Plant and Animal Populations

Plant and Animal Populations

Methods in Demography

Thomas A. Ebert

Department of Biology
San Diego State University
San Diego, California

ACADEMIC PRESS

San Diego   London   Boston   New York   Sydney   Tokyo   Toronto

This book is printed on acid-free paper.

Copyright © 1999 by ACADEMIC PRESS

All Rights Reserved.
No part of this publication may be reproduced or transmitted in any form or by any
means, electronic or mechanical, including photocopy, recording, or any information
storage and retrieval system, without permission in writing from the publisher.

Academic Press
a division of Harcourt Brace & Company
525 B Street, Suite 1900, San Diego, California 92101-4495, USA
http://www.apnet.com

Academic Press
24-28 Oval Road, London NW1 7DX, UK
http://www.hbuk.co.uk/ap/

Library of Congress Catalog Card Number: 98-86238

International Standard Book Number: 0-12-228740-1

PRINTED IN THE UNITED STATES OF AMERICA
98 99 00 01 02 03 QW 9 8 7 6 5 4 3 2 1

For Kay and Tim and Chris

Contents

Preface

...if I'd 'a' knowed what a trouble it was to make a book I wouldn't 'a' tackled it....

Mark Twain,
Huckleberry Finn, 1884

Arrangement of birth and death statistics for purposes of gaining insight into the current and potential future state of human populations dates to John Graunt (1662) and Edmund Halley (1693) but detailed analysis of the interactions of birth and death and understanding of the importance of age-specific rates really started with the work of Alfred Lotka during the early 1900's. Combining the schedules of births and deaths to estimate a rate of population growth was, at least in part, slow to enter standard ecological work because calculations never were simple. Trial-and-error methods using the Euler equation, which involved the use of tables of exponentials, made demographic analysis rather off-putting. Furthermore, although P. H. (George) Leslie developed the matrix method of combining survival and reproductive schedules, the application of his approach was long in coming mostly, I suspect, because of the daunting aspects of analysis with just pencil and paper.

Mainframe computers opened up avenues of analysis, although methods for trial-and-error solutions, at least for ecologists, required discovering branches of mathematics that were both strange and wonderful. I first looked at a numerical analysis textbook in 1967 or '68. A student at the University of Hawaii gave it to me because he was dropping out of school. I couldn't understand what the subject was about and, accordingly, couldn't see what one would do with numerical analysis even though I had read the 1965 paper by Augustus Fabens on the von Bertalanffy growth model, which uses the Newton-Raphson algorithm, and had started

to use this model for describing growth. Discovering that numerical analysis and computers went together was one of those "AHA' s" you hear about now and again. For me, this was about 1971 after I had left the University of Hawaii and was teaching at San Diego State.

The advent of pocket calculators in the mid to late 1970's began to open the possibilities for doing things using paper and pencil that were ever so tedious before. Gone were tables of logs and exponentials and, for many real problems, a programmable calculator with batteries could free one to do calculations while sitting in the library(!) Programs for matrix analysis existed, but they certainly were not widely circulated. Some of this lack might have been associated with computer use in ecology at the time. The International Biological Program (IBP), at least in the US, seemed to dominate computer applications and ecosystem simulation represented the intellectual high ground. Systems were driven from the bottom up, as we would now say, and so models were focused on primary production. Dynamics of plant and animal populations would be emergent properties of primary production.

The microchip of the 1980's probably has done more for demographic studies than any other single device. The general program for matrix analysis given in this book was first developed by taking a number of FORTRAN programs in Pennington (1965) and Kuo (1965) and translating them into AppleSoft BASIC for the Apple II+ computer. I had decided during the early 1980's that I wanted to use matrix techniques for modeling populations and having a personal computer that I could play with at home made developing programs something of a recreational activity.

Publication in 1982 of Hal Caswell's paper in American Naturalist (vol. 120, pp 317-339) showed me a way that I could combine individual growth and survival models and analyze the population consequences of small changes in, for example, K and S_∞ of the Brody-Bertalanffy equation, by using matrix techniques. In retrospect, Caswell's 1982 paper probably is the real beginning of this book and so it is fitting that this book ends with a chapter that focuses on some of the themes of Caswell's 1982 paper.

Notes for this book grew as a consequence of a graduate seminar in 1983 and then as a graduate course at San Diego State University, which, since 1988, has been taught every 3rd semester as Advanced Topics in Population and Community Ecology. The graduate body at San Diego State primarily is masters level students but with a small number of doctoral candidates. In general, students who take my class are in their first or second year of graduate school. All have had an introductory course in ecology at some university but most introductory ecology courses seem to contain very little demography. Students bring some recognition of birth and death schedules to the course but most have never heard of, or have suppressed, knowledge about analysis. Many bring various degrees of math anxiety including what seem to be advanced degrees.

The course I have been teaching has various goals. The first, and most important from my perspective, is that it has served as the device for helping me learn how to analyze and interpret data. For students, the course probably has been most important as a means of building confidence with an ever growing body of the ecological literature. In this respect it is much like the old Charles Atlas body building course that was advertised on the back covers of comic books when I was a kid in Merrill, Wisconsin. In these ads, following isometric exercises, a 97 pound weakling would be transmogrified into an awesome hunk. No longer would big guys kick sand in his face and both his physical and Darwinian fitness increased. An important aspect of my graduate course, and of this book, is that once through it there are fewer authors of ecological papers who are able to bedazzle or baffle with mathematical sand. It will be possible to see whether authors have considered assumptions of their models and to get past the demographic argot and number juggling in order to discover and understand what frequently are wonderful biological stories.

I have focused on methods of analysis because I believe that understanding methods is the best route for understanding and application of more general ideas. Clearly, the next steps are books by Hal Caswell (1989) or Shripad Tuljapurkar (1990) on matrix models or the various books that focus on life history evolution such as those of Brian Charlesworth (1980), Steve Stearns (1992), Derek Roff (1992) and Eric Charnov (1993).

There are various software packages that can do everything that I present in this book. The most general and versatile systems probably are Mathematica and Maple but it also would be possible to assemble several packages such as MatLab and a statistical package with a nonlinear regression module such as SYSTAT as well as a package to decompose size-frequency distributions such as MIX or MULTIFAN. Not everyone can afford all of these packages and, moreover, not everyone has a computer that is compatible with all of them. My goal is to make the methods accessible and to this end I provide programs with this book. I have elected to use BASIC because I think that it still is the most accessible language; FORTRAN and C compilers are rather pricey. Programs are available over The Web at **http://www.sci.sdsu.edu/Cornered_Rat/** together with directions for obtaining a freeware BASIC interpreter. Code for all programs also is printed in appropriate chapters and so could be entered by hand.

I have written BASIC in the most plain vanilla form. All lines are numbered and I have avoided such "modern" touches as WHILE WEND. BASIC has never been standardized but I have tried to write to fit most if not all versions *except* for the bugbear of input-output commands that involve external devices. All programs run with Chipmunk BASIC (the freeware) as well as with Microsoft Quick BASIC but there may be problems with other versions of BASIC with respect to input and output commands involving disk access. I have used an end of file command in a few programs and this may cause some BASIC versions to have fits.

In February 1985 at a workshop in Mazzara del Vallo, Sicily, I was part of a group that was supposed to evaluate software for length-based methods in fisheries research. We were to include "friendliness" as a category for consideration and John Pope, an English fisheries biologist, suggested that we use a scale with "Bangkok massage parlor" at one end and "cornered rat" at the other. I was charmed by the scale of friendliness and several years later the fictitious company Cornered Rat Software was formed. All programs in this book are Cornered Rat products. Error trapping in programs is minimal; some mistakes in data entry will be fatal and you may have to start over including in some cases rebooting. On the up side, this means that the programs are fairly short and much easier to follow and so if you want to modify them you will find it much easier than if the programs were cluttered with hand-holding baggage.

Graduate students have found my course to be very useful and my hope is that the book will be found equally instructive for upper division undergraduates. I also hope that it will be useful to academic colleagues as well as those in resource management or environmental consulting who wish to do population analysis.

Thomas A. Ebert

Acknowledgments

A number of colleagues have been generous in supplying me with raw data from their publications or data yet to see the white literature: Lee McClenaghan, John Mauchline, Paul Zedler and, in particular, John Rae whose data set on a rare cactus form the basis for discussing calculation of confidence limits in Chapter 10. Growth and size data for red sea urchins were gathered in collaboration with Steve Schroeter and John Dixon as well as biologists from resource agencies along the Pacific coast: Pete Kalvass (California Department of Fish and Game), Neil Richmond (Oregon Department of Fisheries and Wildlife), Alex Bradbury (Washington Department of Fish and Wildlife), and Doug Woodby (Alaska Department of Fish and Game).

Data analysis was done, mostly, with the BASIC programs that are supplied with this book although I also used SYSTAT 5.2 for some nonlinear regression problems. Data analysis was done and this book was written and eventually composed as camera-ready on a rather cranky Mac SE/30. Writing and composition were done with Microsoft Word 5.1a. Most figures were created using Cricket Graph 1.3.1 and then brushed, burnished, buffed, and shined using Aldus FreeHand 3.0; some figures were created directly in FreeHand.

Various chapters in this book were read and critiqued by Lee McClenaghan, Gordon Fox, Peter Frank, Kaius Helenurm, Mike Russell, Steve Schroeter and, in particular, Bill Hazen, who read the entire manuscript. Over the years, students in my graduate class found various problems with all sorts of things that I did. Students of the spring class of 1998 (Jon Ball, Jennifer Campbell, Loanne Doan, Michelle Duggan, Megan Johnson, Mary Ann Tiffany, and Mark Tucker) were particularly helpful because they had the penultimate camera-ready version for lecture notes and so functioned, among other things, as my copy editors. I considered all comments but didn't always follow advice. I hope that I made correct choices concerning suggestions; errors that remain, however, are mine.

1

Projection from a Life Table

*P*opulation growth is less like it used to be than ever it has been before.

J. Cohen,

How Many People Can the Earth Support, 1995.

INTRODUCTION

Population ecology is based on two attributes of living organisms: birth and survival. The relationship between births and deaths in determining population change is intuitive but systematic study and analysis of tables really didn't begin until the 17th century when bills of mortality and records of births were used to describe human populations (Graunt 1662, Halley [of comet fame] 1693a,b). The details of the schedules of birth and survival determine whether a population increases, decreases or remains the same size over some period of time. What makes examination of such simple traits so fascinating is the richness of how additions and losses of individuals are scheduled.

Additions of new individuals can be by sexual or asexual reproduction. Asexual reproduction includes parthenogenesis, which is development from an unfertilized egg, and is found in many groups of marine and freshwater invertebrates (Lynch 1985, Ghiselin 1987, Gomez *et al.* 1995) and insects (*e.g.* Templeton 1982), as well as reptiles (Cuellar 1977, Vanzolini *et al.* 1978). Many plants produce seeds asexually following a modified meiosis or from somatic cells in the ovarian wall (*e.g.* Solbrig and Simpson 1974, Kaur *et al.* 1978, Ford 1981) and there are species that produce seeds both sexually and asexually (*e.g.* Michaels and Bazzaz 1986). Other modes of asexual

reproduction include fission or fragmentation found in species that also reproduce sexually such as, for example, some sea stars, brittle stars and sea cucumbers (*e.g.* Ottesen and Lucas 1982, Emson and Mladenov 1987, Hendler 1991) or ascidians (*e.g.* Stoner 1989). Many Cnidarians can reproduce by fission or budding (*e.g.* Sebens 1982, Stocker 1991, Fabricius 1995) as well as sexually. Various plants produce root sprouts (*e.g.* Stohlgren and Rundel 1986) or tillers (*e.g.* Fetcher and Shaver 1983, Bartlett and Noble 1985, Eriksson 1988, Cain 1990, Carlsson and Callaghan 1991, Busso and Richards 1995).

Many species combine several different modes of adding new individuals (*e.g.* Caswell and Werner 1978, Watkinson and White 1985, Hoagland *et al.* 1988) and some include long dormant stages such as seed banks (*e.g.* Templeton and Levin 1979, Pake and Venable 1996) or propagule pools (Hairston and De Stasio 1988, Hairston and Cáceres 1996). The greatest complexity of life cycles probably is found in various parasites (Morand *et al.* 1995), parasitic fungi, and red algae (Searles 1980, Klinger 1993). With such a wealth of reproductive modes, it should be no surprise to discover that even though the "schedule of births" sounds simple, it can be, in fact, very complex.

Losses of individuals, the schedule of survival, can include massive mortality of newborn or germinated stages or, on the other hand, have high survival of new individuals, which frequently is associated with parental care. Many species show senescence; that is, the rate of mortality increases with age as it does in humans and other mammals, birds, at least some fishes, and some plants (*cf.* Medawar 1957, Williams 1957). In other species senescence does not seem to occur or is negligible so that the probability of surviving each year is not dependent on age. Some trees, such as redwoods and bristlecone pines, seem to be

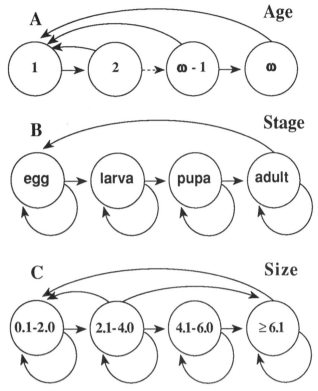

FIGURE 1.1 A , Age-structured life cycle with transitions (survival probabilities) from one age to the next and return transitions to age 1, which are reproductive contributions; ω is the final age-class and no individuals survive past this age. **B**, Stage-structured life-cycle for an insect; during a time period, individuals may remain in a stage or transfer to the next stage. **C**, Life-cycle structured by size classes showing some of the possible transfers including shrinking to a smaller size and skipping a size class.

physiologically immortal and survival probabilities may not change with age. Survival rates of some animals also may be independent of age. There are studies that suggest that at least some sponges, cnidarians, arthropods, molluscs, echinoderms, and chordates have survival probabilities of post-juveniles that are independent of age (reviewed by Finch 1990).

Schedules of births and deaths in animals and plants range from lives that are completed within a few weeks to generations measured in hundreds of years. All of these different life cycles "work" in the sense that organisms with long and organisms with short schedules all continue to leave offspring that live long enough, in turn, to leave successful offspring. Moreover, species with different life cycles coexist in, what seems to us, the same environment; there are long-lived and short-lived plants and animals living together in the same hectare of desert or temperate evergreen forest or coral reef.

Life cycles of organisms are very diverse and many factors may be associated with changes in schedules of births and survivors, including changes in the physical and

chemical environment, predators, competitors, and pathogens. Some species, such as humans and other mammals and many bird species, have cycles in which transitions for survival and reproduction are predictably related to age (Figure 1.1A). These predictable age-specific survival rates are what makes the cost of life insurance change with age. We, as well as many other organisms, also have quite narrow limits on the number of offspring produced at each reproductive episode. The usual litter size is one for humans as well as for elephants, whales, and penguins. It is two for doves and four for armadillos. Large women tend to have one baby for each pregnancy and this is the same number that very small women have. Furthermore, it is much more likely that a woman age 24 will have a baby than will a woman age 44. For humans, and many other creatures, size is not important but age is.

Many species, such as insects, have life cycles with definite stages. Survival as well as reproductive contributions of an individual depend more on stage than on age. The graph of such a cycle (Figure 1.1B) shares some features with age-structured graphs such as a single over all direction to flow. Pupae either remain as pupae or turn into adults or they die. They cannot turn back into larvae or eggs.

In certain types of stage-structured life cycles, individuals *can* transfer back and forth between stages. For example, many species have individuals that are in breeding condition as well as individuals that have not reproduced for one or more years. Being in breeding condition places an individual in a stage that is different from the stage of nonbreeding individuals. Over a period of time, individuals may transfer back and forth between these two stages.

A third general type of life cycle is an extension of stage-structure: namely, size (Figure 1.1C). For many species of trees and shrubs, many fish, reptiles, and marine invertebrates, survival and reproduction are more dependent on size than on age. Big trees produce more seeds than small trees even though a big tree may be younger than a small individual of the same species. Transfers within a size-structured life cycle can be very complex because over some fixed period of time, such as one year, some individuals may remain in a size class, others may grow to the next size class or skip a size class and jump ahead into a larger size category. Also, some individuals may shrink one or more classes. Size-classified life cycles can have many connections.

A final point about graphs of life cycles is that there is no reason why they must be restricted to the three general types shown in Figure 1.1. Ages, stages, and sizes could be mixed together in a single graph, other ways of classifying creatures could be used, or individual graphs could be linked together into spatially structured populations.

A variety of models can be used to explore schedules of births and survivors. These models provide not only tools for practical purposes such as helping life insurance companies make a great deal of money by betting with you, or someone else, on when and how you are going to die but also for other practical applications such as management of natural resources like forests and fish, conservation of wildlife, and preservation of rare and endangered species.

There also are more theoretical reasons for studying the schedules of birth and survival because these schedules have a genetic basis and so are subject to natural selection just as are morphological, physiological, or many behavioral traits. It is an important insight to realize that not only *number* of offspring may be selected for but the *scheduling* of the births and subsequent survival can be adjusted by natural selection. Schedules are shaped by the constraints of evolutionary history and the opportunities provided by current genetic variation as well as other forces, such as effects of small population size, that modify phenotypes.

The combination of the schedules of birth and survival rates is called an organism's life history and the quantitative analysis of the evolution of these schedules continues to grow in the ecological and evolutionary literature from a beginning in about 1954 (Cole 1954) to, more recently, Charlesworth (1980), Tuljapurkar (1990), Stearns (1992), and Roff (1992). A study of schedules of births and survivors links physiology and population genetics with large scale, multispecies systems; demographic studies provide windows into the large black boxes of communities and ecosystems.

LIFE TABLES

A life table is a schedule of probabilities of things that may happen to individuals with particular attributes. The most general choice for attributes is age but it is possible to use some other criteria such as instars for insects or vegetative states for plants. Usually, the entries in a life table are various ways of viewing survival; some authors (*e.g.* Stearns 1992), however, include reproductive schedules as part of a life table. Most of what I present in the following chapters includes both survival and fecundity and so I use the more generous definition of life table to include not only survival but also fecundity and, in some cases, size. I use "life table" in the sense of "life history table". To begin, I focus on survival, move on to fecundity, and then combine these to explore population growth.

A pattern of survival usually is described as though we were watching a group of new born (hatched, germinated, fertilized) individuals decline in numbers over time. A group of individuals such as this is called a *cohort* although cohort also can have many other meanings. A cohort really is just a collection of individuals that share some trait or group of traits for purposes of a study. All freshmen who enter a university in the fall are a cohort and have an academic survival rate. This cohort could be separated into men and women and so we could study academic survival of a male cohort and academic survival of a female cohort. For right now, a cohort will be a collection of new individuals all age 0.

A number of statistics of reproduction and survival or mortality form a life table. They are all related and different types of data can be used to calculate the same values. The variable name x is the attribute that is used to classify individuals. In some cases x will mean age and in other cases it may mean stage or size. For the following statistics, x appears as a subscript. Time units are whatever is appropriate for the organism being studied: parts of a day for yeast or some rotifers, days or weeks for microcrustaceans, years for large mammals, or decades for many trees. Some of the statistics are defined as follows:

m_x is the number of female offspring produced by a female with attribute x and can be concentrated at one point during a time interval as a pulse, or continuously over the interval t to t+1;

p_x is the probability that an individual with attribute x survives one time period and so is conditional on an individual being alive at the beginning of the time interval;

l_x is the probability that a new individual (age, stage or size 0) is alive at the *beginning* of age, stage or size x;

n_x is the number of individuals in state x.

With x = age,

$$p_x = \frac{n_{x+1}}{n_x} \qquad (1.1)$$

and, with x = stage or size,

$$p_x = \frac{\sum_{i=0}^{\omega} n_{x+i}}{n_x}, \qquad (1.2)$$

where x+i refers to individuals in all stages or sizes that survived one time unit from the original number n_x in stage or size x. For example, if there are 100 individuals in size class x ($n_x = 100$), one year later there may be 50 still in size class x, 10 that transferred to the next size class (x+1) and are still alive, and 20 that grew all the way through class x+1 and so transferred to size class x+2 and are still alive. The survival rate, p_x, is

$$p_x = \frac{50+10+20}{100} \text{ or } 0.80 \text{ yr}^{-1}.$$

Some individuals stayed in size class x and died, some transferred to size class x+1 and died before the next census and some that transferred to size class x+2 died during the year. The total number that died was 20 (i.e. 100 - 80) but we don't know which size class they had reached at the time of death.

The probability that a new individual in age, stage, or size 0 survives to age, stage or size x is determined by dividing all numbers of individuals alive at x, which is n_x, by the original number, n_0:

$$l_x = \frac{n_x}{n_0}. \qquad (1.3)$$

Note: l_0 is always equal to 1.0 and l_x is the cumulative survival from 0 to x, which is

$$l_x = \prod_{i=0}^{x-1} p_i \qquad \text{for } x > 0 \qquad (1.4)$$

and

$$p_x = \frac{l_{x+1}}{l_x}. \qquad (1.5)$$

Life cycles of substantial complexity are explored over the next several chapters; however, we start with a simple age structured model and build from there.

A simple model of unlimited population growth is an exponential

$$N_t = N_0 e^{rt} \qquad (1.6)$$

or

$$N_t = N_0 \lambda^t,$$

which, for a single time interval (t=1), is

$$N_{t+1} = N_t \lambda, \qquad (1.7)$$

where

N_t = number of individuals at time t,
N_0 = original or starting number of individuals,
e^r or λ = population growth rate per unit of time.

Differences in r or λ result in different numbers of individuals through time. If $r > 0$, that is $\lambda > 1.0$, the population increases through time and the larger the value of r or λ, the more rapid the increase. If $r < 0$, that is $\lambda < 1.0$, the population decreases through time and if $r=0$ ($\lambda = 1.0$) the population remains unchanged.

The instantaneous growth rate of a population growing according to Equation 1.6 is

$$\frac{dN}{dt} = rN \qquad (1.8)$$

or

$$r = \frac{dN_t/dt}{N_t}. \qquad (1.9)$$

"Little r" is the *population growth rate per individual* (Equation 1.9) and frequently is called the *intrinsic rate of natural increase*; λ is the finite growth rate. Little r and λ are used for comparing the relative health of populations. These summary parameters, r and λ, are used in analyzing populations as part of resource management and conservation of species and also are central to understanding life history evolution.

There are time units (days, months, years), t, associated with r or λ so you should think of these as $r\,t^{-1}$ or $\lambda\,t^{-1}$. All this means is if the time units are in years then r or λ are appropriate for time measured in years or, if you wish to calculate r or λ using years but then make predictions using time units in months you can obtain the appropriate value of r by dividing $r\,yr^{-1}$ by 12 or $ln\lambda\,yr^{-1}$ by 12. For example, if $r\,yr^{-1} = 0.2$ then $r\,month^{-1} = 0.017$ or $\lambda\,yr^{-1} = 1.22$ and $\lambda\,month^{-1} = 1.017$

POPULATION PROJECTION

The purpose of this section is to show how schedules of birth and survivorship interact to determine population growth. The data in Table 1.1 are for a short-lived species with a high reproductive rate. Assume that reproduction occurs once each year rather than continuously and is concentrated at the time when a female enters the next age class. For example, a female who has just attained the age of 1 year, with a probability of 0.24, will reproduce and have 20 female offspring. The consequences of continuous reproduction are presented in a later section.

Changes in numbers in each age class as well as change in total number can be illustrated by assuming some arbitrary initial conditions and then following the fate of the population by applying the rates given in Table 1.1.

I have selected 1000 newborn individuals to start the population but obviously some other number could be selected and with some other mix of age classes. During the first time interval these 1000 small animals survive at a rate appropriate for the age interval 0 to 1, which is 0.240 (Table 1.1). At the beginning of the next interval, there are 240 individuals remaining out of the original 1000, that is 1000 × 0.240. Each of these 240 females reproduces at a rate of 20 females per female so there are 4800 newborn females (240 × 20). The population now consists of 4800 age 0 and 240 age 1 females plus an unspecified number of males.

During the next time interval, $t_1 \rightarrow t_2$, the 4800 age 0 females survive at a rate of 0.240 and the 240 age 1 females survive at a rate of 0.242. At the beginning of the next time

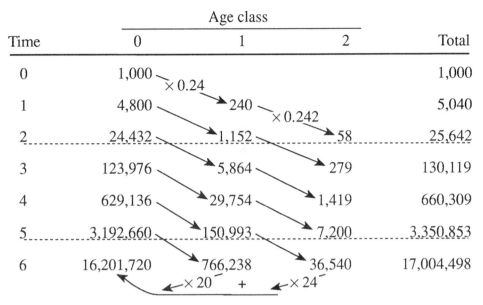

FIGURE 1.2 Projection of a population starting with 1000 age-0 individuals and applying the transitions given in Table 1.1; probability of living from age 0 to age 1, p_0, is 0.24; probability of living from age 1 to age 2, p_1, is 0.242; age-1 females produce 20 female offspring (m_1); and age-2 females produce 24 female offspring (m_2).

TABLE 1.1 Survival, l_x, Conditional Survival, p_x, and Fecundity, m_x, for a Short-Lived Creature

Age (x)	l_x	p_x	m_x
0	1.000	0.240	0
1	0.240	0.242	20
2	0.058	0.000	24
3	0.000		

TABLE 1.2 Fraction of Individuals in Each Class Based on Projection Using Numbers Given in Figure 1.2

	Age class			
Time	0	1	2	r
0	1.0000			
1	0.9524	0.0476		1.6174
2	0.9528	0.0449	0.0023	1.6268
3	0.9528	0.0451	0.0021	1.6242
4	0.9528	0.0451	0.0021	1.6242
5	0.9528	0.0451	0.0021	1.6243
6	0.9528	0.0451	0.0021	1.6243

interval there are 1152 1-year-old females (4800 × 0.240) and 58 2-year-old females (240 × 0.242). Each 1-year-old female has 20 female offspring (1152 × 20 = 23,040) and each 2-year-old female has 24 female offspring (58 × 24 = 1392) giving a total of 24,432 age 0 females. The results of continuing this process for several more years are shown in Figure 1.2.

Several features of population growth can be shown with the expanding population in Figure 1.2 by exploring some of the relationships among age classes and between pairs of years. For successive estimates of population size, population growth rate, λ or e^r, is just the number after one year divided by the original number

$$e^r = \frac{N_{t+1}}{N_t} \tag{1.10}$$

and

$$r = ln\left(\frac{N_{t+1}}{N_t}\right) \tag{1.11}$$

The *fraction* of the population in each age class can be calculated by dividing the number in a class by the total population size. For t = 5, the total population is 3,350,853 so the fraction in age class 0 is 3,192,660/3,350,853, the fraction in age-class 1 is 150,993/3,350,853 and the fraction in age-class 2 is 7,200/3,350,853, which are 0.95279, 0.04506, and 0.00215 respectively. Table 1.2 shows these fractions for all years as well as estimates of r for each year by using a modification of Equation 1.11 (Equation 1.12).

$$r = ln\left(\frac{N_t}{N_{t-1}}\right). \tag{1.12}$$

Notice that the proportions become fixed and that the successive estimates of r settle at a single value. The fixed proportions in each age class collectively are called the

TABLE 1.3 Calculation of Overall Birth and Death Rates for the Creature Described by Table 1.1

Part A Numbers of Individuals in Each Age Class

Time	Age class 0	1	2	Total
0	1.0000			1.0000
1	4.8000	0.2400		5.0400
2	24.4339	1.1520	0.0581	25.6445
3	123.9736	5.8641	0.2788	130.1161
4	629.1323	29.7537	1.4191	660.3052
5	3192.6443	150.9918	7.2004	3350.8378
6	16201.6525	766.2346	36.5400	17004.4266
7	82218.2209	3888.3965	185.4288	86292.0465
8	417231.2583	19732.3726	940.9920	437904.6246
9	2117315.6143	100135.4997	4775.2341	2222226.3462

Part B Total Births and Survivors

Time	Age 1 × 20	Age 2 × 24	Total Births	Survivors
0				
1	4.8000	0.0000	4.8000	0.2400
2	23.0400	1.3944	24.4344	1.2101
3	117.2820	6.6912	123.9732	6.1429
4	595.0740	34.0584	629.1324	31.1728
5	3019.8360	172.8096	3192.6456	158.1922
6	15324.6920	876.9600	16201.6520	802.7746
7	77767.9300	4450.2912	82218.2212	4073.8253
8	394647.4520	22583.8080	417231.2600	20673.3646
9	2002709.9940	114605.6184	2117315.6124	104910.7338

Note: population simulated until a stable-age distribution was attained at a tolerance of 1×10^{-9}.

stable-age distribution. Whenever a population has a fixed schedule of births and survivors, it will always converge on the stable-age distribution.

Population growth rate can be found by determining the ratio of successive population sizes or successive numbers in each age class. For example, the growth rate between $t = 5$ and $t = 6$ is 17,004,498/3,350,853 or 5.0747, which is λ. The growth rate, λ, between $t = 4$ and $t = 5$ is 3,350,853/660,309 or 5.0747, the same rate, which gives $r = 1.6243$ (*i.e. ln*(5.0747) = 1.6243).

Once a stable-age distribution has been achieved, it should be clear that not only does the entire population grow according to Equation 1.6 but each age class also changes according to Equation 1.6. For example, if just age class 0 is used from $t = 5$ to 6, then 16,201,720/3,192,660 = 5.0747, which is the same rate one gets by using the entire population.

Because the proportions are fixed, it is possible to dissect r into birth and loss rates per individual. The number of individuals at $t+1$ is equal to the number that survived from t to $t+1$ plus the number born at time $t+1$ (Equation 1.13).

Table 1.3. shows deaths and births for the small creatures from Table 1.1.

$$N_{t+1} = \text{survivors}_{t \to t+1} + \text{births}_{t+1}. \qquad (1.13)$$

The survival and birth rates can be combined with an initial total number, N_t, after a stable-age distribution has been attained, to determine number at N_{t+1}

$$N_{t+1} = \underbrace{N_t \text{ (survival rate)}}_{\text{survivors from } t} + \underbrace{N_t \text{ (survival rate)(birth rate)}}_{\text{born at } t+1} \qquad (1.14)$$

and Equation 1.14 can be rearranged by factoring out N_t (survival rate)

$$N_{t+1} = N_t(\text{survival rate})(1 + \text{birth rate}). \qquad (1.15)$$

Birth rate at time $t+1$ is the total number of births divided by the adults that produced them; that is, the survivors from the total population at time t. Survivors are the number of newborn individuals at time t multiplied by the appropriate survival rate, p_0, plus the number of age-1 individuals multiplied by p_1, the probability of surviving from t to t+1. For example, at t_4, survivors come from t_3,

$$\text{survivors} = \underbrace{123.9736 \times 0.24}_{\text{age } 0 \to 1} + \underbrace{5.8641 \times 0.242}_{\text{age } 1 \to 2} = 31.1728.$$

$$= 29.7537 + 1.4919 = 31.1728$$

These survivors will reproduce so the total number of births is:

$$\text{births} = \underbrace{29.7537 \times 20}_{\text{age } 1} + \underbrace{1.4191 \times 24}_{\text{age } 2} = 629.1324.$$

The birth rate is the number of births divided by the number of individuals not including the newborns

$$\text{birth rate} = \frac{629.1324}{31.1728} = 20.1821 \text{ females per female yr}^{-1}.$$

The survival rate is the number surviving from t to t+1 divided by the number at the beginning of the time period. For example, at time period 4, individuals have survived from the beginning of time 3 to the beginning of time 4

$$\text{survival rate} = \frac{31.1728}{130.1161} = 0.2396 \text{ yr}^{-1}.$$

Equation 1.15 can be rearranged by dividing both sizes by N_t,

$$\frac{N_{t+1}}{N_t} = (\text{survival rate})(1 + \text{birth rate}), \qquad (1.16)$$

or

$$\lambda = e^r = (\text{survival rate})(1 + \text{birth rate}). \qquad (1.17)$$

For the small creatures from Table 1.1,

$$\lambda = 0.2396 \times (1 + 20.1821) = 5.075231$$

or

$$r = ln\lambda = 1.6243,$$

which is the value that was obtained by projection in Table 1.2.

Little r is the sum of the instantaneous birth rate per individual (b) and instantaneous survival rate per individual (s)

$$r = b+s, \qquad (1.18)$$

so

$$r = ln(1 + \text{birth rate}) + ln(\text{survival rate}) \qquad (1.19)$$

TABLE 1.4 The Stable-Age Distribution for the Creature in Table 1.1 Together with Products Used to Calculate Annual Birth and Survival Rates

Age	Fraction, c_x	p_x	m_x	$c_x p_x$	$c_x m_x$
0	0.9528	0.24	0	0.22867	-
1	0.0451	0.242	20	0.01090	0.90122
2	0.0021	0.0	24		0.05157
Total	1.0000			0.23957	0.95279

or

$$r = 3.0531 - 1.42878 = 1.6243$$

with

$$s = -1.4278$$

and

$$b = 3.0531.$$

The annual survival rate, S, is e^s or

$$S = e^{-1.4278} = 0.2398 \text{ yr-1}$$

and annual birth rate, B, is $e^b - 1$ or

$$B = e^{3.0531} - 1 = 21.1809 - 1$$
$$= 20.1809 \text{ females per female yr}^{-1}.$$

The values for both annual birth rate and annual survival rate should look somewhat familiar. The annual birth rate, B, is between the values of m_1 and m_2 but not just the average. Also, the annual survival rate, S, is in the range of p_0, p_1, and p_2 but, again, not just the average of these three probabilities. B is obtained from m_x values and S from p_x values by incorporating terms of the stable-age distribution, c_x (Table 1.4) and Equations 1.20 and 1.21. Using c_x and m_x values, the annual birth rate, B, is,

$$B = \frac{\sum_{x=1}^{\omega} c_x m_x}{\sum_{x=1}^{\omega} c_x}; \qquad (1.20)$$

or, using values from Table 1.4,

$$B = \frac{0.9528}{0.0472} = 20.1821 \text{ females female}^{-1}.$$

The annual survival rate, S, is

$$S = \sum_{x=0}^{\omega} c_x p_x; \qquad (1.21)$$

or, using values from Table 1.4,

$$S = 0.2396 \ yr^{-1}.$$

These are the same rates as were obtained by using the actual total numbers of births and survivors. What should be clear is that population growth rate, either as λ or e^r, is a statistic that summarizes the details of specific rates of survival and reproduction of the age classes of a population. The population growth rate has two parts, birth and survival, and these two parts can be derived from knowledge of age-specific rates and terms of the stable-age distribution. All of these rates and summary statistics can be derived by simulation of population growth and, for some purposes, such results may be all that are required. Projection can be done using a rather simple computer program such as **PROJECTION.BAS** presented at the end of this chapter, or exactly the same results can be obtained using any of various spreadsheets such as Excel or Lotus 1-2-3 (*e.g.* Starfield and Bleloch 1986, Norton 1994). Whichever approach is used, the process will aid in building intuition concerning the relationships between schedules of birth and survival in determining changes in populations.

PROBLEMS

1. A life table for gray foxes (Table 1.5) is given by Michod and Anderson (1980).
   a. Calculate the values for p_x.
   b. Start with some arbitrary numbers in each age class and project the growth of the population until you have attained a stable-age distribution.
   c. What are the terms of the stable-age distribution?
   d. What are r and λ?
   e. What are the instantaneous birth and survival rates, b and s?

TABLE 1.5 Life table for Female Gray Foxes

Age	l_x	m_x
0	1.000	0
1	0.300	2.08
2	0.086	2.21
3	0.029	2.34
4	0.014	2.34

Note: age in years; data from Michod and Anderson (1980).

2. Survival and reproductive data for a freshwater snail, *Viviparus georgianus* are provided by Buckley (1986). Note in Table 1.6 that p_x values are provided rather than l_x.

a. Calculate the values for l_x.
b. There is some reason to believe that the m_x include both males and females. Compare the estimates of r using the m_x schedule (Table 1.6) and a schedule using 1/2 of each m_x value.
c. Calculate the annual birth and survival rates, B and S, for a population that has assumed a stable-age distribution.

TABLE 1.6 Life Table for *Viviparus georgianus*

Age	p_x	m_x
0	0.321	0
1	0.639	0
2	0.565	1.240
3	0.307	4.800
4	0.0	8.350

Note: age in years; data from Buckley (1986).

3. Table 1.7 shows survivorship and fecundity data for female gray squirrels in North Carolina (Barkalow, Hamilton and Soots 1970).
   a. Calculate the terms of the stable-age distribution.
   b. What is the annual population growth rate, λ, and the growth rate per individual, r?
   c. What are b and s? B and S?
   d. Compare how rapidly you attained a stable age distribution for gray squirrels with how rapidly you arrived at a stable-age distribution using the gray fox data in problem 1. Why do you suppose there is a difference?

TABLE 1.7 Life Table for Female Gray Squirrels

Age	l_x	m_x
0	1.000	0
1	0.253	1.28
2	0.116	2.28
3	0.089	2.28
4	0.058	2.28
5	0.039	2.28
6	0.025	2.28
7	0.022	2.28

Note: age in years; data from Barkalow, Hamilton and Soots (1970).

4. Table 1.8 shows survivorship and fecundity for sand smelt (*Atherina boyeri*) in the English Channel Assume that survivorship means survival from the previous age to the age the table (*e.g.* 0.0016 for age 1 is p_0 or survival from hatching to age 1). Also, assume that fecundity means total egg production.

a. Create a new table with l_x and m_x values.

b. Estimate r and the terms of the stable-age distribution.

c. Project the population using three different starting conditions such as all individuals in age-class 1, all in age-class 2 and equal numbers in 1 and 2.

TABLE 1.8 Life table for sand smelt (*Atherina boyeri*) in the English Channel.

Age	1	2	3
Fecundity	1423	4567	8718
Survivorship	0.0016	0.1082	0.1082

Note: table from Henderson and Bamber (1987) derived from data presented by Turnpenny *et al.* (1981).

5. It is very unlikely that the small creature presented in Table 1.1 actually could have such a large value of r or λ otherwise we would be overrun by them. It is more likely that in the long run λ would be close to 1.0 (*i.e.* r close to 0). What value of p_0 for the small creature would give $\lambda = 1.0$? [*Hint:* Solve this problem by trial and error. Change p_0, run the simulation, get an answer and then change p_0 and try again.]

BASIC PROGRAM

The following program, **PROJECTION.BAS**, written in BASIC determines r, birth rate and survival rate, and the terms of the stable-age distribution by projection.

```
10 dim p(20),m(20),l(20),t(500),n(20),
   live(500),birth(500),surv(500)
20 print "A fine product from Cornered
   Rat Software©"
30 ro = 10000
40 input "Are data from a file (F) or
   from the keyboard (K)? (F/K): ";f$
50 if f$ = "F" or f$ = "f" then goto 250
60 if f$ <> "k" and f$ <> "K" then goto
   40
70 input "How many age classes (e.g.
   ages 0, 1 and 2 are 3 classes): ";x
80 x = x-1
90 for i = 0 to x
100 print "l(";i;") = ",
110 input l(i)
120 print "m(";i;") = ",
130 input m(i)
140 next i
150 input "Do you want to save the data
    file? (Y/N)";f$
160 if f$ = "N" or f$ = "n" then goto
    370
170 input "Name of file for data: ";f$
180 open f$ for output as #9
190 print #9,x
200 for i = 0 to x
210 print #9,l(i);",";m(i)
220 next i
230 close #9
240 goto 370
250 input "Name of file with data: ";f$
260 open f$ for input as #9
270 input #9,x
280 for i = 0 to x
290 input #9,l(i),m(i)
300 next i
310 close #9
320 l(x+1) = 0
330 print "x = ";x
340 for i = 0 to x
350 print l(i),m(i)
360 next i
370 omega = 500
380 l(x+1) = 0
390 print "Enter the initial numbers in
    each age class"
400 for i = 0 to x
410 print "n(";i;") = ",
420 input n(i)
430 next i
440 t(0) = 0
450 for i = 0 to x
460 t(0) = t(0)+n(i)
470 next i
480 for i = 0 to x
490 p(i) = l(i+1)/l(i)
500 next i
510 for j = 1 to omega
520 for i = x to 1 step -1
530 n(i) = n(i-1)*p(i-1)
540 live(j) = live(j)+n(i)
550 next i
560 n(0) = 0
570 for i = 1 to x
580 n(0) = n(i)*m(i)+n(0)
590 next i
600 for i = 0 to x
610 t(j) = t(j)+n(i)
620 next i
630 birth(j) = n(0)/(t(j)-n(0))
640 surv(j) = (t(j)-n(0))/t(j-1)
650 r = log(t(j)/t(j-1))
```

```
660 if abs(r-ro) < 1.000000E-05 then
    goto 690
670 ro = r
680 next j
690 print "t","Population
    size","r","survival rate","birth
    rate"
700 i = 0
710 print i,t(i)
720 for i = 1 to j
730 r = log(t(i)/t(i-1))
740 print i,t(i),r,surv(i),birth(i)
750 next i
760 print "Terms of the stable age
    distribution"
770 print "age","fraction"
780 for i = 0 to x
790 print i,n(i)/t(j)
800 next i
810 input "Would you like to save output
    to a file? (Y/N): ";f$
820 if f$ <> "Y" and f$ <> "y" then goto
    980
830 input "File for output: ";f$
840 open f$ for output as #9
850 print #9,"t","population
    size","r","survival rate","birth
    rate"
860 i = 0
870 print #9,i,t(i)
880 for i = 1 to j
890 r = log(t(i)/t(i-1))
900 print #9,i,t(i),r,surv(i),birth(i)
910 next i
920 print #9,"Terms of the stable age
    distribution"
930 print #9,"age","fraction"
940 for i = 0 to x
950 print #9,i,n(i)/t(j)
960 next i
970 close #9
980 end
```

The program **PROJECTION.BAS** projects an initial age distribution using l_x and m_x values until a stable-age distribution has been attained, which is accepted when $|r_i-r_{i-1}| < 1.0 \times 10^{-5}$ where r_i is the current estimate of r_i and r_{i-1} is the previous estimate. When suitable convergence has been attained, the terms of the stable-age distribution are printed.

Data can be entered at the keyboard or from a file. File construction includes a value for the oldest non-0 age class followed by l_x and m_x values for each age class. For example, the data file for the small beast in Table 1.1 is

```
2
1,0
0.24,20
0.05808,24
```

The meaning of each line is

2	oldest age class is 2 so there are 3 age classes
1,0	$l_0 = 1$ and $m_0=0$
0.24,20	$l_1 = 0.24$ and $m_1 = 20$
0.058,24	$l_2 = 0.058$ and $m_2 = 24$

Initial conditions must be provided for each age class. Prompts are provided for n(0), n(1), etc. For example, you might want to start with 10 individuals in age class 0 without any individuals in any of the other age classes.

$n(0) = 10$
$n(1) = 0$
$n(2) = 0$

The total population size at time 0 is calculated by summing all of the initial terms in lines 440 to 470.

```
440 t(0) = 0
450 for i = 0 to x
460 t(0) = t(0)+n(i)
470 next i
```

The survivors from one time period to the next, n(i), together with total survivors, live(j), are calculated in lines 520 to 550..

```
520 for i = x to 1 step -1
530 n(i) = n(i-1)*p(i-1)
540 live(j) = live(j)+n(i)
550 next i
```

The total number of new age-0 individuals is calculated in lines 560 to 590.

```
560 n(0) = 0
570 for i = 1 to x
580 n(0) = n(i)*m(i)+n(0)
590 next i
```

Birth and survival rates are calculated in lines 600 to 640.

```
600 for i = 0 to x
610 t(j) = t(j)+n(i)
```

```
620 next i
630 birth(j) = n(0)/(t(j)-n(0))
640 surv(j) = (t(j)-n(0))/t(j-1)
```

and r is determined in line 650.

```
650 r = log(t(j)/t(j-1))
```

Lines 660 to 680 test to see whether the new value for r is very close to the value calculated in the previous iteration and if the absolute difference is less then 1.0×10^{-5} then it is declared that it is good enough and the program moves on to line 690. If the absolute difference is greater than 1.0×10^{-5} then the new value of r becomes the old value of r, called ro in line 670, and the program continues.

```
660 if abs(r-ro) < 1.000000E-05 then
    goto 690
670 ro = r
680 next j
```

The rest of the program does various things to output results and calculate terms of the stable-age distribution in lines 780 to 800.

```
780 for i = 0 to x
790 print i,n(i)/t(j)
800 next i
```

Results first are shown on the screen and then, if you wish, sent to a file on a disk.

2

Additional Ways of Combining Survival and Fecundity

It may be true that demographers know too much.

R. H. MacArthur,
Quarterly Review of Biology, 1960.

INTRODUCTION

Although it is possible to estimate population growth rate per individual, r, and terms of the stable-age distribution by simple projection as was done in Chapter 1, other ways of looking at survival and birth schedules provide additional insights into the process of population growth and the evolution of life-histories.

NET REPRODUCTIVE RATE PER GENERATION

A commonly used technique for estimating r makes use of l_x and m_x values to calculate the net reproductive rate per generation, R_0, and generation time, T, and then uses these two parameters to estimate r

$$n_T = R_0 n_0, \qquad (2.1)$$

which means that if you have n_0 individuals, one generation later you will have n_T. Equation 2.1 really is the same as Equation 1.6 and so R_0 also can be written

$$R_0 = e^{rT} \qquad (2.2)$$

or

$$r = \frac{ln R_0}{T}. \qquad (2.3)$$

Growth rate per individual, r, is the $\log_e$ of the net growth rate per generation divided by the mean generation length.

Picture n_0 newborn females. At age x the surviving females ($n_0 l_x$) will produce m_x new females per female. If there is just one such reproductive event, then the number of individuals that start the next generation will be $n_0 l_x m_x$ and so $R_0 = l_x m_x$ The length of the generation, T, would be however long it took to get from one cohort of newborn females to the next cohort of newborns. This might be one year for annual species such as Saturnid moths or many years for such creatures as periodic cicadas. Species where $R_0 = l_x m_x$ have non-overlapping generations; it is necessary, however, to be careful with this definition because many species called "annual", such as "annual weeds", actually may have substantial seed banks in the soil and so flowering stages in any particular year may be of many ages from few to many years old; time in the seed bank has to be counted and the "annual" weed in fact may be 1 or 10 or 50 years old.

When there are overlapping generations, R_0 is not just a single $l_x m_x$ value but the sum of the $l_x m_x$ values for the entire life table from 0 to the final age class, ω,

$$R_0 = \sum_{x=0}^{\omega} l_x m_x. \qquad (2.4)$$

It should be apparent that not only is R_0 the net growth rate per generation but also the expected lifetime reproduction of a newborn female ($\Sigma l_x m_x$). Similar R_0 values can be obtained by having very poor survival of new-born females and large values for m_x, like oysters, or very good survival for newborns and small values for m_x like humans. There also can be many terms to the summation of $\Sigma l_x m_x$, like redwood trees, or just a few or one, like sea hares. Identical values for R_0 would mean that the growth rates *per generation* would be the same, which leads us to the problem of estimating mean generation time, T.

AGE OF MOTHERS DISTRIBUTION

Picture all reproduction as concentrated at one point each generation. The concentration would be at the average age of mothers and so to get mean generation time it is necessary to estimate the average age of mothers, which means, in effect, asking each new-born female "How old is your mom?" This isn't the same as getting the average age of females that are reproducing because, for many species, fecundity changes with age. Think of the number of eggs produced by a female sturgeon at its first reproduction compared with egg production in a 100 year-old fish. So, out of n_0 newborn females, what fraction survive to age x? The answer is l_x. And how many offspring will be contributed by the survivors at each age class? The answer is $l_x m_x$. In graphical form (Figure 2.1), age is on the x-axis and $l_x m_x$ is on the y-axis.

Determining the centroid of a function between "a" and "b", such as in Figure 2.1, is treated in calculus. A centroid has both $\overline{x}$ and $\overline{y}$ components but all we want is $\overline{x}$, which is the average age of mothers or an estimate of T

$$\overline{x} \text{ or } T = \frac{\displaystyle\int_a^b x l_x m_x dx}{\displaystyle\int_a^b l_x m_x dx}. \qquad (2.5)$$

Because of the way life tables are constructed (*e.g.* Table 2.1), summations are used as approximations of the integrals

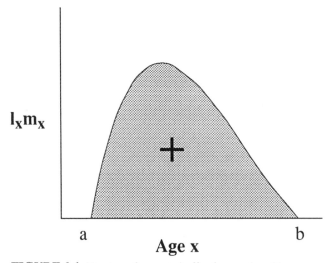

FIGURE 2.1 Number of expected offspring produced by a new-born female as a function of age. The position of "+" indicates the center of mass or centroid where the mean age of mothers, $\overline{x}$, is the generation time T or age at which all reproduction can be considered to be concentrated.

TABLE 2.1 Life Table for a Short-Lived Creature

Age	l_x	m_x	$l_x m_x$	$x l_x m_x$
0	1.000	0	0	0
1	0.240	20	4.800	4.800
2	0.058	24	1.392	2.784
Total			6.192	7.584

$$T = \frac{\displaystyle\sum_{x=0}^{\omega} x l_x m_x}{\displaystyle\sum_{x=0}^{\omega} l_x m_x}. \qquad (2.6)$$

Equation 2.6 really should be quite familiar in that sense that it has the same form as the formula used in calculating a mean from grouped data. In both cases there is an x and a frequency of x, which is f. For calculating a mean, fx is the frequency of occurrence of x values and the sum of the frequencies is n. Look at Figure 2.1 again and consider that $l_x m_x$ *is* a frequency; $l_x m_x$ is the number of females born (m_x) to the fraction of females that survived to age x, which is l_x. Therefore, $l_x m_x$ *is* the same as f and $\Sigma l_x m_x$ or R_0 is the sum of these frequencies or is like n when determining a mean. One difference is that $\Sigma l_x m_x$ does not have to be an integer. The data for our small creature in Chapter 1 (Table 1.1) provides an example of how data are manipulated to estimate R_0, T, and r (Table 2.1).

Using the sums in Table 2.1, the net reproductive rate per generation is

$$R_0 = \Sigma l_x m_x = 6.1920$$

and so an estimate of generation time is

$$T = \frac{\sum x l_x m_x}{\sum l_x m_x} = 7.584/6.192 = 1.225 \text{ years.}$$

With R_0 and T, Equation 2.3 can be used to calculate r

$$r = \frac{ln R_0}{T} = \frac{ln 6.192}{1.225}$$

or

$$r = 1.489 \text{ yr}^{-1},$$

which is *not* the same as r obtained in Chapter 1 where r was 1.6242 yr^{-1}. Equation 2.3 is widely used to estimate r yet we didn't get the same result that we obtained in Chapter 1. A reasonable question is, "why the difference?" We could correctly conclude that r was about 1.5 from both methods and this might be good enough for certain purposes; two different but equally valid approaches, however, should have provided identical answers. We will return to this problem later in the chapter after exploring the use of the Euler equation to estimate r.

DERIVATION OF THE EULER EQUATION

A common method used to estimate r is known as the Euler equation because it was derived by the Swiss mathematician Leonhard Euler (1707 - 1783). Euler (1760) used fixed schedules of survival and births and estimated population growth, λ, by trial and error. Euler used age-specific rates for survival but not births. Rather, he used an overall population birth rate, B. Population growth with fixed schedules was introduced into ecology by Alfred J. Lotka (1880 - 1949) and so sometimes the Euler equation is called the Lotka equation (Lotka 1907a,b). The following section provides a derivation of the Euler equation for estimating r from a table of l_x and m_x values.

The number of individuals that are born (hatched, spawned) during the current reproductive season will be called b_0. They can be considered to be concentrated at a single point, which is "right now." We can express b_0 in terms of how many individuals were born one reproductive interval ago (b_1)

$$b_0 = b_1 e^{r \cdot 1}. \tag{2.7}$$

This really is just Equations 1.6 or 2.2 and is another way of saying that all age classes grow at the same rate *once a population has reached a stable-age distribution*.

Equation 2.7 can be turned around and one can ask, "Relative to the number born 'right now', how many were born one time unit ago?" The answer is

$$b_1 = \frac{b_0}{e^{r \cdot 1}}. \tag{2.8}$$

Taking the reciprocal of a value with an exponent is the same as just changing the *sign* of the exponent and so Equation 2.8 can be written

$$b_1 = b_0 e^{-r \cdot 1}. \tag{2.9}$$

It should be clear that we can calculate the number born x-time units ago in exactly the same manner and so

$$b_x = b_0 e^{-rx}. \tag{2.10}$$

Of the individuals born x time units ago, how many have survived to the present? The answer is found by multiplying the number born x-time units ago by their survival rate for x time units, which is l_x

$$l_x b_x \text{ or } l_x b_0 e^{-rx}.$$

Each of these survivors will contribute to current reproduction by having m_x female offspring. We can show this contribution of individuals in each age group that was born x time units ago and has survived to the present with a series of terms, each of which is

$$l_x m_x b_0 e^{-rx}.$$

All of these individual terms can be summed to give the total number of new born individuals, b_0

$$b_0 = \sum l_x m_x b_0 e^{-rx}. \tag{2.11}$$

Because b_0 is a constant, it can be moved outside of the summation, Σ,

$$b_0 = b_0 \sum l_x m_x e^{-rx} \tag{2.12}$$

and then both sides can be divided by b_0

$$1 = \sum l_x m_x e^{-rx}, \tag{2.13}$$

which is the discrete Euler or Lotka equation.

Given l_x and m_x, the only unknown is r and so it would seem that solving for r would be a straightforward exercise in algebra. Unfortunately, r can not be isolated on one side of the equation and so a simple analytical solution is not

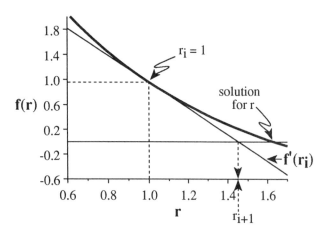

FIGURE 2.2 f(r), using Equation 2.15, as a function of r for data in Table 2.1; $f'(r)$ was calculated at r = 1 and so $f'(r)$ is tangent to f(r) at r = 1; bold line is the value of the function, Equation 2.15, plotted for different values of r but with fixed $l_x m_x$ so there is just one point, r = 1.62, where f(r) = 0, which is the solution for r.

possible. Trial and error works and it is possible to select a value of r, calculate each term in Equation 2.13, and then see whether the sum is equal to 1.0 and, if it is, the correct calculated. After a few guesses, it usually is possible to begin to zero in on better and better solutions; there are, however, some rather more efficient ways of approaching the problem which fall in a branch of mathematics called numerical analysis. An old, but still very good, algorithm was proposed by Isaac Newton (1642 - 1727) and Joseph Raphson (1648 - 1715).

THE NEWTON -RAPHSON METHOD

The Newton-Raphson method, or sometimes just Newton's method, is an efficient algorithm for finding r and is of very general applicability in cases where the solution of an equation cannot be achieved analytically.

Figure 2.2 presents the idea of the Newton-Raphson method. The first step is to express the function of interest so it is equal to zero

$$f(r) = 0. \qquad (2.14)$$

For the Euler equation, just subtract 1 from both sides,

$$f(r) = \Sigma l_x m_x e^{-rx} - 1, \qquad (2.15)$$

and the correct solution for r is when f(r) = 0. When some arbitrary value of r is selected, r_i, so $f(r_i) \neq 0$, a better value for r, r_{i+1}, lies on a line that is tangent to $f(r_i)$. A line that is

TABLE 2.2 Application of the Newton-Raphson Algorithm to Determine r with the Euler Equation

Iteration number 1, r = 1.0

Age (x)	$l_x m_x$	$l_x m_x e^{-rx}$	$-x l_x m_x e^{-rx}$
0	0	0	0
1	4.800	1.7658	-1.7658
2	1.392	0.1884	-0.3768
Σ		1.9542	-2.1426

r_{i+1} = 1 - (1.9542 - 1)/-2.14257 = $\boxed{1.4453}$

Iteration number 2, r = 1.4453

Age (x)	$l_x m_x$	$l_x m_x e^{-rx}$	$-x l_x m_x e^{-rx}$
0	0	0	0
1	4.800	1.1312	-1.1312
2	1.392	0.0773	-0.1546
Σ		1.2085	-1.2858

r_{i+1} = 1.4453 - (1.2085 - 1)/-1.2858 = $\boxed{1.6075}$

tangent to $f(r_i)$ has a slope that is the derivative of $f(r_i)$, which is $f'(r_i)$, at r_i

$$f'(r) = -\Sigma x l_x m_x e^{-rx}. \qquad (2.16)$$

The equation of the straight line that passes through the point $(r_i, f(r_i))$ is

$$f(r_i) = f'(r_i)r_i + c. \qquad (2.17)$$

A better estimate of r, that is r_{i+1}, has coordinates $(r_{i+1}, 0)$ and the straight line equation is

$$0 = f'(r)r_{i+1} + c. \qquad (2.18)$$

What we want to know is what we should use for r_{i+1} given r. One way of doing this is to combine Equations 2.17 and 2.18 by subtracting Equation 2.18 from Equation 2.17,

$$f(r_i) - 0 = f'(r_i)r_i - f'(r_i)r_{i+1} + c - c. \qquad (2.19)$$

A bit of rearrangement now will produce

$$r_{i+1} = r_i - \frac{f(r_i)}{f'(r_i)}, \qquad (2.20)$$

which in the Newton-Raphson algorithm. Table 2.2 shows several of the steps in applying Equation 2.20, which can be continued until any degree of convergence has been attained. The final result will be 1.6242 just as was obtained in Chapter 1. A BASIC program of the Newton-Raphson algorithm for estimating r, **EULER.BAS**, is given at the end of this chapter.

CALCULATING r FROM GENERATION TIME

Having established that r is equal to 1.6242, it is interesting to return to generation time, T, and the problem of why estimating r using Equation 2.3 provided an estimate of r that was quite different; that is, 1.488. May (1976) showed that the shape for the xl_xm_x *vs.* x distribution (Figure 2.1) influences how close the estimate of r from Equation 2.3 approaches the true value of r, such as we obtained in Chapter 1 or by using the Euler equation (Equation 2.13). To make the distinction clear, let's call the r calculated from Equation 2.3 r_c, and the generation time calculated from Equation 2.6 T_c, then, as May showed, the differences between r and r_c and between T and T_c can be explained by the shape and symmetry of the "age of mothers" or net maternity distribution (Figure 2.1). What is needed is a mathematical way of describing the shape of a distribution.

Moments, μ_p, and cumulants, κ_p, (κ is the Greek kappa) can be used to describe shape and symmetry of distributions. The pth moment about the mean, $\overline{x}$, of a distribution is

$$\mu_p = \frac{\sum f(x_i - \overline{x})^p}{N}. \qquad (2.21)$$

The first moment μ_1 is always zero [since $\sum(x_i - \overline{x}) = 0$] and when p = 2, μ_2 is the variance. Using moments is not the only way of calculating a set of descriptive constants for distributions. Kendall and Stuart (1977) discuss another series of constants, called cumulants, which are the ones that will be used to reconcile r from the Euler equation with r from $ln R_0/T$ because higher cumulant terms become smaller more rapidly than do higher moment terms and hence become less important in the estimation of r (Lotka 1932). The relationship between moments, μ, and cumulants, κ, is

$$\exp\left\{\kappa_1 r + \frac{\kappa_2 r^2}{2!} + \dots \frac{\kappa_\omega r^\omega}{\omega!}\right\} = 1 + \mu_1' r + \frac{\mu_2' r^2}{2!} + \dots \frac{\mu_\omega' r^\omega}{\omega!} \quad (2.22)$$

$$\kappa_2 = \mu_2 \qquad \text{(variance),} \qquad (2.23)$$

$$\kappa_3 = \mu_3 \qquad \text{(skew),} \qquad (2.24)$$

$$\kappa_4 = \mu_4 - 3\mu_2^2 \qquad \text{(kurtosis),} \qquad (2.25)$$

$$\kappa_5 = \mu_5 - 10\mu_3\mu_2, \qquad (2.26)$$

and $\qquad \kappa_6 = \mu_6 - 15\mu_4\mu_2 - 10\mu_3^2 + 30\mu_2^3. \qquad (2.27)$

Cumulants up to κ_{10} are provided by Kendall and Stuart (1977 vol. 1, p 73) although it seems unlikely that one would ever want to use cumulants past κ_6 - if even that far.

Skewness, κ_3, measures the asymmetry of the two sides of a distribution and kurtosis, κ_4, is a measure of the flatness or peakedness of the distribution. Equation 2.28 extends the correction for r provided by May (1976) based on the shape of the age-of-mothers distribution out past variance, skew and kurtosis to the 6th cumulant, κ_6.

$$r_c = r\left(1 - \frac{r\kappa_2}{2!T_c} + \frac{r^2\kappa_3}{3!T_c} - \frac{r^3\kappa_4}{4!T_c} + \frac{r^4\kappa_5}{5!T_c} - \frac{r^5\kappa_6}{6!T_c} + \dots\right) \quad (2.28)$$

In writing a computer program to calculate moments and cumulants, it probably is best to use the definition formula for moments and to use the following substitutions in Equation 2.21

$\qquad l_xm_x$ for f,

$\qquad R_0$ for N because $N = \sum f$ and R_0 is $\sum l_xm_x$,

and

$$T_c = \frac{\sum x l_xm_x}{\sum l_xm_x} = \frac{\sum l_xm_x}{R_0} = \overline{x}.$$

Moments for the age-of-mothers distribution (Figure 2.1) can be calculated by

$$\mu_p = \frac{\displaystyle\sum_{x=0}^{\omega} l_xm_x (x_i - T_c)^p}{R_0} \qquad (2.29)$$

and cumulants can then be determined using Equations 2.23 to 2.27. Finally, cumulants can be used in Equation 2.28 to reconcile the estimates of r obtained from Equation 3.3 and the Euler equation.

So, back to our small beast in Table 2.1. If l_xm_x (Figure 2.1) is normally distributed with respect to x, then just the second moment term in Equation 2.28 can be used to calculate r_c from the r calculated using the Euler equation or by simple projection as in Chapter 1. However, if the expected distribution of l_xm_x as a function of x is *not*

TABLE 2.3 Estimation of r for Our Small Beast (Table 2.1) Using Different Numbers of Terms in Equation 2.30

	r yr^{-1}	T years
start	1.4886	1.225
2 terms	1.6925	1.077
3	1.6199	1.126
4	1.6177	1.127
5	1.6270	1.121
6	1.6245	1.122
Euler	1.6242	1.123

Note: Each higher term means that all lower terms were included; the starting value of r, rc, is 1.4886; the final value using the Euler equation is 1.6242; T is generation time using $R_0 = 6.1920$.

normal, then additional terms, such as terms for skewness, kurtosis, etc. are needed (Equations 2.24-2.27).

The difficulty with Equation 2.28 is that one wants to get r from r_c but it is not possible to isolate r on one side of the equation and solve for it analytically. The solution for this is to use the Newton-Raphson algorithm and so the first step is to write Equation 2.28 so it is equal to 0. The first six terms of the expansion using cumulants

$$f(r) = r - \frac{r^2\kappa_2}{2!T_c} + \frac{r^3\kappa_3}{3!T_c} - \frac{r^4\kappa_4}{4!T_c} + \frac{r^5\kappa_5}{5!T_c} - \frac{r^6\kappa_6}{6!T_c} - r_c \quad (2.30)$$

and

$$f'(r) = 1 - \frac{2r\kappa_2}{2!T_c} + \frac{3r^2\kappa_3}{3!T_c} - \frac{4r^3\kappa_4}{4!T_c} + \frac{5r^4\kappa_5}{5!T_c} - \frac{6r^5\kappa_6}{6!T_c}. \quad (2.31)$$

In fact, of course, Equation 2.30 is not entirely correct because it could be improved by using additional terms of κ but it provides a much better estimate of r than does Equation 2.3.

The procedure is to use Equation 2.20, start with an initial estimate of r, that is r_c, and iterate to obtain an ever better estimate of r. Obviously it is not necessary to use all of the terms in Equation 2.31 and Table 2.3 shows the consequences of using increasing numbers of terms from just variance, κ_2, up to κ_6 in the estimate of r. What this means is that the Newton-Raphson method was used first with

$$f(r) = r - \frac{r^2\kappa_2}{2!T_c} - r_c \quad (2.32)$$

and

$$f'(r) = 1 - \frac{2r\kappa_2}{2!T_c}, \quad (2.33)$$

and then, with three terms,

TABLE 2.4 Calculation of Terms of the Stable-Age Distribution for Data from Table 2.1 Given r = 1.6242.

Age	l_x	$l_x e^{-rx}$	c_x
0	1.000	1.0000	0.9528
1	0.240	0.0473	0.0451
2	0.058	0.0022	0.0021

$$f(r) = r - \frac{r^2\kappa_2}{2!T_c} + \frac{r^3\kappa_3}{3!T_c} - r_c \quad (2.34)$$

and

$$f'(r) = 1 - \frac{2r\kappa_2}{2!T_c} + \frac{3r^2\kappa_3}{3!T_c}, \quad (2.35)$$

and so on.

Although the estimate of r is very close to the correct value when 6 terms are used, it is different at the 4th decimal place, 1.6245 *vs.* 1.6242. For our small beast, additional terms would have to be used with the age-of-mothers distribution in order to achieve the same accuracy as obtained with the Euler equation.

The point of all of this has been to reconcile the estimation of r using generation time with the estimation of r from the Euler equation and to indicate the nature of the problem. Knowing just the mean sometimes is insufficient for one's desires though possibly sufficient for one's needs.

THE STABLE-AGE DISTRIBUTION

The terms of the stable-age distribution, c_x, are calculated in much the same manner as they were in Chapter 1 except for a substitution using l_x and m_x in the equation (Table 2.4)

$$c_x = \frac{\text{Number in an age class}}{\text{Total population}}.$$

What is the number in any particular age class, x? It is the number born x time units ago (e^{-rx}) times the survival rate up to the present, l_x. The total population is the sum of these terms

$$c_x = \frac{l_x e^{-rx}}{\Sigma l_x e^{-rx}} \quad (2.36)$$

A glance back at Chapter 1 will show that the terms of the stable-age distribution are the same as obtained by population projection.

TABLE 2.5 Reproductive Value, v_x, and Residual Reproductive Value, v_x*, for Our Beast in Table 2.1

Age	l_x	m_x	$l_x m_x e^{-rx}$	$\sum l_t m_t e^{-rt}$	e^{rx}/l_x	v_x	v_x*
0	1.00	0	0.0	1.0	1.0000	1.0	1.0000
1	0.24	20	0.9459	1.0	21.1430	21.1430	1.1430
2	0.058	24	0.0541	0.0541	443.9438	24.0000	0.0000

Note: r = 1.6242 yr^{-1}

THE REPRODUCTIVE-VALUE DISTRIBUTION

A final column in a life table that can be calculated, after r has been estimated, is the reproductive-value distribution, v_x, which was developed by R. A. Fisher (1890 - 1962) in 1930. The reproductive value provides a measure of the present *and future* contribution of a female age x to r. Fisher posed the question in terms of the present value of future offspring. The question he asked was to what extent persons of a particular age, on average, contribute to the ancestry of future generations? He then continued (Fisher 1930, p 27),

" The question is one of some interest, since the direct action of Natural Selection must be proportional to this contribution."

The larger the reproductive value, the greater the contribution.

In words, the reproductive value, v_x, is the number of female offspring produced by a females age x and older divided by the number of females that are age x right now

$$v_x = \frac{\text{females produced right now by females age x and older}}{\text{total number of females age x}}.$$
(2.37)

The terms of Equation 2.37 are

females produced right now by females age x and older =

$$\sum_{t=x}^{\omega} e^{-rt} l_t m_t$$

and

total number of females a = $l_x e^{-rx}$.

Substituting into Equation 2.37 and doing a bit of rearranging produces

$$\frac{v_x}{v_0} = \frac{e^{rx}}{l_x} \sum_{t=x}^{\omega} e^{-rt} l_t m_t,$$
(2.38)

where v_x is the reproductive value at age x and v_0 is the reproductive value at birth, which always turns out to be 1.0 because when x = 0, l_0 =1, $e^{r\cdot 0}$ = 1.0, and

$$\sum_{t=0}^{\omega} e^{-rt} l_t m_t = 1$$

from Equation 2.13. Placing v_0 in Equation 2.38 really doesn't seem to have any other purpose than to show consistency with the past. Throw it out, if you like; it won't be missed.

The reproductive value has two parts: current contribution to reproduction, m_x, and the future contribution, which is called the *residual reproductive value*, v_x*. The residual value is found by subtracting m_x from v_x

$$v_x^* = v_x - m_x.$$
(2.39)

For purposes of computing v_x it is best to reformulate Equation 2.39 slightly to make use of the two parts, v_x* and m_x, and to formulate the calculation so it can be done recursively starting with the final v_x value and proceeding *backwards* to x = 0

$$v_x = m_x + \frac{v_{x+1} l_{x+1}}{l_x e^{r((x+1)-x)}}.$$
(2.40)

Multiplying r by ((x+1)-x) in the numerator may at first seem a bit strange because shouldn't it always be just equal to 1? It will equal 1 only as long as elements in the life table are separated by just one time unit but it is possible to have differences that are not just 1; it depends on how the table is constituted.

If time intervals are equal to 1, then Equation 2.40 is simplified to

$$v_x = m_x + \frac{v_{x+1} l_{x+1}}{l_x e^r}$$
(2.41)

or

$$v_x = m_x + \frac{v_{x+1} p_x}{e^r}.$$
(2.42)

Notice that by starting with the *last* age, ω, it is very easy to calculate v_ω because it always will be equal to the final m_x value, m_ω. Once the final v_x value, v_ω, is known, the penultimate v_x value, that is $v_{\omega-1}$ can be calculated using Equations 2.40 to 2.42. One proceeds in this fashion all the way back to v_0, which always will turn out to be 1.0 (Table 2.5).

NOTES ON BIRTH AND SURVIVAL RATES

Recall from Chapter 1 that the finite birth-rate, B, is

$$B = \frac{\# \text{born}_t}{\# \text{survivors}_{t-1 \; t \to t}}. \tag{2.43}$$

When the stable-age distribution has been achieved, it is possible to express B in terms of the stable-age distribution. First,

$$\frac{\# \text{born}_t}{\text{total}_t} = c_0, \tag{2.44}$$

which is the fraction of the population that is age 0. The survivors from t-1 to t are just the total population size at time t minus the number born at t

$$\# \text{survivors}_{t-1 \; t \to t} = \text{total}_t - \# \text{born}_t$$

or, the fraction of the population that are survivors is

$$\frac{\text{total} - \# \text{born}}{\text{total}} = 1 - \frac{\# \text{born}}{\text{total}} \tag{2.45}$$

or

$$\text{survivors} = 1 - c_0.$$

Therefore, substituting Equations 2.44 and 2.45 into 2.43 gives

$$B = \frac{\dfrac{\# \text{born}}{\text{total}}}{\dfrac{\text{total} - \# \text{born}}{\text{total}}} = \frac{c_0}{1 - c_0}. \tag{2.46}$$

Recall that

$$b = ln(1+B) = 1 + \frac{c_0}{1 - c_0} = \frac{1}{1 - c_0}$$

so:

$$\text{✳☞} \quad b = \frac{ln\left(\dfrac{1}{1 - c_0}\right)}{\Delta x}, \tag{2.47}$$

where Δx is the time interval between values in the life-table. Generally Δx will be equal to 1 such as 1 day or 1 year but other values are possible such as 0.5 days or 3 months.

The calculation of s, the survival-rate per individual, also can be done by referring to the stable-age distribution, as was done in Chapter 1. You can express a population, N, with a stable-age distribution as $N\Sigma c_x$, which means that you can multiply each term of the stable-age distribution by N to get the numbers in that age class. Now, if the population grows for 1 time period all of the individuals in an age class shift to the next age class and so the actual numbers in the age class will change depending on the population growth rate e^r, which is what was done in Chapter 1. At the start, the population is

$$N \sum_{x=0}^{\omega} c_x$$

and one time unit later it is

$$e^r N \sum_{x=1}^{\omega} c_x.$$

[*n.b.* x started with 0 for the original population and starts with 1 for the population one time period later.] Survival would be based on the number in age classes >0 after one time unit divided by the original population size, which means that age 0 individuals are included at the beginning but not after 1 time period. Consequently, survival rate, S, is

$$S = \frac{e^{r\Delta x} N \displaystyle\sum_{x=1}^{\omega} c_x}{N \displaystyle\sum_{x=0}^{\omega} c_x}. \tag{2.48}$$

N's cancel and

$$\sum_{x=0}^{\omega} c_x = 1.0,$$

which gets rid of the denominator, and so

$$s = \frac{ln\left(e^{r\Delta x} \displaystyle\sum_{x=1}^{\omega} c_x\right)}{\Delta x}. \tag{2.49}$$

TABLE 2.6 Life Table Values for a Marine Worm *Streblospio benedicti* Reared at 20°C and a Salinity of 34‰

Age	l_x	m_x	Age	l_x	m_x	Age	l_x	m_x
0	1.0000	0	13	0.0101	97.82	26	0.0056	35.17
1	0.4123	0	14	0.0101	0.00	27	0.0056	41.83
2	0.1700	0	15	0.0095	34.64	28	0.0050	32.60
3	0.1105	0	16	0.0084	70.56	29	0.0034	0
4	0.0718	0	17	0.0078	20.78	30	0.0017	0
5	0.0467	0	18	0.0078	61.56	31	0.0017	0
6	0.0303	0	19	0.0078	112.33	32	0.0011	0
7	0.0197	25.23	20	0.0072	215.33	33	0.0011	0
8	0.0197	32.62	21	0.0072	217.0	34	0.0006	0
9	0.0191	54.08	22	0.0072	231.89	35	0.0006	0
10	0.0168	70.36	23	0.0072	209.44	36	0.0006	0
11	0.0135	89.40	24	0.0072	171.78	37	0.0006	0
12	0.0106	107.30	25	0.0072	49.56	38	0.0006	0

Note: ages are in weeks; data from Levin *et al.* (1987).

S is the finite survival rate and s is the survival rate per individual (*cf.* Gulland 1969). Finally, note that

$$\sum_{x=1}^{\omega} c_x = 1 - c_0 \qquad (2.50)$$

and so

☞ $$s = \frac{ln\left[e^{r\Delta x}(1 - c_0)\right]}{\Delta x} \qquad (2.51)$$

which is similar to the derivation presented by Caughley (1977).

The time increment, d, is important in estimating both b and s and it must be a constant. Population growth rate per individual, r, can be estimated with different time intervals between lx values: x does not have to increment by a fixed amount; however, if r must be decomposed into b and s then a fixed interval is required.

The data for our small creature (Table 1.1) can be used to illustrate the application of equations 2.47 and 2.51. Using Equation 2.47,

$$b = 3.053143$$

and using Equation 2.51

$$s = -1.42895.$$

Therefore,

$$r = 3.053143 - 1.428951 = 1.624191.$$

At this point you probably have sufficient details about estimating r and the meanings of b and s that you would be able to calculate any of them from life table data such as Table 1.1.

BIRTH-FLOW POPULATIONS

When there is no defined reproductive season, reproductive data may require transformation before population growth rate and other parameters can be estimated. In birth-pulse populations, reproduction is concentrated at the point where individuals enter a new age class. The l_x values are appropriate for the m_x values because the probability of surviving to a particular age matches exactly the time at which reproduction takes place.

In birth-flow populations, individuals are constantly being added to an age class and also are constantly leaving the class. Furthermore, reproduction is spread across a time period from x to x+1 and females exactly age x will be different from females that are about to enter the age x+1. How one proceeds with analysis depends on how data were gathered. A common approach to gathering data for many small organisms is to follow individual females under laboratory conditions and to count the number of offspring every day or few days or every week and also to document survival. Table 2.6 shows results of such a study using a marine worm *Streblospio benedicti* (Levin *et al.* 1987).

The l_x values in Table 2.6 have their usual meaning; that is, the survival of a newborn individual to the beginning of an age class x to x+1. The m_x values are the number of female offspring produced by each female that enters the age class x and survives for the entire period x to x+1. This is an important point. The m_x values in the table are not correct for females that enter the age interval and then die sometime along the way from x to x+1 and so a correction

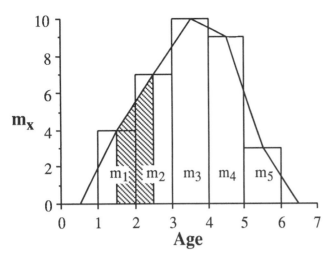

FIGURE 2.3 m_x values, the bars, determined for females that lived during the period x to x+1; the line connects the midpoints and adjusts for changes in fecundity during x to x+1; the shaded area, the trapezoid, is the adjusted m_x value that would be concentrated at x = 2. Note: we still have not adjusted for deaths of females during the interval x+.5 to x+1.5.

must be included. A second correction is needed because when a female enters the age class her fecundity may be quite different from a female that is about to leave the age class. We need to establish points at which reproduction can be considered to be concentrated and also to adjust m_x for female survival during an interval.

Where should we consider reproduction to be concentrated? There are various solutions and here is one. The important thing is to be consistent so that reproduction is concentrated in the middle of the interval over which m_x is adjusted. Figure 2.3 shows the relationships between m_x values and adjustments to account for changes in fecundity during a time interval. Reproduction is centered at exact ages (birthdays) such as 1, 2, 3, etc.

Figure 2.3 shows exactly what intervals are used, how areas for m_x are calculated, and where they must be concentrated for purposes of further analysis. First of all, lines are drawn that connect the midpoints of the bars for m_x, the total number of female off-spring produced from x to x+1. The area, which is m_x adjusted for changes in fecundity during the time interval is the area of the shaded trapezoid,

$$M_x = \frac{(m_{x-1} + m_x)}{2} \qquad (2.52)$$

except for the fact that females die during the interval, which can be corrected by considering that if n females are alive at the beginning of an interval, at the end of the interval there will be n times the survival rate. What is needed is the survival rate that covers the time-interval of the shaded trapezoid in Figure 2.3; that is, from the midpoint

of one interval, x-1 to x, to the midpoint of the interval x to x+1.

For the interval x-1 to x, the probability of survival from birth to the midpoint, x-0.5 is approximately

$$l_{x-0.5} = \frac{l_{x-1} + l_x}{2}.$$

For the interval x to x+1, the probability of survival from birth to the midpoint, x+0.5 is approximately

$$l_{x+0.5} = \frac{l_x + l_{x+1}}{2}.$$

Therefore, the probability, p_x, of surviving from x-0.5 to x+0.5 is

$$p_x = \frac{\dfrac{l_x + l_{x+1}}{2}}{\dfrac{l_{x-1} + l_x}{2}} = \frac{l_x + l_{x+1}}{l_{x-1} + l_x} \qquad (2.53)$$

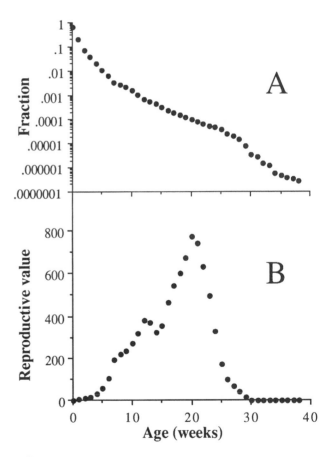

FIGURE 2.4 Analysis of *Streblospio benedicti* with planktonic development (Levin *et al.* 1987); **A,** stable-age distribution; **B,** Reproductive-value distribution; data were analyzed using a birth-flow model.

and so we can finish adjusting the value of m_x that will be concentrated at x

$$m_x^* = \frac{(m_{x-1} + p_x m_x)}{2}. \qquad (2.54)$$

One final point concerning intervals is the use of "x-0.5". This correction, 0.5, is appropriate if x increments by 1. Obviously it would have to be changed if the increment was something else and there are different ways of handling this. If all of the intervals are the same size, such as 3 years, then one can just call them interval 1, interval 2, interval 3 etc. and make the correction for actual time units at the end of the analysis. However, if intervals are of different sizes, then the actual units would have to be used for calculating the correct value for p in Equation 2.54.

So, finally, here is the analysis for *Streblospio benedicti* in Table 2.6. The value for r is 0.1895 week^{-1} or $\lambda = 1.2086$ week^{-1} when data were analyzed assuming flow-through births. These values are smaller than if pulsed births were assumed. With pulsed births, r = 0.2063 or $\lambda = 1.2292$. The differences are not great, but consider the consequences. Over a period of one year the differences suddenly seem much greater. If you started with 100 worms that followed the rules provided in Table 2.6, and if you did the analysis using pulsed births, then the estimated female population after one year would be 4,559,713. Using the birth-flow model, the population would be 1,903,434. The birth-pulse assumption would overestimate the population by more than 2 times. You might start losing money if you were running a worm farm with errors like that. The numbers after one year were calculated using the exponential equation. For example, for the birth-flow model:

$$1,903,434 = 100e^{0.1895 \times 52}.$$

The stable age distribution (Figure 2.4A) shows that most of the population would be concentrated in the young age classes and there would be few individuals in the old non-reproductive ages. The reproductive value for *Streblospio benedicti* with planktonic development (Figure 2.4B) shows a typical pattern for a species where numbers of offspring produced during a time interval increase with age, which, generally, is due to increased size of females. For *S. benedicti*, the maximum reproductive value for the data set that was analyzed, is at 20 weeks, which is 13 weeks after the age at first reproduction.

The great difference in reproductive value between new worms in the plankton and females age 20 weeks, provides a hint that selection for adult characteristics of mature worms would possibly be more likely than selection of traits in the planktonic stages.

TABLE 2.7 Life Table for *Gammarus lawrencianus*

Age	Pivotal age	Time period	l_x	L_x	m_x	M_x
0	2.5	0	1.00	0.99	0.0	0.0
5	7.5	1	0.98	0.96	0.0	0.0
10	12.5	2	0.94	0.93	0.0	0.0
15	17.5	3	0.92	0.90	0.0	0.0
20	22.5	4	0.88	0.81	0.0	0.0
25	27.5	5	0.74	0.71	0.04	0.02
30	32.5	6	0.68	0.64	2.0	1.02
35	37.5	7	0.60	0.56	4.6	3.3
40	42.5	8	0.52	0.26	10.7	7.65

Note: age 0 is five days after release from the brood pouch; m_x is the number of 1-day-old offspring (estimated as 1/2 the total number) collected per female during the 5-day interval and multiplied by 0.76, the survival rate from age 1 day to 5 days; data from (Doyle and Hunte 1981).

PROBLEMS WITH TIME INTERVALS

Life table data for an amphipod (*Gammarus lawrencianus*) maintained for 26 generations in a laboratory environment (Doyle and Hunte 1981) are shown in Table 2.7. It is necessary to understand how data were gathered in order to select the appropriate means for estimating population growth rate, and terms of the stable-age and reproductive-value distributions. Doyle and Hunte followed six cohorts of amphipods grown in 1.5 liter containers. Cohorts were started from broods that were one day old but counts were not started until animals were five days old so the l_x distribution starts with animals that were five days old but starting time still is called time 0. This is no problem as long as everything else in the life table is appropriately adjusted. Cohorts were censused every five days to determine survival but when reproduction began, new individuals were removed every day so "new individuals" really means juveniles that were released from brood chambers during a 24 hour period. All of these are called one-day old even though some might have been released moments before the daily census. An assumption is that survival for the first day is perfect and so the 24-hour census is all new recruits that females released from their brood chambers during one day. Survival rate for the next four days, to the start of the life table, is 0.76. Total counts of new juveniles during each 5 day period were first divided in half, assuming a 50:50 sex ratio, and then further reduced by multiplying by 0.76 before being entered in the m_x column. The m_x column also includes loss of females during the 5-day period and so Equation 2.54 would not be appropriate for analysis.

The life table begins with 5-day-old juveniles rather than numbers of eggs and m_x values have been adjusted so they

TABLE 2.8 Analysis of *Gammarus lawrencianus* Using Different Times When Reproduction is Viewed as Concentrated

Model	R_0	T_c days	r day^{-1}	Generation time
$L_x m_x$	6.666	38.58	0.0496	38.21
$l_x M_x$	6.666	37.44	0.0511	37.12

Note: all parameters have units of days; $L_x m_x$ uses pivotal ages and $l_x M_x$ uses ages; data from Doyle and Hunte (1981).

TABLE 2.9. Stable-Age and Reproductive-Value Distributions for *Gammarus lawrencianus*s Determined with Two Different Models

$R_0 = \Sigma l_x M_x$			$R_0 = \Sigma L_x m_x$		
Age (x)	c_x	v_x	Age (x)	c_x	v_x
0	0.2787	1.0	2.5	0.2815	1.1436
5	0.2116	1.3175	7.5	0.2130	1.5115
10	0.1572	1.7734	12.5	0.1610	1.9999
15	0.1191	2.3395	17.5	0.1215	2.6488
20	0.0883	3.1579	22.5	0.0853	3.7722
25	0.0575	4.8486	27.5	0.0584	5.5159
30	0.0409	6.7845	32.5	0.0410	7.7863
35	0.0280	8.4350	37.5	0.0280	8.4759
40	0.0188	7.65	42.5	0.0101	10.7

Note: data from Doyle and Hunte (1981).

are appropriate for a life table that defines 5-day old animals as entering the table at time = 0 . A life table can begin at any age one chooses; all that is required is that the table be consistent, which means that if the table begins when a brood is released, then m_x values must be based on numbers released from the pouch and not fertilized eggs or some other stage. For example, with birds m_x may be newly laid eggs, in which case the life table must begin with eggs and not nestlings or fledglings but the table could begin with fledglings. All that is required is that m_x and t=0 match and the amphipod data illustrate this problem in defining m_x.

With the data gathered by Doyle and Hunte (1981),*Gammarus lawrencianus* i is best modeled as a birth-flow population. The important key to selection of this method is that gathering data on production of new individuals over a 5-day period implicitly *included* the loss of reproductive females due to death and so differs from the example of *Steblospio benedicti* (Table 2.6) where m_x values were only for worms that survived for an entire week.

There are at least two ways of adjusting the life table so that reproduction can be viewed as concentrated at a particular point. One would be to adjust l_x values so they match the midpoint of a time interval over which m_x was determined and other would be to adjust m_x values to match entry into an age class. It would seem both approaches should yield the same results and so it is instructive to

compare them. The first approach is adjustment of l_x to the midpoint of an age class where m_x can be considered to be concentrated each five day period. Survival of a newborn female to the midpoint of an age interval, the pivotal age, is L_x

$$L_x = \frac{l_x + l_{x+1}}{2}. \tag{2.55}$$

If m_x can be considered to be concentrated at the midpoint of a time interval, the Euler equation can be used with column of L_x and m_x and pivotal ages in Table 2.7 to estimate r. Similarly, m_x could be concentrated around an age x by calculating M_x

$$M_x = \frac{m_{x-1} + m_x}{2}. \tag{2.56}$$

and so the Euler equation could be used with l_x and M_x columns together with age x in Table 2.7 to estimate r.

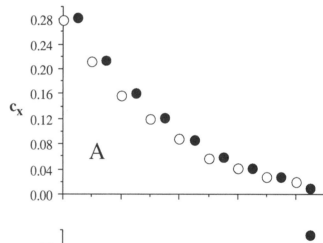

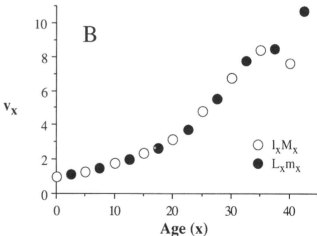

FIGURE 2.5 Analysis of *Gammarus lawrencianus* using two different models for matching survival and reproductive schedules: Equation 2.55 for survival (solid circles) and Equation 2.56 for reproduction (open circles); **A**, stable-age distribution, c_x; **B**, reproductive-value distribution, v_x.

Results are summarized in Table 2.8 and it is clear that estimates of r are not the same for the two models. Estimates of the net reproductive rate per generation, R_0, are the same but initial estimates of generation times, T_c, differ, which means that the selection of times at which reproduction is concentrated were not equivalent. This is not surprising considering that Equations 2.54 and 2.55 are *approximations*. Estimates of r day^{-1} are close for the two methods and possibly the best estimate would be the mean of the two estimates or r = 0.050 day^{-1}.

Use of pivotal ages associated with L_x values is important only for reproductive ages; nonreproductive classes can have l_x values and then L_x values can be used with pivotal ages. The estimated value of r is 0.0496 day^{-1}, the same as when pivotal ages are used for all values in the table. Obviously this has to be true because the various sums that are used all include $l_x m_x$ or $L_x m_x$. Where m_x is 0, the term is of no importance. Also, it is possible to do the analysis using time periods (Table 2.7) rather than actual ages. When this is done, r must be divided by 5 to obtain r day^{-1} and generation time must be multiplied by 5 to get time in days.

The stable-age (c_x) and reproductive-value (v_x) distributions (Table 2.9) show that use of the two different ways of matching survival and fecundity provide similar distributions of c_x and v_x although shapes are not identical. In particular, the reproductive values for the oldest individuals decrease when $l_x M_x$ is used but increase with $L_x m_x$. Comparisons are presented in Figure 2.5.

The estimated value for r = 0.05 day^{-1} may seem quite small but actually is very large with extrapolated to one year by multiplying by 365

$$r = 18.25 \text{ yr}^{-1},$$

which is a finite growth rate, λ, of

$$\lambda = 84,000,000 \text{ yr}^{-1}.$$

If you started with 100 females distributed so that they fit the stable age distribution, in one year you would have 8.4×10^9 female amphipods, a frightening prospect indeed. The point of doing this estimate is to suggest that the measurements made in the laboratory probably are not ones that could be sustained for any length of time without some external intervention. In the study conducted by Doyle and Hunte (1981), all juveniles were removed from cultures started with 5-day-old animals and so the cultures never were limited by increasing density. The high value of r, in fact, probably could be maintained as long as harvesting was continued. Without harvesting, the life table would change so r would approach 0 or become negative. A life table always is for some stated conditions, which may be in the laboratory or in the field. If in the field, conditions are the

prevailing ones for the period of the study and at least implicit is that a somewhat or a very different table might be obtained at some other time or place. There is, of course, no reason why a population could not have characteristics appropriate for unlimited growth in the field. All that would be required would be periodic catastrophes so that population growth would be punctuated by loss of a substantial fraction of the population without modifying the basic l_x and m_x schedules.

USE OF r IN ANALYSIS OF LIFE HISTORIES

Birch (1948) in a classic paper on the use of r, presented a life table for the rice weevil *Sitophilus* (=*Calandra*) *oryzae* starting at the age of first reproduction; l_x began with eggs and so the first l_x value in the table represents survival from a newly laid egg to the pivotal age of 4.5 weeks (Table 2.10).

Birch used the Euler equation (Equation 2.13) to estimate r as equal to 0.762wk^{-1}. Imagine how this would have been done in the 1940's — no computers, no pocket calculators and mechanical calculators that could not do much more than add and subtract. Birch used a book of math tables that contained exponentials and provides tips on how to use the tables to squeeze out additional significant digits.

Birch explored an interesting relationship between the length of the immature stage and r and demonstrated that if the immature phase is short, then fewer eggs would be needed in order to achieve the same value or r. He used a very short version of his life table and included only four reproductive age classes. Because we can use a computer program that will calculate r very easily, we can use all of the age classes and test to see whether Birch's conclusions still seem correct. The procedure is to shift the age of first reproduction and then adjust m_x in order to achieve a value of r equal to 0.762wk^{-1}.

TABLE 2.10 Life Table for the Rice Weevil *Sitophilus* (=*Calandra*) *oryzae*

Age x	l_x	m_x	Age x	l_x	m_x
4.5	0.87	20.0	12.5	0.59	11.0
5.5	0.83	23.0	13.5	0.52	9.5
6.5	0.81	15.0	14.5	0.45	2.5
7.5	0.80	12.5	15.5	0.36	2.5
8.5	0.79	12.5	16.5	0.29	2.5
9.5	0.77	14.0	17.5	0.25	4.0
10.5	0.74	12.5	18.5	0.19	1.0
11.5	0.66	14.5			

Note: m_x is egg number/2 assuming 50% males and is concentrated at a pivotal age in weeks; data from Birch (1948).

TABLE 2.11 Total Age-Specific Egg Production
Required to Maintain r=0.762 wk^{-1}

Age	Eggs/week	Eggs/week	Eggs/week
1.5	4.07		
2.5	4.67		
3.5	3.05		
4.5	2.54	**40**	
5.5	2.54	**46**	
6.5	2.84	**30**	183.49
7.5	2.54	**25**	211.01
8.5	2.95	**25**	137.61
9.5	2.24	**28**	114.68
10.5	1.93	**25**	114.68
11.5	0.51	**29**	128.44
12.5	0.51	**22**	114.68
13.5	0.51	**19**	133.03
14.5	0.81	**5**	100.92
15.5	0.20	**5**	87.16
16.5		**5**	22.93
17.5		**8**	22.93
18.5		**2**	22.94
19.5			36.70
20.5			9.17
Total	31.91	314	1440.37

Note: actual age of first reproduction is at 4.5 weeks and column of observed fecundities is shown in **bold**; data from Birch (1948).

The evaluation of Birch's (1948) life-history analysis can be done by modifying the program that uses Newton's method to calculate r and this is shown at the end of this chapter. Age-specific egg production (Table 2.11) is similar to the table produced by Birch (1948, his Table 6) but there are a few minor differences. When 6.5 weeks is the age at first reproduction, the production for 6.5 weeks is 183.5 eggs. Birch gives 117 eggs. Considering that egg production for 7.5 weeks is 211.0 (Table 2.11) compared with 204 given by Birch, the value of 117 eggs for 6.5 weeks probably is a typographical error in his paper. A second difference between Table 2.11 and Birch's Table 6 is the total number of eggs produced. Table 2.11 includes the entire life table so it should not be surprising to see that totals are much greater than those given by Birch. What is important is that Birch's conclusions do not have to be changed by the analysis presented here.

Shifts in the age at first reproduction have profound effects on population growth. From a life history stand point, one would expect selection to favor ever decreasing the age at first reproduction; however, such a decrease may carry other burdens such as small size. As Birch points out, *Sitophilus* (=*Calandra*) *oryzae* has an adult size that matches the size of the food that is eaten by the larvae. Resources of a single rice grain are just sufficient to provide for one larva. The food resource fixes the length of development and the size at metamorphosis which, in turn, fixes the size of the adult and so places a constraint on how many eggs can be produced. A delay in age at first reproduction may result in a larger size but with a burden of having to produce large numbers of offspring in order to compensate for the delay. A delay in *S. oryzae* would require either a shift to a grass with larger seeds or a major leap in larval feeding that would include more than one seed. The concepts of trade-offs and constraints in life histories remain important in evolutionary ecology and even though there are just two vectors, l_x and m_x, how natural selection combines these schedules continues to provide wonderful puzzles.

GENERAL COMMENTS

Birth and survival schedules interact to determine whether numbers in a population of organisms increase, decrease or remain the same. There are additional consequences. If these schedules remain fairly constant over time, a population will tend towards assuming a fixed proportion of individuals in each age class, which is the stable-age distribution. Also, the relative contributions of females to current and future population growth can be calculated, which is the reproductive value of females in each age class. Populations may have pulsed or continuous reproduction and this temporal scheduling modifies the manner in which data must be analyzed.

The Euler equation is one convenient model for combining survival and reproductive data for purposes of estimating population growth as well as stable-age and reproductive-value distributions but there are other ways of combining survival and reproduction. Population growth can be determined using descriptive statistics of the age-of-mothers distribution; however, this approach has not been developed.

PROBLEMS.

1. Use the data on gray foxes from Chapter 1 and the BASIC program for the Newton-Raphson method (called **EULER.BAS**) to complete the following:
   a. Calculate r, the terms of the stable-age distribution and the reproductive-value function using the Euler equation. Show the results of each iteration.
   b. Graph the expected life-time female production of a newborn female gray fox ($l_x m_x$ *vs.* x). What is the $\overline{X}$ (T_c) component of the centroid of the distribution? What is the variance?
   c. Calculate b and s and check to be sure that b + s = r.

2. Brousseau and Baglivo (1988) provide a life table for the soft-shell clam *Mya arenaria* at Westport, CT (Table 2.12).

  a. Calculate r. Is it equal to 0.0 as the authors claim?

  b. What is the initial estimate of generation time, T_c, and how does this compare with your final estimate?

  c. Calculate the terms of the stable age and reproductive value vectors. Graph the reproductive value distribution. How does its shape compare with the distribution we obtained for *Gammarus lawrencianus* in Table 2.9?

TABLE 2.12 Life Table for *Mya arenaria*

Age	l_x	m_x
0	1.0	
1	1.0599×10^{-7}	1485615
2	6.7728×10^{-8}	2989512
3	4.3630×10^{-8}	3867846
4	2.8905×10^{-8}	5192053
5	1.9309×10^{-8}	5892286
6	1.2734×10^{-8}	6464925
7	8.1780×10^{-9}	6478050
8	5.0115×10^{-9}	6842903
9	3.0024×10^{-9}	8390977
10	1.5075×10^{-9}	8390977

Note: age in years; data from Brousseau and Baglivo (1988).

TABLE 2.13 Life Table for the West Indian Manatee *Trichechus manatus*

Age	l_x	Age	l_x
1-6	0.955	31-36	0.540
7-12	0.860	37-42	0.345
13-18	0.830	43-48	0.144
19-24	0.780	49-54	0.027
25-30	0.690	55-60	0

Note: age in years; data from Packard (1985).

3. Very nearly everyone I know likes manatees. They are not easy to study and the data assembled by Packard (1985) for the West Indian manatee *Trichechus manatus* (Table 2.13) must be viewed as preliminary. Females begin to reproduce at age 7 and litter size was assumed to be 1. The interbirth interval generally is 2 years. The proportion of mature females breeding was selected as 0.61 so m_x for a 6 year period should be $3 \times 0.5 \times 0.61$ or 0.915 (remember that life tables are for *females*).

  a. What is r for the manatee life table?

  b. Calculate the terms of the stable age and reproductive value vectors for manatees. Compare the shape of the reproductive value distribution with *Gammarus lawrencianus* in Table 2.9 with respect to when the peak occurs relative to the age at first reproduction.

  c. What is the generation time *in years*? *Hint:* remember that the values in the table are for intervals of six years.

4. Hughes and Roberts (1981) present a life table for *Littorina rudis* (Table 2.14), a marine snail, in Aber saltmarsh near Bangor, North Wales.

  a. The authors claim that the generation time is 24 months. Do you agree?

  b. The authors show (Figure 2 on page 259 of their paper) that although there were substantial fluctuations in density from 1974 through 1979, the overall trend was r=0. Does their life table match the trend?

  c. Graph m_x as a function of age, x. When spawning takes place at different times of the year, when do new snails enter the population? At once? Some time later? How might time of entry of new individuals to the population influence your estimate of generation time in part a?

TABLE 2.14 Life Table for *Littorina rudis* in Aber Saltmarsh, N. Wales

Age (months)	l_x	m_x
0	1.0	0
3	0.04	0
6	0.025	0
9	0.01569	0
12	0.00983	3.5
15	0.00615	24.0
18	0.00385	12.5
21	0.00241	8.0
24	0.00151	15.5
27	0.00095	78.0
30	0.00059	30.5
33	0.00037	17.5
36	0.00023	31.0
39	0.00015	137.5
42	0.00009	49.0
45	0.00006	26.5
48	0.00004	45.0

Note: data from Hughes and Roberts (1981).

5. Strijbosch and Creemers (1988) studied a lizard, *Lacerta vivipara*, from 1976-1982 at Bergen in the Netherlands. They provide estimates of juveniles (1st year class), subadults (2nd year class) and adults (3+ year classes) (Figure 2.5). The numbers of juveniles varied between 152 and 65 but the adults were much more constant and varied from a high of 74 to a low of 40. There doesn't seem to be a major population trend of increasing or decreasing and so r should be rather close to 0.0.

TABLE 2.15 Life Table for a Lizard, *Lacerta vivipara*

Age	l_x	m_x	Age	l_x	m_x
0	1000	0.00	4	57	4.88
1	424	0.08	5	10	6.50
2	308	2.94	6	7	6.50
3	158	4.13	7	2	6.50

Note: age in years; data from Strijbosch and Creemers (1988).

a. Does the life table (Table 2.15) support Figure 2.5? Is r very close to 0?
b. Examine your calculated terms for the stable-age distribution and compare these with Figure 2.5. Is the graph consistent with the stable-age distribution?

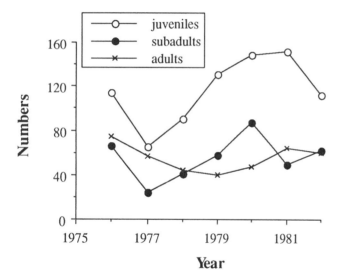

FIGURE 2.5 Population estimates for the lizard *Lacerta vivipara* on the de Hamert reserve at Bergen, Netherlands; data from Strijbosch and Creemers (1988).

TABLE 2.16 Life Table for the 1978 Cohort of the Cactus Finch *Geospiza scandens* on Isla Daphne Major

Age	l_x	m_x	Age	l_x	m_x
0	1.000	0.0	7	0.089	0.0
1	0.434	0.051	8	0.056	0.0
2	0.367	0.667	9	0.056	2.20
3	0.333	1.500	10	0.033	0.0
4	0.322	0.655	11	0.022	0.0
5	0.300	5.500	12	0.0	...
6	0.178	0.687			

Note: age in years; data from Grant and Grant (1992).

6. Grant and Grant (1992) studied Darwin's finches on Isla Daphne Major, Galápagos, from 1975-1991. Over the long run, one would expect r=0 and Grant and Grant (1992) present a graph of numbers of nesting pairs that suggest that *Geospiza scandens*, the cactus finch, had its up and downs but without any obvious trend of increase or decrease. A life table for a cohort should be a sample of all cohorts and so should tend towards r = 0. Should we consider the 1978 cohort (Table 2.16) to be typical in the sense of r=0?

TABLE 2.17 Life Table for *Daphnia pulex*

Day	l_x	m_x	Day	l_x	m_x
0	1.00	0	21	0.15	5.667
3	1.00	0	24	0.15	1.667
6	0.65	0	27	0.15	6.333
9	0.50	0	30	0.15	7.333
12	0.25	0	33	0.15	5.000
15	0.15	0	36	0.05	
18	0.15	0			

Note: data from Paloheimo and Taylor (1987).

7. Various species of the genus *Daphnia* have been used to study population dynamics for many years. Paloheimo and Taylor (1987) report rearing *D. pulex* at a concentration of 5,000 cells of *Chlamydomonas reinhardtii* per milliliter and determining l_x and m_x (Table 2.17). Recall that under good conditions *Daphnia* spp. are parthenogenetic and all offspring are female. The m_x column should be read as the number of offspring produced during the interval x to x+1 by females that survived the entire period. Do the analysis using a birth-flow model.
   a. Graph the m_x distribution *vs.* age and show where mid-points occur and where you are creating the trapezoids that approximate m_x for the mid-points.
   b. Produce a table to show the new values for m_x both before and after you have adjusted for deaths during a time interval.
   c. Calculate r, b, and s. How do these compare with estimates assuming a birth-pulse model?
   d. Determine the stable age and reproductive value distributions.
   e. What is generation time.
   f. Calculate the variance, skew and kurtosis of the age-of-mothers distribution and show how incorporation of each of these adjusts the estimate of r?
   g. What would be the effect on r of delaying age of first reproduction by one time interval; that is, from the interval 21-24 days to 24-27 days? How much would reproduction have to be increased to bring r back to the original value?

TABLE 2.18 Life Table for Waterbuck in the Umfolozi Game Reserve in South Africa

Age	l_x	m_x	Age	l_x	m_x
0	1000	0	7	125.5	0.42
1	193.0	0	8	111.3	0.42
2	180.8	0.25	9	89.0	0.42
3	180.8	0.42	10	64.6	0.42
4	174.2	0.42	11	17.9	0.42
5	166.1	0.42	12	0.0	
6	145.8	0.42			

Note: age in years; data from Melton (1983).

8. The waterbuck (*Kobus ellipsiprymnus*) has been studied in the Umfolozi Game Reserve in South Africa (Melton 1983). Survival was determined from estimates of ages at death using a sample of 95 skulls (Table 2.18) and are presented starting with a cohort of 1000 and so to calculate r you must first divide all lx values in the table by 1000. Melton discusses the problem of getting l_x from a death assemblage. If you know r then you should be able to correct estimates of l_x. A problem, of course, is that a short-term estimate of population change probably is not the same as r. He shows that "corrected" and "uncorrected" estimates give about the same values.

   a. Melton calculated r = -0.13 yr^{-1} and a generation time of 5.55 years. Do you agree with these estimates?
   b. What change in first year survival would be required to make r=0.0? [*Hint*: when you change first year survival the entire rest of the l_x table must be changed.]
   c. What would be the effect on r of leaving juvenile survival as given but make adult survival, starting at age 2, 100%; that is, keep l_x 180.8 from 2 to 11?

9. The following life table (Table 2.19) was reconstructed from Figures 7 and 8 in Dumont and Sarma (1995) for the predatory rotifer *Asplanchna girodi* at a prey density 2000 prey ml^{-1}. The prey was another rotifer *Anuraeopsis fissa*. This is how data were gathered:

"Each test cup consisted of 1 ml of medium with a defined prey density and one neonate A. girodi (ca 1 h old). Observations, made at 12 h intervals, included the number of neonates born and the number of dead adults. After each 24 h interval, the surviving adults were transfer to fresh embryo cups containing an appropriate food density. Neonates and dead adults at each observation were discarded. The experiment was continued until the last individual had died."

TABLE 2.19 Life Table for *Asplanchna girodi* at a Prey Density of 2000 ml^{-1}

Age	l_x	m_x	Age	l_x	m_x
0.0	1	0	2.5	1.00	5.5
0.5	1	0	3.0	0.32	3.0
1.0	1	0.5	3.5	0.32	1.5
1.5	1	3.6	4.0	0.17	0
2.0	1	4.0			

Note: age in days; data from Dumont and Sarma (1995).

   a. Select the most appropriate model for estimating r, b and s, explain why you selected it, and then estimate these parameters. Be sure to include appropriate time units in your answer.
   b. Redo the analysis using population projection program **PROJECTION.BAS**. Be very sure that you construct your data file correctly. Present your results and compare them with your results in part a. If results are different, explain why they are different in the sense of showing how you can turn the results from one method into the results of the other.

TABLE 2.20 Life Table for Domestic Sheep, *Ovis aries*, in New Zealand

Age	l_x	m_x	Age	l_x	m_x
0	1.0	0	6	0.626	0.543
1	0.845	0.045	7	0.532	0.502
2	0.824	0.391	8	0.418	0.468
3	0.795	0.472	9	0.289	0.459
4	0.755	0.484	10	0.162	0.433
5	0.699	0.546	11	0.060	0.421

Note: data from Hickey (1960, 1963).

10. Caughley (1967) presents survival and reproductive data for domestic sheep, *Ovis aries*, in New Zealand (Table 2.20) based on data in Hickey (1960, 1963); ages are in years.

   a. Calculate r, b, and s. What are the *annual* growth, birth and survival rates for the population? [*Hint*: remember the relationship between b and the annual birth rate per female.]
   b. It certainly looks as though females might survive for longer than just 11 years. What would be the effect on r of 4 more age classes all with l_x values of 0.060 and m_x values of 0.421. Add for age 12, $l_{12} = 0.06$, $m_{12} = 0.421$, for age 13 $l_{13} = 0.06$, $m_{13} = 0.421$ etc.

TABLE 2.21 Life Table for the Human Louse, *Pediculus humanus*

Age	l_x	m_x		Age	l_x	m_x		Age	l_x	m_x	
0.0	1.000	0	⎫	22.5	.749	0.934	⎫	45.5	.178	2.303	⎫
0.5	.993	0		23.5	.748	1.628		46.5	.161	2.176	
1.5	.978	0		24.5	.729	2.212		47.5	.136	2.208	
2.5	.964	0		25.5	.718	2.363		48.5	.125	2.371	
3.5	.949	0	⎬ eggs	26.5	.700	2.527		49.5	.112	2.380	
4.5	.935	0		27.5	.678	2.733		50.5	.101	2.708	
5.5	.920	0		28.5	.665	2.617		51.5	.089	2.394	
6.5	.906	0		29.5	.640	2.688		52.5	.072	2.434	
7.5	.891	0		30.5	.609	2.825		53.5	.051	2.722	
8.5	.877	0	⎭	31.5	.579	2.687		54.5	.036	2.895	
9.5	.868	0	⎫	32.5	.553	2.877	⎬ adults	55.5	.033	1.824	⎬ adults
10.5	.859	0		33.5	.511	2.389		56.5	.021	2.636	
11.5	.850	0		34.5	.469	2.649		57.5	.017	5.056	
12.5	.840	0		35.5	.439	2.670		58.5	.015	4.000	
13.5	.831	0		36.5	.397	2.817		59.5	.015	1.688	
14.5	.822	0		37.5	.356	2.923		60.5	.011	2.833	
15.5	.813	0	⎬ larvae	38.5	.328	3.029		61.5	.010	2.200	
16.5	.804	0		39.5	.299	2.737		62.5	.010	2.200	
17.5	.795	0		40.5	.267	2.940		63.5	.009	1.500	
18.5	.786	0		41.5	.261	2.721		64.5	.008	2.125	
19.5	.776	0		42.5	.235	2.835		65.5	.006	2.166	
20.5	.767	0		43.5	.216	2.539		66.5	.002	2.000	⎭
21.5	.757	0	⎭	44.5	.189	2.640	⎭				

Note: x is age in days; data from Evans and Smith (1952)

c. Assume that you are a rancher interested in harvesting your sheep and also getting the flock to grow as rapidly as possible. Use the reproductive-value distribution to justify which sheep to slaughter. Show how killing 50% of these age classes changes r.

d. Now assume that you work for the New Zealand government and are charged with reducing sheep in a national park. Use the reproductive-value distribution to justify a management plan that will focus on just one age class. How will changing the survival rate of this one age class by 50% change r? [*Hint*: if you change one l_x value you change all values that come after it]

11. Evans and Smith (1952) determined survival and fecundity for the human louse, *Pediculus humanus* (Table 2.21). The l_x values are survival to the *midpoint* of the time interval and m_x is total egg production divided by 2 and so m_x is female eggs assuming an equal sex ratio. This is what Evans and Smith say about their rearing technique:

"The lice were kept in uncrowded colonies on small (1.5 x 1.5 inches) patches of woolen cloth in glass beakers, which were removed from the cabinet for about 15 or 20 minutes in the early morning and for a similar period in the late afternoon to allow the insects to feed on the arms or legs of volunteer hosts. Counts of surviving individuals in each colony were made daily, and dead lice were removed as well as all eggs produced during the preceding 24-hour period. Each day's egg production was kept separately and formed the basis of a subcolony of which a number were kept for observations of larval development."

Evans and Smith used the data presented in Table 2.21 with no further transformations or adjustments, and the Euler equation to estimate r = 0.111 day^{-1}.

a. Analyze the data gathered by Evans and Smith to obtain the best estimate of r, b and s. In order to do this you must decide whether birth-flow or birth-pulse is the better model for lice and then you must decide where time intervals must be centered and if and how m_x must be adjusted. Be sure to write a justification for why you did the analysis in a particular way. This is the same as asking whether you think that Evans and Smith did their analysis in the best manner.

TABLE 2.22 Life Table for the Crayfish
Orconectes virilis

Age	l_x	m_x	Age	l_x	m_x
0	1.000	0.00	2.5	0.0088	0.00
0.5	0.075	0.00	3.0	0.0014	53.50
1.5	0.028	0.00	3.5	0.0013	0.00
2.0	0.017	41.35			

Note: age in years; data from Momot (1967).

12. Crayfish, *Orconectes virilis*, were studied by Momot (1967) using mark-recapture (Table 2.22). Age, x, is in years and marks the beginning of the time interval so, for example, 0.5 means from 0.5 to 1.5, etc.

    a. Estimate r first by using the data exactly as presented in the table and then by restructuring the table so not only can you get r but also b and s.

    b. What are the *finite* growth, birth and survival rates *per year*?

    c. Justify why you selected a particular model for your analysis.

BASIC PROGRAMS

EULER.BAS

The program **EULER.BAS** uses the Euler equation and calculates r using the Newton-Raphson method.

```
10 dim m(100),l(100),c(100),v(100),
   t(100),mc(100)
20 print "Estimating population growth
   per individual using the Euler/Lotka
   equation"
30 print " "
40 print "Another fine product from
   Cornered Rat Software ©"
50 print " "
60 print "                        T. A.
   Ebert      1996, revised 1998"
70 print " "
80 input "Are data from a file (F) or
   from the keyboard (K)? (F/K): ";f$
90 if f$ = "F" or f$ = "f" then goto 320
100 if f$ <> "k" and f$ <> "K" then goto
    80
110 input "How many age classes (e.g.
    ages 0, 1 and 2 would be 3): ";x
120 x = x-1
130 for i = 0 to x
140 print "age(";i;")=",
```

```
150 input t(i)
160 print "l(";t(i);") = ",
170   input l(i)
180 print "m(";t(i);") = ",
190   input m(i)
200 next i
210 if l(x) = 0 then let x = x-1
220 input "Do you want to save the data
    file? (Y/N)";f$
230 if f$ = "N" or f$ = "n" then goto
    440
240 input "Name of file for data: ";f$
250 open f$ for output as #9
260 print #9,x
270 for i = 0 to x
280 print #9,t(i);",";l(i);",";m(i)
290 next i
300 close #9
310 goto 440
320 input "Name of file with data: ";f$
330 open f$ for input as #9
340 input #9,x
350 print "x = ";x
360 for i = 0 to x
370 input #9,t(i),l(i),m(i)
380 print t(i),l(i),m(i)
390 next i
400 if l(x) = 0 then let x = x-1
410 close #9
420 print " "
430 print " "
440 lm1 = 0
450 lm2 = 0
460 input "Is for analysis for
    continuous births (C) or pulsed
    births (P)? C/P ";a$
470 if a$ = "P" or a$ = "p" then goto
    600
480 l(x+1) = 0
490 m(x+1) = 2*m(x)-m(x-1)
500 if m(x+1) < 0 then let m(x+1) = 0
510 t(x+1) = t(x)+(t(x)-t(x-1))
520 for i = 1 to x
530 p = (l(i)+l(i+1))/(l(i-1)+l(i))
540 mc(i) = (m(i-1)+p*m(i))/2
550 next i
560 for i = 0 to x
570 m(i) = mc(i)
580 print i,m(i)
590 next i
600 rem Calculate the sums of lxmx and
    xlxmx
610 for i = 0 to x
620 lm1 = lm1+l(i)*m(i)
```

```
630 lm2 = lm2+t(i)*l(i)*m(i)
640 next i
650 rem Estimate and initial value for r
660 t = lm2/lm1
670 r = log(lm1)/t
680 print "R0=";lm1,"T=";t,"r = ";r
690 for j = 1 to 100
700 f = 0
710 fp = 0
720 for i = 0 to x
730 w = l(i)*m(i)*exp(-r*t(i))
740 f = f+w : fp = fp+t(i)*w
750 next i
760 fp = -fp
770 f = f-1
780 r1 = r-f/fp
790 print f,fp,r,r1
800 if abs(r1-r) < 1.000000E-07 then
    goto 840
810 r = r1
820 print j,r
830 next j
840 print " "
850 r = r1
860 print "Final r = ";r;"  after ";j;"
    iterations"
870 print "exp(r) = ";exp(r)
880 print
890 print "The Stable Age Distribution
    and Reproductive Value Function"
900 d = 0
910 for i = 0 to x
920 d = d+l(i)*exp(-r*t(i))
930 next i
940 for i = 0 to x
950 c(i) = l(i)*exp(-r*t(i))/d
960 next i
970 v(x+1) = 0
980 l(x+1) = 0
990 m(x+1) = 0
1000 for i = x to 0 step -1
1010 v(i) = m(i)+l(i+1)/l(i)*exp(-
     r*(t(i+1)-t(i)))*v(i+1)
1020 next i
1030 for i = 0 to x
1040 print t(i),c(i),v(i)
1050 next i
1060 d = t(x)-t(x-1)
1070  w = 0
1080 for i = 1 to x
1090 dx = t(i)-t(i-1)
1100 if abs(d-dx) <> 0 then let w = 1
1110 next i
1120 b = log(1/(1-c(0)))/d
```

```
1130 s = log(exp(r*d)*(1-c(0)))/d
1140 t = log(lm1)/r
1150 if w = 1 then print "Unequal time
     intervals   "
1160 if w = 1 then goto 1180
1170 print "b = ";b;"  s = ";s;"  r =
     ";b+s
1180 print "Generation time = ";t
1190 input "Filename for output: ";f$
1200 open f$ for output as #9
1210 print #9,"Final r = ";r;"  after
     ";j;" iterations"
1220 print #9,"exp(r) = ";exp(r)
1230 print #9,"The Stable Age
     Distribution and Reproductive Value
     Function"
1240 for i = 0 to x
1250 print #9,t(i),c(i),v(i)
1260 next i
1270 if w = 1 then print #9,"Unequal
     time intervals   "
1280 if w = 1 then goto 1300
1290 print #9,"b = ";b;"  s = ";s;"  r =
     ";b+s
1300 print #9,"Generation time = ";t
1310 close #9
1320 end
```

Instructions for EULER.BAS.

Data can be entered either at the keyboard or from a file, which can be created using a word processor and then *saved as a text file*. If you don't save your data as a text file, when you run **EULER.BAS** and access the data file, the program may fill the screen with strings of ?????????? until you kill the program, which you can do using "Control ."; that is, hold down the Control key and then depress period (.). Be careful! Always save data files as text files.

Here is an example of how a life table is written as an appropriate file for **EULER.BAS**. The data, once again, are our small creature presented in Table 2.1 and repeated here:

Age (x)	l_x	m_x
0	1.000	0
1	0.240	20
2	0.058	24

The first value in a data file should be the number of age classes *not counting 0*. For example, the maximum age-class starts at 2 and so the first value in the data file is 2. This should be followed by rows of data with each row containing "age", "lx", and "mx". Data values should be separated by commas

```
2
0,1 ,0
1,0.24 ,20
2,0.058 ,24
```

R_FROM_RC.BAS

The program **R_FROM_RC.BAS** calculates r from the shape of the age-of-mothers distribution using up to a 6th order cumulant. The data format is the same as for **EULER.BAS**.

```
10 dim m(100),l(100),t(100),u(10),
   k(10), fc(10)
20 print "Estimating population growth
   per individual using moments of the
   age-of-mothers distribution"
30 print " "
40 print "Another fine product from
   Cornered Rat Software ©"
50 print " "
60 print "                         T. A.
   Ebert      1996"
70 print " "
80 input "Are data from a file (F) or
   from the keyboard (K)? (F/K): ";f$
90 if f$ = "F" or f$ = "f" then goto 300
100 if f$ <> "k" and f$ <> "K" then goto
    80
110 input "How many age classes (e.g.
    ages 0, 1 and 2 would be 3): ";x
120 x = x-1
130 for i = 0 to x
140 print "age(";i;")=",
150   input t(i)
160 print "l(";t(i);") = ",
170 input l(i) : print "m(";t(i);") = ",
180 input m(i)
190 next i
200 input "Do you want to save the data
    file? (Y/N)";f$
210 if f$ = "N" or f$ = "n" then goto
    410
220 input "Name of file for data: ";f$
230 open f$ for output as #9
240 print #9,x
250 for i = 0 to x
260 print #9,t(i);",";l(i);",";m(i)
270 next i
280 close #9
290 goto 410
300 input "Name of file with data: ";f$
310 open f$ for input as #9
320 input #9,x
330 print "x = ";x
340 for i = 0 to x
350 input #9,t(i),l(i),m(i)
360 print t(i),l(i),m(i)
370 next i
380 close #9
390 print " "
400 print " "
410 r0 = 0
420 t1 = 0
430 for i = 0 to x
440 r0 = r0+l(i)*m(i)
450 t1 = t1+t(i)*l(i)*m(i)
460 next i
470 print "R0 = ";r0;
480 tc = t1/r0
490 rc = log(r0)/tc
500 print "rc (mean) = ";rc;"    Tc =
    ";tc
510 for i = 2 to 6
520 u(i) = 0
530 k(i) = 0
540 next i
550 for j = 2 to 6
560 for i = 0 to x
570 u(j) = u(j)+l(i)*m(i)*(t(i)-tc)^j
580 next i
590 next j
600 for j = 2 to 6
610 u(j) = u(j)/r0
620 next j
630 k(2) = u(2)
640 k(3) = u(3)
650 k(4) = u(4)-3*u(2)*u(2)
660 k(5) = u(5)-10*u(3)*u(2)
670 k(6) = u(6)-15*u(4)*u(2)-
    10*u(3)*u(3)+30*u(2)^3
680 fc(1) = 1
690 for j = 2 to 6
700 fc(j) = fc(j-1)*j
710 next j
720 for j = 2 to 6
730 fc(j) = fc(j)*tc
740 next j
750 r = rc
760 input "Enter the highest-order term
    you wish to use (from 2 to 6) ";wj
770 for k = 1 to 100
780 f = 0
790 fp = 0
800 for j = 2 to wj
810 f = f+(-1)^(j+1)*r^j*k(j)/fc(j)
820 fp = fp+(-1)^(j+1)*j*r^(j-
    1)*k(j)/fc(j)
```

```
830 next j
840 fp = 1+fp
850 f = f+r-rc
860 r1 = r-f/fp
870 r = r1
880 if abs(r1-r) < 1.000000E-07 then
    goto 910
890 print k,r
900 next k
910 print " Final r = ";r
```

BIRCH.BAS

The evaluation of Birch's (1948) life-history analysis can be done by modifying the **EULER.BAS** program that uses Newton's method to calculate r. The following program is single purpose and was written solely to explore Birch's data. It would be possible to make it more general by adding lines for getting data from a file but a general program was not the intent. **BIRCH.BAS** makes adjusting both age at first reproduction and m_x values a simple matter of changing two constants, F and A, or :f: and "a" in the code. Each m_x value is divided by F before r is estimated. The constant A scales the time of first reproduction. Both l_x and m_x values are initialized in lines 50-200 by starting with an index of 1 even though in the original data the first l_x and m_x values are for x = 4.5. As long as x is just used as an index it doesn't matter whether it starts at 0 or 1 or 4.5. The actual *value* of x is important only when it is used in a calculation. For example, the values of $l_x m_x$ use x only as an *index* to keep appropriate l_x and m_x values together and the actual value of any x is unimportant. On the other hand, values of $x l_x m_x$ require a correct value for the x that is multiplied by $l_x m_x$. In the program, lines 450, 540, and 560 all use the actual value of x and so the constant A is added to x to scale the data for different ages of first reproduction. Values of f must be found by trial and error and reasonable values are suggested in line 270. The iterations to arrive at reasonable values of F are shown in Table 2.23.

TABLE 2.23 Trial and Error Search for Values of F that Result in r = 0.762.

Iteration	A = 1.5		A = 6.5	
	f	r	f	r
1	10	0.7544	0.34	0.7004
2	9	0.8028	0.25	0.7429
3	9.5	0.7778	0.20	0.7739
4	9.6	0.7730	0.23	0.7544
5	9.8	0.7636	0.22	0.7606
6	9.85	0.7613	0.215	0.7638
7	9.84	0.7617	0.218	0.7619

```
10 rem Recalculation of Birch's (1948)
   results
20 dim m(20),l(20)
30 x = 15
40 rem x is the number of age-classes
   with m(x) not equal to 0.
50 l(1) = 0.87 : l(2) = 0.83
60 l(3) = 0.81 : l(4) = 0.8
70 l(5) = 0.79 : l(6) = 0.77
80 l(7) = 0.74 : l(8) = 0.66
90 l(9) = 0.59 : l(10) = 0.52
100 l(11) = 0.45 : l(12) = 0.36
110 l(13) = 0.29 : l(14) = 0.25
120 l(15) = 0.19
130 m(1) = 20 : m(2) = 23
140 m(3) = 15 : m(4) = 12.5
150 m(5) = 12.5 : m(6) = 14
160 m(7) = 12.5 : m(8) = 14.5
170 m(9) = 11 : m(10) = 9.5
180 m(11) = 2.5 : m(12) = 2.5
190 m(13) = 2.5 : m(14) = 4
200 m(15) = 1
210 rem tm is a counter for life-time
    total number of eggs
220 input "Want to continue? (Y/N) :";f$
230 if f$ = "N" or f$ = "n" then goto
    690
240 rem A is the time adjustment for age
    at first reproduction
250 input "Age at first reproduction :
    ";a
260 print "f adjusts m(x) when age at
    first reproduction changes."
270 print "Use f=1 for first
    reproduction = 4.5 weeks, 9.84 for
    1.5 weeks,"
280 print "and .218 for 6.5 weeks"
290 input "f = ";f
300 a = a-1
310 tm = 0
320 for i = 1 to x
330 m(i) = m(i)/f
340 tm = tm+m(i)
350 next i
360 for i = 1 to x
370 print i+a,m(i)*2
380 next i : print " "
390 print "total # eggs: ";tm*2
400 lm1 = 0
410 lm2 = 0
420 rem Calculate the sums of lxmx and
    xlxmx
```

```
430 for i = 0 to x                          580 fp = -fp
440 lm1 = lm1+l(i)*m(i)                      590 f = f-1
450 lm2 = lm2+(i+a)*l(i)*m(i)                600 r1 = r-f/fp
460 next i                                   610 if abs(r1-r) < 1.000000E-05 then
470 rem Estimate and initial value for r         goto 650
480 t = lm2/lm1                              620 r = r1
490 r = log(lm1)/t                           630 print j,r
500 for j = 1 to 100                         640 next j
510 f = 0                                    650 r = r1
520 fp = 0                                   660 print "Final r = ";r;"  after ";j;"
530 for i = 0 to x                               iterations"
540 w = l(i)*m(i)*exp(-r*(i+a))              670 print "exp(r) = ";exp(r)
550 f = f+w                                  680 goto 30
560 fp = fp+(i+a)*w                          690 end
570 next i
```

3

Descriptive Statistics of Life Tables

It has never occurred to me that counting millions of leaves and swarms of bees could make money. Who could possibly be interested in how many branches there are in a tree, how many birds in a flight that crosses the sky?

Malba Tahan,
The Man Who Counted:
A Collection of Mathematical Adventures, 1972

INTRODUCTION

In Chapters 1 and 2, and in all chapters that follow, life tables are presented as the starting point for further analysis and it may seem as though gathering such data is a routine or trivial task. In fact, analysis usually is the easy part and gathering the data for survival and reproductive schedules is far more difficult. Sometimes this is because of the large numbers of organisms that must be studied and sometimes because many years must be included. Not infrequently, very interesting or economically important or rare and endangered species have life cycles with stages that are difficult to observe and require delightfully clever innovative studies before life tables can be constructed.

CONFIDENCE LIMITS

Two major points of this chapter are to show how confidence limits can be attached to life table estimates and to provide examples of how survival and reproductive data are gathered for a variety of organisms. Using n_x as the numbers in state x,

p_x is the conditional probability that an individual with attribute x survives one time period, which is

$$p_x = \frac{n_{x+1}}{n_x} \qquad (3.1)$$

or

$$p_x = \frac{l_{x+1}}{l_x}; \qquad (3.2)$$

q_x is the conditional probability an individual with attribute x dies during the time interval t to t+1;

$$q_x = 1 - p_x \qquad (3.3)$$

or

$$q_x = 1 - \frac{l_{x+1}}{l_x} \quad \text{or} \quad \frac{l_x - l_{x+1}}{l_x};$$

and

l_x is the probability a new individual (age, stage, or size 0) is alive at the beginning of age, stage, or size x,

$$l_x = \frac{n_x}{n_0}, \qquad\qquad\qquad (3.4)$$

which is appropriate if a cohort of new individuals is followed and individuals leave the study only through death and not by emigration or dropping out for some other reason. Data for which survival times have not been observed for all individuals are called censored.

The value of l_0 always is equal to 1.0 and l_x is the cumulative survival from 0 to x

$$l_x = \prod_{i=0}^{x-1} p_i \qquad\qquad \text{for } x > 0, \qquad (3.5)$$

which is an appropriate way for dealing with censored survival data but requires adjustments of numbers alive to account for losses to causes other than death. Also, it may be important to separate deaths due to hunting, for example, from deaths due to natural causes. Equation 3.1 can be modified to

$$p_x = \frac{n_{x+1}}{n_x - k}, \qquad\qquad\qquad (3.6)$$

where k is the number of individuals that leave at time x for reasons other than those appropriate for the study. The adjustment k is suitable for many purposes; however, if individuals leave the study over the entire period x to x+1, then a better correction is number leaving, k, divided by 2 (*cf.* Collett 1994).

The probability that a new individual (age, stage, or size 0) dies during the time period x to x+1 is d_x

$$d_x = l_x - l_{x+1}, \qquad\qquad\qquad (3.7)$$

which is the same as the probability of living to the start of the period (l_x) multiplied by the probability of dying during the interval x to x+1,

$$d_x = q_x \prod_{i=0}^{x-1} p_i . \qquad\qquad\qquad (3.8)$$

If x means age, Equations 3.8 or 3.9 are appropriate for estimating d_x. If there are no censored survival times, Equation 3.9 could be used but if the data contain censored times, Equation 3.8 should be used.

$$d_x = \frac{n_x - n_{x+1}}{n_0} \qquad\qquad\qquad (3.9)$$

For stage-structured analysis,

$$d_x = \frac{n_x - \sum_{i=0}^{\omega} n_{x+i}}{n_0}, \qquad\qquad (3.10)$$

where $\sum_{i=0}^{\omega} n_{x+i}$ is the summation of all surviving individuals that have remained in category x or have transferred to some other category. This summation is not all of the individuals in category x+1, just those that transferred from x during one time interval.

The life table statistics p_x, q_x, l_x and d_x all are *estimates* and so require statements of statistical properties (*cf.* Lee 1992, Fox 1993, Collett 1994). Equations 3.5 and 3.8 show calculations based on products of p_x and so to obtain statistics based on sums, an obvious transformation is to take logarithms of survival rates so variances can be estimated. A logarithmic transformation of Equations 3.5 and 3.8 gives

$$ln l_x = \sum_{i=0}^{x-1} ln p_i \qquad\qquad\qquad (3.11)$$

and

$$ln d_x = ln(1 - p_x) + \sum_{i=0}^{x-1} ln p_i) \qquad (3.12)$$

The variance of $ln l_x$ is

$$\text{var}[ln l_x] = \sum_{i=0}^{x-1} \text{var}[ln p_i]. \qquad\qquad (3.13)$$

Taylor series approximations to variance (Collett 1994) lead to

$$\text{var}(l_x) \approx l_x^2 \sum_{j=0}^{x-1} \frac{q_j}{n_j p_j} \qquad\qquad (3.14)$$

and

$$\text{var}(l_x) = \frac{l_x(1 - l_x)}{n_0}. \qquad\qquad (3.15)$$

The general variance estimate for d_x, appropriate for either censored or uncensored data, is

$$\text{var}(d_x) \approx \frac{d_x^2}{b} \left[\sum_{j=0}^{x-1} \frac{q_j}{n_j p_j} + \frac{p_x}{n_x q_x} \right] \qquad (3.16)$$

but if there are no censored survival times, a simpler estimate is

$$var(d_x) = \frac{d_x(1 - d_x)}{n_0}.$$ (3.17)

The variance estimate for p_x or q_x is

$$var(p_x \text{ or } q_x) = \frac{p_x q_x}{n_x}.$$ (3.18)

Standard errors, se, are the square roots of the variance estimates and the approximate 95% confidence intervals are ±1.96 se (Zar 1974). In Equation 3.16, b is the width of the time interval, which, in many cases, will be equal to 1.0.

Problems with the estimates of variance, associated standard errors, and 95% confidence limits are evident when Equations 3.14-3.18 are used. One problem is that it is possible to obtain 95% confidence intervals with one limit that is less than 0 or greater than 1.0. One approach to obtaining confidence limits that do not extend outside [0,1] is to use transformed variables with a range (-∞, +∞), obtain a confidence interval, and then transform back. The resulting confidence interval will not extend outside [0,1].

Collett (1994) shows two transformations that keep confidence limits within [0,1]. Using l_x, for example, the logistic transformation is $ln(l_x/(1 - l_x))$ and the double logarithmic transformation is $ln(-ln(l_x))$. With a *ln-ln* transformation, Equation 3.14 is

$$var[ln\{-ln\, l_x\}] \approx \frac{1}{[ln\, l_x]^2} \sum_{j=0}^{x-1} \frac{q_j}{n_j p_j}.$$ (3.19)

The square root of Equation 3.19 is the standard error, se, and so the 95% confidence limits can be found using Equation 3.20.

$$l_x^{\exp(\pm 1.96se)}.$$ (3.20)

A second problem with Equations 3.14 to 3.18 occurs when an estimated probability is equal to 1 or 0. In such cases, regardless of sample size n, the variance will be 0 and so the 95% confidence intervals will be [0,0] or [1,1], which is perfectly acceptable when n is very large but does not seem correct when n is 50 or fewer. A way of calculating the 95% confidence intervals that addresses both the problem of probability estimates of 0 or 1 and the problem of keeping confidence limits within the interval [0,1] is to use profile likelihood intervals (Verzon and Moolgavar 1988, Lebreton *et al.* 1992).

The profile likelihood interval is based on the relationship between -2ln(L) and the chi-square distribution where ln(L)

is the log-likelihood of a life-table statistic such as p_x or l_x. The 95% interval is all values of θ that satisfy the inequality

$$-2[y\,ln(p_x)+(n-y)(1-p_x)]+2[y\,ln(\theta)+ (n-y)ln(1- \theta)] \leq 3.8416,$$ (3.21)

where y is the number of "successes," such as survivors, out of the total starting number n. The value 3.8416 is the chi-square value at 0.05 with 1 degree of freedom. Two end values of θ satisfy the inequality of the 95% interval, [θ_1, θ_2] and the trick, of course, is finding these two values. It is rather easy when p_x is 0 or 1 because one of the end values will be 0 or 1 and so there is just one value of θ to find. If the estimated value of p_x = 1, then θ_2 = 1 and

$$\theta_1 = e^k,$$

where

$$k = -\frac{3.8416}{2n}.$$

For example, if n = 20 and p_x was estimated as 1.0, the 95% confidence interval would be [0.908, 1] because

$$k = -\frac{3.8416}{2 \times 20} = -0.09604$$

and

$$\theta_1 = e^{-0.09604} = 0.908.$$

If the estimate of p_x was 0, then the 95% confidence interval would be [0, 0.092]. The value of k is calculated as above with 20 observations and

$$\theta_2 = 1 - e^{-0.09604} = 1 - 0.908 = 0.092.$$

If p_x is some value other than 1 or 0, then both θ_1 and θ_2 must be estimated. One way of doing this is by trial and error using the Newton-Raphson algorithm introduced in Chapter 2. The algorithm is

$$\theta_{i+1} = \theta_i - \frac{F(\theta)}{F'(\theta)}.$$ (3.22)

Inequality 3.21 is written as an equality and rearranged so that F(θ) is equal to 0 and F$'$(θ) is the derivative of the function with respect to θ

$$F(\theta) = -2[y\,ln(p_x) + (n-y)(1-p_x)] + 2[y\,ln(\theta) + (n-y)ln(1 - \theta)] - 3.8416 = 0$$ (3.23)

TABLE 3.1 Comparison of 95% Confidence Intervals for Survival Using the Binomial (Equation 3.15, 3.17 or 3.18), *ln(-ln)* **Transformation (Equation 3.20, and Profile Likelihood (Equation 3.21)**

Survivors (y)	y/20	Binomial)	*ln(-ln)* Transformation	Profile Likelihood 95%
20	1.0	[1.0, 1.0]	[1.0, 1.0]	[0.908, 1.0]
19	0.95	[0.854, 1.046]	[0.695, 0.993]	[0.798, 0.997]
15	0.75	[0.560, 0.940]	[0.500, 0.887]	[0.538, 0.902]
10	0.50	[0.281, 0.719]	[0.271, 0.692]	[0.291, 0.709]
5	0.25	[0.060, 0.440]	[0.091, 0.448]	[0.100, 0.462]
1	0.05	[-0.046, 0.146]	[0.003, 0.205]	[0.003, 0.202]
0	0.00	[0.0, 0.0]	[0.0, 0.0]	[0.0, 0.092]

Note: n = 20 for all estimates and y is the number of survivors (= "successes") out of 20.

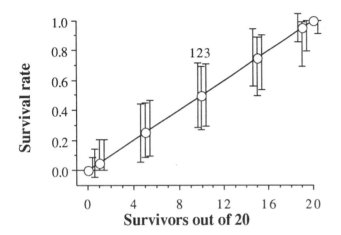

FIGURE 3.1 Comparison of 95% confidence intervals using three different methods; confidence intervals for methods are offset slightly for better definition; the left confidence interval (1) is the binomial (Equation 3.15, 3.17 or 3.18); the middle interval (2) uses a *ln(--ln)* transformation (Equation 3.20); and the right interval (3) uses profile likelihood estimation (Equation 3.21); data are from Table 3.1.

and

$$F'(\theta) = 2\left[\frac{y}{\theta} - \frac{n-y}{1-\theta}\right]. \qquad (3.24)$$

If a starting value of θ is selected that is close to 1, such as 0.99999, the algorithm will converge on the upper limit, θ_2, and if a starting value that is close to 0 is selected, such as 0.00001, the lower limit, θ_1, will be found. Table 3.1 shows a comparison of 95% confidence intervals for survival values using the binomial (Equations 3.15, 3.17, 3.18), logarithmic transformation (Equations 3.19 and 3.20), and profile likelihood intervals (Equations 3.21 to 3.23). For easier comparison, results are shown in Figure 3.1.

Several features are evident in comparing the 95% confidence intervals using the binomial, *ln(-ln)*) transformation, and profile likelihood estimates. First, the sizes of the intervals, that is upper minus lower, are fairly similar for middle values from about 5 to 15 survivors (Figure 3.1). The major differences are at 0 and 20 and at 1 and 19. For descriptive purposes, any of the three methods would provide a sense of the confidence that should be placed in particular estimates. The profile likelihood intervals seem to be the best of the three methods because they do not extend outside [0,1] and they provide confidence intervals for estimate of 0 and 1, which seems better than having intervals of zero size regardless of the value of n as are provided by the other methods.

ESTIMATING SURVIVAL

A wide variety of ways of gathering the basic data for life table construction exist and the method used depends on the nature of the organism and available time and money. One distinction in construction is based on how cohort survival is assembled. The most obvious way is to start with a collection of age 0 individuals and follow them until the last member of the cohort is dead. This is horizontal construction of survival data in the sense that one follows along with the cohort through time. Vertical construction is accomplished by taking a slice through an existing population and determining the ages of all individuals. If the population has a stable-age distribution and is stationary (r = 0 or the equivalent λ = 1.0), then the relative proportions in each age class can be used to determine survival rates. A modification of this vertical slice through a population is to determine ages of all individuals, tag or map them, and then determine survival over some reasonable period. For many creatures this would be for a period of not less than one year. Additional modifications of this vertical approach are used for organisms that are classified by something other than age, size for example, and survival is determined on a size basis rather than on an age basis. The following examples show several ways of estimating survival both horizontally and vertically.

**TABLE 3.2 Seedling Survival of *Cercidium microphylla*, Paloverde, in a
557 Square Meter Plot at the Desert Laboratory, Tucson, Arizona**

Year	1910	1911	1912	1913	1914	1915	1916	1917
1910	**542**	62	35	16	4	3	3	2
1911		**122**	49	6	1	0	0	0
1912			**151**	24	3	3	3	2
1913				**34**	5	2	2	2
1914					**7**	2	2	2
1915						**0**	0	0
1916							**29**	5
1917								**38**

Note: numbers are based on mapped individuals; boldface numbers are the number of new seedlings; data from Shreve (1917).

**TABLE 3.3 Annual Survival Rates for Six Cohorts of Paloverde (*Cercidium microphylla*)
in a 557 Square Meter Plot at the Desert Laboratory, Tucson, Arizona**

	Cohort					
Age	1910	1911	1912	1913	1914	1916
0-1	0.1144	0.4016	0.1589	0.1471	0.2857	0.1724
	[0.089, 0.143]					
1-2	0.5645	0.1224	0.1250	0.4000	1.0000	
	[0.440, 0.683]					
2-3	0.4571	0.1667	1.0000	1.0000	1.0000	
	[0.300, 0.621]					
3-4	0.2500	0.0000	1.0000	1.0000		
	[0.085, 0.489]					
4-5	0.7500		0.6667			
	[0.278, 0.984]					
5-6	1.0000					
	[0.527, 1.0]					
6-7	0.6667					
	[0.161, 0.977]					

Note: age in years; numbers are from Table 3.2; 95% confidence intervals are included just for the 1910 cohort; data from Shreve (1917).

A Desert Tree, Paloverde, in Arizona

Sessile organisms such as plants, barnacles, and corals are particularly amenable to analysis because survival can be determined by mapping (*e.g.* Harper 1977). If a cohort can be followed from the time of germination or settlement, then the decline in numbers over time defines the age specific survival rates. Deserts tend to have open vegetation and so lend themselves very well to documenting plant germination and following the fates of individual seedlings.

Shreve (1917) mapped new seedlings of a variety of perennial species from 1909 to 1917 in a 557 m^2 plot near the Desert Laboratory at Tucson, Arizona. Table 3.2 shows annual seed germination and subsequent survival for a small tree, *Parkinsonia microphylla*, paloverde, which now is called *Cercidium microphylla*, starting in 1910, when Shreve was certain that small plants were new seedlings. These data can be used to estimate age specific survival

rates, p_x, which are the fraction of individuals in a particular age class, n_x, that survive until the next year, n_{x+1}. In 1910 there were 542 seedlings so n_0 is 542. In 1911 there were only 62 plants remaining from the original 542 so 62 is n_{0+1} or n_1. First year survival was

$$p_0 = \frac{62}{542} = 0.1144 \text{ yr}^{-1}$$

with a 95% confidence interval of [0.089, 0.143] using Equation 3.21. With such a large value for n, 542, the binomial (Equation 3.18) gives nearly identical results, [0.086, 0.141]. The probability of dying during the interval 0 to 1, q_0, was

$$q_0 = 1 - 0.1144 = 0.8856 \text{ yr}^{-1}$$

with a 95% confidence interval of [0.857, 0.911].

3. Descriptive Statistics of Life Tables

TABLE 3.4 Survivors (S) and Deaths (D) During Age Intervals x to x+1 for Paloverde, *Cercidium microphylla*, at the Desert Laboratory, Tucson, Arizona

	Age class														
	0	1		2		3		4		5		6		7	
Year		D	S	D	S	D	S	D	S	D	S	D	S	D	S
1910	542	480	62	27	35	19	16	12	4	1	3	0	3	1	2
1911	122	73	49	43	6	5	1	1	0						
1912	151	127	24	21	3	0	3	0	3	1	[2]				
1913	34	29	5	3	2	0	2	0	[2]						
1914	7	5	2	0	2	0	[2]								
1915	0														
1916	29	24	[5]												
n_{x+1}		738	147	94	48	24	24	13	9	2	5	0	3	1	2
n_x	885		142		48		22		7		3		3		

Note: a box is drawn around survivors in 1917 to indicate that they must be removed from analysis before the next p_x value is calculated; data from Shreve 1917.

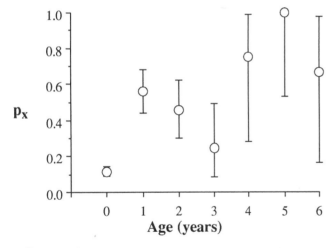

FIGURE 3.2 Age-specific survival of paloverde, (*Cercidium microphylla*) in a 557 square meter plot at the Desert Laboratory, Tucson, Arizona (Shreve 1917); error bars are 95% profile likelihood intervals (Equation 3.21).

Calculation of p_1 for the 1910 cohort is based on 62 at the beginning of the sampling year in 1911 and 35 individuals one year later in 1912,

$$p_1 = \frac{35}{62} = 0.5645 \text{ yr}^{-1},$$

with a 95% confidence interval of [0.440, 0.683]. Death rate for plants during the age interval 1 to 2 is q_1,

$$q_1 = 1 - 0.5645 = 0.4355 \,[0.317, 0.569] \text{ yr}^{-1}.$$

Annual survival rates for all annual cohorts of paloverde were calculated (Table 3.3) using Equation 3.6 and data from Table 3.2. Year to year variation can be reduced by pooling all of the cohorts of *Cercidium microphylla* (Table 3.4). This approach can be justified for the paloverde data because of the very small sample sizes for all but survival from age 0 to age 1 (Table 3.2).

Survival of individuals shown in Table 3.4 is complicated because a new cohort was added each year and followed just to 1917. It is unknown whether age 0 individuals in 1916 that survived to 1917 died during the next year and so although the survivors are appropriately used to estimate survival from age 0 to age 1, they must be removed when estimating survival from age 1 to age 2: survival times are censored. This is the meaning of the rows n_{x+1} and n_x in Table 3.4. Some individuals must be removed from the n_x group each time because the particular cohort reached 1917, and even though some were still alive, they should not be included for estimating the next p_x value (Equation 3.6).

Paloverde survival (Figure 3.2) shows three patterns common to studies of mapped individuals (Table 3.5). First is the general increase in survival with age, which is true for many species: young stages suffer higher mortality than do later stages. Second is the variability caused by differences in environmental conditions from year to year. Third is the increase in the confidence limits with age. This increase is a reflection of the decreasing numbers of individuals available to estimate annual survival, which is a common problem in studies where horizontal construction is used.

Equation 3.14 can be used to estimate the 95% confidence intervals for l_x, the probability that a newborn is alive at time x, and Equation 3.16 can be used for d_x, the probability that a newborn dies during the interval x to x+b, where b usually will be equal to 1. The steps in calculation are shown in Tables 3.6 and 3.7 and results are plotted in Figure 3.3. The variance for l_0 always is 0 because l_0 is 1 by

TABLE 3.5 Annual Survival Probabilities for Paloverde, *Cercidium microphylla*, Tucson, Arizona

Age	n_x	n_{x+1}	p_x	binomial 95% interval	profile likelihood 95% interval
0	885	147	0.166	[0.141, 0.191]	[0.142, 0.192]
1	142	48	0.338	[0.260, 0.416]	[0.264, 0.418]
2	48	24	0.500	[0.359, 0.641]	[0.361, 0.639]
3	22	9	0.409	[0.204, 0.615]	[0.222, 0.616]
4	7	5	0.714	[0.380, 1.050]	[0.350, 0.946]
5	3	3	1.000	[1.000, 1.000]	[0.527, 1.000]
6	3	2	0.667	[0.133, 1.200]	[0.161, 0.977]

Note: binomial 95% confidence interval is from Equation 3.13 and the profile likelihood 95% confidence interval is from Equation 3.14 and the Newton-Raphson algorithm (Equation 3.15); data from Shreve (1917).

TABLE 3.6 Calculation of the Variance for l_x for Paloverde, Tucson, Arizona, 1910-1917, Using Equation 3.14

Age	n_x	p_x	q_x	l_x	$\dfrac{q_x}{n_x p_x}$	$\sum\limits_{j=0}^{x-1} \dfrac{q_j}{n_j p_j}$	l_x^2	var(l_x)	se(l_x)	95% interval
0	885	0.1661	0.8339	1.0000	0.0057		1	0	0.0000	[1.000, 1.000]
1	142	0.3380	0.6620	0.1661	0.0138	0.0057	0.027590	0.000157	0.0125	[0.142, 0.191]
2	48	0.5000	0.5000	0.0561	0.0208	0.0195	0.003152	0.000061	0.0078	[0.041, 0.071]
3	22	0.4091	0.5909	0.0281	0.0657	0.0403	0.000788	0.000032	0.0056	[0.017, 0.039]
4	7	0.7143	0.2857	0.0115	0.0571	0.1060	0.000132	0.000014	0.0037	[0.004, 0.019]
5	3	1.0000	0.0000	0.0082	0.0000	0.1631	0.000067	0.000011	0.0033	[0.002, 0.015]
6	3	0.6667	0.3333	0.0082	0.1667	0.1631	0.000067	0.000011	0.0033	[0.002, 0.015]
7	2			0.0055		0.3298	0.000030	0.000010	0.0031	[-0.001, 0.012]

Note: columns for all intermediate terms are shown; p_x and n_x are from Table 3.5; $\sum q_j / n_j p_j$ is a running total of the $q_x / n_x p_x$ column that is displaced by one age category and so calculates sums from j=0 to x-1; var(l_x) is $l_x^2 \sum q_j / n_j p_j$; se(l_x) is $\sqrt{\text{var}(l_x)}$ and the 95% confidence interval is ±1.96se; data from Shreve (1917).

TABLE 3.7 Calculation of the variance of d_x for Paloverde, Tucson, Arizona, 1910-1917, Using Equation 3.16

x	n_x	D_x	d_x	$\dfrac{p_x}{n_x q_x}$	$\sum\limits_{j=0}^{x-1} \dfrac{q_j}{n_j p_j}$	$\sum\limits_{j=0}^{x-1} \dfrac{q_j}{n_j p_j} + \dfrac{p_x}{n_x q_x}$	d_x^2	var(d_x)	se(d_x)	95% interval
0	885	738	0.8339	0.00023		0.00023	0.695386	0.000157	0.0125	[0.809, 0.858]
1	142	94	0.1100	0.00360	0.0057	0.00927	0.012090	0.000112	0.0106	[0.089, 0.131]
2	48	24	0.0281	0.02083	0.0195	0.04030	0.000788	0.000032	0.0056	[0.017, 0.039]
3	22	13	0.0166	0.03147	0.0403	0.07177	0.000275	0.000020	0.0044	[0.008, 0.025]
4	7	2	0.0033	0.35714	0.1060	0.46310	0.000011	0.000005	0.0022	[-0.001, 0.008]
5	3	0	0.0000	-	0.1631	-	0.000000	-	-	-
6	3	1	0.0027	0.66667	0.1631	0.82976	0.000007	0.000006	0.0025	[-0.002, 0.008]
7	2				0.3298					

Note: initial number of plants was 885 and D_x is the number dying during the period x to x+1 so d_x is $D_x/885$ (Table 3.4); all intermediate terms are shown; $\sum q_j / n_j p_j$ is a running total of the $q_x / n_x p_x$ column and so calculates sums from j=0 to x-1; var(l_x) is $d_x^2 [\sum q_j / n_j p_j + p_x / n_x q_x]$; se($d_x$) is $\sqrt{\text{var}(d_x)}$.; data from Shreve (1917).

TABLE 3.8 Number of female cactus finches (*Geospiza scandens*) remaining out of 90 fledglings in 1978; 95% confidence intervals are from Equations 3.10 and 3.13

Year	Age	n	p_x	p_x profile likelihood	l_x	Binomial 95% interval for l_x	l_x profile likelihood
1978	0	90	0.433	[0.334, 0.536]	1.000	-	-
1979	1	39	0.846	[0.713, 0.936]	0.434	[0.331, 0.537]	[0.334, 0.537]
1980	2	33	0.909	[0.781, 0.977]	0.367	[0.267, 0.467]	[0.272, 0.469]
1981	3	30	0.967	[0.861, 0.998]	0.333	[0.236, 0.430]	[0.242, 0.434]
1982	4	29	0.931	[0.802, 0.988]	0.322	[0.225, 0.419]	[0.232, 0.423]
1983	5	27	0.593	[0.405, 0.763]	0.300	[0.205, 0.395]	[0.212, 0.399]
1984	6	16	0.500	[0.269, 0.731]	0.178	[0.099, 0.257]	[0.108, 0.265]
1985	7	8	0.625	[0.290, 0.890]	0.089	[0.030, 0.148]	[0.042, 0.159]
1986	8	5	1.000	[0.681, 1.000]	0.056	[0.008, 0.104]	[0.020, 0.116]
1987	9	5	0.600	[0.199, 0.919]	0.056	[0.008, 0.104]	[0.020, 0.116]
1988	10	3	0.667	[0.161, 0.977]	0.033	[-0.004, 0.070]	[0.008, 0.084]
1989	11	2	0.000	[0.000, 0.383]	0.022	[-0.008, 0.052]	[0.004, 0.067]
1990	12	0			0.000		[0, 0.021]

Note: age in years; data from Grant and Grant (1992).

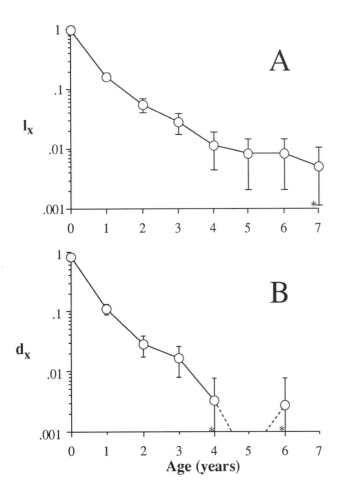

FIGURE 3.3 **A,** Survivorship curve, l_x, for paloverde, (*Cercidium microphylla*), at the Desert Laboratory, Tucson, Arizona based on data from Shreve (1917); * the estimate of the lower 95% value was negative (Table 3.5) and so was arbitrarily truncated. **B,** Probability that a newly germinated seedling will die during the interval x to x+1, d_x; dashed line is for d_5=0.

definition and so does not vary. The final value for d_x comes one time unit before the final l_x value if survival times are censored; that is, there is doubt concerning future survival. For example, for paloverde there were 2 survivors at age 7 (Table 3.6) but it is unknown how many lived from age 7 to 8 and so it would be wrong to assume that both died. Rather than making an assumption about the fates of individuals age 8, it is better not to estimate d_7, the probability of a newborn dying between 7 and 8.

Paloverde is a long-lived shrub and data given in Table 3.2 represent the beginning of a table that could extend out to 200 years or more. Two things are clear: (1) because of the low survival of seedlings, it would be necessary to start each year with samples at least 10 times larger than the ones used by Shreve; and, (2) probably no organization will ever gather 200 years of data on a species, other than humans. In order to study the demography of paloverde, some other approach is needed and a reasonable one would be to classify individuals by size rather than age (Chapter 9). The same idea as used by Shreve would be followed in mapping individuals but rather than classification by age, size could be used and so all individuals in a population would be recorded although numbers still would have to be increased by at least an order of magnitude. Age classification and following cohorts works very well when organisms are not so longevous. Many animals are suitable but many also bring additional problems.

Mobile animals provide challenges for obtaining estimates of survival although the fundamental idea is exactly the same as with plants. The desired data are numbers of individuals of known age at the beginning of a period and numbers present one time period later, which, for many animals, is one year but could be as short as half a day (*e.g.* Dumont and Sarma 1995). It depends on the nature of the beasts. In general, the approach is to keep track of

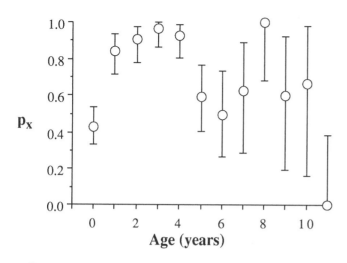

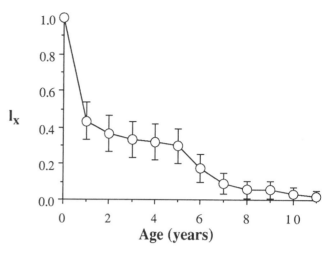

FIGURE 3.4 Age-specific survival rate, p_x , for the 1978 cohort of female cactus finches (*Geospiza scandens*); data from Grant and Grant (1992); error bars are 95% profile likelihood intervals (Equation 3.21).

FIGURE 3.5 Fraction, l_x , of female cactus finches (*Geospiza scandens*) remaining out of 90 fledglings in 1978; 95% confidence intervals calculated using Equation 3.15; data from Grant and Grant (1992).

individuals and count survivors for a period of time; however, an unfortunate attribute of motile creatures is that they can leave study sites. Estimating survival for animals in open populations is not a trivial problem because loss of tagged animals can be due death, emigration, and lost tags (*e.g.* Seber 1982, Lebreton *et al.* 1992).

Cactus Finches in the Galápagos

Estimating survival for motile animals is considerably easier when the population is closed and confined to a relatively small area. A good example is the work of Peter and Rosemary Grant who have been studying Darwin's finches in the Galápagos since 1975 (Grant and Grant 1992). Their work has included tagging all nestlings of particular finch species, which permits them to have a unique mark for every individual. Unique tagging means that the numbers of all birds banded as nestlings or as age 1 immatures can be determined each year when a census is made. The populations are closed and loss of individuals means death; however, not seeing a bird during a season may mean that the bird died before censusing or that, for whatever reason, it was missed even though it was still alive. This appears not to have been a problem with Darwin's finches but, typically, is one in mark-recapture studies. The following data are for one cohort that was followed for 12 years.

The year 1978 was good for the cactus finch *Geospiza scandens* on Isla Daphne Major and 90 new fledgling female finches were added to the population. Table 3.8 shows the decline in numbers over a 12-year period together with annual survival, p_x, and the probability of survival of a

fledgling to age x; that is, l_x. Annual survival probabilities (Figure 3.4) show a rise and then a drop at age 5 during 1983 when there was heavy rainfall and associated extensive breeding. The next three years had low or zero rainfall and very little breeding; survival rates never recovered the high levels attained early in life. The survival data for the 1978 cohort of *Geospiza scandens* (Table 3.8) can be expressed as an l_x distribution (Figure 3.5).

Values of both l_x and p_x for the cactus finch data have a problem. There is a confounding of characteristics of age (x) with environmental (e) differences from year to year. Both l_x and p_x are functions,

$$l_x = f(x,e)$$

and

$$p_x = g(x,e),$$

and these functional relationships will always be true for organisms that we examine. Cactus finches have very highly variable breeding success and some years during the Grants' study were complete failures. A major breeding success in 1975 shows survival of another cohort that differed in certain respects from the 1978 cohort. The significant point is that the survival probabilities reflect not only age but also environmental conditions that prompt breeding; highly successful breeding for the cactus finch is correlated with decreased survival. Trade-offs between survival and breeding are central to life-history analysis and will be touched on very briefly in Chapter 14; however, a detailed discussion lies outside the aims of this book. Books by Stearns (1992) and Roff (1992) present summaries of the data and arguments concerning trade-offs.

TABLE 3.9 Numbers of Gorse Weevils in Different Developmental Stages at Whitford, New Zealand

Days	E	I_1	I_2	I_3	P	A
0	0					
7	26					
14	149					
21	177					
28	417					
35	604					
42	777					
49	1151	0				
56	1447	22	0			
63	1656	72	4			
70	1217	151	20	0		
77	1192	894	45	25		
84	400	618	207	188	0	
91	238	261	521	374	16	0
98	81	89	130	457	65	7
105	30	13	8	324	113	18
112	5	9	5	197	346	67
119	0	3	6	187	372	260
126		12	16	177	324	345
133		0	0	37	230	590
140				0	65	607
147					0	147
154						56
161						0

Note: samples based on 495 gorse pods (*Ulex europaeus*) collected weekly; E=eggs, I_1, I_2 and I_3 are larval instars, P = pupae and A=adults; sampling began on 31 July 1990 and ended on 9 January 1991; data from Hoddle (1991).

TABLE 3.10 Cumulative Stage Distributions of the Gorse Weevil based on Numbers in Table 3.9

Days	E	I_1	I_2	I_3	P	A
0	0					
7	26					
14	149					
21	177					
28	417					
35	604					
42	777					
49	1151	0				
56	1469	22	0			
63	1732	76	4			
70	1388	171	20	0		
77	2156	964	70	25		
84	1413	1013	395	188	0	
91	1410	1172	911	390	16	0
98	829	748	659	529	72	7
105	506	476	463	455	131	18
112	629	624	615	610	413	67
119	828	828	825	819	632	260
126	874	874	862	846	669	345
133	857	857	857	857	820	590
140	672	672	672	672	672	607
147	147	147	147	147	147	147
154	56	56	56	56	56	56
161	0	0	0	0	0	0
Total (T)	18267	8700	6556	5594	3628	2097
$(T \times 7)$: A_j^*	127869	60900	45892	39158	25396	14679

Note: numbers summed from adult to egg for each date and then columns summed to give values of A^*; data from Hoddle (1991).

TABLE 3.11 Cumulative Stage Frequencies for Gorse Weevils Multiplied by Time and Summed to Produce D_j^*

Days	E	I_1	I_2	I_3	P	A
0	0					
7	182					
14	2086					
21	3717					
28	11676					
35	21140					
42	32634					
49	56399	0				
56	82264	1232	0			
63	109116	4788	252			
70	97160	11970	1400	0		
77	166012	74228	5390	1925		
84	118692	85092	33180	15792	0	
91	128310	106652	82901	35490	1456	0
98	81242	73304	64582	51842	7056	686
105	53130	49980	48615	47775	13755	1890
112	70448	69888	68880	68320	46256	7504
119	98532	98532	98175	97461	75208	30940
126	110124	110124	108612	106596	84294	43470
133	113981	113981	113981	113981	109060	78470
140	94080	94080	94080	94080	94080	84980
147	21609	21609	21609	21609	21609	21609
154	8624	8624	8624	8624	8624	8624
161	0	0	0	0	0	0
Total	1481227	924084	750281	663495	461398	278173
$T \times 7$: D_j^*	10368589	6468588	5251967	4644465	3229786	1947211

Note: data from Hoddle (1991).

Life Table for the Gorse Weevil Using Stages

Insects are relatively easy to classify to developmental stage but usually impossible to age (but see Neville 1963). Because they molt when transferring out of a stage, external tags are of limited value in obtaining survival rates. For field populations, numbers of individuals in each stage of development can be gathered relatively easily and a goal is to use such data to estimate survival and stage duration rates.

One method of analysis was developed by Kiritani and Nakasuji (1967) and modified by Manly (1976, 1977, 1985) and is called the Kiritani-Nakasuji-Manly method. The idea behind the method is simple: the area under a stage-frequency curve, A_j, is determined by the number entering a stage, M_j, a survival parameter, φ, and the duration of the stage, a_j. Three assumptions were used in development of the method:

1. Survival rate per unit of time is the same in all stages for the entire sampling period.
2. Sampling is started at the time when individuals begin to enter stage 1 and continues until all individuals are dead.
3. Population losses are due only to mortality and not emigration.

Gorse weevils, *Apion ulicis*, were studied by Hoddle (1991) and data on numbers in each stage were gathered weekly (Table 3.9). The three assumptions seem to be met and so these data can be used to estimate stage duration and survival. Calculations of survival and stage duration are simplified (Manly 1990) by use of cumulative stage and time-stage distributions with A_j^* being the area under the cumulative stage-frequency curve $F_j(t)$ and D_j^* the area under the cumulative time-stage curve, $tF_j(t)$.

Using the primary observational data of Table 3.9, two new tables can be produced. The first (Table 3.10) is the cumulative stage frequency for each sample date starting with adults and summing backward to eggs . This is $h\Sigma F(j)$ and called A_j^*, with h equal to the time units used, in this case 7 days. The second table (Table 3.11) is created from the summations of Table 3.10 multiplied by time: $h\Sigma tF(j)$, which are D_j^*. The initial goal is to estimate the time at which each stage can be viewed as concentrated. This is

TABLE 3.12 Estimates of A_j^*, D_j^*, and B_j^* From the Cumulative Stage-Frequency Distribution (Table 3.11), Stage-Specific Survival, W_j, and Stage Duration, a_j

Stage	A_j^*	D_j^*	B_j^*	W_j	a_j	Σa_j
E	127869	10368589	81.088	0.47627	17.670	17.670
I_1	60900	6468588	106.217	0.75356	6.740	24.410
I_2	45892	5251967	114.442	0.85326	3.780	28.190
I_3	39158	4644465	118.608	0.64855	10.315	38.505
P	25396	3229786	127.177	0.57800	13.059	51.564
A	14679	1947211	132.653			

Note: all rates and durations have units of days; Σa_j is the length of time to entry into the *next* stage; data from Hoddle (1991).

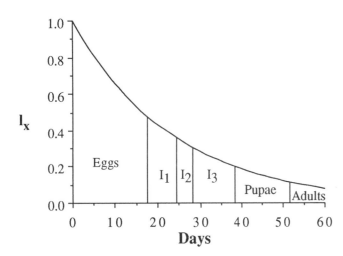

FIGURE 3.6 Survivorship curve, l_x, for the gorse weevil *Apion ulicis* based on data from Hoddle (1991).

like calculating a mean with grouped data or the centroid of a geometric figure. The time at which a stage j is concentrated is called B_j^*,

$$B_j^* = \frac{h\Sigma tF(j)}{h\Sigma F(j)} = \frac{D_j^*}{A_j^*}. \tag{3.25}$$

The next step is to use the values of B_j^*, that is the times at which the mass of a stage is concentrated, to estimate the survival rate. The survival parameter φ (Equations 3.26 and 3.27) is a constant for exponential decay and is similar to the symbol "s" in Chapter 1. The difference in usage is s is negative and so is not used with an attached minus sign; $s = -\varphi$. In studies of many other organisms, φ is called "the instantaneous mortality rate" and a different symbol is used such as M or Z in fisheries studies (*e.g.* Ricker 1975a),

$$A_A^* = A_E^* \, e^{-\varphi t} \tag{3.26}$$

or

$$\varphi = \frac{-ln(A_A^*/A_E^*)}{t}. \tag{3.27}$$

The number of adults, A_A^*, is equal to the number of eggs, A_E^*, times the survival rate, $e^{-\varphi t}$. In order to express survival rate in terms of time, it is necessary to know how many units time t represents in Equations 3.26 and 3.27. An estimate of this is the period between the time where eggs can be viewed as concentrated and the time where adult numbers can be viewed as concentrated. These two points are B_E^* and B_A^*:

$$\varphi = \frac{-ln(A_A^*/A_E^*)}{B_A^* - B_E^*} = \frac{-ln(0.114797)}{51.565} = 0.042 \text{ day}^{-1}.$$

The daily survival rate, p_x, from x to x+1 day is

$$p_x = e^{-\varphi} = e^{-0.04198} = 0.959 \text{ day}^{-1}.$$

The stage-specific survival rate, W_j, is

$$W_j = \frac{A_{j+1}^*}{A_j^*}. \tag{3.28}$$

Assuming survival rate was constant over all stages, the duration of a stage can be calculated from the daily survival rate, p_x, and the stage-specific rate, W_j because

$$p_x^a = W_j, \tag{3.29}$$

where the exponent a is the number of days in the stage. Taking natural logarithms of both sides and rearranging gives

TABLE 3.13 Life Table Based on Human Remains at the Libben Site, Ohio

Age	D_x	d_x	l_x	p_x	Age	D_x	d_x	l_x	p_x
0	226	0.1753	1.0000	0.8247	20	63	0.0489	0.4577	0.8932
1	50	0.0388	0.8247	0.9529	25	78	0.0605	0.4088	0.8520
2	52	0.0403	0.7859	0.9487	30	115	0.0892	0.3483	0.7439
3	43	0.0334	0.7456	0.9552	35	154	0.1195	0.2591	0.5388
4	25	0.0194	0.7122	0.9728	40	97	0.0752	0.1396	0.4613
5	117	0.0908	0.6928	0.8689	45	50	0.0388	0.0644	0.3975
10	94	0.0729	0.6020	0.8789	50	33	0.0256	0.0256	0.0000
15	92	0.0714	0.5291	0.8650					

Note: data from Lovejoy *et al.* (1977).

$$a_j = \frac{ln W_j}{ln p_x} = \frac{-ln W_j}{\varphi} \qquad (3.30)$$

and a summary of calculations is shown in Table 3.12.

An l_x distribution (Figure 3.6) can be calculated based on p_x and the stage duration, a_j. First a running total of a_j is calculated, which is the age at which an individual enters a stage.

The restrictive assumption of constant survival for all life stages probably is not met by most populations although for the example given here, it may be just fine. Life inside a gorse pod may be equally safe for all life stages. The Kiritani-Nakasuji-Manly method of analysis of multicohort stage-frequency data represents a good introduction to a much larger and richer collection of analytical methods (Manly 1990, Wood 1994).

Human Remains at the Libben Site, Ohio

In the following example, data are an assemblage of dead individuals. The Libben site is a human burial ground in the Great Black Swamp of northern Ohio where people lived from about 800 AD to 1100 AD. The site was excavated and a total of 1289 skeletal remains were classified with respect to age at death (Lovejoy *et al.* 1977).

For a collection of remains of dead individuals, the data are D_x values and the objective is to calculate l_x and p_x values. We have to assume that the population producing the death assemblage was neither growing nor declining and that the relative proportions of individuals in the age classes were not changing during the period when remains accumulated; that is, a stable-age distribution with stationary structure had been attained (r=0 or l = 1). Using a death assemblage to create a life table is most easily done first by adding all dead individuals to get a total. This total can be thought of as a cohort that all started together at the same time. They didn't of course, but if the population had not changed in size and the proportions in each age class had been constant, then the results would be the same whether we used a collection of remains or actually did start with a

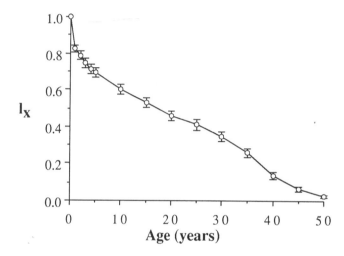

FIGURE 3.7 Survivorship curve determined from 1289 skeletal remains at a human burial site (The Libben site) in the Great Black Swamp of northern Ohio; the site was occupied from about 800 AD to 1100 AD (Lovejoy *et al.* 1977).

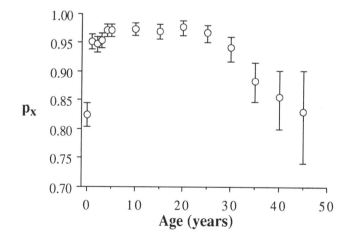

FIGURE 3.8 Age-specific survival, p_x, determined from 1289 skeletal remains at a human burial site (The Libben site) in the Great Black Swamp of northern Ohio; the site was occupied from about 800 AD to 1100 AD (Lovejoy *et al.* 1977); p_x values are adjusted so all are for a time period of one year; 95% confidence intervals were calculated using Equation 3.14.

cohort of age 0 creatures. The assumptions are very important in this regard. The initial number in the cohort, n_0, is

$$n_0 = \Sigma D_x \qquad (3.31)$$

and

$$d_x = \frac{D_x}{\Sigma D_x}. \qquad (3.32)$$

Once this has been done, it is possible to calculate l_x,

$$l_{x+1} = l_x - d_x, \qquad (3.33)$$

and p_x is calculated using Equation 3.2.

Human remains at the Libben site span many years (Table 3.13) although it is uncertain exactly how many. If the burial site was used for several hundred years, it probably is reasonable to believe that year-to-year variability evened out with as many years as good for survival as bad. The survivorship curve probably represents quite accurately the changes in numbers with age (Figure 3.7).

Another way of looking at the Libben Site data is to use age-specific survival, p_x, rather than l_x. A problem with the table is that some time intervals are 1 year and some are 5 years so that if we use Equation 3.1, some p_x values will refer to survival probability for a period of 1 year and some will refer to a period of 5 years. For purposes of comparison, it is best to use the same time periods, either 1 or 5, and not mix them. Let's use 1 year as the time unit. If p_{45} is the survival probability from age 45 to 50 it really represents survival from 45 to 46, 46 to 47, 47 to 48, 48 to 49 and 49 to 50. The probability of surviving from 45 to 47, 2 time units, would be the probability of survival for 1 time unit, 45 to 46 multiplied by the probability of surviving from 46 to 47. If we know survival for a 2-time-unit period, a reasonable guess for survival for 1 time unit would be

survival for 1 time unit = (survival for 2 time units)$^{1/2}$.

For ages where survival is estimated using 5 year periods:

survival for 1 time unit = (survival for 5 time units)$^{1/5}$.

Survival from age 45 to 50, p_{45}, is 0.3975, which is a 5-year period so age-specific survival per year is

$$(0.3975)^{1/5} = 0.8315 \text{ per year.}$$

The age-specific survival of humans at the Libben Site (Figure 3.8) shows a fairly typical series of age-specific survival rates. Following birth, first-year survival is low but if an individual makes it through the first year of life, survival improves substantially and remains about level up to the age of 25 when it begins to turn downward; that is,

annual survival rate drops. Another way of expressing this is that annual mortality rate $(1-p_x)$ increases. This downward turn of the age-specific survival rate is an objective way of defining senescence. Creatures like us humans that have a p_x curve that plunges downward starting at some age are said to become senescent.

European Dippers in Eastern France

A simple example of a mark-recapture study is provided for a bird by Lebreton *et al.* (1992). European Dippers (*Cinclus cinclus*) were banded from 1981 to 1987 in eastern France (Marzolin 1988, G. Marzolin unpublished). Analysis of data is by the Cormack-Jolly-Seber method (Cormack 1964, Jolly 1965, Seber 1965), which has a number of assumptions that must always be considered when the method is applied:

1. the probability of capture is the same for all individuals in the population whether tagged or not.
2. survival probability is the same for all individuals whether tagged or not.
3. emigration is permanent and therefore is the same as mortality.

These assumptions have some curious properties that must be kept in mind. All of the assumptions exclude age and size effects such as differences in behavior associated with age/size or senescence. Assumption 3 can be met if species have home ranges but is much more of a problem if losses due to movement are substantial. In such a case, the survival probability becomes 1.0 minus the probability of loss, where loss has two components: mortality and emigration. Assumptions appear to have been met in the work with dippers.

After tagging, birds were released. Recaptures and additional tagging were done with a time interval of 1 year. Small numbers of birds were not found after 1 year but were recaptured for the first time after 2 years and two birds during the study were not recaptured for the first time until 3 years after initial tagging. Creating a table such as Table 3.14 can be very tedious and hence error prone. I have written a short BASIC program, **JOLLY-SEBER.BAS**, that takes mark-recapture data and creates a summary table. The program is provided at the end of this chapter.

In order to estimate survival from the recapture data, it is necessary to calculate an additional intermediate value, z_i, which is based on the number caught in the ith sample that were last caught on or before the hth sample time (Table 3.15). Note that in Table 3.15 each element in a diagonal is the sum of all values in the column above it in Table 3.14. For example, for i=3 in Table 3.14, the entry for h=1 is 2

TABLE 3.14 Mark-Recapture Data for European Dippers, *Cinclus cinclus*, Banded From 1981 to 1987 in France

Year (h)	R_i	i=2	3	4	5	6	7	Total r_i
1	22	11	2	0	0	0	0	13
2	60		24	1	0	0	0	25
3	78			34	2	0	0	36
4	80				45	1	2	48
5	88					51	0	51
6	98						52	52
m_i		11	26	35	47	52	54	

Note: h = year of tagging; i = year of first recapture; R_i = number with tags that were released in year h; data from Lebreton *et al.* 1992.

TABLE 3.15 Tabulations of Numbers Caught in the ith Sample Last Caught in or before the hth Sample for European Dippers Banded from 1981 to 1987

Year (h)	i=2	3	4	5	6	7	z_i
1	11	2	0	0	0	0	2
2		26	1	0	0	0	1
3			35	2	0	0	2
4				47	1	2	3
5					52	2	2
6						54	

Note: h = year of tagging, i = year of first recapture; data from Lebreton *et al.* 1992; z_i is the number of different animals caught before the ith sample that were not caught in the ith sample but were caught subsequently.

TABLE 3.16 Summary of Intermediate Terms Needed to Estimate Annual Survival Rate, ϕ_i

i	R_i	M_i	m_i	r_i	z_i	ϕ_i	$se(\phi_i)$
1	22	0	0	13	0	0.718	0.155
2	60	15.8	11	25	2	0.435	0.069
3	78	28.167	26	36	1	0.478	0.060
4	80	38.333	35	48	2	0.626	0.059
5	88	52.176	47	51	3	0.599	0.056
6	98	55.769	52	52	2		

Note: some values are collected from Tables 3.14 and 3.15 as well as Equations 3.34 - 3.37.

and for h=2 is 24. In Table 3.15, for h=1 the entry is 2 and for h = 2 it is 26, that is, the sum of 2 and 24.

The total number of marked animals in the population just before sampling is M_i, which is estimated

$$M_i = \frac{R_i z_i}{r_i} + m_i \qquad (3.34)$$

where

R_i = number released from the ith sample after marking,

z_i = number marked before time i that are not caught in the ith sample but are caught subsequently,

r_i = number of the R_i animals that subsequently are recaptured, and

m_i = number of marked animals in the ith sample.

The survival rate ϕ_i is

$$\phi_i = \frac{M_{i+1}}{M_i - m_i + R_i} \qquad (3.35)$$

with an associated variance of

$$var(\phi_i) = \phi_i^2 \left[\frac{(M_{i+1} - m_{i+1})(M_{i+1} - m_{i+1} + R_{i+1})}{M_{i+1}^2} \right.$$

$$\left. \left(\frac{1}{r_{i+1}} - \frac{1}{R_{i+1}}\right) + \frac{M_i - m_i}{M_i - m_i + R_i}\left(\frac{1}{r_i} - \frac{1}{R}\right) + \frac{1 - \phi_i}{M_{i+1}}\right]. \qquad (3.36)$$

The standard error of the estimate of survival is

$$se(\phi_i) = \sqrt{var(\phi_i)}. \qquad (3.37)$$

Table 3.16 collects all of the intermediate terms as an aid to calculation. Calculation of variances and standard errors of the estimate of ϕ_i are best done using a computer program, which is shown at the end of this chapter (**JOLLY-SEBER_VAR.BAS**).

Substantial complications can arise in mark-recapture studies because of the biological attributes of the organisms being studied. A excellent discussion of problems and suggestions is provided by Lebreton *et al.* (1992), who include an evaluation of software packages that can be used in analysis.

Estimating survival rates can be as simple as mapping plants and watching them or as complex as aging and tagging large mammals. Techniques for data manipulation change but the goals are all exactly the same: to obtain the best possible estimates of survival because these estimates will be used in, among other things, models that predict the future states of populations.

TABLE 3.17 Human Fecundity Schedule for the United States Based on 1964 Data

Age (yrs)	Fecundity rate	Age (yrs)	Fecundity rate
0	0.0	27.5	0.438250
1	0.0	32.5	0.253695
7.5	0.0	37.5	0.122180
12.5	0.002115	42.5	0.033730
17.5	0.177815	47.5	0.001970
22.5	0.537085		

Note: data from Keyfitz and Flieger (1968, p. 162); fecundity values are the expected number of female infants produced during a 5-year period by females in the age category x to x+5 years.

TABLE 3.18 Birth Rates of 225 Caribou Cows ≥3 Years Old in the Porcupine Herd

Year	n	no. giving birth	Birth rate
1982	9	8	0.89
1983	23	20	0.87
1984	31	25	0.81
1985	56	43	0.77
1986	42	31	0.74
1987	51	40	0.78
1988	91	76	0.84
1989	74	58	0.78
1990	74	61	0.82
1991	74	55	0.74
1992	78	67	0.86

Note: all cows were radio collared so could be located during calving season; data from Fancy *et al.* (1994).

ESTIMATING FECUNDITY

U. S. Women in 1964

Determining the numbers of offspring produced by a female with stated characteristics such as age or size frequently is remarkably difficult. Part of the problem can be shown by examining a fecundity schedule for U. S. women in 1964 (Table 3.17).

The average number of babies per pregnancy is about 1, so why does the age-specific rate change? The obvious answer, of course, is that the probability of becoming pregnant changes with age. Also, for the U. S. population in 1964, the gender bias was 51.1:48.9 in favor of males, which is about the same world wide. In order to model population growth, it is very important to know not only the number of offspring per reproductive episode and the sex ratio, but also the probability that a female will become pregnant during a stated time interval.

The Porcupine Caribou Herd, 1983 to 1992

The Porcupine caribou herd (*Rangifer tarandus granti*) is in northeastern Alaska and northwestern Canada and was studied from 1983 to 1992 (Fancy *et al.* 1994) to determine demographic transitions in sufficient detail so that a population model could be developed. Reproduction was studied using 225 radio-collared cows aged ≥3 years that were monitored for a total of 603 reproductive attempts. Age-1 cows were never observed to have become pregnant as calves. Also, it was rare, although not impossible, for 2--year olds to have calves. Out of 53 known 2--year olds, 2 gave birth. In general, females first give birth at age 3.

Cows were intensively monitored each year from the end of May to the end of June. Whether a cow had given birth was determined from a combination of symptoms: (1) presence of a live or dead calf; (2) udder distension; and, (3) presence or absence of hard antlers. Any cow that did not show obvious signs of having given birth was relocated one or more times to make sure that pregnancies were not missed. Results (Table 3.18) indicate that, in general, the probability of having a successful pregnancy was between 0.74 and 0.89 with a mean of 0.809.

In order for these birth rates to be useful they must be combined with survival estimates of newborn calves and survival to age 1 year. These values were estimated by Fancy *et al.* (1994).

The New Zealand Nikau Palm.

Seed production was studied in the nikau palm *Rhopalostylis sapida* over a 7-year period (Enright 1992). Nikau palms bear seeds on branched inflorescences and an entire inflorescence may have in excess of 2000 seeds. To determine the total number of seeds produced, the usual practice is not to count all seeds but rather, in some fashion, to subsample. In this case a branchlet was selected and all seeds counted and then this number was multiplied by the number of branchlets in order to obtain the total number of seeds (Table 3.19). Seed production is complicated and

TABLE 3.19 Mean Annual Seed Production in *Rhopalostylis sapida* at Huapai Scientific Reserve, N. Z.

Height (m)	n	Seed-bearing inflorescences per palm	Seeds per palm yr^{-1}
0-2	39	0	0
2-4	32	0.59	283
4-6	60	1.20	685
6-8	20	1.60	1140
8-10	6		866

Note: data based on 7 years of observation (Enright 1992).

TABLE 3.20 Annual Variation in Reproduction of *Rhopalostylis sapida* at Huapai Scientific Reserve, N. Z.

Year	Seeds per palm
1983	367±34
1984	422±61
1985	553±87
1986	1180±132
1987	1154±117
1988	478±76
1989	523±100

Note: values are for mature individuals only (≥2m tall); data from Enright (1992).

numbers of seeds varied with palm size between years and within years over a single flowering season. One aspect of reproduction not evident in the presentation of means is the substantial year-to-year differences, which was substantial (Table 3.20). For analysis of reproduction to be complete, there would have to be estimates of germination rate and survival, which were addressed in a companion paper (Enright and Watson 1992).

Work on *Rhopalostylis sapida* presents the general problems of estimating reproduction in plants. First there is the problem of determining age-specific rates, which, in general, is avoided by turning the analysis into one that focuses on size-specific rates. Second, there is the problem of how to determine the numbers of seeds when numbers are large, which can be solved by subsampling. Third, there is a problem with substantial year-to-year variation. A frequent solution is to ignore the problem by using just mean values. Not infrequently this is a reasonable solution; however,

variation is important in estimating confidence limits for population growth and this topic is presented in Chapter 10.

Although it would seem that measuring reproductive rates in plants should be easier than in animals, as can be seen with the nikau palm, there are interesting and specific difficulties that are as challenging as measurements of reproduction in animals.

A Sea Star, *Linckia multifora*, in Guam

The sea star *Linckia multifora* can reproduce asexually by arm autotomy. An autotomized arm can regenerate a new disc and new arms; individuals that lose arms regenerate new ones (Figure 3.9). A population of *L. multifora* in various stages of regeneration following arm autotomy was studied in Guam (Rideout 1978) and his results are presented in Table 3.21.

Because autotomized arms take roughly one month to become comets, the monthly rate of autotomy is the number of new individuals per individual, m_A, and can be calculated from the monthly number of autotomized arms (A) and the total number of sea stars not including the autotomized arms (T) in the samples

$$m_A = \frac{A}{T} \text{ month}^{-1}. \tag{3.38}$$

For example, the autotomy rate for April 1974 was 0.286 month^{-1}. The mean monthly autotomy rate for 12 months was 0.096 individual^{-1}; that is, each month, on average, about 10% of adult starfish lose an arm and contribute a new individual to the population (Ebert 1996).

TABLE 3.21 Population Structure of *Linckia multifora* on Asan Reef Flat, Guam

Sample	Arms (A)	Comets (B + C)	D	E	F	G	H + I	Total (B through I)
April 1974	6	14	1	1	0	1	4	21
May	9	23	5	3	2	3	20	56
June	8	28	4	0	0	1	14	47
July	1	42	9	2	1	2	27	83
Aug.	10	48	11	2	4	2	33	100
Sept.	1	19	3	1	1	1	11	36
Oct.	0	33	13	0	2	4	12	64
Nov.	13	31	26	5	4	3	26	95
Dec.	0	9	5	0	1	1	4	20
Jan. 1975	6	36	9	3	2	2	13	65
Feb.	9	39	18	4	4	2	20	87
Mar.	4	30	13	3	4	3	7	60

Note: letters refer to Figure 3.9; samples start in 1974 and extend through March 1975; data from Rideout (1978).

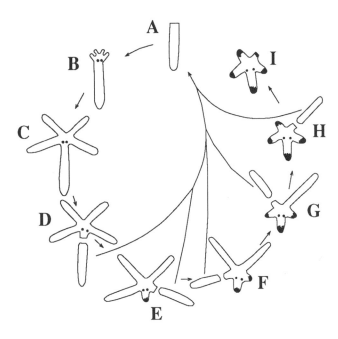

FIGURE 3.9 Asexual life cycle of *Linckia multifora* by arm autotomy; autotomized arms (A) form comets (B) that regenerate (C) and then autotomize the original arm and form a counter-comet (D), which can lose another arm forming a post-counter comet (E) and so on to a disc parent (I) in which all arms regenerated by a comet have been autotomized; disc parents can re-enter the cycle by losing a regenerated arm; (redrawn from Rideout 1978).

GENERAL COMMENTS

Examples in this chapter show how survival and reproduction can be determined and illustrate both the richness and idiosyncratic nature of demographic data. Various groups of organisms may lend themselves to similar techniques but frequently even within closely related animals or plants, differences in the biological details of the life cycle will demand changes in how data must be gathered. A reasonable approach in all studies is first to draw the life cycle for the study organism to ensure that all parts are included such as both sexual and asexual contributions to reproduction. The second step is to decide on how all of the connections can be studied. Often this means breaking the life cycle into parts that later can be combined into a single transmission line in graphs shown in Chapter 1 (Figure 1.1); a reproductive transmission from an adult female that is shown as a single line in Figure 1.1 requires knowing the probability of reproducing, number of offspring produced if reproduction takes place, and survival of new individuals up to the age of 1. What should be clear is that gathering data on survival and reproduction requires a deep understanding of the natural history of the organism that is to be studied.

PROBLEMS

1. Empty shells of the clam *Chione fluctafraga* were collected in San Diego Bay at Bayside Park on October 6, 1979, by students in an introductory ecology class (Table 3.22). Only right shells were used. Many clams have growth lines that have been used in various studies to determine age. An age 0 clam is one that is large enough to be sampled by hand and without special equipment.

TABLE 3.22 Empty Shells of *Chione fluctafraga* Collected at Chula Vista, CA, October 6, 1979

Age	Number
0	15
1	110
2	212
3	134
4	39
5	16
6	4
7	1

a. Determine l_x and p_x for this clam. Calculate 95% confidence limits for the estimates of p_x and l_x.

b. Assume that survival is age-dependent past the age of 1 year. How could you explain any apparent age-dependent patterns?

2. White pine (*Pinus strobus*) was studied by Holla and Knowles (1988) in Ontario, Canada. Ages were determined by counting "growth lines" in cores extracted from the trunks (Table 3.23). In order to use these data to estimate terms of a life table, it is necessary to assume that the population has a stable-age distribution and is stationary ($r = 0$).

a. Determine p_x and l_x together with the 95% confidence intervals and graph the distribution.

TABLE 3.23 Numbers of Live Pine Trees, *Pinus strobus*, in Each Age Class Determined by Growth Rings

Age interval	n_x	Age interval	n_x
0-10	255	100-110	36
10-20	93	110-120	27
20-30	68	120-130	21
30-40	57	130-140	19
40-50	51	140-150	14
50-60	48	150-160	10
60-70	47	160-170	6
70-80	45	170-180	2
80-90	44	180-190	1
90-100	39	190-200	0

Note: data from Holla and Knowles (1988).

TABLE 3.24 Mark-Recapture Data for Merriam Kangaroo Rats, *Dipodomys merriami*

Males: N = 82			
1000000000000000000000000	0000110000000000000000000	0000000000101100000000000	0000000000000000101100000
1000000000000000000000000	0000111110000000000000000	0000000000110110000000000	0000000000000000101000000
1100000000000000000000000	0000010000000000000000000	0000000000111111110000000	0000000000000000001100000
1111100000000000000000000	0000010000000000000000000	0000000000111111111111111	0000000000000000001000000
1111110000000000000000000	0000010000000000000000000	0000000000010000000000000	0000000000000000001000000
1111110111000000000000000	0000010000000000000000000	0000000000010000000000000	0000000000000000001000000
1111111000000000000000000	0000010000000000000000000	0000000000001010000000000	0000000000000000000111111
1111111100000000000000000	0000010000000000000000000	0000000000001100000000000	0000000000000000000111111
1111111110000100000000000	0000011000000000000000000	0000000000011111111111111	0000000000000000000100000
1111111111111110000000000	0000011000000000000000000	0000000000001111111000000	0000000000000000000011111
0100100000000000000000000	0000011000000000000000000	0000000000001100000000000	0000000000000000000011100
0100111000000000000000000	0000001000000000000000000	0000000000001100000000000	0000000000000000000010100
0101000000000000000000000	0000001000000000000000000	0000000000001100000000000	0000000000000000000001100
0110111101000000000000000	0000000101111111111110101	0000000000000100010111000	0000000000000000000000111
0111110000000000000000000	0000000111001110000000000	0000000000000111111111111	0000000000000000000000100
0010111111100000000000000	0000000111111100000000000	0000000000000110000000000	0000000000000000000000100
0001111100011111000000000	0000000011011110000000000	0000000000001000000000000	0000000000000000000000011
0001111000000000000000000	0000000001000000000000000	0000000000000111110000000	0000000000000000000000010
0000100000000000000000000	0000000000100000000000000	0000000000000111000000000	0000000000000000000000001
0000100000000000000000000	0000000000100000000000000	0000000000000111000000000	
0000110000000000000000000	0000000000100000000000000	0000000000000111000000000	

Females N = 104			
1001000000000000000000000	0000100000000000000000000	0000000000100000000000000	0000000000000000101000000
1001000000000000000000000	0000100000000000000000000	0000000001001100000000000	0000000000000000100000000
1010000000000000000000000	0000101000000011111111111	0000000001011100000000000	0000000000000000100000000
1100000000000000000000000	0000111001010000000000000	0000000001100000000000000	0000000000000000010000001
1100000000000000000000000	0000110000000000000000000	0000000000111111111001001	0000000000000000001111000
1110000000000000000000000	0000011111110000000000000	0000000001000000000000000	0000000000000000001101111
1111000000000000000000000	0000011000111111111111111	0000000010000000000000000	0000000000000000001100100
1111000000000000000000000	0000011000000000000000000	0000000000111110000000000	0000000000000000001011110
1111110000000000000000000	0000001000000000000000000	0000000000111000000000000	0000000000000000001000000
0100000000000000000000000	0000001000000000000000000	0000000000100000000000000	0000000000000000001000000
0100000000000000000000000	0000001100000000000000000	0000000000010000000000000	0000000000000000001000000
0110111000000000000000000	0000001110111111111000000	0000000000010000110000000	0000000000000000001000000
0111100100000000000000000	0000001111101111111111111	0000000000011000000000000	0000000000000000000111111
0111110000000000000000000	0000000011111111100000000	0000000000011000000000000	0000000000000000000110100
0111110000000000000000000	0000000110000000000000000	0000000000011000000000000	0000000000000000000110000
0111110000000000000000000	0000000010000000000000000	0000000000011110000000000	0000000000000000000100000
0111111011000111100000000	0000000010000000000000000	0000000000011111111011	0000000000000000000011000
0111111111110000000000000	0000000010000000000000000	0000000000011111111100	0000000000000000000010000
0011011000000100000000000	0000000010000000000000000	0000000000011111111100	0000000000000000000001110
0011000000000000000000000	0000000010000100111111100	0000000000001100000000000	0000000000000000000001000
0001000000000000000000000	0000000010111100000000000	0000000000001000001000	0000000000000000000000010
0001000000000000000000000	0000000011100000000000000	0000000000001000000000000	0000000000000000000000010
0001110000000000000000000	0000000011111100000000000	0000000000001000000000000	0000000000000000000000010
0001111011000111111101011	0000000011111110000000000	0000000000000111110101	0000000000000000000000001
0001111100011111110011111	0000000011111111000000000	0000000000000111000000000	0000000000000000000000001
0001111110011111101010100	0000000000100000000000000	0000000000000101100000000	0000000000000000000000001

Note: sampling was done every month starting in September 1992 and ending in August 1994 at the Barry M. Goldwater Air Force Range, Arizona (unpublished data from Leroy McClenaghan); data structure follows the structure of the dipper data.

TABLE 3.25 Mark-Recapture Data for the Black-Kneed Capsid (*Blepharidopterus angulatus*)

i	1	2	3	4	5	6	7	8	9	10	11	12	13
R_i	54	143	164	202	214	207	243	175	169	126	120	120	-
h													
1		10	3	5	2	2	1	0	0	0	1	0	0
2			34	18	8	4	6	4	2	0	2	1	1
3				33	13	8	5	0	4	1	3	3	0
4					30	20	10	3	2	2	1	1	2
5						43	34	14	11	3	0	1	3
6							56	19	12	5	4	2	3
7								46	28	17	8	7	2
8									51	22	12	4	10
9										34	16	11	9
10											30	16	12
11												26	18
12													35

Note: data from Jolly (1965)

3. The mark-recapture data in Table 3.24 are for the Merriam kangaroo rat *Dipodomys merriami* that was studied for a two year period on the Barry M. Goldwater Air Force Range, Arizona (L. McClenaghan, unpublished data). Samples were gathered every month starting in September 1992 through August 1994.
   a. Determine survival rates for males and females separately.
   b. Would you feel justified in combining data for both sexes? If so, redo the analysis.
   c. How can you explain patterns in the results?

4. Jolly (1965) presented capture-recapture data for an insect, female black-kneed capsid (*Blepharidopterus angulatus*) in an apple orchard. During the study, there were additions of new females both by fresh emergence and by immigration. The following table (Table 3.25) shows the number caught in the ith sample last captured in the hth sample. These data are in the same form as Table 3.14 for dippers. The 13 samples were taken at alternatively 3- and 4-day intervals; that is, the interval between sample 1 and 2 was 3 days and between 2 and 3 it was 4 days.
   a. Estimate daily survival for each period. Plot the results with the estimated standard errors for each point.
   b. What are the daily survival rates for each period?

5. Lesser kestrels (*Falco naumanni*) were studied in southern Spain from 1988 to 1993 by Hiraldo *et al.* (1996). Table 3.26 shows the capture or resighting history for yearlings and adults ≥ 2 years old.

TABLE 3.26 Capture or Resighting Matrix for Tagged Lesser Kestrels in Southern Spain from 1988 to 1993

History	N	History	N	History	N
100000	11	010010	2	001000	13
100100	1	010100	2	001011	1
100111	1	010101	1	001100	12
101000	2	010110	3	001110	4
101111	2	010111	3	001111	9
110000	3	010200	1	001120	1
110010	1	011000	18	000100	22
110100	1	011100	15	000101	4
111000	3	011110	7	000110	15
111010	1	011111	14	000111	12
111100	2	011120	1	000010	23
010000	22	011200	1	000011	28

Note: capture or resighting = 1; no capture or resighting = 2; found dead = 2; data from Hiraldo et al. (1996).

Estimate annual survival rates together with standard errors. Is there evidence that survival rates are changing over time for lesser kestrels? [*Hint:* you will have to expand the data in the table so you have a data entry for each bird.]

6. Hartman (1995) studied eastern moles (*Scalopus aquaticus*) on the United States Department of Energy's Savannah River site in Aiken and Barnwell Counties, South Carolina. Tooth wear patterns were used to age animals and patterns were validated, in part, using tagged individuals. Reproduction is during late winter and early spring and

numbers of individuals in each age class (Table 3.27) are for 101 moles captured during October and November. The data represent a vertical life table and so in order to use them to estimate survival rates, it is necessary to assume that population growth rate, λ, is equal to 1.0.

 a. With an assumed $\lambda = 1.0$, estimate l_x and p_x together with 95% profile likelihood intervals.

 b. Does survival of *Scalopus aquaticus* differ from the European mole *Talpa europaea* based on non overlap of 95% confidence limits? Use the data in Table 3.28.

TABLE 3.27 Numbers of eastern moles, *Scalopus aquaticus*, assigned to age classes based on tooth wear

Age	N	Age	N
0-1	37	3-4	12
1-2	25	4-5	6
2-3	20	6-7	1

Note: animals captured during autumn; data from Hartman (1995).

TABLE 3.28 Numbers of European moles, *Talpa europaea*, assigned to age classes based on tooth wear

Age	N	Age	N
0-1	70	3-4	4
1-2	20	4-5	1
2-3	11	6-7	3

Note: animals captured during autumn; data from Lodal and Grue (1985) and cited in Hartman (1995).

BASIC PROGRAMS

P_LIKEKIHOOD.BAS

The following program calculates the profile likelihood confidence intervals using the Newton-Raphson algorithm (Venzon and Moolgavkar 1988, Lebreton *et al.* 1992)

```
10 dim s(2),p(2)
20 s(1) = 1.000000E-05 : s(2) = 0.99999
30 print "Estimating profile likelihood
   95% intervals"
40 print "Cornered Rat Software   June
   1996 T. A. Ebert"
50 print : input "Do you want to
   continue? (y/n) ";s$
60 if s$ = "n" or s$ = "n" then goto 380
70 input "number of observed successes:
   ";y
```

```
80 input "Total sample size:   ";n
90 x = y/n
100 if y = 0 or y = n then goto 270
110 for i = 1 to 2
120 p = s(i)
130 for j = 1 to 10000
140 f = 2*(y*log(x)+(n-y)*log(1-x))
150   f  =  2*(y*log(p)+(n-y)*log(1-
    p))+3.8416-f
160 fd = 2*(y/p-(n-y)/(1-p))
170 p = p-f/fd
180 if abs(f/fd) <= 1.000000E-06 then
    goto 200
190 next j
200 p(i) = p
210 next i
220 p = y/n
230 s = 1.96*sqr(p*(1-p)/n)
240 print "Binomial [lower 95%, p, upper
    95%]: ";p-s,x,p+s
250 print "Profile likelihood [lower
    95%, p, upper 95%]: ";p(1),x,p(2)
260 goto 50
270 p = exp(-3.8416/(2*n))
280 if y = 0 then goto 320
290 p(2) = 1
300 p(1) = p
310 goto 330
320 p(2) = 1-p : p(1) = 0
330 p = y/n
340 s = 1.96*sqr(p*(1-p)/n)
350 print "Binomial interval [lower 95%,
    p, upper 95%]: ";p-s,x,p+s
360 print "Profile likelihood [lower
    95%, p, upper 95%]: ";p(1),x,p(2)
370 goto 50
380 end
```

JOLLY-SEBER.BAS

The following program creates a table of number caught in the ith sample last captured in the hth sample using data in which each individual is showen as initially marked (1) and then either recaptured (1) or not (0) on each sampling date. For example, a single individual might be listed as 1001, which would mean that it had been marked on the first date, not seen on the second or third dates and then captured or resighted on the 4th date.

```
10 dim a(500,40),a$(500),m(40,40),r(40)
20 rem n is the number of records and k
   is the number of sample dates
30 input "Name of file with data: ";f$
```

```
40 open f$ for input as #1
50 input #1,n,k
60 for i = 1 to n
70 input #1,a$(i)
80 next i
90 close #1
100 for i = 1 to n
110 for j = 1 to k
120 a(i,j) = val(mid$(a$(i),j,1))
130 r(j) = r(j)+a(i,j)
140 next j
150 next i
160 for w = 1 to k-1
170 i = 1
180 if a(i,w) = 0 then goto 260
190 j = w+1
200 if a(i,j) = 0 then goto 230
210 m(w,j) = m(w,j)+1
220 goto 260
230 j = j+1
240 if j = k+1 then goto 260
250 goto 200
260 i = i+1
270 if i = n+1 then goto 290
280 goto 180
290 next w
300 for i = 1 to k-1
310 print r(i),
320 next i
330 print r(k)
340 for i = 1 to k-1
350 for j = 1 to k-1
360 print m(i,j),
370 next j
380 print m(i,k)
390 next i
400 input "Do you want results saved to
    a file? (y/n) ";f$
410 if f$ = "Y" or f$ = "y" then goto
    440
420 if f$ <> "N" and f$ <> "n" then goto
    400
430 goto 570
440 input "File for output: ";f$
450 open f$ for output as #1
460 for i = 1 to k-1
470 print #1,r(i),
480 next i
490 print #1,r(k)
500 for i = 1 to k-1
510 for j = 1 to k-1
520 print #1,m(i,j),
530 next j
540 print #1,m(i,k)
550 next i
560 close #1
570 end
```

For example here are test data for dippers as presented by Lebreton *et al.* (1992) (their Table 2, p 70); 1=capture or resighting, 0=no capture or sighting. There are 22 records with 3 sample dates

```
22, 3
111
110
111
110
100
101
100
110
100
100
110
100
111
100
111
100
100
110
110
100
101
110
```

Output should be

```
22  11  6
0   11  2
0   0   4
```

The top row is R_i, the second row shows that of the 22 birds released on the first date, 11 were captured on the second date, 2 were not seen on the second date but were seen on the third date. The third row shows that of the 11 birds released on the second date, 4 were recaptured on the third date. The data set is sufficiently small that it is easy to verify that these numbers are correct.

JOLLY-SEBER_VAR.BAS.

The following program calculates standard errors for the estimates of survival using the Cormack-Jolly-Seber method. A data file must be created with the first number in the file being the number of time periods used, which is k in the program. The data file must have k lines and each line must contain f, r_i, M_i, m_i and R_i The file for European dippers, shown in Table 3.16, is

6
.718, 13, 0, 0, 22
.435, 25, 15.8, 11, 60
.478, 36, 28.167, 26, 78
.626, 48, 38.333, 35, 80
.599, 51, 52.176, 47, 88
0, 52, 55.769, 52, 98

JOLLY-SEBER_VAR.BAS

```
10 dim ph(30),cmi(30),mi(30),cri(30),
   ri(30)
20 input "file with data: ";f$
30 open f$ for input as #1
40 input #1,k
50 for i = 1 to k
60 input #1,ph(i),ri(i),cmi(i),mi(i),
   cri(i)
70 next i
```

```
80 print "Standard error of estimate of
   survival rate"
90 for i = 1 to k-1
100 a = (cmi(i+1)-mi(i+1))*(cmi(i+1)-
    mi(i+1)+cri(i+1))/(cmi(i+1)^2)
110 a = a*(1/ri(i+1)-1/cri(i+1))
120 b = (cmi(i)-mi(i))/(cmi(i)-
    mi(i)+cri(i))*(1/ri(i)-1/cri(i))
130 var = ph(i)*ph(i)*(a+b+(1-
    ph(i))/cmi(i+1))
140 print i,sqr(var)
150 next i
160 close #1
170 end
```

The output for dippers is the same as shown by Lebreton *et al.* (1992) in their Table 11

Standard errors of survival rates for each time period

1	0.155527
2	0.068865
3	0.059696
4	0.059262
5	0.056068

4

Life Cycle Graphs

The tiniest changes in a polynomial's coefficients can, in the worst case, send its roots sprawling all over the complex plane.

W. H. Press *et al.*,
Numerical Recipes. The Art of Scientific Computing, 1986.

INTRODUCTION

Population growth can be modeled as a life cycle graph (Hubbell and Werner 1979; Caswell 1982b, 1989) and can be analyzed using techniques that are appropriate for other sorts of networks such as electronic circuits, food webs, and successional changes in communities. Graphs have components, such as age or stage or size classes, and there are transitions or flows among components.

THE EULER EQUATION

Understanding life cycle graphs may be somewhat easier by starting with the Euler equation and restricting the way the equation is used to cases with a fixed time unit such as one week or one day or one year. As a first example, we can use our small beast from Chapters 1 and 2. With 3 age classes, 0 to 2, the Euler equation is

$$1 = \sum_{x=0}^{2} l_x m_x e^{-rx},$$

which can be expanded as

$$1 = l_0 m_0 e^{-r0} + l_1 m_1 e^{-r1} + l_2 m_2 e^{-r2} \qquad (4.1)$$

and, because $\lambda = e^r$, we can substitute 1 into Equation 4.1

$$1 = l_0 m_0 \lambda^{-0} + l_1 m_1 \lambda^{-1} + l_2 m_2 \lambda^{-2}. \qquad (4.2)$$

Also, l_x values can be replaced by p_x values because they are the products of p_x values:

$$l_1 = p_0,$$
$$l_2 = p_0 p_1,$$
$$l_3 = p_0 p_1 p_2,$$
etc.

In equation 4.2, the term $l_0 m_0 \lambda^{-0}$ drops out because m_0 always will be 0; there are no creatures where new-born individuals give birth to more individuals at the same instant that they themselves are born. Some creatures are born pregnant (some aphids, some mites) but it always takes some amount of time from birth until new births take place. The important point is that $l_0 m_0 \lambda^{-0}$ always will be 0 and so can be dropped from Equation 4.2, which becomes

$$1 = p_0 m_1 \lambda^{-1} + p_0 p_1 m_2 \lambda^{-2}. \qquad (4.3)$$

TABLE 4.1 Annual Survival Probability, p_x, and Fecundity, m_x, for a Short-Lived Beast

Age	p_x	m_x
0	0.240	0
1	0.242	20
2	0.000	24

Note: age is in years.

Now rearrange to make the equation equal to 0

$$1 - p_0 m_1 \lambda^{-1} - p_0 p_1 m_2 \lambda^{-2} = 0 \qquad (4.4)$$

and clear λ^{-2} by multiplying through by λ^2

$$\lambda^2 - p_0 m_1 \lambda - p_0 p_1 m_2 = 0. \qquad (4.5)$$

Survival and fecundity values for our small beast (Table 4.1) can be used in Equation 4.5 to form a polynomial that will have two roots

$$\lambda^2 - (0.240)(20)\lambda - (0.242)(0.240)(24) = 0$$

or

$$\lambda^2 - 4.8\lambda - 1.39392 = 0.$$

For this particular example, the roots, λ_i, can be found by using the formula for finding roots of a quadratic equation

$$\lambda_i = \frac{-b \pm \sqrt{b^2 - 4ac}}{2a} \qquad (4.6)$$

where

    $a = 1$ (the coefficient of λ^2),
    $b = -4.8$ (the coefficient of λ),

and

    $c = -1.39392$ (the coefficient of λ^0).

Substituting values into Equation 4.6 yields

$$\lambda_i = \frac{4.8 \pm \sqrt{23.04 - 5.5756}}{2}$$

so

    $\lambda_1 = 5.0747$

and

    $\lambda_2 = -0.2747.$

The largest positive root is the population growth rate, λ or e^r. In this case there is just one positive root, $\lambda_1 = 5.0747$, so $r = 1.624$, the same as calculated by projection methods in Chapter 1 and using the Euler equation or cumulants of the age-of-mothers distribution in Chapter 2. The relationship between the Euler equation and the

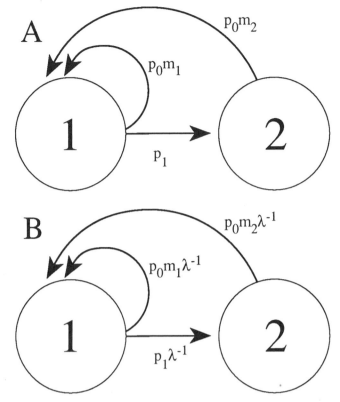

FIGURE 4.1 A, Life cycle graph with two ages; new-born individuals survive for one time period at which time they are age 1; p_x values are survival rates from x to x+1; **B**, Z-transformation of the life cycle graph in A.

characteristic polynomial or characteristic equation that can be formed from it is an important link to understanding the analysis of life cycle graphs.

The single most important rule in drawing life cycle graphs is that lines must be assigned transitions in such a manner that time units are not changed; if transitions are in years the lines must be years or multiples of years and, if multiples of years, then this must be explicitly included in the graph. Actually, it is much easier than it sounds. To begin, a birth-pulse population is presented so it is necessary to decide when reproduction takes place. For a graph, nothing happens "at once". If the time reference for a particular graph is 1 year, then births that take place "now" will take one full year in transit before entering the population and so individuals will be one year old when they enter the graph. There are somewhat different ways of handling this problem but I will start with the convention that the first node (box, circle) of a graph will represent organisms that are age 1.

Transitions that represent births will include not only m_x but also survival for one time interval, p_0, from 0 to age 1,

$$f_x = p_0 m_x. \qquad (4.7)$$

GRAPH ANALYSIS

Our Small Beast

A life cycle graph for our small beast is shown in Figure 4.1A and Caswell (1985, 1989) provides a set of steps for finding the characteristic equation of a such a graph. Actually the steps are very much like expanding the Euler equation and using λ rather than e^r and p_x values rather than l_x values. The steps are

Step 1

Multiply each coefficient by λ^{-1}. This is called a Z-transformation and is the same as using e^{-rx} in the Euler equation.

Step 2

Take products of coefficients around all loops in the graph. This is the same as forming the individual terms of the Euler equation, $l_x m_x e^{-rx}$.

Step 3

Sum these products and set 1 minus this sum equal to zero. The value 1 is the same as in the Euler equation; namely, the total number of new born individuals added to the population right now (Equation 2.11 - 2.13).

Step 4

Clear the equation of the highest λ^{-i} term by multiplying through by λ^i.

Once the characteristic equation has been formed, the final step is to find the roots. The largest positive root, λ_1, is the population growth rate after a stable-age distribution has been attained.

To find the population growth rate for the life cycle in Figure 4.1A, the first step is to perform a Z-transformation (Figure 4.1B), which is just multiplying each transmission or transfer by λ^{-1}. The next step is to multiply terms around each loop. The first loop in Figure 4.1B originates at node 1 and, in one cycle re-enters node 1. The second loop originates at node 1 and passes through node 2 before returning to node 1:

Loop 1 $p_0 m_1 \lambda^{-1}$

and

Loop 2 $p_1 p_0 m_2 \lambda^{-2}$.

Next, sum the loops, subtract the sum from 1 and set equal to 0:

TABLE 4.2 Survival and Fecundity for *Bradybaena fruticum*, a Pulmonate Land Snail

Age	l_x	p_x	m_x	$f_x = p_0 m_x$
0	1.000	0.8190	0.0	0.0
1	0.819	0.5165	0.0	0.0
2	0.423	0.2648	16.69	13.6691
3	0.112	0.0536	6.00	4.914
4	0.006	0.1667	0.0	
5	0.001	0.0	0.0	

Note: ages are in years; f_x is the number of offspring multiplied by the probability that a new-born survives for one year to enter age class 1; that is, $p_0 m_x$. data from Staikou *et al.* (1990).

$$1 - (p_0 m_1 \lambda^{-1} + p_1 p_0 m_2 \lambda^{-2}) = 0$$

or

$$1 - p_0 m_1 \lambda^{-1} - p_1 p_0 m_2 \lambda^{-2} = 0.$$

Because the largest exponent is -2, multiply the equation by λ^2 to form the characteristic equation of the graph

$$\lambda^2 - p_0 m_1 \lambda - p_1 p_0 m_2 = 0, \qquad (4.8)$$

which is the same as Equation 4.5 and so if p_x and m_x values from Table 4.1 would be used, the same roots would be obtained.

A Pulmonate Land Snail *Bradybaena fruticum*

Table 4.2 presents survival and fecundity for a pulmonate land snail *Bradybaena fruticum* in Greece (Staikou *et al.* 1990). Analysis of the life cycle graph of *B. fruticum* (Figure 4.2A) follows the same steps as used for the first example:

Step 1

Z-transform the graph by multiplying each transition by λ^{-1} (Figure 4.2B).

Step 2

Identify the loops and multiply transitions around each loop. The first loop goes from 1 to 2 and back to 1 and the second loop goes from 1 to 2 to 3 and back to 1,

loop 1: $(0.5165\lambda^{-1})(13.6691\lambda^{-1}) = 7.06009\lambda^{-2}$

loop 2: $(0.5165\lambda^{-1})(0.2648\lambda^{-1})(4.914\lambda^{-1}) = 0.67208\lambda^{-3}$.

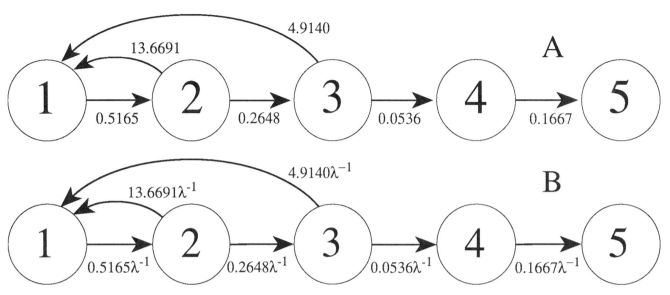

FIGURE 4.2 **A**, Life cycle graph for *Bradybaena fruticum* ; **B**, Z-transform of graph (Staikou *et al.* 1990).

Step 3

Sum the loops, subtract from 1, and set equal to 0

$$1 - (7.06009\lambda^{-2} + 0.67208\lambda^{-3}) = 0$$

or

$$1 - 7.06009\lambda^{-2} - 0.67208\lambda^{-3} = 0.$$

Step 4

Find the highest negative power of λ and clear the equation of this value by multiplying the equation by λ raised to this positive power. In this case -3 is the largest negative power and so multiply the equation by λ^3

$$\lambda^3 - 7.06009\lambda - 0.67208 = 0, \qquad (4.9)$$

which is the characteristic equation of the graph for *Bradybaena fruticum* as given in Figure 4.2.

Step 5

Find the largest, positive root of the characteristic equation.

The problem is that Equation 4.9 is a cubic and so we can not use Equation 4.6 to find the roots. An approach similar to Equation 4.6 to find roots of cubics was published in 1615 by François Viète (p 146 in Press *et al.* 1986); however, because for the moment all we want is the largest, positive, real root, it once again is easier to use the Newton-Raphson algorithm

$$\lambda_{i+1} = \lambda_i - \frac{f(\lambda)}{f'(\lambda)},$$

TABLE 4.3 Application of the Newton-Raphson Method to Find a Root of Equation 4.9 for *Bradybaena fruticum*

Iteration	λ_i	$f(\lambda)$	$f'(\lambda)$	λ_{i+1}
1	5	89.02747	67.93991	3.68961
2	3.68961	23.50642	33.77958	2.99373
3	2.99373	5.022980	19.82717	2.74039
4	2.74039	0.560156	15.46915	2.70416
5	2.70416	0.010426	14.87735	2.70346
6	2.70346	0.000016		

Note: data from Staikou *et al.* (1990).

where

$$f(\lambda) = \lambda^3 - 7.06009\lambda - 0.67208$$

and

$$f'(\lambda) = 3\lambda^2 - 7.06009.$$

If we select an arbitrary initial value of 5 for λ, we can improve our estimate to a reasonable solution in just a few iterations (Table 4.3). At iteration 5, $\lambda_{i+1} = 2.70346$, which is a reasonable solution because if it is used for λ_i in Equation 4.8, $f(\lambda) = 0.000016$. However, how do we know that there isn't a larger positive root that should be taken as the population growth rate? We don't for certain although, usually, if one starts with a fairly large - but not very large - value for λ, the Newton-Raphson method will converge on the largest real, positive root. If you wanted to be sure, you might want to find the other roots. Now that you have one root, you know that a factor of Equation 4.9 is (λ - 2.70346). You could use synthetic division to find a quadratic equation that could be solved for the other two roots. By using synthetic division; that is, dividing $(\lambda^3 - 7.06009\lambda - 0.67208)$ by $(\lambda - 2.70346)$, you would obtain

$$\lambda^2 + 2.70346\lambda + 0.24861 = 1.63 \times 10^{-5},$$

which is close enough to zero to see whether there is a positive, real root larger than 2.70346. Using the quadratic formula, the two roots are

$$\lambda_2 = -2.60814$$
$$\lambda_3 = -0.09532$$

So, $\lambda_1 = 2.703$ really is the largest, positive root and population growth rate when a stable-age distribution has been attained and $r = 0.994$ yr^{-1}. This is the same result that would be obtained using the Euler equation in Chapter 2.

THE STABLE-AGE DISTRIBUTION

Caswell (1985, 1989) provides a set of steps to obtain the stable-age distribution directly from a graph such as Figure 4.2. One difference in notation between Caswell's recipe and the way we have been expressing the stable-age distribution is that Caswell defines the first c_x value as equal to 1. If we would be starting with age 0, then $c_0 = 1$ whereas, so far, we have been defining $\Sigma c_x = 1$. This is no big deal and it obviously is very easy to convert one distribution to the other.
Caswell's steps are:

| Step 1 |

Set c_x for the first node $= 1$

| Step 2 |

Calculate c_x for other nodes by taking the product of the coefficients along the paths of the Z-transformed graph connecting the nodes from n_0 to n_x. Actually, this is exactly the same way as was done with the Euler equation in Chapter 2.

For our small beast (Figure 4.1) the paths are

1. $c_1 = \boxed{1}$

2. $c_2 = p_1\lambda^{-1} = (0.242)(5.0747^{-1}) = \boxed{0.0477}$.

These values really don't look like the terms of the stable-age distribution that were obtained in Chapters 1 and 2. One problem, of course, is that the first term is equal to 1.0 rather than having all terms sum to 1.0. This point is minor

TABLE 4.4 Comparisons of Terms of the Stable-Age Distribution for *Bradybaena fruticum*

Age	Euler	Starting at 1	Graph
0	0.731734	—	—
1	0.221679	1.00000	1.00000
2	0.042351	0.19105	0.19105
3	0.004148	0.01871	0.01871
4	0.000082	0.00037	0.00037
5	0.000005	0.00002	0.00002

Note: "Euler" is from the Euler equation (**EULER.BAS**, Chapter 2) and "Graph" is from Figure 4.4; "Starting at 1" are terms of the stable-age distribution based on the Euler equation divided by 0.221679, which is c_1; data from Staikou *et al.* (1990).

compared with the problem of only two terms rather than the expected three. Age class 0 has been absorbed into the transitions from reproductive females to the first node and because this takes one time period, there is no 0 age-class node; however, the relationship between the number of age 1 and age 2 animals has been preserved. From Chapter 1 (Table 1.5), the fraction of animals in age 1 was 0.04506 and the fraction in age 2 was 0.00215. If the fractions are divided by the fraction in age 1, so that $c_1 = 1$, then $c_2 = 0.0477$, the same proportion as calculated from the life cycle graph. The relative proportions have been preserved.

The terms of the stable-age distribution for *Bradybaena fruticum* (Staikou *et al.* 1990), the snail shown in Figure 4.2, can be calculated using the same rules given above. First c_1 is set equal to 1.0,

$$c_1 = \boxed{1}$$

$$c_2 = \frac{0.5165}{2.70346} = \boxed{0.19105}$$

$$c_3 = \frac{0.5165 \times 0.2648}{2.70346^2} = \boxed{0.01871}$$

$$c_4 = \frac{0.5165 \times 0.2648 \times 0.0536}{2.7346^3} = \boxed{0.00037}$$

$$c_5 = \frac{0.5165 \times 0.2648 \times 0.0536 \times 0.1667}{2.7346^4} = \boxed{0.00002}.$$

These values are in the same proportion as c_x values determined by using the Euler equation (Table 4.4); the column "Starting at 1" uses the terms of the stable-age distribution from the Euler equation and divides all terms, starting at 1, by 0.221679. Terms of the stable-age distribution determined from the life cycle graph are in the same proportion as terms that were calculated by using the Euler equation or from projection.

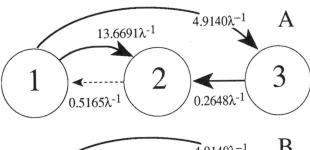

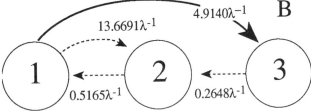

TABLE 4.5 Comparisons of Terms of the Reproductive-Value Distribution for *Bradybaena fruticum*

Age	Euler	Start at 1	Graph
0	1	—	—
1	3.3009	1	1
2	17.2776	5.2342	5.2342
3	6	1.8177	1.8177
4	0	0	0
5	0	0	0

Note: "Euler" is from the Euler equation (**EULER.BAS**, Chapter 2) and "Graph" is from Figures 4.5 and 4.6; "Starting at 1" are terms of the reproductive-value distribution based on the Euler equation divided by 3.3009, which is v_1; data from Staikou *et al.* (1990).

FIGURE 4.3 A, Z-transformed and transposed life cycle graph for *Bradybaena fruticum* (Staikou *et al.* 1990) showing possible paths (solid, bold lines) from age class 1 to age class 2; one path $1 \rightarrow 2$ is current reproduction and $1 \rightarrow 3 \rightarrow 2$ is the future contribution or residual reproduction; **B**, possible path (solid, bold line) from age class 1 to age class 3; only one path is possible, $1 \rightarrow 3$ and is current reproduction; there is no the future contribution or residual reproduction associated with age class 3.

THE REPRODUCTIVE VALUE

The following steps, based on Caswell (1985, 1989), can be used for calculating the terms of the reproductive value distribution from a graph such as Figure 4.1 or 4.2:

Step 1

Transpose the graph; that is, reverse the directions of all arrows.

Step 2

Set the first v_x value equal to 1.

Step 3

Calculate v_x by taking products of coefficients along *all* pathways from n_1 to n_x and summing over all such pathways.

For our small beast in Figure 4.1, once v_1 is assigned the value of 1, there is only one node, age 2, and only one way back to node 1

$$v_1 = \boxed{1}$$

$$v_2 = \frac{5.76}{5.0747} = \boxed{1.13504}.$$

As with the stable-age distribution, these are in the same relative proportion as terms of the stable-age distribution that were determined using the Euler equation in Chapter 2. Starting with age 0, v_1 was 21.143 and v_2 was 24.0. Dividing v_2 by v_1 is 1.1351, which is about the same as determined from analysis of the life cycle graph. The values are not identical because of round-off error in the data files. For the Euler equation l_2 rounded to 0.058 but with $p_2 = 0.242$ and $p_1 = 0.24$, l_2 should have been 0.05808. For practical purposes, the v_x values are the same.

Determining v_x values for *Bradybaena fruticum* (Staikou *et al.* 1990) is done in exactly the same manner:

Step 1

Transpose the graph; that is, change the direction of all arrows.

Step 2

Start at the first node, n_1, and set $v_1 = \boxed{1}$.

Step 3

Starting at node 1, there are two ways to get to node 2: (Figure 4.3A). One path goes directly from $1 \rightarrow 2$ (current reproduction) and the other path goes from $1 \rightarrow 3 \rightarrow 2$ (residual reproductive value)

$$v_2 = \frac{13.669}{2.70346} + \frac{0.2648 \times 4.914}{2.70346^2} = \boxed{5.2342}.$$

Just one possible path goes from node 1 to node 3, $1 \rightarrow 2$ (Figure 4.3B). This is current reproduction and there is no residual

$$v_3 = \frac{4.914}{2.70346} = \boxed{1.8177}.$$

The relative proportions for the v_x distribution have been preserved by making calculations starting at v_1 rather than at v_0 (Table 4.5). If v_x values calculated using the Euler equation are used, then discarding age 0 and dividing the Euler column in Table 4.5 by v_1, 3.3009, will produce the values determined by analyzing the life cycle graph

The proportions of the stable-age, c_x, and reproductive value, v_x, distributions are preserved, which is going to be perfectly good for certain purposes; loss of c_0 and v_0 in the analysis, however, does represent a loss of insight into the contributions of new individuals in terms of relative numbers and a clear distinction between m_x and residual reproductive value. It is possible to obtain all terms of the stable-age distribution from the life cycle graph although this seems not to be done in the ecological literature. The additional term for c_0 can be calculated by starting with m_x values and summing all ways of getting m_x from node 1. For our small beast (Figure 4.1) there is a self-loop at node 1 so the first way of getting an m_x value, m_1, requires no adjustment for population growth, λ^{-1}, or survival probability, p_x. The first term to be included in our estimate is m_1, which is 20. The second way of getting an m_x value is to survive from node 1 to node 2, p_1, which also requires including an adjustment for population growth, and then producing 24 offspring, m_2. This second term is $(0.242)(24)/5.0747$. The c_0 term is

$$c_0 = m_1 + p_1 m_2 \lambda^{-1}$$

or

$$c_0 = 20 + \frac{0.242 \times 24}{5.0747} = \boxed{21.1445}$$

Other terms of the stable-age distribution remain unchanged and so

$$c_0 = 21.1445,$$
$$c_1 = 1.00,$$
and $$c_2 = 0.0477.$$

The sum of these three c_x values is 22.1922 and so the relative proportion in each age class is

$$c_0 = 0.9528,$$
$$c_1 = 0.0451,$$
$$c_2 = 0.0021,$$

which is the same as determined for our small beast in Chapters 1 and 2.

The approach for *Bradybaena fruticum* is the same. Age 2 females each produce 16.69 female offspring but it takes one year to get from age 1 to age 2 so the contribution of age 2 females is $m_2 p_1 \lambda^{-1}$. Age 3 females each produce 6 female offspring but it takes 2 years to get from age 1 to age 3. The contribution is $m_3 p_1 p_2 \lambda^{-2}$. There are just two reproductive classes so

$$c_0 = m_2 p_1 \lambda^{-1} + m_3 p_1 p_2 \lambda^{-2}$$

or

$$c_0 = \frac{16.69 \times 0.5165}{2.7035} + \frac{6.00 \times 0.5165 \times 0.2648}{2.7035^2} = \boxed{3.3009}.$$

All other terms remain unchanged so, from Table 4.4,

$$c_0 = 3.30088,$$
$$c_1 = 1.00,$$
$$c_2 = 0.19105,$$
$$c_3 = 0.01871,$$
$$c_4 = 0.00037,$$
and $$c_5 = 0.00002.$$

The total of these terms is 4.51103 and the *relative* values are

$$c_0 = 0.73173,$$
$$c_1 = 0.22168,$$
$$c_2 = 0.04235,$$
$$c_3 = 0.00415,$$
$$c_4 = 0.00008,$$
and $$c_5 = 0.000004,$$

which are the same as determined using the Euler equation and the Newton-Raphson algorithm (Table 4.4) with a bit of round-off error in c_5.

It is possible to use graphs and calculate terms of the reproductive-value distribution starting with $v_0 = 1$ rather than $v_1 = 1$. The approach is similar to calculating c_0 and starts with m_x values; also, we know that $v_0 = 1.0$ and so what we want to calculate is all v_x values other than v_0. For our small beast (Figure 4.4A), there are two v_x values to calculate and the best way is to start with the final v_x value, which for our beast is v_2. There is just one reproductive input to node 2 and this is m_2 so v_2 is equal to m_2, which makes good sense because there is no residual reproductive value.

The first way of getting to node a is m_1, current reproduction, and the second way, starting with an m_x value, is $m_2 p_1 \lambda^{-1}$ and so

$$v_1 = m_1 + m_2 p_1 \lambda^{-1}.$$

Using values for m_x and p_x from Table 4.1 gives all of the reproductive values:

$$v_0 = \boxed{1.0}$$
$$v_1 = 20 + \frac{24 \times 0.242}{5.0747} = \boxed{21.144}$$
$$v_2 = \boxed{24}$$

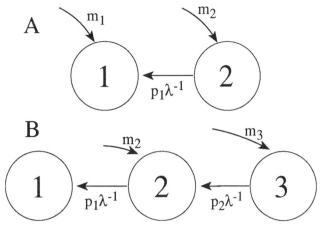

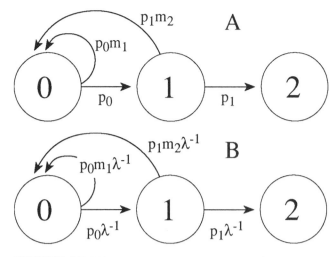

FIGURE 4.4 Life cycle graphs with $f_x = p_0 m_x$ adjusted so the reproductive value can be calculated starting with $v_0 = 1.0$; **A**, our small beast; **B**, the land snail *Bradybaena fruticum* (Staikou *et al.* 1990).

FIGURE 4.5 Life cycle graph with $f_x = p_x m_{x+1}$; **A**, individuals survive and reproduce when they arrive at the next age class; each line represents a transition that takes one time unit; **B**, Z-transformed life cycle graph.

The approach for *Bradybaena fruticum* is exactly the same (Figure 4.4B). Again the starting points are the m_x values. There is just one way of getting to node 3, m_3, and so $v_3 = m_3$. There are two ways of getting to node 2, m_2 and $m_3 p_2 \lambda^{-1}$ and so

$$v_2 = m_2 + m_3 p_2 \lambda^{-1}.$$

Two ways of getting to node 1 are: (1) going from node 2 to node 1; and, (2) starting at node 3, passing through node 2, and then on to node 1

$$v_1 = m_2 p_1 \lambda^{-1} + m_3 p_1 p_2 \lambda^{-2}.$$

Values for p_x and m_x in Table 4.2 can be combined with the estimate of $\lambda = 2.7035$ yr^{-1} (Table 4.3), to calculate terms of the reproductive-value distribution

$$v_0 = \boxed{1.0}$$

$$v_1 = \frac{16.69 \times 0.5165}{2.7035} + \frac{6.00 \times 0.5165 \times 0.2648}{2.7035^2} = \boxed{3.301}$$

$$v_2 = 16.69 + \frac{6.00 \times 0.2648}{2.7035} = \boxed{17.278}$$

$$v_3 = \boxed{6.0}.$$

These v_x values are the same as obtained by using the Euler equation (Table 4.5), which should not be surprising considering that the graphs and the Euler equation really are

exactly the same. The graphs are important because they aid in visualizing the transitions and help to illustrate why particular calculations are made.

Other ways of drawing life cycle graphs preserve age-class 0 as a separate node; however, as will be shown, such graphs have additional problems.

OTHER WAYS OF DRAWING GRAPHS

Each transition takes one time unit and we began by making an adjustment from a node back to node 1 by absorbing age class 0 into the transition; however, there are other ways of making sure each transition has one time unit. It is possible to obtain one unit of time by starting at a node, surviving to the next node, and then reproducing at the next node. This changes the definition of f_x to

$$f_x = p_x m_{x+1}. \tag{4.10}$$

For example, f_0 would be survival from $0 \to 1$, that is p_0, and then reproducing at age 1, at a rate m_1. The life cycle graph for our small beast (Figure 4.5A) illustrates how this shift of view-point changes the graph. Population growth rate and the stable-age and reproductive-value distributions are calculated as before.

Step 1

Perform a Z-transformation (Figure 4.5B).

Step 2

Identify the loops and multiply all transitions around each loop. There are two loops in Figure 4.5B

$$\text{Loop 1: } p_0 m_1 \lambda^{-1}$$

and

$$\text{Loop 2: } p_0 p_1 m_2 \lambda^{-2};$$

Step 3

Sum the loops, subtract the sum from 1 and set equal to 0

$$1 - (p_0 m_1 \lambda^{-1} + p_0 p_1 m_2 \lambda^{-2}) = 0$$

or

$$1 - p_0 m_1 \lambda^{-1} - p_0 p_1 m_2 \lambda^{-2} = 0.$$

Step 4

Find the largest negative exponent, -2 in this case, and clear the equation of negative exponents by multiplying through by $\lambda^{+exponent}$, in this case λ^2,

$$\lambda^2 - p_0 m_1 \lambda - p_0 p_1 m_2 = 0.$$

Substitute values from Table 4.1 to form the characteristic equation

$$\lambda^2 - 4.8\lambda - 1.39392 = 0,$$

which is the same as previously determined; the roots are going to be the same and so the population growth rate, λ, is 5.0747. So far so good. And now the terms of the stable-age distribution, which are determined by following the rules for Z-transformed graphs:

Assign $c_0 = \boxed{1}$

$$c_1 = 0.24\lambda^{-1} = \frac{0.24}{5.0747} = \boxed{0.04729}$$

$$c_2 = (0.24 1^{-1})(0.24 2\lambda^{-1}) = \frac{0.05808}{5.0747^2} = \boxed{0.00225}$$

$$\Sigma c_x = 1.04955.$$

We can change all of these values to match those that we have been using in previous chapters by dividing each c_x by the total, 1.04955

$$c_0 = 0.95279,$$
$$c_1 = 0.04506,$$
$$c_2 = 0.00215,$$

which are the same values for the stable age distribution that we obtained in previous chapters.

TABLE 4.6 Reproductive Value, v_x, and Residual Reproductive Value, v_x^*, for Our Small Beast

x	l_x	m_x	v_x	v_x^*
0	1.00	0	1.0000	1.000
1	0.24	20	21.143	1.143
2	0.058	24	24.000	0.000

Note: r = 1.6242 or λ = 5.0743.

Terms of the reproductive-value distribution follow the same rules as given above. First assign 1 to the first node.

$$v_0 = \qquad\qquad \boxed{1}$$

$$v_1 = \frac{5.808}{5.0747} = \qquad \boxed{1.1445}$$

$$v_2 = \qquad\qquad \boxed{0}$$

These are terms of the *residual* reproductive value, v_x^* (Table 4.6), which makes sense when you consider that the most direct transition from node x back to node 0 does not contain current reproduction, m_x, but rather a term of *future* reproduction, m_{x+1}.

Constituting a graph with $f_x = p_x m_{x+1}$ is perfectly fine for obtaining population growth rate and the terms of the stable-age distribution; however, attempting to obtain the reproductive value distribution results in actually calculating the residual reproductive value.

Another way of constructing a graph for our small beast (Figure 4.6A), which preserves the time constant, is by multiplying m_x by λ; that is

$$f_x = m_x \lambda. \qquad (4.11)$$

When a stable-age distribution has been achieved, all parts of population grow at the same rate; namely, λ. This means that new age-0 individuals that enter the population right now, represent a number that is equal to the number of age 0 individuals produced one time unit ago multiplied by the population growth rate. This construction does not have the intuitive feel to it that the previous two ways of constructing graphs; namely, $f_x = p_0 m_x$ and $f_x = p_x m_{x+1}$; the important time constant of the graph, however, is preserved.

A Z-transformation of the graph (Figure 4.6B) shows the interesting feature of λ being canceled for transitions returning to node 0. The characteristic equation is formed in the familiar manner. There are two loops in Figure 4.6B

Loop 1: $\qquad\qquad p_0 m_1 \lambda^{-1}$

Loop 2: $\qquad\qquad p_0 p_1 m_2 \lambda^{-2}.$

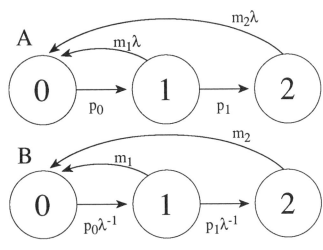

FIGURE 4.6 A, Life cycle graph with $f_x = m_x\lambda$; B, Z-transformed graph

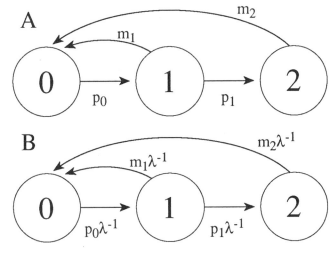

FIGURE 4.7 **A,** *Incorrect way* of constituting a life cycle graph; transitions from 0→1 and 1→2 are completed in 1 time unit but transitions from 1→0 and 2→0 are *immediate*; that is, in 0 time units; **B,** Z-transform for the life cycle graph

Sum the loops, subtract the sum from 1 and set equal to 0

$$1 - (p_0m_1\lambda^{-1} + p_0p_1m_2\lambda^{-2}) = 0$$

or

$$1 - p_0m_1\lambda^{-1} - p_0p_1m_2\lambda^{-2} = 0.$$

Find the largest negative exponent, -2 in this case, and clear the equation of negative exponents by multiplying through by $\lambda^{+exponent}$, λ^2, so

$$\lambda^2 - p_0m_1\lambda - p_0p_1m_2 = 0.$$

Substitute values from Table 4.1 to form the characteristic equation for our small beast

$$\lambda^2 - 4.8\lambda - 1.39392 = 0.$$

This equation is the same one we have obtained for the previous two ways of constituting graphs and so we are assured that we will get the same population growth rate, $\lambda = 5.0747$. The stable-age and reproductive-value distributions can be calculated in the usual manner. For the stable-age distribution, assign 1 to c_0:

$$c_0 = \boxed{1}$$

$$c_1 = 0.24\lambda^{-1} = \frac{0.24}{5.0747} = \boxed{0.04729}$$

$$c_2 = (0.24\lambda^{-1})(0.242\lambda^{-1}) = \frac{0.05808}{5.0747^2} = \boxed{0.00225}$$

This is the same distribution we obtained when we constituted $f_x = p_xm_{x+1}$ and is the same as we obtained previously.

Terms of the reproductive-value distribution follow the same rules as given above. First transpose the Z-transformed graph and assign 1 to v_0:

$$v_0 = \boxed{1},$$

$$v_1 = 20 + \frac{0.242 \times 24}{5.0747} = \boxed{21.144}$$

$$v_2 = \boxed{24}.$$

These, indeed, are the terms of the reproductive value distribution and terms are formed from Figure 4.6B in a very intuitive manner. One path from node 0 to node 1 is m_1, current reproduction, and a second path goes through node 2 and then back to node 0, the future contribution, or residual reproductive value.

There is one final way of assembling a graph (Figure 4.7A). It has been used but *it is an incorrect method*. At first glance Figure 4.7A appears to be the same as Figure 4.6A; however, the m_x values have not been multiplied by λ and so reproduction that takes place must show-up at once at node 0. Values of time are mixed: one unit of time from node 0 to node 1 and one unit of time from node 1 to node 2 but time is equal to 0 for reproduction from node 1 to node 0 and also is 0 from node 2 to node 0.

If the incorrect life cycle graph in Figure 4.7A would be used to obtain the characteristic equation, one would proceed with a Z-transformation (Figure 4.13). The characteristic equation of the graph in Figure 4.7B is

$$\lambda^3 - p_0m_1\lambda - p_1p_0m_2 = 0$$

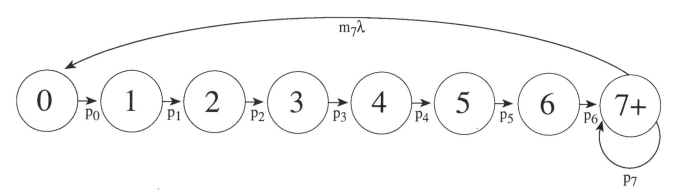

FIGURE 4.8 Life cycle graph for painted turtles, *Chrysemys picta* studied at the E. S. George Reserve of the University of Michigan (Tinkle *et al.* 1981); adult females have the same annual survival probability and same fecundity and so are drawn as a single node with a self loop.

or

$$\lambda^3 - 4.8\lambda - 1.39392 = 0,$$

which is a *cubic* rather than a quadratic equation and so has three roots

$$\lambda_1 = 2.32376,$$
$$\lambda_2 = -0.29579,$$
$$\lambda_3 = -2.02770.$$

Figure 4.7A is not an acceptable way of constituting a life cycle graph; the mixture of time units is the problem.

Three of our methods now agree at least with respect to estimating λ and it is clear that one way of constituting the life cycle graph is wrong:

1. $\boxed{f_x = p_x m_{x+1}}$ Correct method for constituting f_x to obtain a correct λ and stable-age distribution but you will obtain the *residual* reproductive value.

2. $\boxed{f_x = p_0 m_x}$ Correct method for constituting f_x to obtain a correct λ and *relatively* correct stable-age and reproductive-value distributions

3. $\boxed{f_x = m_x \lambda.}$ Correct method for constituting f_x to obtain a correct λ and correct stable-age and reproductive-value distributions

4. Incorrect method for constituting f_x:

.

TABLE 4.7 Life Table for Painted Turtles,
Chrysemys picta,

Age	l_x	m_x	Age	l_x	m_x
0	1.0000	0	7	0.1299	2.8
1	0.6700	0	8	0.0981	2.8
2	0.5092	0	9	0.0746	2.8
3	0.3870	0	10	0.0567	2.8
4	0.2941	0	15	0.0144	2.8
5	0.2235	0	20	0.0009	2.8
6	0.1699	0	35	0.0001	2.8

Note: age is in years; average clutch was 7.55; "≈4% nest twice each year while a *maximum* of 70% reproduce once and 30% not at all." With an assumed 50:50 sex ratio, m_x is ≈2.8; annual survival past age 1 has been smoothed to 0.76 for all remaining age classes; Study conducted at the E. S. George Reserve of the University of Michigan (Tinkle *et al.* 1981).

THE PROBLEM OF SELF LOOPS

Analyses so far all have been for life cycles where all loops pass through a single node, n_0 or n_1. Analysis of many cases is improved if one or more nodes have loops back into the same node rather than a transition either to a node with the next higher number or reproduction back to the first node. Table 4.7 for painted turtles, *Chrysemys picta* (Tinkle *et al.* 1981) is a life cycle that benefits from using a self loop.

It would be possible to estimate population growth rate for painted turtles using the Euler equation but this would require either creating a very long data file of 35 entries or re-scaling the table to some other time units, such as 7 years. When each year is like the previous year for reproductive adults, another way of dealing with the life table is to create a single "adult" node in a life cycle graph (Figure 4.8).

Self loops are a problem because the rules we have been using so far require that all loops pass through a single node,

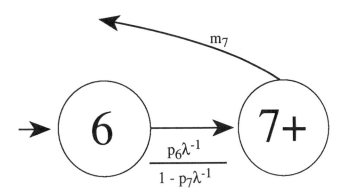

FIGURE 4.9 Detail of transmission from node 6 to node 7+ following a Z-transformation and reduction of the self-loop at node 7+ for painted turtles, *Chrysemys picta*, based on data from Tinkle *et al.* (1981).

$$\frac{a\lambda^{-1}}{1 - b\lambda^{-1}}$$

that is,

$$1 - b\lambda^{-1} \overline{\left) \begin{array}{l} a\lambda^{-1} \quad + \quad ab\lambda^{-2} \quad + \quad ab^2\lambda^{-3} \quad \dots \\ a\lambda^{-1} \\ \underline{a\lambda^{-1} \quad - \quad ab\lambda^{-2}} \\ \qquad\qquad ab\lambda^{-2} \\ \qquad\qquad \underline{ab\lambda^{-2} \quad - \quad ab^2\lambda^{-3}} \\ \qquad\qquad\qquad\qquad ab^2\lambda^{-3} \end{array}\right.} \quad \text{etc}$$

Moving from $node_x$ to $node_{x+1}$ has a probability of a. Moving and staying for one time period has a probability of ab. Moving and staying for two time periods has a probability of ab^2 and so on.

There is just one transmission entering node 7, $p_6\lambda^{-1}$, and so this transmission should be divided by $(1 - p_7\lambda^{-1})$ to eliminate the self loop. A detail of nodes 6 and 7 following a Z-transformation is shown in Figure 4.9. Following the Z-transformation and elimination of the self loop for painted turtles there is just one loop (Figure 4.10A) and so it is easy to form the characteristic equation

such as n_0 or n_1. There are a number of solutions to the problem, one of which is to create a matrix from the graph and proceed with analysis using matrix techniques. We will do this in the next chapter. Additional calculations can be made directly from the graph but correction terms for disjoint loops, such as self loops, can become rather involved (*cf.* Caswell 1982b, 1989) and are deferred to Chapter 7. Dealing with the self loop in Figure 4.14 is fairly easy because the graph can be reduced by absorbing the loop into transmissions coming into node 7. First, perform a Z-transformation; that is, multiply every transmission in Figure 4.14 by λ^{-1}. A self-loop of strength $b\lambda^{-1}$ at node n_j divides the transmission of all incoming arcs by $(1 - b\lambda^{-1})$. Once this is done, the self-loop is eliminated. The logic behind this is provided by Caswell (1989). If an incoming transmission is $a\lambda^{-1}$, then perform the division

$$\frac{p_0p_1p_2p_3p_4p_5p_6m_7\lambda^{-7}}{1 - p_7\lambda^{-1}} = 1,$$

$$p_0p_1p_2p_3p_4p_5p_6m_7\lambda^{-7} = 1 - p_7\lambda^{-1}$$

or

$$\lambda^7 - p_7\lambda^6 - p_0p_1p_2p_3p_4p_5p_6m_7 = 0.$$

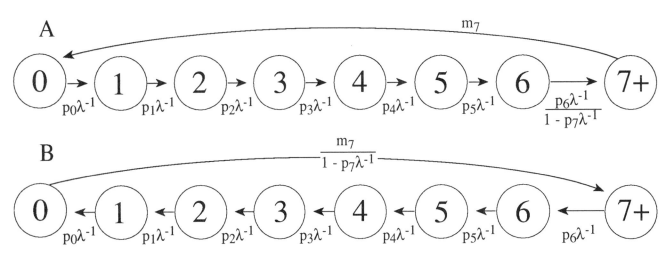

FIGURE 4.10 A, Z-transformed life cycle graph for painted turtles following reduction of the self-loop at node 7+; **B**, life cycle graph transposed for calculation of reproductive values.

TABLE 4.8 Comparisons of Terms of the Stable-Age Distribution for Painted Turtles Using the Euler Equation and the Life Cycle Graph

Age	Graph	Graph, frequency	Euler
0	1.0	0.2964	0.2955
1	0.6429	0.1905	0.1903
2	0.4688	0.1389	0.1390
3	0.3419	0.1013	0.1016
4	0.2493	0.0739	0.0742
5	0.1818	0.0539	0.0542
6	0.1326	0.0393	0.0396
7+	0.3570	0.1058	0.1055
Σ		1.0000	0.9999

Note: 37 age classes were used with the Euler equation; life-cycle graph contained a self loop.

First-year survival, p_0, is 0.67, all other p_x values are 0.76 and all $m_x = 2.8$ so

$$\lambda^7 - .76\lambda^6 - .361505 = 0. \qquad (4.12)$$

The Newton-Raphson algorithm can be used to find the largest, positive, real root, which is 1.0422 yr^{-1}.

The terms of the stable-age distribution are estimated by starting with $c_0 = 1$ and proceeding as was done with our small beast or with the land snail *Bradybaena fruticum*. The only point of difference is for age class 7+, which includes the term for the reduced self loop.

Differences between terms of the stable-age distribution from the life cycle graph and the Euler equation (Table 4.8) can be attributed to truncation of the life table at age 37 whereas with the self loop in the graph, there is no truncation of the life table. Differences are not large and either method could be used for practical purposes such as resource management.

Calculation of terms of the reproductive value in graphs with self loops proceeds in the same manner as presented above and the rule for absorbing a self loop is preserved; namely, dividing all *incoming* loops by $1-p_x\lambda^{-1}$ (Caswell 1989). Figure 4.10B shows the transpose of the graph and removal of the self loop and node 7+.

Approaching the problem of calculation of v_x values by working backwards is exactly the same way that v_x values were calculated using the Euler equation in Chapter 2.

For Figure 4.10B, the v_x value for node 7+ is

$$v_{7+} = \frac{m_7}{1-p_7\lambda^{-1}} = \frac{2.8}{1 - \dfrac{0.76}{1.0422}} = \boxed{10.342},$$

and the rest of the values follow by recursion:

TABLE 4.9 Reproductive Value, v_x, Estimates for Painted Turtles

Age	v_x (Euler)	v_x (graph)
0	1.0	1.0
1	1.555	1.555
2	2.133	2.133
3	2.925	2.915
4	4.011	4.011
5	5.500	5.500
6	7.541	7.542
7+	10.341	10.342

Note: Euler equation estimates used 38 age classes; life cycle graph estimates used 8 age classes with a self loop for age 7+ adults; data from Tinkle *et al.* (1981).

$$v_6 = \frac{v_7 \times p_6}{\lambda} = \frac{(10.342)0.76}{1.0422} = \boxed{7.542},$$

$$v_5 = \frac{v_6 \times p_5}{\lambda} = \frac{(7.542)0.76}{1.0422} = \boxed{5.500},$$

$$v_4 = \frac{v_5 \times p_4}{\lambda} = \frac{(5.500)0.76}{1.0422} = \boxed{4.011},$$

$$v_3 = \frac{v_4 \times p_3}{1} = \frac{(4.011)0.76}{1.0422} = \boxed{2.925},$$

$$v_2 = \frac{v_3 \times p_2}{\lambda} = \frac{(2.925)0.76}{1.0422} = \boxed{2.133},$$

$$v_1 = \frac{v_2 \times p_1}{\lambda} = \frac{(2.133)0.76}{1.0422} = \boxed{1.555},$$

and $v_0 = \dfrac{v_1 \times p_0}{\lambda} = \dfrac{(1.555)0.67}{1.0422} = \boxed{1.0}.$

For comparison, Table 4.9 shows v_x values calculated using the Euler equation and 37 age classes. The value of λ for the Euler analysis was 1.042157 yr^{-1} compared with 1.04264 yr^{-1} from the characteristic equation of the graph in Figure 4.8.

One additional feature of graph analysis can be demonstrated with the painted turtle data. Annual survival rates are the same after age 0 and so Figure 4.10A can be redrawn and nodes 1-6 condensed into a single pre-reproductive or juvenile node (Figure 4.11). Notice that, other than p_0, survival transitions are now called p_j with j standing for juvenile. In this particular case, juvenile and adult survival are the same. Because there are 6 years from age 0 to age 7, p_j must be raised to the 6th power and there are six λ^{-1} terms and so p_j^6 is accompanied by λ^{-6}.

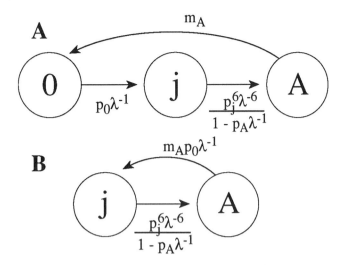

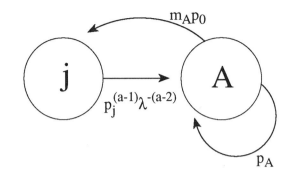

FIGURE 4.11 Z-transformed life cycle graphs for painted turtles combining all pre-reproductive ages with similar survival rates; **A**, $f_x = m_x\lambda$; **B**, $f_x = p_0 mx$; both graphs have exactly the same characteristic equation.

FIGURE 4.12 General life cycle for species with constant juvenile survival, adult survival, and fecundity; "a" is the age at first reproduction.

TABLE 4.10 Life History Data for Three Primate Species

Species	a	m_A	p_0	p_j	p_A
howler monkey					
Alouatta palliata	4	0.48	0.6	0.72	0.82
rhesus monkey					
Macaca mulatta	4	0.62	0.78	0.74	0.78
chimpanzee					
Pan troglodytes	14	0.19	0.73	0.96	0.92

Note: column headings are parameters in Equation 4.14: a = age at first reproduction, m_A = fecundity, p_0 = 1st year survival, p_j = juvenile survival rate yr^{-1}, and p_A = adult survival rate yr^{-1}; data from (Dobson and Lyles (1989).

The characteristic equations for the graphs in Figure 4.11 are formed by multiplication of terms around the single loop and both graphs give the same result:

$$\frac{p_0 p_j^6 m_A \lambda^{-7}}{1 - p_A \lambda^{-1}} = 1$$

or

$$1 - p_A \lambda^{-1} - p_0 p_j^6 m_A \lambda^{-7} = 0$$

and, finally,

$$\lambda^7 - p_A \lambda^6 - p_0 p_j^6 m_A = 0. \qquad (4.13)$$

When 0.76 is used for p_A and p_j, 0.67 for p_0 and 2.8 for m_A, Equation 4.13 is the same as Equation 4.12.

Equation 4.13 can be generalized for all analyses where juvenile survival, adult survival, and fecundity, m_A, all are constant so the general life cycle graph with these characteristics is shown in Figure 4.12.

The characteristic equation of Figure 4.12 is

$$\lambda^a - p_A \lambda^{a-1} - p_0 p_j^{a-1} m_A = 0 \qquad (4.14)$$

with "a" being the age at first reproduction, p_A adult survival, p_j juvenile survival, and m_A fecundity.

Although probably few species fulfill all assumptions of Equation 4.14, it often is the only way that one can gain any insight into the dynamics of populations because some species are sufficiently difficult to study that details of age-specific rates are, at present, impossible to obtain. Table 4.10 shows a summary of life-history data for several primates (Dobson and Lyles 1989) that are suitable for analysis using Equation 4.14. None of the assumptions is strictly true; however, estimates of population growth rates for these species under field conditions probably are the best that can be determined at present. The characteristic equations are

howler monkey $\quad \lambda^4 - 0.821\lambda^3 - 0.107495 = 0,$

rhesus monkey $\quad \lambda^4 - 0.781\lambda^3 - 0.195966 = 0,$

and,

chimpanzee $\quad \lambda^{14} - 0.921\lambda^{13} - 0.081583 = 0.$

Roots of the characteristic equations can be found using the Newton-Raphson algorithm either by hand or with a computer program. For example, with the chimpanzee equation

$$f(\lambda) = \lambda^{14} - 0.92\lambda^{13} - 0.081583$$

and

$$f'(\lambda) = 14\lambda^{13} - 11.96\lambda^{12}$$

so

$$\lambda_{i+1} = \lambda_i - \frac{\lambda^{14} - 0.92\lambda^{13} - 0.081583}{14\lambda^{13} - 11.96\lambda^{12}}.$$

Results of each iteration are shown in Table 4.11 starting with $\lambda = 2.0$. As shown by the chimpanzee data, convergence to reasonable level of tolerance may be quite slow and so the following short computer program probably is a better than using just pencil and paper.

```
10 print "Newton's method"
20 print "Cornered Rat Software  June
   1996 T. A. Ebert"
30 print : input "Do you want to
   continue? (y/n) ";s$
40 if s$ = "N" or s$ = "n" then goto 190
50 input "age at first reproduction, 1st
   year survival,juvenile survival,
   adult survival :";a,p0,pj,pa
60 input "number of female offspring
   produced each year:  ";ma
70 input "Starting value for lambda:
   ";gs
```

```
80 t = ma*pj^(a-1)*p0
90 g = gs
100 for j = 1 to 10000
110 f = g^a-pa*g^(a-1)-t
120 fd = a*g^(a-1)-pa*(a-1)*g^(a-2)
130 g = g-f/fd
140 print g
150 if abs(f/fd) <= 1.000000E-06 then
    goto 170
160 next j
170 print "Final lambda = ";g
180 goto 30
190 end
```

For many species, Equation 4.14 will provide estimates of λ that probably will be too high because the adult self-loop means that adult survival and fecundity never change; there is no senescence.

ANALYSIS WITH AGE FREQUENCY DATA

Michod and Anderson (1980) present a method for estimating r and other demographic parameters when age-frequency data from a single sample are used to estimate survival rates. In general, if just age-frequency data are available, it must be assumed that r = 0 because it is necessary to believe that the relative proportions in each class are the same as the relative proportions of l_x values. However, if one survival transition, such as p_0, is known, then r can be estimated correctly and correct l_x

TABLE 4.11 Iterations, i, to Find the Largest, Positive, Real Root of the Characteristic Equations for Three Primate Species Using the Newton-Raphson Algorithm

	Chimpanzee		Rhesus monkey		Howler monkey	
i	λ_i	λ_{i+1}	λ_i	λ_{i+1}	λ_i	λ_{i+1}
0	2.0	1.865338	2.0	1.577560	2.0	1.578858
1	1.865338	1.740763	1.577560	1.280486	1.578858	1.279279
2	1.740763	1.625648	1.280486	1.093082	1.279279	1.082877
3	1.625648	1.519445	1.093082	1.005393	1.082877	0.979756
4	1.519445	1.421694	1.005393	0.985924	0.979756	0.949230
5	1.421694	1.332056	0.985924	0.985036	0.949230	0.946711
6	1.332056	1.250386	0.985036	0.985035	0.946711	0.946695
7	1.250386	1.176898				
8	1.176898	1.112515				
9	1.112515	1.059550				
10	1.059550	1.022278				
11	1.022278	1.004451				
12	1.004451	1.000894				
13	1.000894	1.000771				
Final λ		1.000771		0.985035		0.946695

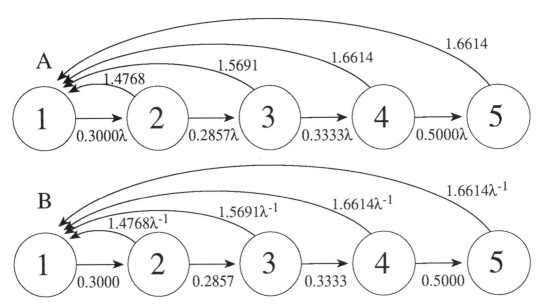

FIGURE 4.13 A, Life cycle graph for gray foxes using p_x values calculated from a stable-age distribution so each p_x has an associated λ; **B**, Z-transformed life cycle graph of gray foxes; multiplying by λ^{-1} cancels λ for all p_x values.

TABLE 4.12 Vertical Life Table for Gray Fox

Age	% in Age	c_x	m_x	p_x	f_x
1	70	0.70	0.00	0.30000l	0
2	21	0.21	2.08	0.28571l	1.4768
3	6	0.06	2.21	0.33333l	1.5691
4	2	0.02	2.34	0.50000l	1.6614
5	1	0.01	2.34	0	1.6614

Note: data from Wood (1958); $p_0 = 0.71$ and $f_x = p_0 m_x$.

values can be calculated. Michod and Anderson (1980) presented their analysis before graphs were used in the analysis of life cycles and what I want to show is that the use of graphs makes the analysis much easier.

If an age-frequency distribution is determined and $r \neq 0$, then the best one can hope for is that at least the age-frequency data are a stable-age distribution. The terms of the stable-age distribution, c_x, are determined using

$$c_x = \frac{l_x \lambda^{-x}}{\Sigma l_x \lambda^{-x}} \qquad (4.15)$$

which means that l_x is

$$l_x = \frac{c_x \Sigma l_x \lambda^{-x}}{\lambda^{-x}} \qquad (4.16)$$

and

$$l_{x+1} = \frac{c_{x+1} \Sigma l_{x+1} \lambda^{-(x+1)}}{\lambda^{-(x+1)}}. \qquad (4.17)$$

Also, because

$$p_x = \frac{l_{x+1}}{l_x},$$

survival probabilities are

$$p_x = \frac{c_{x+1} \lambda}{c_x}. \qquad (4.18)$$

Gray fox data (Wood 1958) used by Michod and Anderson are shown in Table 4.12. An independent estimate of first-year survival, p_0, is used in the analysis and is the key to obtaining a correct estimate of λ and r as well as correct terms of l_x and p_x. Note that the p_x values have been estimated assuming that the observed frequencies are c_x values (Equation 4.16).

Values of p_x and f_x in Table 4.12 can be assembled as a life cycle graph in the usual manner (Figure 4.13A). What is different, is that each p_x term has a λ. When a Z-transformation is done by multiplying all transitions by λ^{-1} (Figure 4.13B), all of the λ's associated with p_x values cancel.

The characteristic equation, based on summing the products around all loops in Figure 4.13B, is

$$0.44360 \, \lambda^{-1} + 0.13449 \, \lambda^{-1} + 0.04746 \, \lambda^{-1} + 0.02373 \, \lambda^{-1} = 1$$

TABLE 4.13 Corrected p_x Values for Gray Foxes Based on the Calculated Value of r

Age	Freq. (c_x)	l_x	Correct p_x
1	0.70	1.0	0.19479
2	0.21	0.19479	0.18551
3	0.06	0.03613	0.21643
4	0.02	0.00782	0.32464
5	0.01	0.00254	0

Note: data from Wood (1958) and following the analysis of Michod and Anderson (1980).

or

$$\lambda = 0.64929 \text{ yr}^{-1}.$$

The value of r, $ln\lambda$, is -0.4319 which, compares favorably with the value of r = -0.433 provided by Micod and Anderson

The values of l_x can now be estimated using Equations 4.16 and 4.17,

$$\frac{l_{x+1}}{l_x} = \frac{c_{x+1}\lambda}{c_x}$$

so

$$l_{x+1} = \frac{c_{x+1}\lambda l_x}{c_x} \qquad (4.19)$$

Using Equation 4.19, all l_x terms can be calculated by starting with $l_x = 1$ (Table 4.13) and then correct p_x values can be calculated. The l_x column should start with l_0 rather than l_1 and this adjustment is shown in Table 4.14. The c_x and l_x columns are very different and the observed age structure (c_x) should not be used to calculate p_x unless at least one survival rate is actually measured.

The use of life cycle graphs makes the analysis proposed by Michod and Anderson (1980) very clear. There also is an important lesson in the analysis concerning how data should be gathered in demographic studies. Specifically, it never would be appropriate to measure age structure and fecundity and then use these data to determine λ or r. This approach is still being used, particularly in plant studies, and does have the beguiling feature that a single field trip to a site can provide all data that seem to be required. Unfortunately, as shown by Michod and Anderson (1980) there is a need for at least one measured transition and so a single trip never will be adequate. A minimum of two trips always will be required so that an initial number and number after one year (or other time unit) can be used to estimate a survival rate. The other approach is to *assume* $\lambda = 1$ and proceed with an analysis that can not include the estimation of λ. There are things that can be done other than estimating λ and some of these are presented in following chapters.

TABLE 4.14 Corrected Life Table for Gray Foxes Starting at Age 0

Age	Freq. (c_x)	l_x	Correct p_x
0	—	1.00	0.71
1	0.70	0.71	0.19479
2	0.21	0.13830	0.18551
3	0.06	0.02566	0.21643
4	0.02	0.00555	0.32464
5	0.01	0.00180	0

Note: $p_0 = 0.71$ yr^{-1}; data from Wood (1958) and following the analysis of Michod and Anderson (1980).

GENERAL COMMENTS

Life cycle graphs have a great virtue in that the visual representation of transitions is intuitively appealing. Nodes and arrows look like age classes with survival to the next age classes and reproduction flowing back to the youngest age. For this reason, if none other, they deserve wide use because they aid in structuring thinking about all parts of a life cycle. A graph forces one to consider all possible transitions and, as will be apparent with more complex cycles, a graph is the best way of organizing both thoughts and data.

Graphs also are more than just pictures of life cycles. They can function as a way of formulating the characteristic polynomial that can be used to calculate the population growth rate, stable-age, and the reproductive-value distributions. In the next chapter, it is shown that a graph can be turned into a matrix and how matrix methods can be used to make calculations.

PROBLEMS

1. Use the following data sets from Chapter 2. Draw the correct life cycle graph and form the correct characteristic equation.
   a. Brousseau and Bagliovo (1988) data on *Mya arenaria*.
   b. Hughes and Roberts (1981) data on *Littorina rudis*.
   c. Strijbosch and Creemers (1988) data on *Lacerta vivipara*.
   d. Grant and Grant (1992) data on *Geospiza scandens*.
   e. Paloheimo and Taylor (1987) data on *Daphnia pulex*.

2. Use the data on *Kobus ellipsiprymnus* (Melton 1983) provided in Chapter 2. Construct a life cycle graph using a self-loop for females age 3 and older and $f_x = m_x\lambda$.. Use the Newton-Raphson algorithm to estimate 1 and determine c_x and v_x using the graph. Compare these results with your

results using the Euler equation. Why are results somewhat different?

3. Turner *et al.* (1970) present an estimated age-specific survival and fecundity schedule for a lizard, *Xantusia vigilis*, (Table 4.15). They begin their table at the age of first reproduction, 3 years.
  a. Determine λ and terms of the stable-age and reproductive-value distributions using the Euler equation.
  b. Create a life cycle graph for this lizard with a self loop for adults. Determine the characteristic equation of the graph and find the largest, positive, real root of the equation using the Newton-Raphson algorithm. *Note*: you can use your estimate from the Euler equation to support the fact that you have found the largest root; don't bother finding all of the roots to justify your answer.
  c. Calculate the stable-age and reproductive-value distributions using the graph.
  d. Compare results from the Euler equation and life cycle graph. Explain any differences you found.

TABLE 4.15 Life Table for *Xantusia vigilis*

Age	l_x	m_x
3	0.45	0.5
4	0.32	0.5
5	0.23	0.5
6	0.16	0.5
7	0.08	0.5
8	0.02	0.5
9	0.001	0.5

Note: data from Turner *et al.* (1970) and adapted from Zweifel and Lowe (1966).

4. Bonnethead sharks (*Sphyrna tiburo*) were studied in Florida by Cortés and Parsons (1996). Several different approached were used to estimate survival for animals older than one year and these different methods provide somewhat different life tables. The method of Hoenig (1983) uses estimates of maximum age to calculate adult survival. Pauly's method relates survival rates of fishes to mean annual temperature. Table 4.16 gives estimated survival and fecundity for sharks in Florida Bay.
  a. Construct a life cycle graph for bonnethead sharks using a self loop for adults.
  b. Estimate λ for with methods of estimating survival of sharks ≥1 years old.

c. Compare estimates of r from the life-cycle analysis with estimates of r using the Euler equation. Why are there differences?

Table 4.16 Life Table for Bonnethead Sharks from Florida Bay Using Two Methods of Estimating Survival

Age	Method 1		Method 2	
	l_x	m_x	l_x	m_x
0	1.000	0	1.000	0
1	0.475	0	0.475	0
2	0.335	0	0.252	0
3	0.236	4.65	0.133	4.65
4	0.166	4.65	0.071	4.65
5	0.117	4.65	0.037	4.65
6	0.083	4.65	0.020	4.65
7	0.058	4.65	0.011	4.65
8	0.041	4.65	0.006	4.65
9	0.029	4.65	0.003	4.65
10	0.020	4.65	0.002	4.65
11	0.014	4.65	0.001	4.65
12	0.010	4.65	<0.001	4.65

Note: method 1 based on Hoenig (1983); method 2 based on Pauly (1980); data from Cortés and Parsons (1996).

5. Powell *et al.* (1996) studied black bears (*Ursus americanus*) in and around Pisgah Bear Sanctuary adjacent to the Pisgah National Forest in North Carolina. A life table for bears (Table 4.17) was prepared using radio-tagged animals. Estimates were for 51 tagged bears for a total of 89 bear years. Estimates of survival are for bears within the sanctuary, bears that were not restricted to the sanctuary, and bears that possibly were poached.
  a. Estimate cub survival by assuming that $\lambda = 1$ for bears in the sanctuary and use this value to estimate λ for sanctuary plus non-sanctuary as well as for data that include poaching.

Table 4.17 Life Table for Black Bears Under Different Levels of Human Disturbance

Age	Sanctuary	Sanctuary + Nonsanctuary	Including Poaching	Fecundity
1	0.73	0.75	0.75	0.00
2	0.83	0.73	0.69	0.04
3	0.67	0.62	0.53	0.64
4	0.57	0.67	0.60	0.78
5	0.67	0.60	0.60	0.71
6+	0.76	0.73	0.58	0.77

Note: data from Powell et al. (1996)

6 The effects of fire and fire exclusion on an annual grass, *Andropogon brevifolius*, were studied in Venezuelan savannas (Canales *et al*. 1994). Plants grow vegetatively for 5 months and produce seeds that spend 7 months in and on the ground before germinating. Here is a table (Table 4.18) for plants in burned and unburned areas with *monthly* survival rates, p_x, starting from germination. Values of f_x are the number of seeds (m_x) multiplied by p_0, which in this case is survival for 7 months and then germination.

a. Compare population growth rate of *A. brevifolius* in burnt and unburnt areas by using a life cycle graph for your analysis. *Hint:* each transition is one month except for f_x which is 7 months. When you do a Z-transformation each *monthly* transition must be multiplied by λ^{-1} and there would have to be 7 of these for f_x so rather than λ^{-1} you would use .... $\lambda^{-?}$.

b. What will be the fate of *A. brefolious* if fires are eliminated?

TABLE 4.18 Life Table for an Annual Grass, *Andropogon brevifolius*

Treatment	p_1	p_2	p_3	p_4	p_5	m_x	p_0	f
Burnt	0.98	0.97	0.98	0.96	0.95	303	0.011	3.33
Unburnt	0.87	0.86	0.86	0.93	0.87	182	0.007	1.27

Note: data from Canales *et al.* (1994).

5

The Leslie Matrix

*T*here must be something in the word 'eigenvalue' that causes some to see red!

J. A. Drake, *Dynamic Aquaria* [review], 1992.

A further 'benefit' ... has been the popularization of the appalling non-words 'eigenvalue' and 'eigenvector', created out of (absurdly partial) translations of the German words Eigenwert *and* Eigenvektor.

I. Grattan-Guinness and W. Ledermann
Matrix theory, 1994

INTRODUCTION

Another procedure for determining population growth rate, the terms of the stable-age distribution, and the reproductive-value function is by manipulation of a transition matrix called the Leslie matrix (Leslie, 1945, 1948). Life cycle graphs and transition matrices are identical and easily can be interconverted. In Chapter 4 it was shown that life cycle graphs were the same as the Euler equation with fixed time units and so the Leslie matrix can be considered to be the same as the Euler equation. A graph is visually more pleasing and it is easier to understand transitions; it will be shown, however, that matrix methods are easier to use particularly as graphs become complex. For example, the self loop that was used for survival of painted turtles in Chapter 4 is easier to use with matrix notation than in a graph.

TABLE 5.1 Changing the Life Cycle Graph in Figure 5.1 into Column and Row Transitions

Row (to) ↓	Column (from)	
	1	2
1	p_0m_1	p_0m_2,
2	p_1	0

Note: transition from node 1 to node 2 means transition from column 1 to row 2.

GRAPH TO MATRIX NOTATION

A reasonable place to start with expressing a graph as a matrix is with the life cycle graph for our small beast in (Figure 5.1). There is a simple rule to follow in creating a transition matrix from a life cycle graph. In a graph (Figure 5.1) transitions are obvious such as *from* node 1 *to* node 2, which is p_1, or *from* node 2 *to* node 1, which is p_0m_2 or f_2. In a matrix, transitions always are *from* the class represented by a column *to* the class represented by a row. It is a simple matter of translating from a node to a node into *from* a column *to* a row (Table 5.1).

The self loop at node 1, p_0m_1, means *from* node 1 *to* node 1, or, in Table 5.1, *from* column 1 *to* row 1. As a matrix, the graph in Figure 5.1 and Table 5.1 is

$$A = \begin{pmatrix} p_0m_1 & p_0m_2 \\ p_1 & 0 \end{pmatrix} \qquad (5.1)$$

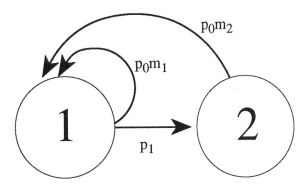

FIGURE 5.1 Life cycle graph with two ages; newborn individuals survive for one time period at which time they are age 1; p_x values are survival rates from x to x+1.

TABLE 5.2 Annual Survival Probability, p_x, and Fecundity, m_x, for a Short-Lived Beast

Age(x)	p_x	m_x
0	0.240	0
1	0.242	20
2	0.000	24

The survival transitions and fecundities for our small beast (Table 5.2) can be used in Equation 5.1 to create a transition matrix (Equation 5.2), which can be used to project a population much in the same manner as we did in Chapter 1.

$$\mathbf{A} = \begin{pmatrix} 4.8 & 5.76 \\ 0.242 & 0 \end{pmatrix} \qquad (5.2)$$

Matrix **A** (Equation 5.2) is an age structured or Leslie matrix and represents two age classes, 1 and 2. As with the life cycle graph (Figure 5.1) age 0 is part of the transition from birth to age 1 so that all transitions have correct time units. Using **A**, we can project a population much as we did in Chapter 2 and this is done using a vector with numbers of individuals in age classes 1 and 2. For example, let's use 240 in age 1 and 0 in age 2

$$\mathbf{B} = \begin{pmatrix} 240 \\ 0 \end{pmatrix} \qquad (5.3)$$

With 240 in age 1 and 0 in age 2 as initial conditions, we can obtain numbers one year later by multiplying of $\mathbf{A} \times \mathbf{B}$

$$\begin{pmatrix} 4.8 & 5.76 \\ 0.242 & 0 \end{pmatrix} \begin{pmatrix} 240 \\ 0 \end{pmatrix} = \begin{pmatrix} 1152 \\ 58.08 \end{pmatrix};$$

that is,

$$4.8 \times 240 + 5.76 \times 0 = 1152,$$

which is the number of individuals age 1 and

$$0.242 \times 240 + 0 \times 0 = 58.08,$$

which is the number age 2.

We now could do matrix multiplication again with our transition matrix and new vector of numbers in age classes 1 and 2, to get numbers in each age class after another year:

$$\begin{pmatrix} 4.8 & 5.76 \\ 0.242 & 0 \end{pmatrix} \begin{pmatrix} 1152 \\ 58.08 \end{pmatrix} = \begin{pmatrix} 5864.14 \\ 278.78 \end{pmatrix};$$

that is,

$$4.8 \times 1152 + 5.76 \times 58.08 = 5864.24$$

and

$$0.242 \times 1152 + 0 \times 58.08 = 278.78.$$

The resulting vector consisting of numbers in each age class can be used to project the population for another year:

$$\begin{pmatrix} 4.8 & 5.76 \\ 0.242 & 0 \end{pmatrix} \begin{pmatrix} 5864.14 \\ 278.78 \end{pmatrix} = \begin{pmatrix} 29753.64 \\ 1419.12 \end{pmatrix}$$

and then another

$$\begin{pmatrix} 4.8 & 5.76 \\ 0.242 & 0 \end{pmatrix} \begin{pmatrix} 29753.64 \\ 1419.12 \end{pmatrix} = \begin{pmatrix} 150991.63 \\ 7200.38 \end{pmatrix}.$$

Look at this last iteration and the ratio between number of individuals in age class 1 initially (29753.64) and number one year later (150991.63). The ratio is 5.0747. Do the same for age class 2; that is,

$$\frac{7200.38}{1419.12}.$$

The ratio is 5.0738. Continue for another iteration

$$\begin{pmatrix} 4.8 & 5.76 \\ 0.242 & 0 \end{pmatrix} \begin{pmatrix} 150991.63 \\ 7200.38 \end{pmatrix} = \begin{pmatrix} 766234.01 \\ 36539.97 \end{pmatrix} \quad (5.4)$$

Now compare the ratios again. For age 1 the ratio is

$$\frac{766234.01}{150991.63},$$

which is 5.0747, and for age 2 the ratio is

$$\frac{36539.97}{7200.38},$$

which also is 5.0747. Both age classes change by a factor of 5.0747 and so we could rewrite Equation 5.4

$$\begin{pmatrix} 4.8 & 5.76 \\ 0.242 & 0 \end{pmatrix}\begin{pmatrix} 150991.63 \\ 7200.38 \end{pmatrix}= 5.0747\begin{pmatrix} 150991.63 \\ 7200.38 \end{pmatrix}$$

This, once again, is a demonstration of applying fixed transitions and achieving a stable-age distribution in which all age classes grow at the same rate, λ or e^r. In matrix notation for our small beast

$$\begin{pmatrix} p_0m_1 & p_0m_2 \\ p_1 & 0 \end{pmatrix}\begin{pmatrix} x_1 \\ x_2 \end{pmatrix}= \lambda\begin{pmatrix} x_1 \\ x_2 \end{pmatrix}$$

or, even more generally,

$$\mathbf{Ax} = \lambda\mathbf{x}, \tag{5.5}$$

which is an appropriate equation for transition matrices of any size,

$$\mathbf{A} = \begin{pmatrix} f_1 & f_2 & f_3 & ... & f_{n-1} & f_n \\ p_1 & 0 & 0 & ... & 0 & 0 \\ 0 & p_2 & 0 & ... & 0 & 0 \\ 0 & 0 & p_3 & ... & 0 & 0 \\ . & . & . & ... & . & . \\ 0 & 0 & 0 & ... & p_{n-1} & 0 \end{pmatrix}.$$

The values of p_x are the survival rates and f_x terms are the effective age-specific reproductive rates, that is, p_0m_x. Equation 5.5 means that projecting the age classes of a population (x_i) using a Leslie matrix is the same as multiplying the age classes by a constant, λ, which is the finite growth rate once a stable-age distribution has been achieved.

It is possible to determine λ from the matrix without having to project the population until a stable age distribution is achieved. The procedure follows the lines of first obtaining the characteristic equation of the matrix and then finding the roots of the characteristic equation. The largest, positive root is the population growth rate.

Starting with Equation 5.5, we can rearrange a bit,

$$\mathbf{Ax} - \lambda\mathbf{x} = 0. \tag{5.6}$$

Factoring out the vector $\mathbf{x}$ gives the matrix equation

$$(\mathbf{A} - \lambda\mathbf{I})\mathbf{x} = 0, \tag{5.7}$$

where $\mathbf{I}$ is the identity matrix, which is a matrix with 1's on the diagonal and 0's everywhere else. The characteristic equation or polynomial of the matrix is the determinant

$$\det(\mathbf{A} - \lambda\mathbf{I}) = 0 \tag{5.8}$$

and, for our small creature, the determinant to evaluate is

$$\det\left[\begin{pmatrix} 4.8 & 5.76 \\ 0.242 & 0 \end{pmatrix} - \lambda\begin{pmatrix} 1 & 0 \\ 0 & 1 \end{pmatrix}\right] = 0$$

or

$$\det\left[\begin{pmatrix} 4.8 & 5.76 \\ 0.242 & 0 \end{pmatrix} + \begin{pmatrix} -\lambda & 0 \\ 0 & -\lambda \end{pmatrix}\right] = 0,$$

which is

$$\begin{vmatrix} 4.8-\lambda & 5.76 \\ 0.242 & -\lambda \end{vmatrix} = 0.$$

Evaluating this particular determinant is quite easy. It is just the product of path a minus the product of path b

$$\det \mathbf{A} = \begin{vmatrix} 4.8-\lambda & 5.76 \\ 0.242 & -\lambda \end{vmatrix}$$

The product of path a is

$$\lambda^2 - 4.8\lambda,$$

and the product of path b is 1.39392. Path a minus path b creates the characteristic polynomial and so

$$\lambda^2 - 4.8\lambda - 1.39392 = 0, \tag{5.9}$$

which is the equation that we obtained in Chapter 4 with the life cycle graph shown in Figure 5.1. There are two roots of this equation, and the largest, positive root is 5.0747.

Before going on to terms of the stable-age and reproductive-value distributions, it is informative to examine the consequences of defining f_x is different ways just as we did with life cycle graphs. If we picture a life cycle with *reproduction first* and *then survival* for one time unit, we have $f_x = m_xp_0$, which is what we have just completed for our small beast. We also can have *survival first* and *then reproduction*, which is Figure 5.2 and has $f_x = p_xm_{x+1}$.

By following the rule of *from* a node *to* a node means *from* a column *to* a row, Figure 5.2 can be expressed as a matrix,

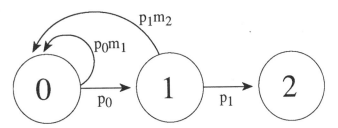

FIGURE 5.2. Life cycle graph with $f_x = p_x m_{x+1}$; individuals survive and reproduce when they arrive at the next age class; each line represents a transition that takes one time unit.

$$A = \begin{pmatrix} p_0 m_1 & p_1 m_2 & 0 \\ p_0 & 0 & 0 \\ 0 & p_1 & 0 \end{pmatrix}$$

and when values from Table 5.1 are substituted for p_x and m_x, we obtain an age-structured or Leslie matrix for our small beast,

$$A = \begin{pmatrix} 4.8 & 5.808 & 0 \\ 0.24 & 0 & 0 \\ 0 & 0.242 & 0 \end{pmatrix}.$$

Once again, the problem is to solve for the population growth rate, λ, which is done by expanding the determinant (Equation 5.10) to create the characteristic polynomial

$$\det(A - \lambda I) = 0. \tag{5.10}$$

For our small creature,

$$\det\left[\begin{pmatrix} 4.8 & 5.808 & 0 \\ 0.24 & 0 & 0 \\ 0 & 0.242 & 0 \end{pmatrix} - \lambda \begin{pmatrix} 1 & 0 & 0 \\ 0 & 1 & 0 \\ 0 & 0 & 1 \end{pmatrix}\right] = 0$$

or

$$\begin{vmatrix} 4.8 - \lambda & 5.808 & 0 \\ 0.24 & -\lambda & 0 \\ 0 & 0.242 & -\lambda \end{vmatrix} = 0 \tag{5.11}$$

The determinant can be expanded in a number of ways; let's use minors.

Minors of a determinant also are determinants and are created by removing equal numbers of rows and columns from the original determinant; $|M_{ij}|$ is created by removing row i and column j. For example, in Equation 5.11 row 2 could be removed so i = 2 and the minor $|M_{21}|$ is formed by removing both row 2 and column 1

$$|M_{21}| = \begin{vmatrix} 4.8-\lambda & 5.808 & 0 \\ 0.24 & -\lambda & 0 \\ 0 & 0.242 & -\lambda \end{vmatrix} = \begin{vmatrix} 5.808 & 0 \\ 0.242 & -\lambda \end{vmatrix}.$$

The other minors that can be formed when row 2 is removed are $|M_{22}|$ and $|M_{23}|$,

$$|M_{22}| = \begin{vmatrix} 4.8-\lambda & 5.808 & 0 \\ 0.24 & -\lambda & 0 \\ 0 & 0.242 & -\lambda \end{vmatrix} = \begin{vmatrix} 4.8-\lambda & 0 \\ 0 & -\lambda \end{vmatrix}$$

and

$$|M_{23}| = \begin{vmatrix} 4.8-\lambda & 5.808 & 0 \\ 0.24 & -\lambda & 0 \\ 0 & 0.242 & -\lambda \end{vmatrix} = \begin{vmatrix} 4.8-\lambda & 5.808 \\ 0 & 0.242 \end{vmatrix}.$$

Evaluating each minor is done by subtracting the product of path b from the product of path a. For example, for $|M_{21}|$

$$|M_{21}| = \begin{vmatrix} 5.808 & 0 \\ 0.242 & -\lambda \end{vmatrix} = -5.808\lambda$$

In similar fashion,

$$|M_{22}| = \lambda^2 - 4.8\lambda$$

and

$$|M_{23}| = 0.242\lambda^2 - 1.1616.$$

Each minor has a plus or minus sign. A signed minor is called a cofactor and is denoted by α_{ij} where i=row number and j=column number

$$\alpha_{ij} = (-1)^{i+j} |M_{ij}|. \tag{5.12}$$

The coefficient $(-1)^{i+j}$ just means that if i+j is even then the sign of the minor is positive but if i+j is odd then the sign is negative. For example, with row 1 and column 1, i+j is 1+1, which is an even number and so the sign of the minor would be positive. With row 1 and column 2, i+j would be 1+2, which is odd, and so the sign of the minor would be negative.

For our small creature,

$$\alpha_{21} = -(-5.808\lambda),$$

$$\alpha_{22} = \lambda^2 - 4.8\lambda,$$

and $\quad \alpha_{23} = 0.242\lambda - 1.1616.$

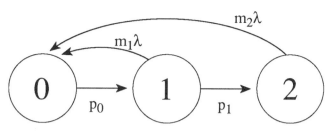

FIGURE 5.3 Life cycle graph with $f_x = m_x\lambda$.

We can now complete the expansion along the second row, a_{2j}, by multiplying each element of the second row by its cofactor and then summing. The determinant of the matrix is

$$\det A = a_{21}\alpha_{21} + a_{22}\alpha_{22} + a_{23}\alpha_{23}$$

$$0.24(5.0808\lambda) - \lambda(\lambda^2 - 4.8\lambda) + 0(0.242\lambda - 1.1616)$$

and so the characteristic polynomial of the matrix is

$$-\lambda^3 + 4.8\lambda^2 + 1.39392\lambda = 0$$

A $-\lambda$ can be factored out so

$$\lambda^2 - 4.8\lambda - 1.39392 = 0$$

which is exactly what we obtained before.

Just as was done with graphs in Chapter 4, a final way of defining f_x, is $f_x = m_x\lambda$. The transition matrix for Figure 5.3, after substitution of values from Table 3.1, is written

$$A = \begin{pmatrix} 0 & 20\lambda & 24\lambda \\ 0.24 & 0 & 0 \\ 0 & 0.242 & 0 \end{pmatrix} \qquad (5.13)$$

and the determinant to evaluate is

$$\det\left[\begin{pmatrix} 0 & 20\lambda & 24\lambda \\ 0.24 & 0 & 0 \\ 0 & 0.242 & 0 \end{pmatrix} - \lambda\begin{pmatrix} 1 & 0 & 0 \\ 0 & 1 & 0 \\ 0 & 0 & 1 \end{pmatrix}\right] = 0$$

or

$$\begin{vmatrix} -\lambda & 20\lambda & 24\lambda \\ 0.24 & -\lambda & 0 \\ 0 & 0.242 & -\lambda \end{vmatrix} = 0$$

The cofactor expansion along row one uses three minors

$$|M_{11}| = \begin{vmatrix} -\lambda & 0 \\ 0.242 & -\lambda \end{vmatrix},$$

$$|M_{12}| = \begin{vmatrix} 0.24 & 0 \\ 0 & -\lambda \end{vmatrix},$$

and $\qquad |M_{13}| = \begin{vmatrix} 0.24 & -\lambda \\ 0 & 0.242 \end{vmatrix}.$

The minors are signed so $|M_{11}|$ and $|M_{13}|$ are positive and $|M_{12}|$ is negative because 1+1 and 1+3 are even and 1+2 is odd.

The signed minors or cofactors are

$$\alpha_{11} = \lambda^2,$$

$$\alpha_{12} = -0.24\lambda,$$

and $\qquad \alpha_{13} = 0.058.$

Now complete the expansion along the first row, a_{1j}, by multiplying each element of the first row by its cofactor and then summing. The determinant of the matrix is

$$\det A = a_{11}\alpha_{11} + a_{12}\alpha_{12} + a_{13}\alpha_{13}$$

or $\qquad -\lambda(\lambda^2) + 20\lambda(0.24\lambda) + 24\lambda(0.058).$

The characteristic polynomial of the matrix is

$$\lambda^2 - 4.8\lambda - 1.39392 = 0$$

and the positive root, λ_1, is 5.0747, just like before.

The stable-age distribution is the eigenvector or latent vector associated with the largest positive root λ_1. The reproductive value function is the latent vector of the transposed transition matrix.

THE STABLE-AGE DISTRIBUTION

Let's start with the transition matrix and f_x defined as $m_x\lambda$, the last example. We now know the value of λ_1 and so it can be substituted back into Equation 5.7

$$\left[\begin{pmatrix} 0 & 20\lambda & 24\lambda \\ 0.24 & 0 & 0 \\ 0 & 0.242 & 0 \end{pmatrix} - 5.0747\begin{pmatrix} 1 & 0 & 0 \\ 0 & 1 & 0 \\ 0 & 0 & 1 \end{pmatrix}\right]\begin{pmatrix} c_0 \\ c_1 \\ c_2 \end{pmatrix} = 0$$

After carrying out the multiplication and combining terms

$$\begin{pmatrix} -5.0747 & 101.494 & 121.7928 \\ 0.24 & -5.0747 & 0 \\ 0 & 0.242 & -5.0747 \end{pmatrix} \begin{pmatrix} c_0 \\ c_1 \\ c_2 \end{pmatrix} = 0$$

three equations are produced,

$$-5.0747c_0 + 101.494c_1 + 121.7928c_2 = 0 \qquad (5.14)$$
$$0.24c_0 - 5.0747c_1 \qquad\qquad = 0 \qquad (5.15)$$
$$0.242c_1 - 5.0747c_2 = 0. \qquad (5.16)$$

An additional constraint must be satisfied. Either

$$c_0 + c_1 + c_2 = 1 \text{ or } c_0 = 1; \qquad (5.17)$$

that is, the terms of the stable-age distribution must sum to 1.0, or the relative number of age 0 is set equal to one. Equation 5.17 can be added or subtracted from one of the equations 5.14 to 5.16; or, 5.17 can be substituted for one of the other equations 5.14 to 5.16. I have chosen substitution and $c_0 = 1$:

$$c_0 \qquad\qquad\qquad = 1$$
$$0.24c_0 - 5.0747c_1 \qquad\qquad = 0$$
$$0.2420c_1 - 5.0747c_2 = 0$$

When $c_0 = 1$ is substituted into the second equation, it is easy to solve for c_1 because

$$c_1 = \frac{0.24 \times 1}{5.0747} = 0.047293$$

and with a solution for c_1, c_2 is

$$c_2 = \frac{0.242 \times 0.047293}{5.0747} = 0.002255.$$

The sum of the c_x values is 1.049548 and so if each value is divided by this sum, terms of the stable-age distribution will sum to 1.0 and so be directly comparable with results from previous chapters:

$$c_0 = \frac{1}{1.049548} = 0.95279,$$

$$c_1 = \frac{0.047293}{1.049548} = 0.04506,$$

and

$$c_2 = \frac{0.002255}{1.049548} = 0.00215.$$

Once again, the results agree with the c_x column obtained using simple projection as shown in Chapter 1, the Euler equation in Chapter 2, and analysis of the life cycle graph in Chapter 4.

THE REPRODUCTIVE-VALUE FUNCTION

Recall from Chapter 4 that the transposed life cycle graph was used to find the reproductive-value vector and transposing a transition matrix is equivalent to transposing a life-cycle graph. Arrows in the life cycle graph were reversed and doing this in a matrix means reversing the direction of transfer. Rather than going *from* a column to a *row*, we want to go *from* a row *to* a column, which is the same as exchanging rows for columns and this is what transposing a matrix does. Starting with Equation 5.7, the matrix equation is

$$(\mathbf{A}' - \lambda\mathbf{I})\mathbf{x} = 0.$$

Note that $\mathbf{A}'$ means the transpose of $\mathbf{A}$.

$$\left[\begin{pmatrix} 0 & 0.24 & 0 \\ 20\lambda & 0 & 0.242 \\ 24\lambda & 0 & 0 \end{pmatrix} - 5.0747 \begin{pmatrix} 1 & 0 & 0 \\ 0 & 1 & 0 \\ 0 & 0 & 1 \end{pmatrix} \right] \begin{pmatrix} v_0 \\ v_1 \\ v_2 \end{pmatrix} = 0.$$

After carrying out the multiplication and combining terms

$$\begin{pmatrix} -5.0747 & 0.24 & 0 \\ 101.494 & -5.0747 & 2.42 \\ 121.7928 & 0 & -5.0747 \end{pmatrix} \begin{pmatrix} v_0 \\ v_1 \\ v_2 \end{pmatrix} = 0.$$

The three equations are

$$-5.0747v_0 + 0.24v_1 \qquad\qquad = 0$$
$$101.494v_0 - 5.0747v_1 + 2.42v_2 = 0$$
$$121.7928v_0 \qquad\qquad - 5.0747v_2 = 0$$

The value of v_0 is 1.0 so it is easy to solve the equations and obtain the following results

$$v_0 = 1.0,$$

$$v_1 = 21.144,$$

and $\quad v_2 = 24,$

which agree with previous estimates of the reproductive-value vector.

If we defined f_x as $p_x m_{x+1}$ then we would obtain the residual reproductive value just as we would with life-cycle graphs in Chapter 4. Similarly, with f_x as $p_0 m_x$ age class 0 is not present but both the stable-age and reproductive-value distributions are *relatively* correct. Defining $f_x = m_x \lambda$ has many attractive features; unfortunately, however, λ is part of transitions in the matrix and this will cause problems when we do additional analyses of transition matrices in later chapters.

For matrices larger than 3×3, solving by hand becomes very tedious and so we will start using a program, **MATRIX.BAS** (shown at the end of this chapter) to help find the coefficients of the characteristic polynomial, all latent roots of the characteristic polynomial, and then solve for the stable-age distribution and reproductive-value vectors.

The program **MATRIX.BAS** accepts data either as a matrix or as coefficients of the characteristic polynomial. Depending on what one wants from an analysis, it may be possible to form the characteristic polynomial from a life-cycle graph, find all roots to ensure that the largest, positive, real one has been found and then to get both the c_x and v_x vectors from the graph. Or, one can get all the pieces one needs by analyzing two separate matrices, one constituted with $f_x = p_0 m_x$ to get λ and a second analysis with $f_x = m_x l$. Additional problems with this approach will be explored in later chapters. My point is that no universal approach will fit all needs and how one casts an analysis will depend on the biological questions that are being asked.

BIRTH-PULSE MATRIX NOTATION

A Viviparous Iguanid Lizard, *Sceloporus jarrovi*

Ballinger (1973) presents a life table (Table 5.3) for a viviparous iguanid lizard, *Sceloporus jarrovi* with x, l_x and m_x having their usual meanings. The first step in analysis is to change l_x and m_x into p_x and f_x (Table 5.4). For p_x from l_x use

$$p_x = \frac{l_{x+1}}{l_x}$$

and, for f_x, use

$$f_x = p_0 m_x.$$

Notice that $p_0 = 0.180$. This is the value to use in order to obtain f_x. The transition matrix will include only ages 1 through 4 because the time constant is one year and so newly hatched lizards appear as age 1 animals after having survived for one year with a probability of 0.180.

TABLE 5.3 Life Table for *Sceloporus jarrovi* from the Chiricahua Mountains, Arizona

Age	l_x	m_x
0	1.000	0
1	0.180	1.2
2	0.086	5.25
3	0.032	5.75
4	0.012	6.00

Note: age in years; data from Ballinger (1973).

TABLE 5.4 p_x and f_x Values for *Sceloporus jarrovi*

Age	p_x	f_x
0	0.180	-
1	0.478	0.216
2	0.372	0.945
3	0.375	1.035
4	0.0	1.080

TABLE 5.5 Stable-Age (c_x) and Reproductive-Value (v_x) Vectors Starting at Age = 1 for *Sceloporus jarrovi*

Age	c_x	v_x
1	1.0000	1.0000
2	0.4961	1.5639
3	0.1915	1.5104
4	0.0745	1.1209

$$\mathbf{A} = \begin{pmatrix} 0.216 & 0.945 & 1.035 & 1.080 \\ 0.478 & 0 & 0 & 0 \\ 0 & 0.372 & 0 & 0 \\ 0 & 0 & 0.375 & 0 \end{pmatrix} \quad (5.18)$$

It is best first to constitute transition matrices with $f_x = p_0 m_x$ in order to avoid problems of suddenly having mixed time constants or having the latent vector associated with $\mathbf{A}'$ containing the *residual* reproductive values. If you see a paper with a transition matrix that begins with age 0, check to see whether f_x values were calculated as $p_x m_{x+1}$; if so, there won't be a problem unless the reproductive value was used for something. In all cases, check very carefully to see how f_x was constituted.

Using **MATRIX.BAS**, the population growth rate, λ, is 0.9635 yr^{-1} for *S. jarrovi* and the stable-age and reproductive-value vectors are shown in Table 5.5. Using $\lambda = 0.9635$ yr^{-1} we can reconstitute the matrix by defining $f_x = \lambda m_x$,

TABLE 5.6 Stable Age (c_x) and Reproductive-Value Vectors for *Sceloporus jarrovi*

Age	c_x	v_x
0	1.0000	1.0000
1	0.1868	5.3530
2	0.0927	8.3715
3	0.0358	8.0851
4	0.0139	6.0000

Note: data from Ballinger (1973).

TABLE 5.7 Survival and Fecundity of the Columbian Ground Squirrel *Citellus* (= *Spermophilus*) *columbianus*

Age	l_x	m_x
0	1.000	0
1	0.338	0.222
2	0.257	1.344
3	0.176	1.536
4	0.095	2.500
5	0.041	2.000

$$A = \begin{pmatrix} 0 & 1.1562 & 5.0586 & 5.5403 & 5.7812 \\ 0.180 & 0 & 0 & 0 & 0 \\ 0 & 0.478 & 0 & 0 & 0 \\ 0 & 0 & 0.372 & 0 & 0 \\ 0 & 0 & 0 & 0.375 & 0 \end{pmatrix}$$

and λ is still 0.9635 yr^{-1} but the stable-age and reproductive-value vectors now include age 0 (Table 5.6).

The Montane Columbian Ground Squirrel

A population of montane Columbian ground squirrels, *Citellus* (= *Spermophilus*) *columbianus* was studied by Zammuto (1987) at 1675m in the Rocky Mountains in southwestern Alberta, Canada. The survival and fecundity estimates are presented in Table 5.7.

The ground squirrel population is treated as birth pulsed and so reproduction is viewed as concentrated on the day a female enters an age class. For example, when a female is exactly 1 year old she has, on average, 0.222 female offspring and this number increases up to a maximum at age 4. Values of m_x reflect both differences in numbers of offspring and the probability that a female reproduces. There are several different ways of forming the transition matrix based on differences in defining f_x and two of these, $f_x = p_0 m_x$ and $f_x = p_x m_{x+1}$, are shown in Table 5.8. The Leslie matrix using $f_x = p_x m_{x+1}$ is

TABLE 5.8 Two ways of Constituting f_x from m_x and p_x for Columbian Ground Squirrels

Age	m_x	p_x	$f_x = p_0 m_x$	$f_x = p_x m_{x+1}$
0	0	0.338	—	0.075
1	0.222	0.760	0.075	1.021
2	1.344	0.685	0.454	1.052
3	1.536	0.540	0.519	1.350
4	2.500	0.432	0.845	0.864
5	2.000	0.000	0.676	0.0

Note: age in years; data from Zammuto (1987).

TABLE 5.9 Stable-Age Distribution c_x, Residual Reproductive Value (v_x^*) and Reproductive Value (v_x) for Columbian Ground Squirrels

Age(x)	c_x	c_x^*	v_x^*	v_x
0	1.00000	0.52625	1.0000	1.0000
1	0.33685	0.17727	2.7468	2.9688
2	0.25513	0.13426	2.2832	3.6272
3	0.17417	0.09166	1.8088	3.3448
4	0.09373	0.04932	0.8610	3.3610
5	0.04035	0.00545	0.0	2.0000
Σ	1.90023			

Note: v_x calculated as $v_x^* + m_x$; data from Zammuto (1987).

$$A = \begin{pmatrix} 0.075 & 1.021 & 1.052 & 1.305 & 0.864 & 0 \\ 0.338 & 0 & 0 & 0 & 0 & 0 \\ 0 & 0.760 & 0 & 0 & 0 & 0 \\ 0 & 0 & 0.685 & 0 & 0 & 0 \\ 0 & 0 & 0 & 0.540 & 0 & 0 \\ 0 & 0 & 0 & 0 & 0.432 & 0 \end{pmatrix}$$

The dominant latent root, λ, for this matrix is 1.003 yr^{-1} so, because $\lambda = e^r$, r=0.003 yr^{-1}. The stable-age and reproductive-value vectors are shown in Table 5.9. Because $f_x = p_x m_{x+1}$ was used for **A**, the eigenvector of the transposed transition matrix was the *residual* reproductive value and so m_x was added to each v_x^* to obtain v_x, the reproductive value at age x.

Self Loops in Matrix Notation: Painted Turtles

Data for painted turtles, *Chrysemys picta* (Tinkle *et al.* 1981) were used in Chapter 4 to illustrate self loops in graphs (Table 5.10, and can be represented as a matrix (Equation 5.19). The first row has the single f_x value, which is $p_0 m_x$ for turtles age 7 and older. Survival probabilities are the sub-diagonal of the matrix.

TABLE 5.10 Life Table for Painted Turtles

Age	p_x	m_x	f_x
0	0.67	0	
1	0.76	0	
2	0.76	0	
3	0.76	0	
4	0.76	0	
5	0.76	0	
6	0.76	0	
7+	0.76	2.8	1.876

Note: $f_x = p_0 m_x$; data gathered at the E. S. George Reserve, University of Michigan (Tinkle *et al.* 1981).

$$\mathbf{A} = \begin{pmatrix} 0 & 0 & 0 & 0 & 0 & 0 & 1.876 \\ 0.76 & 0 & 0 & 0 & 0 & 0 & 0 \\ 0 & 0.76 & 0 & 0 & 0 & 0 & 0 \\ 0 & 0 & 0.76 & 0 & 0 & 0 & 0 \\ 0 & 0 & 0 & 0.76 & 0 & 0 & 0 \\ 0 & 0 & 0 & 0 & 0.76 & 0 & 0 \\ 0 & 0 & 0 & 0 & 0 & 0.76 & 0.76 \end{pmatrix} \quad (5.19)$$

self loop transition.

The self-loop transition for age 7 and older makes sense because all transitions are *from* a column *to* a row, which in this case is from column 7 to row 7. Self loops are easy to handle using matrix notation.

Using **MATRIX.BAS** provides the following characteristic equation for painted turtles

$$\lambda^7 - 0.76\lambda^6 - 0.361505 = 0,$$

which is the same as was obtained in Chapter 4 using graph analysis. The dominant latent root is $\lambda = 1.04216$, which can be used to form $f_x = \lambda m_x$ and permit age 0 to be included as a column in the transition matrix

$$\mathbf{A} = \begin{pmatrix} 0 & 0 & 0 & 0 & 0 & 0 & 0 & 2.918 \\ 0.67 & 0 & 0 & 0 & 0 & 0 & 0 & 0 \\ 0 & 0.76 & 0 & 0 & 0 & 0 & 0 & 0 \\ 0 & 0 & 0.76 & 0 & 0 & 0 & 0 & 0 \\ 0 & 0 & 0 & 0.76 & 0 & 0 & 0 & 0 \\ 0 & 0 & 0 & 0 & 0.76 & 0 & 0 & 0 \\ 0 & 0 & 0 & 0 & 0 & 0.76 & 0 & 0 \\ 0 & 0 & 0 & 0 & 0 & 0 & 0.76 & 0.76 \end{pmatrix} \quad (5.20)$$

TABLE 5.11 Stable-Age and Reproductive-Value Vectors for Painted Turtles

Age (x)	$c_x(1)$	$c_x(2)$	v_x
0	1.000	0.296	1.000
1	0.643	0.190	1.555
2	0.469	0.139	2.133
3	0.342	0.103	2.925
4	0.249	0.074	4.011
5	0.182	0.054	5.500
6	0.133	0.037	7.542
7+	0.357	0.106	10.34
Total	3.375	1.000	

Note: stable-age (1) uses $c_0=1$ (**MATRIX.BAS**) and Stable-age (2) uses $\sum c_x = 1$ obtained from **EULER.BAS**; data from Tinkle *et al.* (1981).

The characteristic equation (Equation 5.20) is

$$\lambda^8 - 0.76\lambda^7 - 0.376748 = 0$$

and the dominant latent root is 1.042163 which is the same as obtained for Equation 5.19 except at the 6th decimal place. The stable-age and reproductive-value vectors are shown in Table 5.11 and are, again, the same as obtained using the Euler equation or by graphical analysis.

BIRTH-FLOW MATRIX NOTATION

A Microcrustacean, *Daphnia pulex*

There is nothing particularly new for birth-flow models; the important ideas were presented in Chapter 2 with the Euler equation. Both survival, p_x, and births, m_x, need to be adjusted so that they reflect the fact that individuals are being added continuously during a time period and, as a consequence, individuals just being added have characteristics appropriate for age x whereas individuals that are nearly ready to enter age x+1 have characteristics more appropriate for x+1.

For the time period 0 to 1, some new-born individuals have to survive the entire period whereas some that are born just before a new time period is attained, have to survive for just a short period of time. On average, new-born individuals must survive just 1/2 time period and so f_x values are averaged m_x values multiplied by survival for 1/2 of a time period. The time adjustment is $\sqrt{l_1/l_0}$ This is the approach used by Levin and Caswell (1987) and recommended by Caswell (1989). Survival is calculated using Equation 5.21,

TABLE 5.12 Life table for *Daphnia pulex*

Age	l_x	m_x	Age	l_x	m_x
0	1.00	0	21	0.15	5.667
3	1.00	0	24	0.15	1.667
6	0.65	0	27	0.15	6.333
9	0.50	0	30	0.15	7.333
12	0.25	0	33	0.15	5.000
15	0.15	0	36	0.05	
18	0.15	0			

Note: m_x is the total number of females produced by females that survived the entire 3-day period; data from Paloheimo and Taylor (1987).

TABLE 5.13 Values of p_x and f_x for *Daphnia pulex*

period	l_x	m_x	p_x	f_x
1	1.00	0	0.8250	0
2	0.65	0	0.6970	0
3	0.50	0	0.6522	0
4	0.25	0	0.5333	0
5	0.15	0	0.7500	0
6	0.15	0	1.0000	2.8335
7	0.15	5.667	1.0000	3.6670
8	0.15	1.667	1.0000	4.0000
9	0.15	6.333	1.0000	6.8330
10	0.15	7.333	1.0000	6.1665
11	0.15	5.000	0.6667	2.5000
12	0.05	0	0.2500	0

Note: calculations of p_x and f_x based on Equations 5.21 and 5.22: time period is 3 days; data from Paloheimo and Taylor (1987).

$$p_x = \frac{l_x + l_{x+1}}{l_x + l_{x-1}} \tag{5.21}$$

and appropriate f_x values are

$$f_x = \sqrt{l_1}\left(\frac{(m_x + p_x m_{x+1})}{2}\right) \tag{5.22}$$

Birth-flow analysis is shown using data for a microcrustacean, *Daphnia pulex*, (Paloheimo and Taylor 1987) (Table 5.12). Values of m_x are the total number of offspring produced during the time interval for females that survived the entire interval. Under good conditions *Daphnia* spp. are parthenogenetic and so all offspring are female.

For further analysis, I use a time period of three days so Table 5.13, which contains the p_x and f_x values, has periods 1, 2, 3, etc. and these mean the periods 3-6, 6-9, and 9-12 days. Values of p_x were calculated using Equation 5.21 and f_x values were calculated using Equation 5.22. Values of p_x and f_x in Table 5.13 were assembled as a transition matrix (Equation 5.23) and the characteristic equation is

$$\lambda^{11} - .425025\lambda^5 - .550050\lambda^4 - .6\lambda^3 - 1.02495\lambda^2 - .924975\lambda^2 - .375 = 0.$$

The population growth rate, λ_1, is 1.174 *per 3 days*. The population growth rate per day would be calculated by first determining r,

$$r = ln(1.173954) = 0.160377 \text{ per 3 days.}$$

Growth rate per individual, r, per day is

$$\frac{r}{3} = \frac{0.160377}{3} = 0.05346 \text{ day}^{-1},$$

and λ day^{-1} is

$$\lambda = e^{0.05346} = 1.055 \text{ day}^{-1}.$$

The stable-age and reproductive-value distributions are shown in Figure 5.4.

$$\mathbf{A} = \begin{pmatrix}
0 & 0 & 0 & 0 & 0 & 2.833 & 3.667 & 4.000 & 6.833 & 6.166 & 2.500 & 0 \\
0.825 & 0 & 0 & 0 & 0 & 0 & 0 & 0 & 0 & 0 & 0 & 0 \\
0 & 0.697 & 0 & 0 & 0 & 0 & 0 & 0 & 0 & 0 & 0 & 0 \\
0 & 0 & 0.652 & 0 & 0 & 0 & 0 & 0 & 0 & 0 & 0 & 0 \\
0 & 0 & 0 & 0.533 & 0 & 0 & 0 & 0 & 0 & 0 & 0 & 0 \\
0 & 0 & 0 & 0 & 0.750 & 0 & 0 & 0 & 0 & 0 & 0 & 0 \\
0 & 0 & 0 & 0 & 0 & 1.000 & 0 & 0 & 0 & 0 & 0 & 0 \\
0 & 0 & 0 & 0 & 0 & 0 & 1.000 & 0 & 0 & 0 & 0 & 0 \\
0 & 0 & 0 & 0 & 0 & 0 & 0 & 1.000 & 0 & 0 & 0 & 0 \\
0 & 0 & 0 & 0 & 0 & 0 & 0 & 0 & 1.000 & 0 & 0 & 0 \\
0 & 0 & 0 & 0 & 0 & 0 & 0 & 0 & 0 & 1.000 & 0 & 0 \\
0 & 0 & 0 & 0 & 0 & 0 & 0 & 0 & 0 & 0 & 0.667 & 0
\end{pmatrix} \tag{5.23}$$

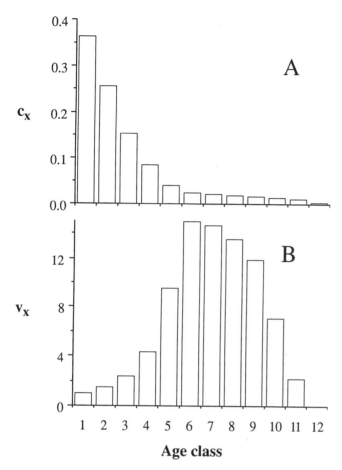

FIGURE 5.4. Stable-age (c_x) and reproductive-value (v_x) distributions for *Daphnia pulex*; data from Paloheimo and Taylor (1987); analysis based on a birth-flow model.

In Table 5.13, the m_x and f_x value for age class 12 is 0; that is, no reproduction during this final stage of life. The result is an entire column of 0's in matrix **A**. Obviously this column could be eliminated along with the final row in **A** and the population growth rate and reproductive-value distribution would be unchanged. The stable-age distribution would be very slightly different because there is a small contribution of this final age class to the population age structure (Figure 5.4). Because the population growth rate would be unchanged, it would, in general, be a good idea to eliminate all age classes that are past reproductive age unless there is some other reason for including them.

GENERAL COMMENTS

This chapter concludes the basic techniques that can be used with age structured populations to estimate population growth rate and the terms of the stable-age and reproductive-value vectors. We have explored simple projection using measures of survival and fecundity, use of the Euler equation, life-cycle graphs, and matrix methods. All of these methods are equivalent. Simple projection can be very useful if, for example, survival or fecundity measures change from iteration to iteration. Spreadsheet software is useful for this (*e.g.* Norton 1994). If time intervals are variable so a single time unit can not can be applied to all transitions, the Euler equation is very useful. Life-cycle graphs provide the clearest view of the meanings of transitions and the calculations that are performed and introduce the useful idea of self loops. Matrices will turn out to be the most useful because graphs become difficult to analyze as they become complex but matrix methods are as easy for complex life cycles as for simple age structured models.

In the next four chapters we explore additional aspects of population models using graphs and matrices, examining all those other latent roots (eigenvalues) to see what they might tell us about dynamics. We also look at how small changes in different parts of a life cycle influence population growth, and then construct more complex life cycles based on stages and sizes and mixtures of ages and stages and sizes. All of this is done one step at a time.

PROBLEMS

1. Use the data for gray foxes (Michod and Anderson 1980) from Chapter 2.
   a. Construct a table showing p_x and f_x values using $f_x = p_0 m_x$
   b. Assemble p_x and f_x to form a transition matrix, **A**.
   c. Form a vector, **n**, with numbers of individuals in each age class; use $n_0 = 531$, $n_1 = 257$, $n_2 = 54$, $n_3 = 32$, $n_4 = 1$. Project the population one year by multiplying **A** by **n**. Has the population achieved a stable age distribution?

2. Use matrix techniques to determine λ and r as well as the stable-age distribution and the reproductive value function for the following life-table (Table 5.14). Work the problem *by hand* (don't use **MATRIX.BAS** except to check your answer) so when you are finished you have 3 ages represented: 0 through 2.

TABLE 5.14 Life Table of an Imaginary Small Beast.

Age(x)	l_x	m_x
0	1.000	0
1	0.312	12
2	0.011	8

3. This is the same problem as problem 2 except an additional age class has been added (Table 5.15). Determine λ and the terms of the stable-age and reproductive-value vectors. Solve this problem "by hand" but use **MATRIX.BAS** to check your results.

TABLE 5.15. Life Table for a Beast Similar to the One in Table 5.14

Age(x)	l_x	m_x
0	1.000	0
1	0.312	12
2	0.011	8
3	0.005	3

4. The meaning of m_x and the first row of a transition matrix are very important in estimating λ, the stable-age distribution, and the reproductive value function. How have the following authors defined f_x: Boyce (1977), Carey (1982), Green (1979), Wright and van Dyne (1981), Buckley (1986), McPeek and Kalisz (1993), and Horvitz and Schemske (1995)?

TABLE 5.16 Life Table for Grizzly Bears in Yellowstone National Park

Age (x)	m_x	l_x	Age (x)	m_x	l_x
0	0	1.0000	11	0.3167	0.1542
1	0	0.8900	12	0.3167	0.1419
2	0	0.6497	13	0.3167	0.1306
3	0	0.3638	14	0.3167	0.1201
4	0	0.2765	15	0.3167	0.1021
5	0.3167	0.2544	16	0.3167	0.0868
6	0.3167	0.2340	17	0.3167	0.0738
7	0.3167	0.2153	18	0.3167	0.0590
8	0.3167	0.1981	19	0.3167	0.0472
9	0.3167	0.1822	20	0.3167	0.0378
10	0.3167	0.1677			

Note: age is years; data from Knight and Eberhardt (1985).

5 Grizzly bears were studied in Yellowstone National Park by Knight and Eberhardt (1985) and they assembled the following data based on the years 1974-1982 (Table 5.16). Calculate r, λ, and the terms of the stable-age and reproductive-value vectors using both the Euler equation and the Leslie matrix. You will have to use **MATRIX.BAS** for the Leslie matrix. Note that you will have to use two Leslie matrices, first with $f_x = p_0 m_x$ and then with $f_x = \lambda m_x$. You may have some problems with starting values and may have to work for a while before obtaining a solution.

6. Schaff *et al.* (1987) provide survival and fecundity data for menhaden (*Brevoortia tyrannus*) on the Atlantic coast

(Table 5.17). Estimating p_0 is difficult but a reasonable first guess is that it is whatever value is needed to make $\lambda = 1$ either for average conditions or for the earliest time period; that is, from 1955-60.

a. Estimate p_0 for the period 1955-60 and then use this value to estimate λ for the periods 1965-70 and 1970-79.

b. How much would p_0 have to change in order to have $\lambda = 1$ for all time periods?

TABLE 5.17 Life Table for Menhaden (*Brevoortia tyrannus*)

	1955-60		1965-70		1975-79	
Age (x)	p_x	eggs	p_x	eggs	p_x	eggs
0						
1	0.14	0	0.09	0	0.10	0
2	0.30	125	0.13	128	0.19	143
3	0.37	242	0.08	199	0.16	199
4	0.31	320	0.06	256	0.14	248
5	0.20	378	0.10	296	0.08	291
6	0.30	405	0.10	322	0.08	331
7	0.30	451	0.10	338		
8	0.30	509				

Note: numbers of eggs are the table values $\times 10^3$; data from Schaff *et al.* (1987).

7. Ainley *et al.* (1983) provide survival and fecundity data for Adélie penguins in the Antarctic. Two tables summarize their findings (Tables 5.18 and 5.19).

TABLE 5.18 Fecundity (m_x) of Adélie Penguins

Age (x)	Fledglings breeder^{-1}	% breeding	Female fledglings female^{-1}
3	0.3	17	0.026
4	0.4	49	0.098
5	0.5	68	0.204
6	0.9	82	0.369
7+	1.0	90	0.450

Note: age is years, data from Cape Crozier, 1967-1974 (Ainley *et al.* 1983).

a. Construct a life-cycle graph and then a matrix to model Adélie penguins. You will have to use a self-loop for age 7+ birds.

b. Survival estimates are different for banded *vs.* unbanded birds form age 0 to 1; unbanded birds have a 28% greater chance of survival during this time period. Is λ also 28% greater if you use 0.513 for p_0 rather than 0.369?

TABLE 5.19 Survival Rates of Adélie Penguins at Cape Crozier, 1967-1969

Age	Unbanded p_x	Banded p_x
0	0.513	0.369
1	0.513	0.513
2+	0.894	0.894

Note: data from Ainley *et al.* (1983).

8. Rowley and Chapman (1991) studied the Major Mitchell or Pink Cockatoo, *Cacatua leadbeateri,* in Western Australia and provide data on reproduction and survival (Tables 5.20 and 5.21). They conclude that $\lambda = 1.13$ and that even if the age at first reproduction is raised to 3 years, λ still is greater than 1.0 (i.e. 1.04). Do you agree with their conclusions? Use whatever method seems best for this problem.

TABLE 5.20 Reproduction in the Pink Cockatoo, *Cacatua leadbeateri*, in Western Australia

Year	Brood sizes (both males and females)	Survival to 6 weeks
1977	3.0	1.00
1978	3.3	0.93
1979	3.3	0.70
1980	2.7	0.79
1981	2.4	0.88
1982	2.7	0.73

Note: age at first reproduction is 2 years; data from Rowley and Chapman (1991).

TABLE 5.21 Survivorship of the Pink Cockatoo in Western Australia

Age (months since tagging)	l_x	Age (months since tagging)	l_x
0	1.00	30	0.250
1	0.800	36	0.197
3	0.613	42	0.152
6	0.581	48	0.116
12	0.444	54	0.116
18	0.365	60	0.090
24	0.312	66	0.090

Note: table *starts* at 6 weeks; that is, t=0 at 6 weeks.

9. Keith and Windberg (1978) studied snowshoe hares for a long time in Canada. Sample calculations of average survival from birth to midwinter over a span of years are shown in Table 5.22). Potentially, there are 4 litters/year with 35 days as the average interval between litters. The total annual birth rate (young expected) per adult female

alive on birth date (11 May) is 11.92, which is the sum of the natality column in Table 5.22. Survival for juveniles is shown in Table 5.23 and for adults in Table 5.24. Keith and Windberg estimate an average λ of 0.968 yr^{-1}. Do you agree with their estimate?

TABLE 5.22 Annual Reproduction of Snowshoe Hares

Litter #	pregnancy rate (F)	Litter size (L)	Natality (F × L × S)
1	0.98	2.68	2.63
2	0.86	5.03	4.27
3	0.85	5.03	4.15
4	0.18	5.03	0.87

Note: data are from Rochester, Alberta, during 1970 (Table 4 of Keith and Windberg 1978); 35 day survival rate (S) for females is 0.958.

TABLE 5.23 Survival Estimates for Snowshoe Hares

Year	Survival: birth to midwinter (180 days)	Survival: midwinter to spring (115 days)
1962-63	0.062	0.426
1963-64	0.110	0.300
1964-65	0.311	0.212
1965-66	0.637	0.485
1966-67	0.332	0.579
1967-68	0.314	0.469
1968-69	0.240	0.423
1969-70	0.575	0.638
1970-71	0.281	0.265
1971-72	0.088	0.119
1972-73	0.062	0.394
1973-74	0.108	
1974-75	0.109	0.782

Note: survival from birth is calculated from the average birth date (26 June 1970) to 17 January; data from Keith and Windberg (1978).

TABLE 5.24 Annual Survival Rates of Adult Snowshoe Hares

Year	Survival rate	Year	Survival rate
1962-63	0.12	1969-70	0.41
1963-64	0.13	1970-71	0.53
1964-65	0.09	1971-72	0.31
1965-66	0.12	1972-73	0.12
1966-67	0.29	1973-74	0.29
1967-68	0.36	1974-75	0.11
1968-69	0.30		

Note: study sites at Rochester, Alberta, and on Halls study area 30 km northwest; data from Keith and Windberg (1978).

BASIC PROGRAMS

MATRIX.BAS

```
10 dim a(40,40),ae(40,40),f(40,40),
   fe(40,40),q(41),c(40)
20 dim rr(40),w(40),b(40),id(40),y(40)
30 print " A fine product from Cornered
   Rat Software©"
40 print "                            T.
   A. Ebert, 1996"
50 print
60 print "        Life-cycle analysis"
70 print
80 print
90 iz = 0
100 input "Do you wish to continue?
    (Y/N)   ",pr$
110 if pr$ = "N" or pr$ = "n" then goto
    4000
120 input "Is analysis of a matrix or
    characteristic equation? (M/E)";pr$
130 if pr$ = "M" or pr$ = "m" then goto
    340
140 if pr$ <> "E" and pr$ <> "e" then
    goto 120
150 input "The highest power of the
    polynomial = ";n
160 for i = 1 to n+1
170 print "Coefficient for s^";(n+1-i),
180 input q(i)
190 next i
200 input "Would you like results saved
    to a file?  (Y/N) ";f$
210 if f$ = "N" or f$ = "n" then goto
    250
220 iz = 1
230 input "Name of file for results:
    ";nm$
240 open nm$ for output as #1
250 for i = 2 to n+1
260 q(i-1) = q(i)
270 next i
280 input "Enter initial values for P
    and Q: 0,0 or 1,1 or whatever.
    ",p1,q1
290 nn = n
300 gosub 1930
310 close #1
320 if p1 = -1000 then goto 280
330 goto 90
340 input "Are matrix data from a file
    or the keyboard? (F/K) ",f$
```

```
350 if f$ = "F" or f$ = "f" then goto
    610
360 if f$ = "K" or f$ = "k" then goto
    380
370 goto 340
380 input "  Enter the order of the
    matrix:    ",n
390 for i = 1 to n
400 for j = 1 to n
410 a(i,j) = 0
420 next j
430 next i
440 print "Enter in sequence: Row #,
    Column #, Element value"
450 print "Separate values with commas
    and enter 0, 0, 0 to end."
460 input i,j,a(i,j)
470 if i = 0 then goto 490
480 goto 460
490 input "Name of file for data:  ",f$
500 open f$ for output as #1
510 print #1,n;",";n
520 for i = 1 to n
530 for j = 1 to n
540 if a(i,j) = 0 then goto 560
550 print #1,i;",";j;",";a(i,j)
560 next j
570 next i
580 print #1," 0, 0, 0"
590 close #1
600 goto 730
610 input "Name of file with data: ";d$
620 open d$ for input as #1
630 input #1,n,n
640 for i = 1 to n
650 for j = 1 to n
660 a(i,j) = 0
670 next j
680 next i
690 input #1,i,j,a(i,j)
700 if i = 0 then goto 720
710 goto 690
720 close #1
730 input "Would you like results saved
    to a file?  (Y/N) ";f$
740 if f$ = "N" or f$ = "n" then goto
    800
750 iz = 1
760 input "Name of file for results:
    ";nm$
770 open nm$ for output as #1
780 print #1," The original data file
    was named:  ";d$
790 print #1," "
```

```
800 q(1) = 1
810 k = 1
820 for i = 1 to n
830 for j = 1 to n
840 f(i,j) = a(i,j)
850 next j
860 next i
870 rem This is a return point from 1090
880 q(k+1) = 0
890 for i = 1 to n
900 q(k+1) = q(k+1)+f(i,i)
910 next i
920 fk = k
930 q(k+1) = -q(k+1)/fk
940 for i = 1 to n
950 f(i,i) = f(i,i)+q(k+1)
960 next i
970 if (k-n+1) = 0 then goto 1100
980 for j = 1 to n
990 for ii = 1 to n : c(ii) = f(ii,j)
1000 next ii
1010 for i = 1 to n
1020 f(i,j) = 0
1030 for ix = 1 to n
1040 f(i,j) = f(i,j)+a(i,ix)*c(ix)
1050 next ix
1060 next i
1070 next j
1080 k = k+1
1090 goto 870
1100 q(n+1) = 0
1110 for j = 1 to n
1120 q(n+1) = q(n+1)-a(1,j)*f(j,1)
1130 next j
1140 if q(n+1) = 0 then goto 1200
1150 for i = 1 to n
1160 for j = 1 to n
1170 f(i,j) = -f(i,j)/q(n+1)
1180 next j
1190 next i
1200 print
1210 print "Coefficients of the
     characteristic equation"
1220 print
1230 if iz = 1 then print #1,
     "Coefficients of the characteristic
     equation"
1240 for i = 1 to n+1
1250 if iz = 1 then print #1,i,q(i)
1260 print i,q(i)
1270 next i
1280 print
1290 if iz = 1 then print #1,"   "
1300 for i = 2 to n+1
```

```
1310 q(i-1) = q(i)
1320 next i
1330 input "Enter initial values for P
     and Q: 0,0 or 1,1 or whatever.
     ",p1,q1
1340 nn = n
1350 gosub 1930
1360 if p1 = -1000 then goto 1330
1370 rm = 0
1380 iv = iv-1
1390 for i = 1 to iv
1400 if rr(i) > rm then rm = rr(i)
1410 next i
1420 print " <return> to continue......"
1430 input x$
1440 print
1450 if iz = 1 then print #1,"   "
1460 print "The dominant root is:   ";rm
1470 if iz = 1 then print #1,"The
     dominant root is:   ";rm
1480 print
1490 if iz = 1 then print #1,"   "
1500 for i = 1 to n
1510 for j = 1 to n
1520 f(i,j) = a(i,j)
1530 next j
1540 next i
1550 for i = 1 to n
1560   f(i,i) = f(i,i)-rm
1570 next i
1580 f(1,1) = f(1,1)+1
1590 b(1) = 1
1600 for i = 2 to n
1610 b(i) = 0
1620 next i
1630 nn = n
1640 for i = 1 to n
1650 for j = 1 to n
1660 fe(i,j) = f(i,j)
1670 next j
1680 next i
1690 wv = 0
1700 gosub 2920
1710 nn = n
1720 for i = 1 to n
1730 for j = 1 to n
1740 f(i,j) = fe(i,j)
1750 next j
1760 next i
1770 gosub 3650
1780 wv = 1
1790 gosub 2920
1800 print "Stage/age","Stable
     age/stage","Reproductive value"
```

```
1810 if iz = 1 then print
     #1,"Stage/age","Stable
     age/stage","Reproductive value"
1820 for i = 1 to n
1830 if iz = 1 then print #1,i,w(i),c(i)
1840 print i,w(i),c(i)
1850 next i
1860 print " <return> to
     continue..........  "
1870 input x$
1880 print
1890 if iz = 1 then print #1," "
1900 gosub 3770
1910 close #1
1920 goto 90
1930 rem Bairstow's method for finding
     roots of a polynomial
1940 print : print
     "Root","Real","Imaginary"
1950 if iz = 1 then print
     #1,"Root","Real","Imaginary"
1960 ep = 1.000000E-06
1970 iv = 1
1980 if (nn-1) < 0 then goto 2910
1990 if (nn-1) > 0 then goto 2070
2000 d = -q(1)
2010 e = 0
2020 print nn,d,e
2030 if iz = 1 then print #1,nn,d,e
2040 rr(iv) = d
2050 iv = iv+1
2060 goto 2910
2070 if (nn-2) > 0 then goto 2110
2080 p = q(1)
2090 q = q(2)
2100 goto 2560
2110 p = p1
2120 q = q1
2130 m = 1
2140 b(1) = q(1)-p
2150 b(2) = q(2)-p*b(1)-q
2160 for k = 3 to nn
2170 b(k) = q(k)-p*b(k-1)-q*b(k-2)
2180 next k
2190 l = nn-1
2200 c(1) = b(1)-p
2210 c(2) = b(2)-p*c(1)-q
2220 for j = 3 to l
2230 c(j) = b(j)-p*c(j-1)-q*c(j-2)
2240 next j
2250 cb = c(1)-b(1)
2260 if (nn-3) <> 0 then goto 2290
2270 dn = c(nn-2)*c(nn-2)-cb
2280 goto 2300
2290 dn = c(nn-2)*c(nn-2)-cb*c(nn-3)
2300 if dn = 0 then goto 2890
2310 if (nn-3) <> 0 then goto 2340
2320 dp = (b(nn-1)*c(nn-2)-b(nn))/dn
2330 goto 2350
2340 dp = (b(nn-1)*c(nn-2)-b(nn)*c(nn-
     3))/dn
2350 dq = (b(nn)*c(nn-2)-b(nn-1)*cb)/dn
2360 p = p+dp
2370 q = q+dq
2380 ap = abs(dp)
2390 aq = abs(dq)
2400 sm = ap+aq
2410 if (m-1) < 0 then goto 2910
2420 if (m-1) > 0 then goto 2450
2430 s1 = sm
2440 goto 2470
2450 if (m-5) <> 0 then goto 2470
2460 if (sm-s1) >= 0 then goto 2510
2470 if (sm-ep) <= 0 then goto 2560
2480 if (m-25) = 0 then goto 2540
2490 m = m+1
2500 goto 2140
2510 print "Functions are diverging for
     assumed values of P and Q"
2520 p1 = -1000
2530 goto 2910
2540 print "  The functions are
     converging slowly"
2550 goto 2490
2560 d = -p/2
2570 f = q-p*p/4
2580 if f > 0 then goto 2780
2590 af = abs(f)
2600 e = sqr(af)
2610 t = d
2620 s = e
2630 e = 0
2640 d = t+s
2650 rr(iv) = d
2660 iv = iv+1
2670 print nn,d,e
2680 if iz = 1 then print #1,nn,d,e
2690 nn = nn-1
2700 s = -s
2710 d = t+s
2720 print nn,d,e
2730 if iz = 1 then print #1,nn,d,e
2740 if abs(e) < 1.000000E-08 then
     rr(iv) = d : iv = iv+1
2750 nn = nn-1
2760 if nn > 0 then goto 2850
2770 if nn <= 0 then goto 2910
2780 af = abs(f)
```

```
2790 e = sqr(af)
2800 print nn,d,e
2810 if iz = 1 then print #1,nn,d,e
2820 if abs(e) < 1.000000E-08 then
     rr(iv) = d : iv = iv+1
2830 e = -e : nn = nn-1
2840 goto 2720
2850 for i = 1 to nn
2860 q(i) = b(i)
2870 next i
2880 goto 1980
2890 print "Good grief!  Divided by
     zero. Try new values for P and Q"
2900 p1 = -1000
2910 return
2920 rem Gaussian elimination
2930 b = abs(ae(k,k))
2940 ne = nn+1
2950 for i = 1 to nn
2960 ae(i,ne) = b(i)
2970 for j = 1 to nn
2980 ae(i,j) = f(i,j)
2990 next j
3000 next i
3010 k = 1
3020 for i = 1 to nn
3030 id(i) = i
3040 next i
3050 kk = k+1
3060 ix = k
3070 it = k
3080 for i = k to nn
3090 for j = k to nn
3100 if abs(ae(i,j)-b) <= 0 then goto
     3140
3110 ix = i
3120 it = j
3130 b = abs(ae(i,j))
3140 next j
3150 next i
3160 if (ix-k) <= 0 then goto 3220
3170 for j = k to ne
3180 c = ae(ix,j)
3190 ae(ix,j) = ae(k,j)
3200 ae(k,j) = c
3210 next j
3220 if (it-k) <= 0 then goto 3310
3230 ic = id(k)
3240 id(k) = id(it)
3250 id(it) = ic
3260 for i = 1 to nn
3270 c = ae(i,it)
3280 ae(i,it) = ae(i,k)
3290 ae(i,k) = c
3300 next i
3310 if ae(k,k) = 0 then goto 3630
3320 for j = kk to ne
3330 ae(k,j) = ae(k,j)/ae(k,k)
3340 for i = kk to nn
3350 we = ae(i,k)*ae(k,j)
3360 ae(i,j) = ae(i,j)-we
3370 if abs(ae(i,j)-1.000000E-
     04*abs(we)) >= 0 then goto 3390
3380 ae(i,j) = 0
3390 next i
3400 next j
3410 k = kk
3420 if (k-nn) > 0 then goto 3630
3430 if (k-nn) < 0 then goto 3050
3440 if ae(nn,nn) = 0 then goto 3630
3450 y(n) = ae(nn,ne)/ae(nn,nn)
3460 nm = nn-1
3470 for i = 1 to nm
3480 k = nn-i
3490 kk = k+1
3500 y(k) = ae(k,ne)
3510 for j = kk to nn
3520 y(k) = y(k)-ae(k,j)*y(j)
3530 next j
3540 next i
3550 for i = 1 to nn
3560 for j = 1 to nn
3570 if (id(j)-i) <> 0 then goto 3600
3580 if wv = 0 then w(i) = y(j)
3590 if wv = 1 then c(i) = y(j)
3600 next j
3610 next i
3620 goto 3640
3630 print "Sorry, no unique solution. "
3640 return
3650 rem Transpose of a matrix
3660 for i = 1 to nn
3670 for j = 1 to nn
3680 ae(i,j) = f(j,i)
3690 next j
3700 next i
3710 for i = 1 to nn
3720 for j = 1 to nn
3730 f(i,j) = ae(i,j)
3740 next j
3750 next i
3760 return
3770 rem Sensitivity and elasticity
     analysis
3780 dt = 0
3790 for i = 1 to n
3800 dt = dt+w(i)*c(i)
3810 next i
```

```
3820 print " <v,c> = ";dt
3830 print
3840 if iz = 1 then print #1," "
3850 if iz = 1 then print #1," <v,c> =
     ";dt
3860 print
3870 print
     "Element","Sensitivity","Elasticity
     "
3880 if iz = 1 then print
     #1,"Element","Sensitivity","Elastic
     ity"
3890 for i = 1 to n
3900 for j = 1 to n
3910 if a(i,j) = 0 then goto 3970
3920 ae(i,j) = c(i)*w(j)/dt
3930 f(i,j) = a(i,j)/rm*ae(i,j)
3940 print
     "A(";i;",";j;")",ae(i,j),f(i,j)
3950 if iz = 0 then goto 3970
3960 print
     #1,"A(";i;",";j;")",ae(i,j),f(i,j)
3970 next j
3980 next i
3990 return
4000 end
```

Instructions for Using MATRIX.BAS

MATRIX.BAS is a general matrix program that finds the coefficients of the characteristic equation of a matrix, determines all of the roots of the characteristic equation, and then, using the dominant latent root, calculates the stable-age (stage) and reproductive-value vectors. Sensitivities and elasticities also are calculated for each element of the matrix and these will be introduced in Chapter 7. The program was initially developed using Applesoft BASIC and made extensive use of FORTRAN routines presented in Kuo (1965) and Pennington (1965).

Transition Matrix: Data from the Keyboard

When **MATRIX.BAS** is run, you have a number of options concerning the nature of data and where data are to be entered. The first decision must be made in response to the question:

> Is analysis of a matrix or characteristic equation? (M/E)

First, let's work with a matrix so enter
> M <return>

or m <return>

You will now be given the following prompt:

> Are matrix data from a file or the keyboard? (F/K)

First, let's enter data from the keyboard so type

> K <return>

or k <return>

For illustration, let's use the data for our small beast and use the matrix in which $f_x = p_0 m_x$.
The prompt on the screen will be

> Enter the order of the matrix:

For our example, enter

> 2 <return>

The new prompt will be

> Enter in sequence: Row #, Column #, element value
> Separate values with commas and enter 0, 0, 0 to end

Several items are important in entering your data:

1. You only have to enter data for non-zero elements.
2. Enter all values as a sequence of three numbers *separated by commas*.
3. When you have entered all data you terminate data entry by entering three 0's separated by commas followed by <return>.

It is a good idea to make a table on a sheet of paper that shows row, column, and element. For our example, the table is

Row	Column	Element
1	1	4.8
1	2	5.76
2	1	0.242

and entered as follows

> 1, 1, 4.8   <return>
> 1, 2, 5.76   <return>
> 2, 1, .242   <return>

N.B. Don't forget the commas.

If you make a mistake, you can correct it.

1. If you *have not* hit <return> you can backspace and make the correction.
2. If you *have* hit <return> all you have to do is reenter correct values and <return>

When you are satisfied with data entry enter

0, 0, 0 <return>

The prompt on the screen will be:

Name of file for data:

Enter a name for your matrix such as

Beast.dat   <return>

The data will be saved on the disk and can be used again. You will receive a prompt:

Would you like results saved to a file? (Y/N)

Saving results is useful because you can open the file with a word processor and print the results, edit them, etc. If you enter

Y <return

then you will be asked

Name of file for results:

Enter something reasonable such as

Beast.out <return>

The program now will be off and running. The first results will be

Coefficients of the characteristic equation
Q(1)	=	1.00
Q(2)	=	-4.80
Q(3)	=	-1.393920

Form the characteristic equation by writing the coefficients from Q(1) to the last non-zero Q-value

1   -4.80   -1.39392

The last coefficient, in this case -1.39392, is the coefficient of λ^0, which is equal to 1.0. The value -4.80 is the coefficient for λ^1, and 1 is the coefficient for λ^2. Attaching

these powers of λ to the coefficients and setting the equation equal to 0 completes the characteristic equation

$$\lambda^2 - 4.80\lambda - 1.39392 = 0$$

The next part of the program will find all of the roots of the characteristic polynomial. This is done using a numerical technique called Bairstow's Method (Bairstow 1914 cited in Grattan-Guinnes 1994) which finds all roots of a polynomial by starting with a quadratic factor of the polynomial, $x^2 + px + q$. The coefficients, p and q are entered as reasonable guesses and these are then improved by the Newton-Raphson method. Reasonable guesses include setting both equal to 0 or both equal to 1; however, you can use other values such as 0.5, -3 etc. Also, p does not have to equal q. Usually the program will work with a wide variety of guesses but sometimes it makes a difference and solutions will be found with some starting values but not with others. This is a bit of an art and you will get better at making it all work after a while.

The prompt on the screen will be

Enter initial values for P and Q: 0,0 or 1,1 or whatever.

Let's use 0,0, so enter

0,0 <return>

N.B. Be sure that you include a "," between the zeros.
The output will provide all roots and will include both the real and imaginary parts.

	Real	Imaginary
X(2)	-.27468129	0
X(1)	5.07468129	0

You must search through the roots to find the ones you need for various purposes. Remember that the *largest positive root*, the dominant latent root, *is the population growth rate*. For this particular case, the population growth rate is 5.07468.

The prompt on the screen will be

<return> to continue......

so when you are ready to continue, hit

<return>

The new output will be the stable-age (stage) and reproductive-value vectors. Both of these are adjusted so the first element in the vector is 1.0. Also, the reproductive value may, in fact, be the residual reproductive value; it depends on how you constituted the matrix. The stable-age (stage) and reproductive-value vectors are printed on the screen.

Stage/age	Stable age/stage	Reproductive value
1	1.0000	1.000
2	.0477	1.135

When you are ready, hit

<return>

The final output includes
1. <v,c>, which is the scalar-product of the two eigenvectors, which is used in sensitivity analysis in Chapter 7.
2. sensitivities and elasticities of the matrix elements.

<v,c> = 1.054		
Element	Sensitivity	Elasticity
A(1,1)	.94865	.89730
A(1,2)	.04522	.05135
A(2,1)	1.07675	.05135

The final prompt is

Do you wish to continue? (Y/N)

Anything other than a

Y <return>

or   y <return>

will bounce you out of the program.

Transition Matrix: Data from a File

Let's assume that you entered Y <return> and have arrived at the prompt

Is analysis of a matrix or characteristic equation? (M/E)

Again, let's work with a matrix so enter

M <return>
or   m <return>

You will now be given the following prompt

Are matrix data from a file or the keyboard? (F/K)

This time let's enter data from a file so type

F <return>
or   f <return>

The new prompt will be

Name of file with data:

so enter the name of your file:

Beast.dat   <return>

The next thing that appears on the screen will be the coefficients of the characteristic equation. You are now back in familiar territory. After you have obtained results you, once again, will receive a prompt

Do you wish to continue? (Y/N)

Enter

Y <return>

Finding Roots of a Polynomial

The prompt again will be

Is analysis of a matrix or characteristic equation? (M/E)

Now let's work with finding the roots of a polynomial so enter

E <return>
or   e <return>

The prompt will be

The highest power of the polynomial =

For our small beast, the highest power is 2 so enter

2 <return>

You then will receive prompts for each coefficient:

Coefficient of s^2

enter
1 <return>

Coefficient of s^1

enter

     -4.8 <return>

Coefficient of s^0

enter

     -1.39392 <return>

Would you like results saved to a file? (Y/N)

This time let's say no and so enter

     N <return>
or   n <return>

You now enter the same subroutine for Bairstow's method that was used when you entered elements of a matrix. You receive a prompt:

     Enter initial values for P and Q: 0,0 or 1,1 or whatever.

Let's use 0,0, so enter

     0,0 <return>

The output will include all roots:

	Real	Imaginary
X(2)	-.27468129	0
X(1)	5.07468129	0

This is all you will get if you enter coefficients of a polynomial. You will not get terms of the stable-age or reproductive-value vectors. The next prompt will be

     Do you wish to continue? (Y/N)

General Comments about MATRIX.BAS

MATRIX.BAS is not able to solve all problems and will not perform well with certain matrices. If in doubt, run a projection to obtain λ_1 and terms of the stable-age/stage distribution. In general, the program will perform very well although you may have to work a bit to find reasonable starting values for p and q when you start Bairstow's Method for finding roots. Some matrices will be challenging and others will converge with a wide range of starting values.

Certain problems can be avoided by changing the way f_x is defined. In general, it probably is best to avoid $f_x = p_x m_{x+1}$ (Figure 5.2) *particularly* if the oldest age class has a self loop. If node 2 in Figure 5.2 would have a self loop it would become an absorbing node with no outgoing transmissions. This may, though not absolutely certainly, cause problems for **MATRIX.BAS**.

Problems may manifest themselves as failure to find all roots, which may mean that the program just hangs with a message about functions converging slowly. Sometimes the problems do not show up until the stable-age and reproductive vectors are calculated. Look at the first values for c_x and v_x. They must be equal to 1.000 or very, very close. If they aren't, then the program is having problems with your data set and you should try different values for p and q. If persistent problems occur for a particular data set, examine the life-cycle graph to see whether you have created any loops that have no outgoing transmissions. Try using $f_x = p_0 m_x$. It also is possible there is an error in the data file. If you are have a persistent problem with a data set, try re-entering the data from within the program. Sometimes small errors are nearly impossible to see when you examine a text file. And be patient. **MATRIX.BAS** is a BASIC program running with an interpreted BASIC so it is not going to be like compiled applications running with a math coprocessor. The larger the matrix, the slower the program; for most data sets, however, **MATRIX.BAS** will get the job done.

6

Transient Behavior in Population Growth

*L*atent roots of a matrix — *latent in a somewhat similar sense as a vapour may be said to be latent in water or smoke in a tobacco leaf.*

J. J. Sylvester, *Philosophical Magazine*, 1883

*Y*ou need not just a mechanical understanding, but a real *intuitive grasp of the slippery little suckers.*

H. Caswell, *Matrix Population Models*, 1989

INTRODUCTION

Depending on the fixed rates of survival and fecundity and the initial age structure, a population may converge in a smooth manner to a stable age distribution and population growth rate, λ, or may exhibit oscillations, which may or may not be damped. It is possible to describe this behavior by examining the subdominant latent roots or eigenvalues obtained from the characteristic equation of the population. Possibly the most important lesson that can be learned from the analysis of transient behavior is that when oscillations are observed in populations, external forces need not be the only explanation; fixed schedules of l_x and m_x can produce oscillations, which can be damped or undamped. Under some initial sets of conditions, many populations may decline although the long-term growth trajectory may be positive — or vice versa, where initial increases may be transitory and the long-term trend may be negative. It is important to realize that if just density is estimated in ecological studies, and then used for population projection, general health of a population may be incorrectly judged.

OSCILLATIONS AND CONVERGENCE

Bernardelli's Beast

Bernardelli (1941) presented a matrix for a hypothetical creature (Equation 6.1). Because the creature is found in a paper, it is reasonable to picture it as a psocid or possibly a silverfish; however, because it also is hypothetical, considering it to be a pseudoscorpion probably is best.

$$\mathbf{A} = \begin{pmatrix} 0 & 0 & 6 \\ \frac{1}{2} & 0 & 0 \\ 0 & \frac{1}{3} & 0 \end{pmatrix} \qquad (6.1)$$

The behavior of matrix **A** can be illustrated by starting with an arbitrary number of individuals in each age class and projecting the population by multiplication. After three cycles, we have returned to initial conditions with 1000 individuals in each age class.

$$\begin{pmatrix} 0 & 0 & 6 \\ \frac{1}{2} & 0 & 0 \\ 0 & \frac{1}{3} & 0 \end{pmatrix} \begin{pmatrix} 1000 \\ 1000 \\ 1000 \end{pmatrix} = \begin{pmatrix} 6000 \\ 500 \\ \frac{1000}{3} \end{pmatrix}$$

$$\begin{pmatrix} 0 & 0 & 6 \\ \frac{1}{2} & 0 & 0 \\ 0 & \frac{1}{3} & 0 \end{pmatrix} \begin{pmatrix} 6000 \\ 500 \\ \frac{1000}{3} \end{pmatrix} = \begin{pmatrix} 2000 \\ 3000 \\ \frac{500}{3} \end{pmatrix}$$

$$\begin{pmatrix} 0 & 0 & 6 \\ \frac{1}{2} & 0 & 0 \\ 0 & \frac{1}{3} & 0 \end{pmatrix} \begin{pmatrix} 2000 \\ 3000 \\ \frac{500}{3} \end{pmatrix} = \begin{pmatrix} 1000 \\ 1000 \\ 1000 \end{pmatrix}$$

This is so fascinating, we should do it again; this time, however, with different initial conditions

$$\begin{pmatrix} 0 & 0 & 6 \\ \frac{1}{2} & 0 & 0 \\ 0 & \frac{1}{3} & 0 \end{pmatrix} \begin{pmatrix} 1800 \\ 900 \\ 300 \end{pmatrix} = \begin{pmatrix} 1800 \\ 900 \\ 300 \end{pmatrix}.$$

With a different set of initial conditions, no oscillations are evident. The population size remains the same with a ratio of 6:3:1, which is the same as the stable-age distribution.

One more set of initial conditions will illustrate a useful point. The stable-age distribution is 0.60, 0.30, 0.10 and the farther away initial conditions are from these proportions, the greater the amplitude; however, the period remains equal to three time units

$$\begin{pmatrix} 0 & 0 & 6 \\ \frac{1}{2} & 0 & 0 \\ 0 & \frac{1}{3} & 0 \end{pmatrix} \begin{pmatrix} 300 \\ 900 \\ 1800 \end{pmatrix} = \begin{pmatrix} 10800 \\ 150 \\ 300 \end{pmatrix}$$

$$\begin{pmatrix} 0 & 0 & 6 \\ \frac{1}{2} & 0 & 0 \\ 0 & \frac{1}{3} & 0 \end{pmatrix} \begin{pmatrix} 10800 \\ 150 \\ 300 \end{pmatrix} = \begin{pmatrix} 1800 \\ 5400 \\ 50 \end{pmatrix}$$

$$\begin{pmatrix} 0 & 0 & 6 \\ \frac{1}{2} & 0 & 0 \\ 0 & \frac{1}{3} & 0 \end{pmatrix} \begin{pmatrix} 1800 \\ 5400 \\ 50 \end{pmatrix} = \begin{pmatrix} 300 \\ 900 \\ 1800 \end{pmatrix}$$

It can be seen from these three different sets of initial conditions that the absence of oscillation when starting conditions matched the stable-age distribution would not

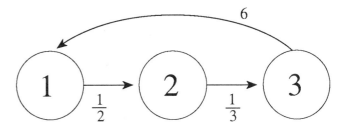

FIGURE 6.1 Life-cycle graph of Equation 6.1, the matrix presented by Bernardelli (1941).

represent a stable point. Rather, any small deviation from the stable-age distribution would introduce oscillations and these oscillations would not be damped over time.

A life cycle graph (Figure 6.1) is an easy way of constituting the characteristic equation for Bernardelli's matrix (Equation 6.2) by following the steps given in Chapter 4.

$$\lambda^3 - 1 = 0 \qquad\qquad (6.2)$$

When we projected a population using Equation 6.1, it returned to its initial state, which means that over the long run, the population is not changing in size; it returns to the same state every three cycles. Because of this, a good guess for one of the roots of Equation 6.2 is $\lambda = 1$, which means that a factor of Equation 6.2 probably is $\lambda - 1$. If we divide Equation 6.2 by $\lambda - 1$, we discover that it divides with no remainder and so it is, indeed, a factor

$$\begin{array}{r} \lambda^2 + \lambda + 1 \\ \lambda - 1 \overline{\smash{\big)}\ \lambda^3 + 0 + 0 - 1} \\ \underline{\lambda^3 - \lambda^2} \\ \lambda^2 + 0 \\ \underline{\lambda^2 - \lambda} \\ \lambda - 1 \\ \underline{\lambda - 1} \end{array}.$$

Now that we know that $\lambda^2 + \lambda + 1$ is a factor of Equation 6.1, we can use the formula for finding roots of a quadratic equation:

$$\lambda_i = \frac{-b \pm \sqrt{b^2 - 4ac}}{2a},$$

where for $\lambda^2 + \lambda + 1$,

a = 1 (the coefficient of λ^2)

b = 1 (the coefficient of λ^1)

c = -1 (the coefficient of λ^0)

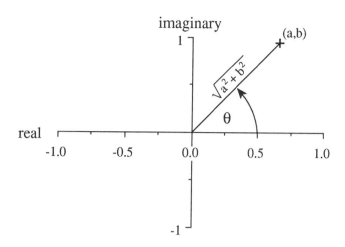

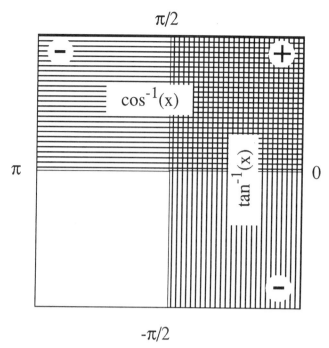

FIGURE 6.2 Relationship between the real and imaginary coefficients, a and b, of a complex eigenvalue and the angle θ in the complex plane.

$$\lambda_i = \frac{-1 \pm \sqrt{1-4}}{2} = \frac{-1 \pm \sqrt{-3}}{2}.$$

It now is clear that two roots are complex. If we use $i = \sqrt{-1}$, then the three roots of Equation 6.2 are

$$\lambda_1 = 1,$$

$$\lambda_2 = -\frac{1}{2} - \frac{\sqrt{3}}{2} i,$$

and

$$\lambda_3 = -\frac{1}{2} + \frac{\sqrt{3}}{2} i.$$

Roots 2 and 3 form a complex conjugate pair of roots that differ only by the sign of the imaginary part of the root. Roots with imaginary parts always occur as conjugate pairs. The general form of these roots is

$$\lambda = a \pm bi. \tag{6.3}$$

It is useful to picture a and b as coordinates in the complex plane (Figure 6.2). Periods of oscillation are based on how many time periods it takes for θ to make one circuit around the origin. If θ is very small then it will take a long time to make a circuit and, consequently, the period of oscillation is long. If θ is large, then just a short time is needed to make one entire circuit and so the period of oscillation is short. In the complex plane the angle between λ and the point a+bi is θ and so

$$\theta = \tan^{-1}(b/a) \tag{6.4}$$

FIGURE 6.3 Ranges (i.e. values of θ in radians) for $\tan^{-1}(x)$ and $\cos^{-1}(x)$ showing overlap in the first quadrant but non-overlap in quadrants 2 and 4; for the inverse tan function, $x = b/a$ and for the inverse cos function, $x = a/\sqrt{a^2 + b^2}$); quadrant 1 is positive and quadrants 2 and 4 are negative for the domains of the inverse functions.

or
$$\theta = \cos^{-1}\left(a/\sqrt{a^2 + b^2} \right). \tag{6.5}$$

Finding θ is a bit of a problem because the domain (*i.e.* possible values for b/a) of the inverse tangent function is all real numbers but the range (*i.e.* possible values for θ) is the open interval $-\pi/2$ to $+\pi/2$, which is -1.5707963 to +1.5707963. The domain of the inverse cosine function is from -1 to +1 and the range is the closed interval $[0, \pi]$. The overlap of the ranges for these two inverse functions is shown in Figure 6.3.

The ranges of inverse tangent and cosine functions are different and these differences are important in calculating periods of oscillation. Figure 6.3 shows that the ranges of the two functions overlap in the first quadrant and so when b/a is *positive*, the two inverse functions give identical angles. When b/a is *negative* then one no longer is in the range of the inverse tan function. Consequently, if b/a is negative, one should first take the inverse tangent of the value and then add π. It is important to remember that a and b are the real parts of complex roots and the trigonometric functions are in *radians*, not *degrees*. The modulus of the conjugate pair is

$$|\lambda| = \sqrt{(a+bi)(a-bi)},$$

which is
$$|\lambda| = \sqrt{a^2 + b^2} \qquad (6.6)$$

or the hypotenuse of the triangle formed by the origin and the point (a, b) (Figure 6.2). For matrix **A**,

$$|\lambda_2| = \sqrt{\left(-\frac{1}{2}\right)^2 + \left(\frac{\sqrt{3}}{2}\right)^2} = 1.$$

The rate of convergence to stable structure is proportional to the ratio between the dominant real eigenvalue and the modulus of the largest subdominant eigenvalue. This ratio is called the *damping ratio* (Caswell 1986) and the symbol is a Greek rho, ρ

$$\rho = \frac{\lambda_1}{|\lambda_2|}. \qquad (6.7)$$

The larger the value of ρ, the more rapid the convergence. Also, when a population is deformed so that it is moved away from a stable-age distribution, the damping ratio is an indication of how rapidly it will return to stable-age structure.

For matrix **A**, the damping ratio is equal to 1 and convergence does not take place. It should be noted that λ_2 does not have to be complex. It can be either positive or negative; only its magnitude is important in the damping ratio.

When subdominant eigenvalues are complex or negative, oscillations can occur and the period of oscillation can be estimated (*e.g.* Keyfitz 1968, Caswell 1986). The period of oscillation, P, is

$$P = \frac{2\pi}{\tan^{-1}\left(\dfrac{b}{a}\right)} \qquad \frac{b}{a} > 0 \qquad (6.8)$$

and
$$P = \frac{2\pi}{\pi + \tan^{-1}\left(\dfrac{b}{a}\right)} \qquad \frac{b}{a} < 0 \qquad (6.9)$$

or
$$P = \frac{2\pi}{\cos^{-1}\left(\dfrac{a}{|\lambda_i|}\right)}. \qquad (6.10)$$

An example of application of Equations 6.8-6.10 can be shown using Equation 6.1 where we know that the period of oscillation is 3 time units. The two sub-dominant eigenvalues have real and imaginary parts,

$$a = -\frac{1}{2}$$

and
$$b = \frac{\sqrt{3}}{2}$$

so
$$\frac{b}{a} = -\sqrt{3}$$

or, approximately, -1.73205, which is negative and so Equation 6.9 should be used instead of Equation 6.8 or, as an alternative, Equation 6.10 could be used. First with Equation 6.10,

$$P = \frac{2\pi}{\cos^{-1}\left(-\dfrac{1}{2}\right)}$$

or   P = 3 time units.

If Equation 6.9 is used then

$$P = \frac{2\pi}{\pi + \tan^{-1}(-1.73205)}$$

or   P = 3 time units;

however, if Equation 6.8 would be used instead of 6.9 or 6.10,

$$P = \frac{2\pi}{\tan^{-1}(-1.73205)} = 6 \text{ time units,}$$

which is incorrect.

A period of 3 units is the same conclusion we arrived at by projection and is approximately equal to the generation time (Lotka 1945, Caswell 1986) or the average age of mothers (Keyfitz 1968), which makes good sense for Equation 6.1 because generations do not overlap and it takes 3 time units from birth to reproduction.

Major oscillations are determined by the negative or conjugate roots with the largest modulus, $|\lambda_i|$; however, all negative or conjugate pairs describe oscillatory behavior and all roots contain information concerning damping. It is possible to describe damping ratios for other roots

$$\rho_{i-1} = \frac{\lambda_1}{|\lambda_i|} \qquad (6.11)$$

and to use Equations 6.8 or 6.9 to describe other periods of oscillation.

The relationships among real and imaginary parts, eigenvalues, and convergence rate and oscillatory behavior is given by the following rules (Caswell 1989):

1. If $|\lambda_i| < 1$, its contribution to population growth decays.
   If $\lambda_i > 0$ then the decay is smooth

If $\lambda_i < 0$ or complex then the decay is with oscillations.

2. If $|\lambda_i| > 1$, its contribution to population growth increases.

   If $\lambda_i > 0$ then the increase is smooth

   If $\lambda_i < 0$ or complex then the increase is with oscillations.

3. If $\lambda_i = 1$ then its contribution to population growth remains unchanged through time.

   If $\lambda_i = -1$ then the population will show undamped oscillations.

Roots that have no imaginary part require an additional comment. Equation 6.10 can be used to calculate θ when a, the real part, is positive or negative, and b, the imaginary part of the root, is 0. In Equation 6.10, if a is positive, then the coordinate "a" is the same as λ and so $a/|\lambda_i| = 1$ and $\cos^{-1}(1) = 0$, which means that the period of oscillation (Equation 6.10) is undefined but you can see what the limit is as $a/|\lambda_i|$ approaches 1 from quadrant 1 in Figure 6.2. The $\cos^{-1}$ function becomes smaller and smaller and so the period of oscillation gets larger and larger and approaches infinity. For positive roots with just real parts, there is no period of oscillation.

On the other hand, if $\lambda_i < 0$ and there is no imaginary part of the root, then $a/|\lambda_i| = -1$ and $\cos^{-1}(-1)$ is π. The period of oscillation, P, is

$$P = \frac{2\pi}{\pi} \text{ or 2 time units,}$$

which is a very convenient result because you can get an immediate sense of how the population will grow by looking at the relative magnitudes of the dominant latent root and the other roots. If there is a negative root that is large relative to the dominant root, then you know at once that the population will grow with oscillations at a period equal to 2 time units.

Our Small Beast

Analysis of data for our small beast provides another example of calculation of the damping ratio and periods of oscillation. In Chapter 4, all of the roots of the characteristic equation were found by using the quadratic formula. These are presented again in Table 6.1.

The damping ratio is

$$\rho = \frac{\lambda_1}{|\lambda_2|} = \frac{5.0747}{0.2747} = 18.473$$

TABLE 6.1 Roots of the Characteristic Equation for a Small Creature

Root	Real	Imaginary
1	5.0747	0.00
2	-0.2747	0.00

Note: there are no complex roots.

TABLE 6.2 Roots of the Characteristic Equation for a Gray Squirrel Population in North Carolina

Root	Real (a)	Imaginary (b)	Modulus, $|\lambda|$
1	1.04218	0.0	
2,3	0.34099	$\pm 0.50711i$	0.61110
4,5	-0.16129	$\pm 0.58022i$	0.60222
6,7	-0.53887	$\pm 0.25495i$	0.59513

Note: the modulus, $|\lambda|$, is $\sqrt{a^2 + b^2}$; data from Barkalow *et al.* (1970).

and the length of time, t, it would take for λ_1 to have 1000 times the influence on growth as λ_2 is

$$t = \frac{ln1000}{ln\rho} = \frac{6.907755}{2.91631} = 2.4 \text{ years}$$

or 10000 times the influence in 3 years. It is clear that the importance of λ_2 disappears very rapidly, which is consistent with our simulation of population growth in which convergence to a stable-age distribution was very rapid. The large damping ratio would have permitted us to predict this without the simulation.

The negative root, -0.2747, will produce a damped oscillation with a period of 2 years; however, the very large damping ratio means that its influence diminishes rapidly to negligible importance.

Gray Squirrels

A third example of the application of analysis of transient behavior is provided using the data from gray squirrel data (Barkalow *et al.* 1970) given in Chapter 1. Using Bairstow's method, all of the roots of the characteristic equation were found and are given in Table 6.2.

The damping ratio for the gray squirrel data will be very similar using the modulus for any of the complex conjugate pairs because all are approximately equal to 0.6. Using the largest modulus, 0.611098, the damping ratio, ρ, is

$$\rho = \frac{\lambda_1}{|\lambda_2|} = \frac{1.0422}{0.6111} = 1.705$$

TABLE 6.3 Calculation of the Period of Oscillation, P, for a Gray Squirrel Population

Roots	b/a	$a/|\lambda_i|$	$\tan^{-1}(b/a)$	$\cos^{-1}(a/|\lambda_i|)$	P
2,3	1.487158	0.558003	0.978819	0.978819	6.42 yr
4,5	-3.597332	-0.267828	*-1.299658	1.841934	3.41
6,7	-0.473115	-0.903937	*-0.441901	2.699684	2.33

Note: remember to use radians and *NOT* degrees; a and b are the real parts of complex roots; $|\lambda_i|$ is the modulus; * means that π must be added before P is calculated; data from Barkalow *et al.* (1970).

The damping ratio is much smaller than the one calculated for our small creature and so we would expect that a gray squirrel population with the characteristics we have been given would converge on a stable-age distribution more slowly. Simulation of the population in Chapter 1 showed that this was true. We can calculate how long it would take for λ_1 to have 1000 times the influence on growth as λ_2

$$t = \frac{ln1000}{ln\rho} = \frac{6.907755}{0.53378} = 13 \text{ years,}$$

which is about six times slower than our small beast.

The complex roots in Table 6.2 indicate that as the gray squirrel population converges to a stable age distribution, it will show oscillations (Table 6.3).

To illustrate oscillations in the squirrel population, I started with 100 new born squirrels and applied the rules we used in Chapter 1; I also could have used matrix multiplication instead. The population growth rate, λ_1, eventually was 1.0422 and so r = 0.0413. In my simulation, N = 255.46 at t=36 years and so a reasonable general growth equation can be calculated that *should* be correct if we started with a stable age distribution (*Note*: first solve for N_0 given N=255.46 at t=36). The resulting growth equation is

$$N_t = 57.67e^{0.0413t}. \qquad (6.12)$$

The next step in analysis was to estimate N from 0 to 36 and then determine the residuals; that is, the observed (from simulation) - calculated (from Equation 6.12). The residuals are plotted in Figure 6.4.

In Figure 6.4, major low points occur at 2, 8 and 15 years for an average of about 6.5 years, which is close to the calculated period of oscillation for the largest complex root of 6.42 years.

The Pulmonate Snail *Bradybaena fruticum*

One final example should help in building intuition concerning the roles of the subdominant roots. Data for a pulmonate snail *Bradybaena fruticum* in Greece (Staikou *et al.* 1990) are given in Table 6.4. The authors adjusted m_x

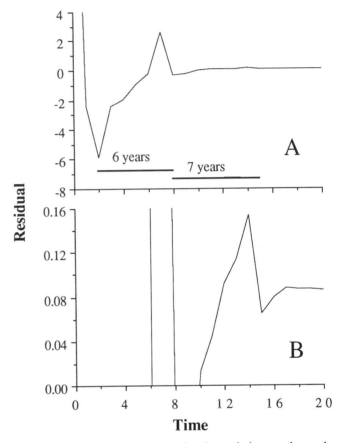

FIGURE 6.4 Residuals for simulated population number and number estimated from Equation 6.12; A and B differ in the scale used for the residuals in order to show damped oscillations over the first 20 years.

TABLE 6.4 Life Table for *Bradybaena fruticum*

Age (yr.)	l_x	m_x
0	1.000	0
1	0.819	0
2	0.423	16.69
3	0.112	6.00
4	0.006	0
5	0.001	0

Note: data from Staikou *et al.* (1990).

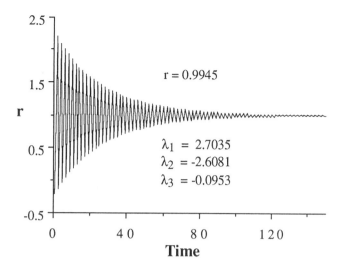

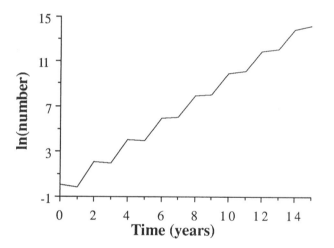

FIGURE 6.5 Convergence of a projected population of *Bradybaena fruticum* to a stable-age distribution and fixed value of r; the 3 roots were obtained from analysis of a transition matrix; data from Staikou *et al.* (1990).

FIGURE 6.6 Population growth of *Bradybaena fruticum* based on data presented by Staikou *et al.* (1990).

values first by a reduction of 34.5% because this is the fraction that do not hatch and then by an additional 30% due to mortality just after hatching. An additional complication is that only 70% of females reproduce when they are 2 years old and just 3% of females reproduce when 3 years old.

Use of the program **PROJECTION.BAS** from Chapter 1 shows that convergence to a stable-age distribution takes hundreds of iterations when starting with all individuals in age class 1. Figure 6.5 shows this slow convergence to r with damped oscillations. Because there are no complex roots, the only possible periods of oscillation are associated with negative roots and these all have periods of exactly 2 years, which is what is shown in Figures 6.5 and 6.6.

The damping ratio, ρ, for roots λ_1 and λ_2 is 1.0365. and so the length of time, t, it would take for the contribution of λ_1 to be 1000 times greater than the contribution of λ_2 is

$$t = \frac{ln\,1000}{ln\,1.0365} = 193 \text{ years.}$$

Convergence to a stable-age distribution is very slow; indeed, the life-table of *Bradybaena fruticum* is such that this snail probably never achieves a stable-age distribution and always will show year to year fluctuations due to the intrinsic schedules of its life history.

THE 𝕮𝖔𝖒𝖕𝖑𝖊𝖆𝖙 PROJECTION EQUATION

Recall that the Leslie matrix can be used to project a population, one cycle at a time

$$\mathbf{n}(t+1) = \mathbf{A}\mathbf{n}(t). \tag{6.13}$$

After a stable-age distribution has been attained, the transition matrix can be replaced by the dominant latent root, λ_1

$$\mathbf{n}(t+1) = \lambda_1 \mathbf{n}(t).$$

Furthermore, the number of individuals can be predicted to any time in the future by raising λ_1 to a power that is the number of time periods or cycles from now to the future time

$$\mathbf{n}(t) = \lambda_1^{\,t}\mathbf{n}(0).$$

Once again, however, the initial age distribution, $\mathbf{n}(0)$, *must* be the stable-age distribution times some constant.

It is possible to find a solution to the projection equation that does not require a stable-age distribution. The solution makes use not only of the dominant latent root λ_1 but also of all other roots of the characteristic equation. Additional information includes all latent vectors associated with **A** and the transpose of **A**, **A'**, as well as a set of constants, g_i, that are determined from initial conditions.

A worked example will help to show how terms are calculated. Once again, let's use our small beast, which has the following transition matrix

$$\mathbf{A} = \begin{pmatrix} 4.8 & 5.76 \\ 0.242 & 0 \end{pmatrix}$$

and

$$\lambda^2 - 4.8\lambda - 1.39392 = 0,$$

which has two roots,

$$\lambda_1 = 5.0747$$

and $\lambda_2 = -0.2747.$

Having found the roots, we now must find the associated latent vectors. With **A**, the stable-age distribution was found in Chapter 5 using λ_1,

$$\left[\begin{pmatrix} 4.8 & 5.76 \\ 0.242 & 0 \end{pmatrix} - 5.0747\begin{pmatrix} 1 & 0 \\ 0 & 1 \end{pmatrix}\right]\begin{pmatrix} x_1 \\ x_2 \end{pmatrix} = 0. \quad (6.14)$$

After carrying out the multiplication and combining terms

$$\begin{pmatrix} -.2747 & 5.76 \\ 0.242 & -5.0747 \end{pmatrix}\begin{pmatrix} x_1 \\ x_2 \end{pmatrix} = 0 \quad .$$

Two equations are produced:

$$-0.2747x_1 + 5.76x_2 = 0,$$
and $$0.242x_1 - 5.0747x_2 = 0.$$

An additional constraint must be satisfied, $x_1 = 1$, and the solution of these equations yields a vector, c_1, that is the stable-age distribution

$$x_1 = 1$$

$$x_2 = 0.0477.$$

A second vector, c_2, is associated with the second eigenvalue and is calculated in the same manner. All that is different is that λ_2 replaces λ_1 in Equation 6.14

$$\left[\begin{pmatrix} 4.8 & 5.76 \\ 0.242 & 0 \end{pmatrix} - (-0.2747)\begin{pmatrix} 1 & 0 \\ 0 & 1 \end{pmatrix}\right]\begin{pmatrix} x_1 \\ x_2 \end{pmatrix} = 0.$$

After carrying out the multiplication and combining terms

$$\begin{pmatrix} 5.0747 & 5.76 \\ 0.242 & 0.2747 \end{pmatrix}\begin{pmatrix} x_1 \\ x_2 \end{pmatrix} = 0,$$

two equations are produced;

$$5.0747x_1 + 5.76x_2 = 0,$$
and $$0.242x_1 + 0.2747x_2 = 0.$$

Once again, an additional constraint must be satisfied $x_1 = 1$, and the solution of these equations yields a vector, c_2:

$$x_1 = 1$$

$$x_2 = -0.8810.$$

The initial age distribution, $\mathbf{n}(0)$ is a linear function of the eigenvectors

$$\mathbf{n}(0) = g_1\mathbf{c}_1 + g_2\mathbf{c}_2 + \dots g_s\mathbf{c}_s$$

and the vector of constants, **g**, is obtained from the matrix equation

$$\mathbf{n}(0) = \mathbf{Cg},$$

where **C** is a matrix consisting of the eigenvectors and **g** is the vector of constants. To use our small creature, let's assume that we start with 1000 in age class 1 and none in age class 2,

$$\mathbf{n}(0) = \begin{pmatrix} 1000 \\ 0 \end{pmatrix}.$$

Accordingly,

$$\begin{pmatrix} 1 & 1 \\ 0.0477 & -0.8810 \end{pmatrix}\begin{pmatrix} g_1 \\ g_2 \end{pmatrix} = \begin{pmatrix} 1000 \\ 0 \end{pmatrix},$$

and the set of simultaneous equations is

$$g_1 + g_2 = 1000$$
$$0.0477g_1 - 0.8810g_2 = 0,$$

which yields

$$g_1 = 948.6379$$
$$g_2 = 51.3621.$$

Now, finally, we can assemble the eigenvalues, eigenvectors, and constants into a solution to the projection equation

$$\mathbf{n}(t) = \sum_{i=1}^{s} g_i\lambda_i^t\mathbf{c}_i. \quad (6.15)$$

For $t = 1$,

$$\mathbf{n}(1) = (948.638)(5.075)\begin{pmatrix} 1 \\ 0.048 \end{pmatrix} + (51.362)(-0.2747)\begin{pmatrix} 1 \\ -0.881 \end{pmatrix}$$

TABLE 6.5 Population Projection using Equation 6.13,

$$n(t+1) = An(t), \text{ and Equation 6.15, } n(t) = \sum_{i=1}^{s} c_i \lambda_i^t w_i$$

Time	Equation 6.13		Equation 6.15	
	n(1)	n(2)	n(1)	n(2)
0	1,000	0	1,000	0
1	4,800	242	4,800	242
2	24,434	1,162	24,434	1,162
3	123,970	5,913	123,973	5,914
4	629,130	30,002	629,133	30,009
5	3,192,600	152,250	3,192,657	152,290

or $\quad n(1) = \begin{pmatrix} 4800 \\ 242 \end{pmatrix}.$

If we wished to determine the numbers of individuals in each age class at time = 5, each latent root would be raised to the 5th power

$$n(5) = (948.64)(5.075)^5 \begin{pmatrix} 1 \\ 0.048 \end{pmatrix} + (51.362)(-0.275)^5 \begin{pmatrix} 1 \\ -0.881 \end{pmatrix}$$

$$(6.16)$$

which is

$$n(5) = \begin{pmatrix} 3,192,657 \\ 152,290 \end{pmatrix}.$$

There is an important item to notice in Equation 6.15, which is evident in Equation 6.16. Because the latent roots are being raised to ever higher powers, the largest latent root will become ever more dominant in the sense that its contribution to **n** will become an ever larger fraction of contributions of all roots. Eventually, its contribution becomes so large that all other roots can be ignored. At this point, the population has attained a stable-age distribution; that is, just c_1 has to be considered. Table 6.5 shows that Eq 6.15 actually provides the same results as Eq 6.13. There is a certain amount of creeping round-off error but, basically, the two methods provide the same answers.

The Bernardelli (1940) example, with complex roots, requires vectors with complex numbers. The three vectors that are required are

$$c_1 = \begin{pmatrix} 1 \\ 0.5 \\ 0.1666667 \end{pmatrix},$$

$$c_2 = \begin{pmatrix} 1+0i \\ -0.25 + 0.433013i \\ -0.083333 - 0.144338i \end{pmatrix},$$

and $\quad c_3 = \begin{pmatrix} 1+0i \\ -0.25 - 0.433013i \\ -0.083333 + 0.144338i \end{pmatrix}.$

Also, the vector **g** will have complex entries. For example, if initial conditions are

$$n(0) = \begin{pmatrix} 1000 \\ 1000 \\ 1000 \end{pmatrix}$$

then

$$\begin{pmatrix} 1 & 1 & 1 \\ 0.5 & -0.25+0.4330i & -0.25-0.4330i \\ 0.1667 & -0.0833-0.1443i & -0.0833+0.1443 \end{pmatrix} \begin{pmatrix} g_1 \\ g_2 \\ g_3 \end{pmatrix} = \begin{pmatrix} 1000 \\ 1000 \\ 1000 \end{pmatrix}$$

and

$$g_1 = 3000.0$$

$$g_2 = -1000.0 + 1154.7i$$

$$g_3 = -1000.0 - 1154.7i$$

It is very unlikely that you would ever *really* want to use the **Compleat** population growth equation; calculations were fairly involved for our small beast with just a 2×2 matrix and no complex roots. Calculations become rather tedious when complex roots are present and so, in general it would be so much easier just to use matrix multiplication and project a population for as long as one wanted in order to determine population size and numbers in each class. However, I think that it is good to go through the calculations a few times in order to help build confidence in understanding what all of those latent roots are doing; why λ_1 is called the "dominant" root and why the relative sizes of the roots determine how rapidly a population converges to a stable configuration.

PROBLEMS

1. Leonardo of Pisa, a.k.a. Fibonacci (c. 1180 - c. 1250), proposed what is now known as his rabbit problem in which one starts with a pair of rabbits, one male and one female. After one month, this pair produces another pair and will continue to produce a new pair every month thereafter. Each new pair will begin reproduction at the age of 2 months and then will produce a new pair every month thereafter. Clearly, Leonardo's parents never let him have

real rabbits for pets because, though he was very good with numbers, he seems to have been a bit dim concerning lagomorph biology. Using current terminology, all p_x values are equal to 1, including p_0, and $m_x = 1$ because the pair that is produced has one male and one female and m_x is the number of female offspring per female.

a. Draw the rabbit problem as a life cycle graph using 2 nodes (1 month and 2+ months). Note that the node for 2+ has a self loop.
b. Write the characteristic equation and determine the two roots. *Note:* the self loop for 2+ should be absorbed in the transition from 1 to 2+ (see Chapter 4 for how to do this).
c. Write the transition matrix for Fibonacci's rabbits and then write the 𝕮𝖔𝖒𝖕𝖑𝖊𝖆𝖙 growth equation with starting conditions of 1 and 0 for numbers in age class 1 and age class 2+.
d. Using the 𝕮𝖔𝖒𝖕𝖑𝖊𝖆𝖙 growth equation, determine numbers of females (or pairs) at times 1 through 6. Notice that you have produced a sequence of Fibonacci numbers. Also, just incidentally, the population growth rate, λ_1, is the so called 'golden ratio'.

2. The 1978 projection matrix for *Gemma gemma* (Weinberg *et al.* 1986) includes $p_0 = 0.156$ and $f_x = p_0 m_x$. The matrix begins with age 1 and the time interval is 3 months.

$$
A = \begin{pmatrix}
0 & 0 & 0 & 0 & 2.29 & 0 & 3.01 & 4.66 \\
1.000 & 0 & 0 & 0 & 0 & 0 & 0 & 0 \\
0 & 0.480 & 0 & 0 & 0 & 0 & 0 & 0 \\
0 & 0 & 0.310 & 0 & 0 & 0 & 0 & 0 \\
0 & 0 & 0 & 1.000 & 0 & 0 & 0 & 0 \\
0 & 0 & 0 & 0 & 0.578 & 0 & 0 & 0 \\
0 & 0 & 0 & 0 & 0 & 0.717 & 0 & 0 \\
0 & 0 & 0 & 0 & 0 & 0 & 0.024 & 0
\end{pmatrix}
$$

Describe convergence and oscillatory behavior of the matrix.

a. How rapidly will *Gemma gemma* converge to a stable-age distribution?
b. What are the periods of oscillation?
c. Do the periods of oscillation make sense for an organism that must fit an annual cycle? Explain your answer.

3. van Aarde (1987) provides data on age-specific survival and fecundity for the Cape porcupine (*Hystrix africaneaustralis*) based on animals at the Tussen-die-Riviere Game Farm (Table 6.6). Data are for 1977/78. Age

classes 3 to 10 declined at a constant rate of 0.0516 yr⁻¹. Determine the damping ratio and periods oscillation for growth of the Cape porcupine population.

TABLE 6.6 Life Table for the Cape Porcupine (*Hystrix africaneaustralis*)

Age	l_x	m_x	Age	l_x	m_x
0	1.0	0	6	0.2578	0.66
1	0.836	0	7	0.2062	0.66
2	0.464	0.48	8	0.1547	0.66
3	0.4124*	0.66	9	0.1031	0.66
4	0.3609	0.66	10	0.0516	0.66
5	0.3093	0.66			

Note: ages are in years; data from van Aarde (1987).

4. Use data from previous chapters to explore transient behavior. Use the following data and determine damping ratios and periods of oscillation.

a. Buckley (1986) for a freshwater snail, *Viviparus georgianus* (Chapter 1).
b. Caughley (1967) for domestic sheep, *Ovis aries* (Chapter 2).
c. Canales *et al.* (1994) for *Andropogon brevifolius* (Chapter 4).
d. Knight and Eberhardt (1985) for grizzly bears (Chapter 5).
e. Schaff *et al.* (1987) for menhaden, *Brevoortia tyrannus* (Chapter 5).
f. Ainley *et al.* (1983) for Adélie penguins (Chapter 5).
g. Rowley and Chapman (1991) for the Pink Cockatoo (Chapter 5).

5. The life table for the pulmonate land snail *Helicella* (*Xerothracia*) *pappi* in Macedonia, Greece (Table 6.7) is modified from Lazaridou-Dimitriadou (1995).
a. Determine transient behavior for *Helicella pappi*. First determine all latent roots and then calculate and interpret the damping ratio and periods of oscillation.
b. How long will it take for λ_1 to have 1000 time the influence on growth as λ_2?

TABLE 6.7 Life table for the Pulmonate Land Snail *Helicella* (*Xerothracia*) *pappi*

Age (years)	l_x	m_x
0	1.000	0
1	0.560	0
2	0.130	18
3	0.101	7
4	0.001	0

Note: modified from data presented by Lazaridou-Dimitriadou (1995).

Table 6.8 Life Table for Northern Sea Lions,
***Eumetopias jubatus*, Near Marmot Island, Alaska**

Age	Annual Survival	Fecundity
0	0.776	0
1	0.776	0
2	0.776	0
3	0.868	0.105
4	0.879	0.267
5	0.888	0.286
6	0.893	0.315
7	0.898	0.315
8	0.874	0.315
9	0.899	0.315
10	0.893	0.315
11	0.896	0.315
12	0.895	0.315
13+	0.895	0.315

Note: maximum observed age was 31 years; data from Calkins and Pitcher (1982).

6. Northern sea lions (*Eumetopias jubatus*) were studied in the vicinity of Marmot Island, Alaska, starting in 1975. The life table (Table 6.8) is from data gathered by Calkins and Pitcher (1982) and presented by York (1994). Survival from age 0 to 3 years was 0.478 and so the annual survival for each of the first three years is $0.478^{1/3}$. For animals between the ages, x, of 13 and 31, annual survival rate, p_x, was described by

$$p_x = 0.895^{x-12}$$

a. Construct a life-cycle graph for northern sea lions.
b. Create a matrix from life cycle graph, form the characteristic equation and find all roots.
c. Are there apparent periods of oscillation? If so, what are they?
d. If the population described by Table 6.8 would be moved away from a stable-age distribution, how rapidly would it return to a stable configuration?

7

Sensitivity Analysis

"Who's overstuffed?" said Pooh.

B. Hoff, *The Te of Piglet*, 1992

INTRODUCTION

A life cycle is the result of the history of a species and the local conditions under which the species was studied. The historical development of the species will include features that are the direct products of natural selection, features that evolved under selective pressures other than those now applied to the feature (the "exaptations" of Gould and Vrba 1982), and constraints of various sorts that include laws of physics and lack of genetic variation. Traits also change within the context of other traits and so there is covariation due to trade-offs among growth, reproduction, and survival.

SENSITIVITY AND ELASTICITY

Analysis of a life-cycle in terms of how sensitive population growth rate, λ, is to changes in terms of the transitions matrix or terms in the life-cycle graph, provides insight into what parts of the life-cycle should be under the most intense selection pressure or, from a management stand point, which parts should be emphasized in conservation or pest control. Population growth rate, λ, is a reasonable measure of fitness and how λ changes with the changes in a trait has been called the "selective pressure" on the trait (Emlen 1970). The change in λ with change in a

trait a_i, $\partial\lambda/\partial a_i$, is the partial derivative of λ with respect to a_i and is called the sensitivity of λ to changes in a_i.

Different elements, a_i, of the life-cycle graph, such as age-specific survival probabilities (p_x) and fecundities (m_x), contribute differently to population growth rate λ. The different contributions can be measured as the proportional sensitivity or "elasticity" (Caswell *et al.* 1984, de Kroon *et al.* 1986), which has an advantage over sensitivity, $\partial\lambda/\partial a_i$, because the elements a_i have different units. All p_x values are probabilities and so can not exceed 1.0, but fecundities, m_x, are not bounded by 1.0 and can be much larger. Elasticity, e_i, is defined

$$e_i = \frac{\partial(ln\lambda)}{\partial(ln a_i)} \qquad (7.1)$$

and so because

$$\partial(ln\lambda) = \left[\frac{1}{\lambda}\right]\partial\lambda, \qquad (7.2)$$

and

$$\partial(ln\ a_i) = \left[\frac{1}{a_i}\right]\partial a_i \qquad (7.3)$$

or

$$e_i = \left[\frac{1}{\lambda}\right]\partial\lambda \div \left[\frac{1}{a_i}\right]\partial a_i,$$

and so elasticity, e_i, is

$$e_i = \frac{a_i}{\lambda}\frac{\partial\lambda}{\partial a_i}. \qquad (7.4)$$

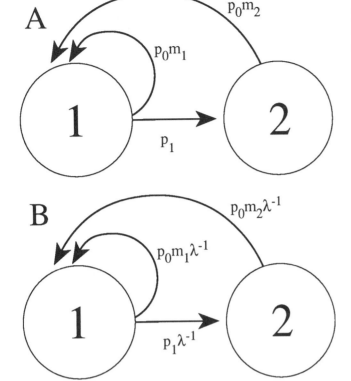

FIGURE 7.1 **A**, Life-cycle graph using $f_x = p_0 m_x$; **B**, graph following Z-transformation; that is, multiplication of each transition by λ^{-1}.

The way to think of elasticity is, for example, if $e_i = 0.5$ the implication is that a 10% change in a_i will result in a 5% change in λ; that is, $0.5 \times 10\%$. The larger the elasticity, the greater will be the change in λ per change in a_i. In evaluating the evolutionary consequences of different elasticities, *if all other things are equal*, then selection should favor increased investment of resources in attributes that produce the greatest positive change in λ. For conservation of a population, emphasis should be placed on preserving or increasing transitions with large elasticities. In pest management, control would be best if focused on transitions with large elasticity values and harvesting a resource would be best if concentrated on stages with small elasticities.

SENSITIVITY ANALYSIS USING A LIFE CYCLE GRAPH

Our Small Beast

Here, once again, is our small beast from previous chapters to demonstrate how calculations of sensitivity are performed. Let's use $f_x = p_0 m_x$ (Figure 7.1A).

Recall how the characteristic equation is formed from a life-cycle graph. First multiply each transition by λ^{-1} (Figure 7.1B). Then multiply transitions for each path around the graph, sum them, and set the sum equal to 1.0

$$p_0 m_1 \lambda^{-1} + p_0 p_1 m_2 \lambda^{-2} = 1.$$

Finally, rearrange and clear the equation by multiplying through by λ^2

$$\lambda^2 - p_0 m_1 \lambda - p_0 p_1 m_2 = 0, \tag{7.5}$$

which is the characteristic equation of the life-cycle graph. We want to calculate $\frac{\partial \lambda}{\partial a_i}$, where $a_i = p_0, p_1, m_1,$ or m_2. These partials can be obtained directly from the characteristic equation, which can be written

$$f(\lambda, a_i) = \lambda^2 - p_0 m_1 \lambda - p_0 p_1 m_2. \tag{7.6}$$

The function $f(\lambda, a_i) = 0$ defines λ implicitly as differentiable with respect to a_i,

$$\frac{\partial f(\lambda, a_i)}{\partial a_i} + \frac{\partial f(\lambda, a_i)}{\partial \lambda} \frac{\partial \lambda}{\partial a_i} = 0,$$

and so

$$\frac{\partial \lambda}{\partial a_i} = - \frac{\partial f(\lambda, a_i)/\partial a_i}{\partial f(\lambda, a_i)/\partial \lambda}. \tag{7.7}$$

Equation 7.7 can be used to find all of the partials of λ with respect to a_i. The first step is to find all partials of Equation 7.6 with respect to a_i and λ; that is, $\partial f(\lambda, a_i)/\partial a_i$ and $\partial f(\lambda, a_i)/\partial \lambda$

$$\frac{\partial f(\lambda, a_i)}{\partial p_0} = - m_1 \lambda - p_1 m_2, \tag{7.8}$$

$$\frac{\partial f(\lambda, a_i)}{\partial p_1} = - p_0 m_2, \tag{7.9}$$

$$\frac{\partial f(\lambda, a_i)}{\partial m_1} = - p_0 \lambda, \tag{7.10}$$

$$\frac{\partial f(\lambda, a_i)}{\partial m_2} = - p_0 p_1, \tag{7.11}$$

and $\quad \dfrac{\partial f(\lambda, a_i)}{\partial \lambda} = 2\lambda - p_0 m_1. \tag{7.12}$

Recall the life-table elements for our small beast (Table 7.1). Partials can be calculated using Equation 7.7 and the data in Table 7.1

TABLE 7.1 Life Table for Our Small Beast

Age(x)	p_x	m_x
0	0.240	0
1	0.242	20
2	0.000	24

TABLE 7.2 Elasticities Calculated Using Equation 7.4 and Partials Calculated in Equations 7.13 - 7.16.

Component	$\dfrac{\partial \lambda}{\partial a_i}$	e_i
$p_0.$	20.0587	0.9486
$p_1.$	1.0768	0.0513
$m_1.$	0.2277	0.8973
$m_2.$	0.0109	0.0513

$$\frac{\partial \lambda}{\partial p_0} = \boxed{\frac{m_1 \lambda + p_1 m_2}{2\lambda - p_0 m_1}} = 20.0587, \quad (7.13)$$

$$\frac{\partial \lambda}{\partial p_1} = \boxed{\frac{p_0 m_2}{2\lambda - p_0 m_1}} = 1.0768, \quad (7.14)$$

$$\frac{\partial \lambda}{\partial m_1} = \boxed{\frac{p_0 \lambda}{2\lambda - p_0 m_1}} = 0.2277, \quad (7.15)$$

$$\text{and} \quad \frac{\partial \lambda}{\partial m_2} = \boxed{\frac{p_0 p_1}{2\lambda - p_0 m_1}} = 0.0109. \quad (7.16)$$

Elasticities for components of the life-cycle graph can be calculated using Equation 7.4 (Table 7.2). For example, the elasticity value for p_0 is

$$e_i = \frac{a_i}{\lambda} \frac{\partial \lambda}{\partial a_i} = \frac{0.240}{5.0747} \times 20.058698 = \boxed{0.9486}.$$

The elasticities show how life table attributes influence population growth rate. For example, a 10% change in either p_0 or m_1 would change λ by about 9%; that is, $10\% \times 0.95 = 9.5\%$ for p_0 and $10\% \times 0.90 = 9\%$ for m_1. Similarly, a 10% change in either p_1 or m_2 would change λ by only 0.5%. Note that p_0 is part of both f_1 and f_2; that is, $p_0 m_1$ and $p_0 m_2$, and so it should not be surprising to discover in Table 7.2 that changing *both* m_1 and m_2 by 10% would have the same effect as changing p_0 by 10%. The sum of the elasticities of m_1 and m_2 is 0.9486, the same as the elasticity for p_0. Relationships among elasticity values also includes the fact that the sum of incoming elasticities equals the sum of outgoing elasticities from a node (van Groenendael *et al.* 1994)

TABLE 7.3 Changes in λ with 10% Changes in Specific Values of a_i to Show the Relationship between Elasticities of Life-Table Values and Change in λ

a_i	original value	increase of 10%	new λ	% increase in λ	elasticity
p_0	0.24	0.264	5.5560	9.484%	0.9490
p_1	0.242	0.2662	5.1006	0.510	0.0513
m_1	20.0	22.0	5.5320	9.011	0.8973
m_2	24.0	26.4	5.1006	0.510	0.0513

Note: original value of λ is 5.0747.

An important question, of course, is whether any of the elasticities actually show what they are supposed to show: namely, the proportional change in λ given a specific change in a_i. These calculations are shown in Table 7.3 and were done by multiplying each a_i value (p_0, p_1, m_1, m_2) by 10% and then adding this amount to the appropriate a_i value. For example, 10% of p_0 is 0.10×0.24 or 0.024 which, when added to 0.24 gives the 10% increase in p_0. As shown in Table 7.3, this 10% increase gives a new value of 0.264 for p_0. When an a_i was changed by 10%, other values were kept at their original values and the % increase in λ was calculated

$$\% \text{ increase in } \lambda = \frac{\text{new } \lambda - 5.0749}{5.0749} \times 100\%. \quad (7.17)$$

The elasticity function is not linear and so can not be extrapolated out to large changes in life-table values; with small changes, however, the elasticity values are quite good (Table 7.3).

For our small beast, the best evolutionary strategy would be to increase the number of offspring at first reproduction or improve early survival. Such a strategy would be much better than increasing longevity or increasing the number of offspring produced at the second breeding season. Basically, the best strategy might be to become an annual. What is not part of present analysis is information concerning year to year variability in parameters or covariation in, for example, reproduction and survival. These are important topics that will receive some attention in following chapters; however, a wider discussion is provided by Caswell (1989), Stearns (1992) and Roff (1992).

A Fresh Water Snail, *Viviparus georgianus*

Survival and reproductive data for a freshwater snail, *Viviparus georgianus*, have been taken from Buckley (1986) and are shown in Table 7.4. The life-cycle graph that would

TABLE 7.4 Life Table for a Freshwater Snail
Viviparus georgianus

Age	p_x	m_x
0	0.321	0
1	0.639	0
2	0.565	1.24
3	0.307	4.80
4	0.0	8.35

Note: data from Buckley (1986); age is in years.

**TABLE 7.5 Coefficients of the Characteristic
Polynomial for *Viviparus georgianus***

Powers of λ	Coefficient
λ^4	1.0
λ^3	0.0
λ^2	-0.2543
λ^1	-0.5563
λ^0	-0.2971

Note: data from Buckley 1986; note that the coefficient for λ^3 is 0.

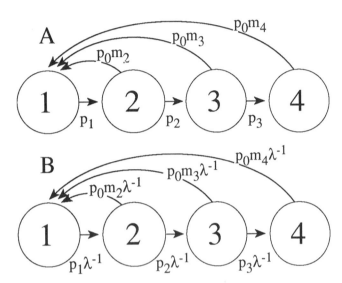

FIGURE 7.2 A, Life cycle appropriate for *Viviparus georgianus* or other species with age at first reproduction equal to 2 and with a maximum life of 4 years; **B,** .Z-transformed life cycle.

be appropriate for *V. georgianus* is shown in Figure 7.2A with four nodes and f_x defined as $p_0 m_x$.

The steps for constituting the characteristic equation are first to do a Z-transformation (Figure 7.2B), identify the loops and multiply the transitions around each loop.

There are three loops in Figure 7.2:

loop 1: $p_0 p_1 m_2 \lambda^{-2}$,

loop 2: $p_0 p_1 p_2 m_3 \lambda^{-3}$,

and   loop 3: $p_0 p_1 p_2 p_3 m_4 \lambda^{-4}$.

Next, sum the loops, subtract from 1, set equal to 0

$$1 - p_0 p_1 m_2 \lambda^{-2} - p_0 p_1 p_2 m_3 \lambda^{-3} - p_0 p_1 p_2 p_3 m_4 \lambda^{-4} = 0$$

and clear the negative exponents by multiplying through by λ^4 to form the characteristic equation,

$$\lambda^4 - p_0 p_1 m_2 \lambda^2 - p_0 p_1 p_2 m_3 \lambda - p_0 p_1 p_2 p_3 m_4 = 0. \quad (7.18)$$

Notice that there is no λ^3 term. Equation 7.18 is a function of λ and a_i where a_i is p_0, p_1, p_2, p_3, or m_4

$$f(\lambda, a_i) = \lambda^4 - p_0 p_1 m_2 \lambda^2 - p_0 p_1 p_2 m_3 \lambda - p_0 p_1 p_2 p_3 m_4$$

and so Equation 7.7 can be used to find all of the partials of λ with respect to a_i. The first step is to find all partials of the function with respect to a_i and λ; that is, $\partial f(\lambda, a_i)/\partial a_i$ and $\partial f(\lambda, a_i)/\partial \lambda$

$$\frac{\partial f(\lambda, a_i)}{\partial p_0} = - p_1 m_2 \lambda^2 - p_1 p_2 m_3 \lambda - p_1 p_2 p_3 m_4, (7.19)$$

$$\frac{\partial f(\lambda, a_i)}{\partial p_1} = - p_0 m_2 \lambda^2 - p_0 p_2 m_3 \lambda - p_0 p_2 p_3 m_4, (7.20)$$

$$\frac{\partial f(\lambda, a_i)}{\partial p_2} = - p_0 p_1 m_3 \lambda - p_0 p_1 p_3 m_4, \quad (7.21)$$

$$\frac{\partial f(\lambda, a_i)}{\partial p_3} = - p_0 p_1 p_2 m_4, \quad (7.22)$$

$$\frac{\partial f(\lambda, a_i)}{\partial m_2} = - p_0 p_1 \lambda^2, \quad (7.23)$$

$$\frac{\partial f(\lambda, a_i)}{\partial m_3} = - p_0 p_1 p_2 \lambda, \quad (7.24)$$

$$\frac{\partial f(\lambda, a_i)}{\partial m_4} = - p_0 p_1 p_2 p_3, \quad (7.25)$$

and

$$\frac{\partial f(\lambda, a_i)}{\partial \lambda} = 4\lambda^3 - 2 p_0 p_1 m_2 \lambda - p_0 p_1 p_2 m_3. \quad (7.26)$$

When numbers from Table 7.4 are substituted into Equation 7.18, the dominant root λ_1 can be found either by using the

TABLE 7.6 Summary of Sensitivity and Elasticity Calculations for a Freshwater Snail *Viviparus georgianus*

Life-Table Element	Sensitivity $\frac{\partial\lambda}{\partial a_i}$	Elasticity $\frac{a_i}{\lambda}\frac{\partial\lambda}{\partial a_i}$
p_0	1.0663	0.3309
p_1	0.5357	0.3309
p_2	0.4618	0.2522
p_3	0.2894	0.0859
m_2	0.0656	0.0787
m_3	0.0358	0.1664
m_4	0.0106	0.0859

Note: data from Buckley (1986).

Newton-Raphson algorithm or by using **MATRIX.BAS**, which permits entering the coefficients of a polynomial (Table 7.5) to find all roots.

The estimate of the largest latent root, λ_1, is 1.0343 and so it is possible to proceed with the calculations of all sensitivities and elasticities using Equation 7.7 and Equations 7.19-7.25 for the numerators and Equation 7.26 for the denominator, $\partial f(\lambda,a_i)/\partial\lambda$

$$\frac{\partial\lambda}{\partial p_0} = -\frac{-p_1m_2\lambda^2 - p_1p_2m_3\lambda - p_1p_2p_3m_4}{4\lambda^3 - 2p_0p_1m_2\lambda - p_0p_1p_2m_3} \quad (7.27)$$

$$= -\frac{-3.565673}{3.34389705} = 1.0663,$$

$$\frac{\partial\lambda}{\partial p_1} = -\frac{-p_0m_2\lambda^2 - p_0p_2m_3\lambda - p_0p_2p_3m_4}{4\lambda^3 - 2p_0p_1m_2\lambda - p_0p_1p_2m_3} \quad (7.28)$$

$$= -\frac{-1.7912071}{3.34389705} = 0.5357,$$

$$\frac{\partial\lambda}{\partial p_2} = -\frac{-p_0p_1m_3\lambda - p_0p_1p_3m_4}{4\lambda^3 - 2p_0p_1m_2\lambda - p_0p_1p_2m_3} \quad (7.29)$$

$$= -\frac{-1.54419}{3.34389705} = 0.4618,$$

$$\frac{\partial\lambda}{\partial p_3} = -\frac{-p_0p_1p_2m_4}{4\lambda^3 - 2p_0p_1m_2\lambda - p_0p_1p_2m_3} \quad (7.30)$$

$$= -\frac{-0.9677}{3.34389705} = 0.2894,$$

$$\frac{\partial\lambda}{\partial m_2} = -\frac{-p_0p_1\lambda^2}{4\lambda^3 - 2p_0p_1m_2\lambda - p_0p_1p_2m_3} \quad (7.31)$$

$$= -\frac{-0.219447}{3.34389705} = 0.0656,$$

$$\frac{\partial\lambda}{\partial m_3} = -\frac{-p_0p_1p_2\lambda}{4\lambda^3 - 2p_0p_1m_2\lambda - p_0p_1p_2m_3} \quad (7.32)$$

$$= -\frac{-0.119872}{3.34389705} = 0.0358,$$

and

$$\frac{\partial\lambda}{\partial m_4} = -\frac{-p_0p_1p_2p_3}{4\lambda^3 - 2p_0p_1m_2\lambda - p_0p_1p_2m_3} \quad (7.33)$$

$$= -\frac{-0.035579}{3.34389705} = 0.0106.$$

The sensitivities $\frac{\partial\lambda}{\partial a_i}$ and elasticities $\frac{a_i\partial\lambda}{\lambda\partial a_i}$ are summarized in Table 7.6.

Based on the elasticity calculations, changes in survival would change λ much more than would similar percent changes in fecundity. Notice that because p_0 is part of all transitions with m_x values, the sum of the elasticities of m_1, m_2, and $m_3 = 0.331$, the same as the elasticity of p_0.

Primates

A third example of using a life-cycle graph to calculate sensitivities and elasticities directly from the characteristic equation uses data on primates (Dobson and Lyles 1989) from Chapter 4 (Table 7.7), where analysis was presented as a general life-cycle graph containing just two nodes, juveniles, j, and adults, A. An assumption was that juvenile survival, adult survival, and fecundity all were constant and, with these assumptions, the characteristic equation of the life cycle is

$$\lambda^a - p_A\lambda^{a-1} - p_0p_j^{a-1}m_A = 0 \quad (7.34)$$

where

a = age at first reproduction,
p_A = adult survival rate,
p_0 = first-year survival rate,
p_j = juvenile survival rate,
and
m_A = annual fecundity.

It is possible to proceed with Equation 7.34 in the same way as was done with Equations 7.5 or 7.18. First it is reasonable to express Equation 7.34 as a function of λ and a_i where a_i represents, a, p_A, p_0, p_j and m_A

TABLE 7.7 Life History Data for Three Primate Species

Species	a	m_A	p_0	p_j	p_A	λ yr^{-1}
Howler monkey *Alouatta palliata*	4	0.48	0.6	0.72	0.82	0.9467
Rhesus monkey *Macaca mulatta*	4	0.62	0.78	0.74	0.78	0.9850
Chimpanzee *Pan troglodytes*	14	0.19	0.73	0.96	0.92	1.0008

Note: column headings are parameters in Equation 7.34: a = age at first reproduction, m_A = fecundity, p_0 = 1st year survival, p_j = juvenile survival rate yr^{-1}, and p_A = adult survival rate yr^{-1}; (Dobson and Lyles 1989).

$$f(\lambda, a_i) = \lambda^a - p_A\lambda^{a-1} - p_0 p_j^{a-1} m_A$$

so

$$\frac{\partial f(\lambda, a_i)}{\partial \lambda} = a\lambda^{a-1} - (a-1)p_A\lambda^{a-2} \qquad (7.35)$$

$$\frac{\partial f(\lambda, a_i)}{\partial a} = \lambda^a ln(\lambda) - p_A\lambda^{a-1} ln(\lambda) - ln(p_j)p_j^{a-1} p_0 m_A \qquad (7.36)$$

$$\frac{\partial f(\lambda, a_i)}{\partial p_A} = -\lambda^{a-1} \qquad (7.37)$$

$$\frac{\partial f(\lambda, a_i)}{\partial p_0} = -p_j^{a-1} m_A \qquad (7.38)$$

$$\frac{\partial f(\lambda, a_i)}{\partial p_j} = -(a-1)p_j^{a-2} p_0 m_A \qquad (7.39)$$

$$\frac{\partial f(\lambda, a_i)}{\partial m_A} = -p_j^{a-1} p_0. \qquad (7.40)$$

The partials, $\frac{\partial \lambda}{\partial a_i}$, can be calculated by dividing Equations 7.36 - 7.40 by Equation 7.35 as shown in Equation 7.7:

$$\frac{\partial \lambda}{\partial a} = \boxed{-\frac{\lambda^a ln(\lambda) - p_A\lambda^{a-1} ln(\lambda) - ln(p_j)p_j^{a-1} p_0 m_A}{a\lambda^{a-1} - (a-1)p_A\lambda^{a-2}}}, \qquad (7.41)$$

$$\frac{\partial \lambda}{\partial p_A} = \boxed{-\frac{-\lambda^{a-1}}{a\lambda^{a-1} - (a-1)p_A\lambda^{a-2}}}, \qquad (7.42)$$

TABLE 7.8 Sensitivity and Elasticity Values for Populations of Three Primate Species

a_i	Howler monkey $\partial\lambda/\partial a_i$	e_i	Rhesus monkey $\partial\lambda/\partial a_i$	e_i	Chimpanzee $\partial\lambda/\partial a_i$	e_i
p_0	0.151	0.096	0.162	0.128	0.054	0.039
p_j	0.377	0.286	0.512	0.384	0.534	0.512
p_A	0.713	0.618	0.616	0.485	0.488	0.449
m_A	0.188	0.096	0.204	0.128	0.207	0.039
a	-0.025	0.105	-0.036	0.147	0.002	0.023

Note: data from Dobson and Lyles (1989) are presented in Table 7.7.

$$\frac{\partial \lambda}{\partial p_0} = \boxed{-\frac{-p_j^{a-1} m_A}{a\lambda^{a-1} - (a-1)p_A\lambda^{a-2}}}, \qquad (7.43)$$

$$\frac{\partial \lambda}{\partial p_j} = \boxed{-\frac{-(a-1)p_j^{a-2} p_0 m_A}{a\lambda^{a-1} - (a-1)p_A\lambda^{a-2}}}, \qquad (7.44)$$

and

$$\frac{\partial \lambda}{\partial m_A} = \boxed{-\frac{-p_j^{a-1} p_0}{a\lambda^{a-1} - (a-1)p_A\lambda^{a-2}}}. \qquad (7.45)$$

The calculations in Equations 7.41 - 7.45 are sufficiently tedious and complicated that I prefer to program the steps. The following short BASIC program does the calculations and prints both the sensitivities and elasticities.

```
10 print "A program to calculate
   sensitivities and elasticities for
   Equation 7.34"
20 input "p0: ";p0
30 input "pj: ";pj
40 input "pA: ";pa
50 input "mx: ";mx
60 input "age at first reproduction: ";a
70 input "Population growth rate:   ";lm
80 dl = a*lm^(a-1)-(a-1)*pa*lm^(a-2)
90 da = -(lm^a*log(lm)-pa*lm^(a-
   1)*log(lm)-pj^(a-
   1)*mx*p0*log(pj))/dl
100 dpa = -(-(lm^(a-1)))/dl
110 dpj = -(-(a-1)*pj^(a-2)*mx*p0)/dl
120 dp0 = -(-(pj^(a-1)*mx))/dl
130 dmx = -(-(pj^(a-1)*p0))/dl
140 ea = da*a/lm
150 epa = dpa*pa/lm
```

```
160 epj = dpj*pj/lm
170 ep0 = dp0*p0/lm
180 emx = dmx*mx/lm
190 print
195 print "d(lambda)/d(ai)","elasticity"
200 print
210 print "p0 ";dp0,ep0
220 print "pj ";dpj,epj
230 print "pA ";dpa,epa
240 print "mx ";dmx,emx
250 print "a ";da,ea
260 end
```

Sensitivity and elasticity calculations for the primate data shown in Table 7.7 are given in Table 7.8. The sign of the sensitivity values indicates the direction of change in λ when a_i values are changed. The sensitivity of the age at first reproduction, a, has a negative sign which means that if the age of first reproduction is *increased*, then λ would *decrease*. On the other hand, if p_0, p_j, p_A or m_A are *increased* then λ would *increase*. Examination of the elasticities for the primate populations suggest a direction for conservation. All three species-populations would benefit from programs that would increase survival, particularly for juveniles and adults. Increasing infant survival would be much less important. With limited resources for conservation, more good could be done with respect to increasing λ by focusing on the life stages with the largest elasticities.

SENSITIVITY ANALYSIS USING A TRANSITION MATRIX

Although it is possible to calculate sensitivities and elasticities from a life-cycle graph and its associated characteristic equation, it also is possible to obtain such measures using matrix techniques (Caswell 1989).

We can ask how fitness changes with changes in the individual elements of the transition matrix

$$\frac{\partial \lambda}{\partial a_{ij}} = \frac{v_i c_j}{<v,c>} \qquad (7.46)$$

The terms of the reproductive value function are v_i, the elements of the stable-age distribution are c_j, and $<v, c>$ is the scalar product of the stable-age, c, and reproductive-value, v, vectors

$$<v, c> = v_1 c_1 + v_2 c_2 + \ldots v_n c_n \qquad (7.47)$$

The vectors should be adjusted so $v_1 = c_1 = 1$.

TABLE 7.9 Stable-Age and Reproductive-Value Distributions for Our Small Beast when $f_x = p_0 m_x$

Age(x)	v_x	c_x	$v_x c_x$
1	1.0000	1.0000	1.0000
2	1.1350	0.0477	0.0541
$<v, c> =$			1.0541

TABLE 7.10 Elasticities of Matrix Elements When $f_x = p_0 m_x$

element	name	value	$\dfrac{\partial \lambda}{\partial a_{ij}}$	e_{ij}
a_{11}	f_1	4.8	0.9486	0.8973
a_{12}	f_2	5.76	0.0452	0.0513
a_{21}	p_1	0.242	1.0768	0.0513
	p_0	0.24	20.059	0.9486

Note: the elasticity for p_0 is the sum of the elasticities for f_1 and f_2; sensitivity, $\partial \lambda / \partial p_0$ can be calculated using Equation 7.4.

f_x Defined as $p_0 m_x$

Using $f_x = p_0 m_x$ the transition matrix for our small beast is

$$A = \begin{pmatrix} 4.8 & 5.76 \\ 0.242 & 0 \end{pmatrix} \qquad (7.48)$$

from which the stable-age and reproductive vectors can be obtained (Table 7.9). For specific elements of matrix A (Equation 7.48)

$$p_1 = a_{21}$$

$$\frac{\partial \lambda}{\partial p_1} = \frac{1.13504 \times 1.0}{1.054131} = \boxed{1.076755}, \qquad (7.49)$$

$$f_1 = a_{11}$$

$$\frac{\partial \lambda}{\partial f_1} = \frac{1.0 \times 1.0}{1.054131} = \boxed{0.94865}, \qquad (7.50)$$

and

$$f_2 = a_{12}$$

$$\frac{\partial \lambda}{\partial f_2} = \frac{1.0 \times 0.04769}{1.054131} = \boxed{0.04524}. \qquad (7.51)$$

The elasticities are shown in Table 7.10. Notice that the elasticities are the same as those calculated from the life

TABLE 7.11 Stable-Age and Reproductive-Value Vectors for $f_x = p_x m_{x+1}$

Age(x)	v_x	c_x	$v_x c_x$
0	1.0000	1.0000	1.0000
1	1.1445	0.0473	0.0541
2	0.0000	0.0022	0.0000
$\langle v, c \rangle =$			1.0541

Note: v_x is the *residual* reproductive vector.

TABLE 7.12 Elasticities of Matrix Elements for $f_x = p_x m_{x+1}$

Element	Name	Value	$\frac{\partial \lambda}{\partial a_{ij}}$	e_{ij}
a_{11}	f_0	4.8	0.9487	0.897
a_{12}	f_1	5.808	0.0449	0.051
a_{21}	p_0	0.24	1.0853	0.051
a_{32}	p_1	0.242	0.0	0.0

TABLE 7.13 Stable-Age (c) and Reproductive-Value (v) Vectors for $f_x = m_x \lambda$

x	c	v
1	1.0000	1.0000
2	0.0473	21.1445
3	0.0023	24.0000

TABLE 7.14 Sensitivities and Elasticities for $f_x = m_x \lambda$

Component	$\frac{\partial \lambda}{\partial a_{ij}}$	e_i
f_1	0.0230	0.4605
f_2	0.0011	0.0264
p_0	10.2937	0.4868
p_1	0.5526	0.0264

cycle graph but only one sensitivity matches: namely for p_1, which reflects the fact that f_x values are products of $p_0 m_x$ and both of these are separate in Table 7.2. The elasticities for f_1 and f_2 match the elasticities for m_1 and m_2 in Table 7.2. The elasticity for p_0 can be calculated by summing all elasticities that contain p_0, namely f_1 and f_2 and the sensitivities for m_x can be calculated by using Equation 7.4 and rearranging to solve for $\partial \lambda / \partial m_x$ given the elasticities, the values of m_x, and λ.

f_x Defined as $p_x m_{x+1}$

As was shown in Chapter 5, the transition matrix for our beast can be constructed using $f_x = p_x m_{x+1}$

$$A = \begin{pmatrix} 4.8 & 5.808 & 0 \\ 0.24 & 0 & 0 \\ 0 & 0.242 & 0 \end{pmatrix} \qquad (7.52)$$

which yields stable-age and reproductive-value vectors that start at age 0 and go to age 2 (Table 7.11). Equation 7.46 can be used to calculate sensitivities and elasticities (Table 7.12).

Entries for sensitivities and elasticities in Table 7.12 do not match the results in Tables 7.2 or 7.10. As was presented in Chapter 5, a problem with using $f_x = p_x m_{x+1}$ is that the *residual* reproductive value is calculated rather than the reproductive value and so all v_x elements are off by the value for current reproduction, m_x. Also, all elasticities are

shifted up so that what is given as the elasticity for f_1 really should be for f_2 and values for p_0 actually are correct for p_1. However, if a shift is done, then the elasticities are correct; for example, 0.897 is the correct elasticity for m_1 and 0.051 is correct for both m_2 and p_1. The elasticity for p_0 can be calculated as the sum of the appropriate elasticities, which are f_0 and f_1 in this case. All of the shifting that is required makes this approach clumsy and error prone and so if elasticities are needed in an analysis, defining f_x as $p_x m_{x+1}$ can not be recommended.

f_x Defined as $m_x \lambda$

With $f_x = m_x \lambda$ and $\lambda = 5.0747$ the transition matrix is

$$A = \begin{pmatrix} 0 & 101.4936 & 121.7924 \\ 0.24 & 0 & 0 \\ 0 & 0.242 & 0 \end{pmatrix} \qquad (7.53)$$

This matrix produces a somewhat different characteristic equation than previously seen for our small beast

$$\lambda^3 - 24.3585\lambda - 7.0737 = 0 \qquad (7.54)$$

Note that there is λ^3 and no λ^2.

Equation 7.54 has three roots rather than just two; the dominant root, however, still is the correct one. The three roots are

$$\lambda_1 = 5.0747,$$

$$\lambda_2 = -0.2914,$$

and

$$\lambda_3 = -4.7833.$$

TABLE 7.15. Comparison of Sensitivities and Elasticities Calculated from the Life-Cycle Graph (A) with Values in Table 7.14 (B)

Component	$\frac{\partial\lambda}{\partial a_{ij}}$ (A)	$\frac{\partial\lambda}{\partial a_{ij}}$ (B)	Relative e_i Table 7.14	e_i
m_1.	0.2277	0.0230	0.4605	0.8973
m_2.	0.0109	0.0011	0.0264	0.0513
p_0.	20.0587	10.2937	0.4868	0.9486
p_1.	1.0768	0.5526	0.0263	0.0513
Σ			1.0000	1.9486

Note: relative e_i is $e_i/\Sigma e_i$; values in the e_i column are from graph analysis (Table 7.2) as well as relative e_i multiplied by 1.9486.

Most things look just fine, such as the stable-age (**c**) and reproductive value (**v**) vectors (Table 7.13); however, the sensitivities and elasticities look a bit strange (Table 7.14), which should not be surprising because the matrix methods that are being used calculate elasticities so that $\Sigma e_i = 1.0$ (*cf.* Messerton-Gibbons, 1993).

In Table 7.14, elasticities sum to 1.0; however, in Table 7.2, elasticities summed to 1.9486. The elasticity values in Table 7.14 are *relatively* correct and can be made *absolutely* correct by multiplying them by 1.9486 (Table 7.15). Sensitivities, $\partial\lambda/\partial a_i$, are more complicated. Sensitivities of λ with respect to changes in p_x values from Table 7.14 can be corrected by multiplying them by 1.9486, the same value as used to correct elasticities; however, for m_x values, it is necessary to apply a correction not only because *relative* elasticities were calculated but also because f_x was $m_x\lambda$ and so the additional λ (5.0747) must be accounted for. For example, in Table 7.14, the sensitivity of λ to changes in m_2 is

$$\frac{\partial\lambda}{\partial m_2} = 0.0011.$$

The correct value is

$$\frac{\partial\lambda}{\partial m_2} = 0.0011 \times 1.9486 \times \lambda$$

or

$$\frac{\partial\lambda}{\partial m_2} = 0.0011 \times 1.9486 \times 5.0747 = \boxed{0.0109},$$

which is the value determined from graph analysis. Similarly, for m_1

$$\frac{\partial\lambda}{\partial m_1} = 0.02302 \times 1.9486 \times 5.0747 = \boxed{0.2276},$$

which agrees with the value from graph analysis but with a difference of 0.0001 due to round-off error in calculations.

The use of a "correction factor" to adjust elasticities when $f_x = m_x\lambda$ is worth a bit more exploration. The *relative* values for sensitivity and elasticity that were calculated using $f_x = m_x\lambda$ could be converted to correct values *if* the appropriate conversion factor could be found. This is possible by using some sensitivities or elasticities from $f_x = p_0 m_x$.

If analysis first is conducted using $f_x = p_0 m_x$ and sensitivities calculated, which will be correct for values of p_x, and then the analysis redone using $f_x = m_x\lambda$, it is possible to use sensitivities or elasticities of p_x values to obtain the correction factor. For example in Table 7.15, the sensitivity values for p_1 are 1.077 and 0.5526 for $f_x = p_0 m_x$ and $f_x = m_x\lambda$ respectively. The ratio of these two calculations in the appropriate correction factor

$$\frac{1.07675}{0.55257} = 1.94862$$

and so the elasticities for the elements of the life-cycle graph, estimated using the matrix with $f_x = m_x\lambda$, are

TABLE 7.16 Comparison of Sensitivities and Elasticities for Our Small Hypothetical Creature using Three Different Methods

	$f_x = p_0 m_x$			Life-cycle graph			$f_x = m_x\lambda$	
	$\frac{\partial\lambda}{\partial a_{ij}}$	e_i		$\frac{\partial\lambda}{\partial a_{ij}}$	e_i		$\frac{\partial\lambda}{\partial a_{ij}}$	e_i
$p_0 m_1$	0.9487	0.897	m_1	0.2277	**0.8973**		0.0230	0.4605
$p_0 m_2$	0.0452	0.051	m_2	0.0109	**0.0513**		0.0011	0.0264
p_1	$\boxed{1.0768}$	0.051	p_1	1.0768	**0.0513**		$\boxed{0.5526}$	0.0264
			p_0	20.0587	**0.9486**		10.2937	0.4868

Note: a conversion factor for $f_x = m_x\lambda$ was obtained by using the sensitivity for p_1 with $f_x = p_0 m_x$; the two values that were used are shown in boxes; correct elasticities are shown in **boldface**.

m_1 $0.46047 \times 1.94862 = 0.8973,$

m_2 $0.02635 \times 1.94862 = 0.0514,$

p_1 $0.02635 \times 1.94862 = 0.0514,$

and
p_0. $0.48682 \times 1.94862 = 0.9486.$

Using $f_x = m_x\lambda$ creates problems in the sensitivity and elasticity analysis and so probably should not be used for this purpose.; using $f_x = p_0 m_x$ is best. The elasticity for p_0 can be obtained by summing all elasticities that contain p_0 and sensitivities for m_x can be obtained by using Equation 7.4.

Matrix analysis of *Viviparus georgianus*

The life-table data (Table 7.4) for *Viviparus georgianus* can be assembled as a transition matrix, **A**, and analyzed using matrix methods. Values of f_x are defined as $p_0 m_x$ and so

$$A = \begin{pmatrix} 0 & 0.3980 & 1.5408 & 2.6804 \\ 0.639 & 0 & 0 & 0 \\ 0 & 0.565 & 0 & 0 \\ 0 & 0 & 0.307 & 0 \end{pmatrix}. \quad (7.55)$$

Table 7.17 shows the sensitivity and elasticity values and as was shown for our small beast, it is possible to obtain the same sensitivities and elasticities from matrix methods that were obtained directly from the characteristic equation that can be formed from the life-cycle graph. Calculating the elasticity for a parameter that is part of several matrix elements can be done by summing the elasticities for all of the matrix elements that have the parameter as part of their element value. In particular, when f_x is defined as $p_0 m_x$, the elasticity for p_0 will be the sum of the elasticities for all of the f_x elements. When elasticity is known, sensitivity of p_0 can be calculated using Equation 7.4.

GENERAL COMMENTS

Elasticity values provide insight for understanding the relative importance of different transitions in a life cycle and can be useful in understanding life history evolution as well as aiding in formulating strategies for conservation, resource management, and pest control. These aspects are explored in following chapters.

TABLE 7.17 Sensitivity Analysis of Matrix Equation 7.54 for *Viviparus georgianus* using $f_x = p_0 m_x$;

Element	Name	Sensitivity	Elasticity
a_{12}	$p_0 m_2$	0.2044	0.0787
a_{13}	$p_0 m_3$	0.1117	0.1664
a_{14}	$p_0 m_4$	0.0331	0.0859
	p_0	1.0663	0.3309
a_{21}	p_1	0.5357	0.3309
a_{32}	p_2	0.4618	0.2522
a_{43}	p_3	0.2894	0.0859

Note: data based on Buckley (1986).

PROBLEMS

1. Jensen (1971) constructed a Leslie matrix for brook trout.

$$A = \begin{pmatrix} 0 & 0 & 37 & 64 & 82 \\ 0.06 & 0 & 0 & 0 & 0 \\ 0 & 0.34 & 0 & 0 & 0 \\ 0 & 0 & 0.16 & 0 & 0 \\ 0 & 0 & 0 & 0.08 & 0 \end{pmatrix}$$

a. Calculate elasticities for elements of the matrix.
b. Based on the calculated elasticity of p_1, what should be the effect of a 20% increase in p_1?
c. Check your prediction by increasing p_1 by 20% and determining λ. Has it changed according to expectation?

2. Return to the gray squirrel data of Barkalow *et al.* (1970) given in Table 1.8.

a. What elements of the transition matrix have the greatest influence on λ?
b. Draw a life-cycle graph from the gray squirrel data. Use general symbols such as m_1, m_2, p_0, p_1, etc. to constitute the characteristic equation. Determine the partials for $\dfrac{\partial f(\lambda, a_i)}{\partial p_0}$, $\dfrac{\partial f(\lambda, a_i)}{\partial p_1}$ and $\dfrac{\partial f(\lambda, a_i)}{\partial m_1}$ then calculate $\dfrac{\partial \lambda}{\partial p_0}$, $\dfrac{\partial \lambda}{\partial p_1}$ and $\dfrac{\partial \lambda}{\partial m_1}$. Calculate elasticities for these partials and then compare them with elasticities obtained in part a.

3. Determine elasticities for the following data sets; a through e are from Chapter 2 and f is from Chapter 5. Be sure to include the elasticity for p_0.

a. Brousseau and Bagliovo (1988) data on *Mya arenaria*.
b. Strijbosch and Creemers (1988) data on *Lacerta vivipara*.
c. Grant and Grant (1992) data on *Geospiza scandens*.

d. Paloheimo and Taylor (1987) data on *Daphnia pulex*.

e. Melton (1983) data on *Kobus ellipsiprymnus*.

f. Knight and Eberhardt (1985) data on *Ursus arctos horribilis*.

4. Bustamante (1996) explored the risks of extinction of a captive bearded vultures (*Gypaetus barbatus*) held in zoos in Europe and used to repopulate the Alps. The questions was whether current rates of release to the wild endangered the captive breeding population. The average demographic values based on captive birds from 1978 to 1993 are shown in Table 7.18.

a. Assemble a life cycle graph following the primate example (Equation 7.34) but keep juveniles (0 - 1 year) as a separate node.

b. Determine elasticities for all transitions.

c. On average, could a population of 39 breeding females sustain removal of one juvenile female per year for release back into the wild? How about two juvenile females per year?

Table 7.18 Summary of Demographic Values for Bearded Vultures Held in European Zoos.

Age at first reproduction : 7 years		
Age after which adults do not reproduce: 31 years		
Sex ratio at hatching: 50:50		
Fecundity rates		
Maximum clutch size: 4 eggs		
Females laying:	0 eggs	28.14%
	1 egg	21.82%
	2 eggs	46.16%
	3 eggs	3.48%
	4 eggs	0.39%
Mortality rates		
Juveniles (0-1 year)*		65.9%
Immatures (2-6 years)		1.11%
Adults (7-31 years)		3.33%

Note: *includes 46.76% hatching success, 78.93% nestling survival, and 92.43% post-fledging survival; data from Bustamante (1996).

8

Stage-Structured Demography

*A*ll the world's a stage

> W. Shakespeare, *As You Like It*, Act II vii, 139.

INTRODUCTION

Many organisms have survival and reproductive rates that are more dependent on the stage they are in than on their age. For example, most trees, fish, and marine invertebrates produce offspring more as a function of size than of age. Also, it is common in various studies to be able to recognize stages but virtually impossible to identify ages; for example, insect instars usually are easy to classify but ages are difficult though, at least in some cases, not impossible to determine (*e.g.* Neville 1963). The Leslie matrix was extended by Lefkovitch (1965) to represent organisms grouped by stages and so any demographic transition matrix that does not use ages often is referred to as a Lefkovitch matrix

$$A = \begin{pmatrix} s_1 & f_2 & f_3 & \ldots & f_n \\ g_1 & s_2 & 0 & \ldots & 0 \\ 0 & g_2 & s_3 & \ldots & 0 \\ \vdots & \vdots & \vdots & \vdots & \vdots \\ 0 & 0 & \ldots & g_{n-1} & s_n \end{pmatrix} \quad (8.1)$$

Transitions labeled f_x again mean numbers of female offspring that have been adjusted in some fashion such as $f_x = p_0 m_x$ or $m_x \lambda$. The transition s_x is the probability of

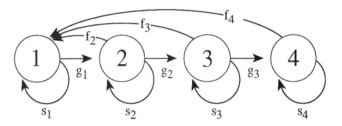

FIGURE 8.1 Life-cycle graph for a stage-structured population showing transitions between successive stages (g_x), transitions of retention within a stage (s_x), and adjusted reproduction rates (f_x); all arcs take one time period to complete.

staying in the same class and g_x is the probability of *transferring* to the next class. The life-cycle graph that corresponds to Equation 8.1 is shown in Figure 8.1.

COMPARISON OF AGE- AND STAGE-STRUCTURE: MEDFLIES

It is instructive to create a stage-structured graph using age-structured data because it shows the meaning of the transitions. Carey (1982) provides data for the Mediterranean fruit fly (*Ceratitis capitata*) taken from Shoukry and Hafez (1979) and these age-structured data (Table 8.1) are used as the basis of a stage-structured analysis.

The life-table begins at x=2 and egg survival is 1.0 (Carey 1982) so the length of the egg stage is 3 days because otherwise l_x would be 1.00 for larvae on day 4. Other survival rates for the stages can be obtained from Table 8.1

TABLE 8.1 Life Table of the Mediterranean Fruit Fly *Ceratitis capitata*, **at 25° C**

Stage	Age (x)	l_x	m_x	Stage	Age (x)	l_x	m_x
Egg	2	1.00	0.00	Adult	40	0.25	25.96
				Adult	42	0.24	25.69
Larva	4	0.83	0.00	Adult	44	0.23	25.42
Larva	6	0.69	0.00	Adult	46	0.22	24.63
Larva	8	0.58	0.00	Adult	48	0.21	23.69
Larva	10	0.48	0.00	Adult	50	0.20	22.63
				Adult	52	0.19	21.56
Pupa	12	0.40	0.00	Adult	54	0.18	20.23
Pupa	14	0.39	0.00	Adult	56	0.17	18.77
Pupa	16	0.38	0.00	Adult	58	0.16	17.44
Pupa	18	0.37	0.00	Adult	60	0.15	15.31
Pupa	20	0.34	0.00	Adult	62	0.14	13.71
Pupa	22	0.32	0.00	Adult	64	0.13	12.11
				Adult	66	0.11	10.38
Adult	24	0.32	0.00	Adult	68	0.10	9.05
Adult	26	0.31	0.00	Adult	70	0.08	7.46
Adult	28	0.31	2.41	Adult	72	0.06	6.39
Adult	30	0.30	7.06	Adult	74	0.04	5.06
Adult	32	0.29	16.51	Adult	76	0.03	4.00
Adult	34	0.28	22.36	Adult	78	0.02	2.80
Adult	36	0.27	25.69	Adult	80	0.02	1.20
Adult	38	0.26	26.49				

Note: m_x is the total number of eggs/2 and so is the number of female on the assumption of a 1:1 sex ratio; table taken from Carey (1982) and based on Shoukry and Hafez (1979).

TABLE 8.2. Summary Statistics for the Mediterranean Fruit Fly *Ceratitis capitata*

Stage	Days in stage	Survival per day, p	Fraction leaving	Fraction staying	g_x	s_x
Egg	3	1.0	0.5	0.5	0.5	0.5
Larva	9	0.90321	0.111111	0.888889	0.100356	0.802853
Pupa	12	0.981576	0.083333	0.916667	0.081798	0.899778
Adult	56	0.951695	0.0	1.0	0.0	0.951695

Note: the eggs take 3 days to hatch but the first day's survival is part of the transit of eggs from the adult node to the egg node and so just 2 days are spent at the node; fraction leaving is 1/days and fraction staying is 1-fraction leaving; g_x = fraction leaving × survival; s_x = fraction staying × survival.; data from Carey (1982) and based on Shoukry and Hafez (1979).

by dividing initial $l_{x+\Delta x}$ by l_x. For example, the survival of the entire pupal stage is

$$\frac{l_{24}}{l_{12}} = 0.32/0.40 = \boxed{0.80}.$$

Notice that l_{24} is used for the last instant of the pupal stage because, by the definition of l_x values, it must be the first instant of the adult stage. The length of the pupal stage is 12 days and so the survival rate *per day* is

$$\sqrt[12]{0.80} = \boxed{0.9816 \text{ day}^{-1}}.$$

The larval stage is 9 days long and the survival rate for the entire period is 0.40 with a *per day* survival of

$$\sqrt[9]{0.40} = \boxed{0.9032 \text{ day}^{-1}}.$$

The adult phase of the life cycle lasts for 56 days and the survival over this period is

$$l_{80}/l_{24} = \frac{0.02}{0.32} = \boxed{0.0625},$$

which gives a daily survival rate of

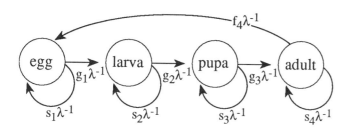

FIGURE 8.2 A Z-transformed graph in which all transitions have been multiplied by λ^{-1} in preparation for constituting the characteristic equation.

$$\sqrt[56]{0.062} = \boxed{0.9517 \text{ day}^{-1}}.$$

If all of the individuals in a particular stage truly are identical, then the fraction that moves to the next node is the reciprocal of the residence time multiplied by the survival rate (Table 8.2). Summing the m_x column in Table 8.1 gives the total number of eggs produced by an average female during a span of 56 days, which is 411 and so the expected number of eggs per day per female is 411/56 or 7.33928, which, multiplied by the expected survival rate for eggs (1.0) gives

$$f_x = \boxed{7.3393}.$$

The g_x, s_x, and f_x values can be assembled into a transition matrix **A** (Equation 8.2)

$$\mathbf{A} = \begin{pmatrix} 0.5 & 0 & 0 & 7.33928 \\ 0.5 & 0.80285 & 0 & 0 \\ 0 & 0.10036 & 0.89978 & 0 \\ 0 & 0 & 0.08180 & 0.95169 \end{pmatrix} \quad (8.2)$$

which, by using **MATRIX.BAS**, has a characteristic equation

$$\lambda^4 - 3.15433\lambda^3 + 3.66994\lambda^2 - 1.85888\lambda + 0.31362 = 0. \quad (8.3)$$

A second approach to formulating the characteristic equation is to work directly from the life-cycle graph. To formulate the characteristic equation, recall (Chapter 4) that the first step is to perform a Z-transformation of the graph by multiply each transition by λ^{-1} as shown in Figure 8.2.

The characteristic equation is obtained by setting the determinant of the Z-transformed graph equal to zero and accounting for the problems introduced by the disjoint loops (Caswell 1989), which are loops that share no nodes. In

Figure 8.2 these are the self-loops; that is, the s_x transitions for staying in the same node.

$$\det (\mathbf{A}) = 1 - \sum_i L_i + \sum_{i,j} {}^*L_iL_j - \sum_{i,j,k} {}^*L_iL_jL_k + \ldots \quad (8.4)$$

The first summation, $\sum_i L_i$, is the sum of all of the loops in the graph. First the major loop passes through all of the nodes and then there are all of the self-loops

$$\sum_i L_i = g_1g_2g_3f_4\lambda^{-4} + (s_1 + s_2 + s_3 + s_4)\lambda^{-1}. \quad (8.5)$$

The summations marked with asterisks (*) in Equation 8.4 add or subtract pairs, triplets, quadruples, etc. of all disjoint loops

pairs

$$\sum_{i,j} {}^*L_iL_j = (s_1s_2 + s_1s_3 + s_1s_4 + s_2s_3 + s_2s_4 + s_3s_4)\lambda^{-2}, \quad (8.6)$$

triples

$$\sum_{i,j,k} {}^*L_iL_jL_k = (s_1s_2s_3 + s_1s_2s_4 + s_1s_3s_4 + s_2s_3s_4)\lambda^{-3}, \quad (8.7)$$

and quadruples

$$\sum_{i,j,k,m} {}^*L_iL_jL_kL_m = s_1s_2s_3s_4\lambda^{-4}. \quad (8.8)$$

Equations 8.5 through 8.8 can be assembled according to Equation 8.4 to form the characteristic equation of Figure 8.2

$$1 - g_1g_2g_3f_4\lambda^{-4}$$
$$- (s_1 + s_2 + s_3 + s_4)\lambda^{-1}$$
$$+ (s_1s_2 + s_1s_3 + s_1s_4 + s_2s_3 + s_2s_4 + s_3s_4)\lambda^{-2}$$
$$- (s_1s_2s_3 + s_1s_2s_4 + s_1s_3s_4 + s_2s_3s_4)\lambda^{-3}$$
$$+ s_1s_2s_3s_4\lambda^{-4} = 0. \quad (8.9)$$

Now multiply through by λ^4 and rearrange a bit to turn Equation 8.9 into the usual form of the characteristic equation.

$$\lambda^4 - (s_1 + s_2 + s_3 + s_4)\lambda^3$$
$$+ (s_1s_2 + s_1s_3 + s_1s_4 + s_2s_3 + s_2s_4 + s_3s_4)\lambda^2$$
$$- (s_1s_2s_3 + s_1s_2s_4 + s_1s_3s_4 + s_2s_3s_4)\lambda$$
$$+ s_1s_2s_3s_4 - g_1g_2g_3f_4 = 0. \quad (8.10)$$

TABLE 8.3 Expansion of Equation 8.15

$$\lambda^2 - s_1\lambda - s_2\lambda + s_1s_2$$
$$\lambda^2 - s_3\lambda - s_4\lambda + s_3s_4$$

$$\lambda^4 - s_1\lambda^3 - s_2\lambda^3 + s_1s_2\lambda^2$$
$$- s_3\lambda^3 + s_1s_3\lambda^2 + s_2s_3\lambda^2 - s_1s_2s_3\lambda$$
$$- s_4\lambda^3 + s_1s_4\lambda^2 + s_2s_4\lambda^2 - s_1s_2s_4\lambda$$
$$s_3s_4\lambda^2 - s_1s_3s_4\lambda - s_2s_3s_4\lambda + s_1s_2s_3s_4.$$

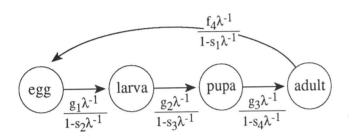

FIGURE 8.3 Self loops, s_x, in Figure 8.1, have been eliminated at each node by dividing $1 - s_x\lambda^{-1}$ into the arc that comes into the node; transit from one node to the next takes one day; each arc has been multiplied by λ^{-1}; $f_4 = $ eggs/female·day^{-1} times p_0.

The values for s_x, g_x and f_x can be substituted into Equation 8.10 to obtain the coefficients. For example, if the values from Table 8.2 are used together with $f_x = 7.33928$, the following equation is obtained:

$$\lambda^4 - 3.154326\lambda^3 + 3.669938\lambda^2 - 1.858882\lambda + 0.313624 = 0, \quad (8.11)$$

which is the same as Equation 8.3 and shows that, indeed, the matrix formulation and the life-cycle graph provide identical results.

The life-cycle graph can be somewhat simplified. Recall (Chapter 4) that self-loops divide all incoming loops by $1 - a_t\lambda^{-1}$ (Figure 8.3). To form the characteristic equation, loops are identified and arcs are multiplied around loops and then all loops are summed and set equal to one. Our Medfly example has just one loop

$$\frac{g_1\lambda^{-1}}{1 - s_2\lambda^{-1}} \times \frac{g_2\lambda^{-1}}{1 - s_3\lambda^{-1}} \times \frac{g_3\lambda^{-1}}{1 - s_4\lambda^{-1}} \times \frac{f_4\lambda^{-1}}{1 - s_1\lambda^{-1}} = 1. \quad (8.12)$$

To simplify Equation 8.12, note that, for example,

$$\frac{g_1\lambda^{-1}}{1 - s_2\lambda^{-1}} = \frac{\frac{g_1}{\lambda}}{\frac{\lambda - s_2}{\lambda}} = \boxed{\frac{g_1}{\lambda - s_2}}$$

so Equation 8.12 can be rewritten

$$\frac{g_1}{\lambda - s_2} \times \frac{g_2}{\lambda - s_3} \times \frac{g_3}{\lambda - s_4} \times \frac{f_4}{\lambda - s_1} = 1$$

or

$$g_1g_2g_3f_4 = (\lambda - s_1)(\lambda - s_2)(\lambda - s_3)(\lambda - s_4). \quad (8.13)$$

The next step is to expand the right side of Equation 8.13

$$(\lambda - s_1)(\lambda - s_2) = \lambda^2 - s_1\lambda - s_2\lambda + s_1s_2 \quad (8.14)$$

and

$$(\lambda - s_3)(\lambda - s_4) = \lambda^2 - s_3\lambda - s_4\lambda + s_3s_4 . \quad (8.15)$$

Now multiply Equation 8.14 by Equation 8.15 to complete the expansion as shown in Table 8.3. Summing and factoring with respect to λ produces

$$\lambda^4 - (s_1+s_2+s_3+s_4)\lambda^3$$
$$+ (s_1s_2+s_1s_3+s_1s_4+s_2s_3+s_2s_4+s_3s_4)\lambda^2$$
$$- (s_1s_2s_3+s_1s_2s_4+s_1s_3s_4+s_2s_3s_4)\lambda$$
$$+ s_1s_2s_3s_4 - g_1g_2g_3f_4 = 0. \quad (8.16)$$

If values of s_x, g_x, and f_x from Table 8.2 are substituted into Equation 8.16, the following characteristic equation is obtained

$$\lambda^4 - 3.15433\lambda^3 + 3.66994\lambda^2 - 1.85888\lambda + 0.31362 = 0, \quad (8.17)$$

which is the same as Equation 8.3 and Equation 8.11. **MATRIX.BAS** can be used either with the transition matrix (Equation 8.2) or with the characteristic equation (Equations 8.11 or 8.17) to obtain all of the roots. The dominant root, λ_1, is 1.2343. For comparison, Carey (1982) calculated the rate of increase per day (λ) to be 1.115.

The difference between the stage structured analysis and age-structured analysis is substantial and can be attributed to a number of factors. First, the stage-structured approach is able more correctly to make use of events that take place at times between time intervals in Table 8.1: it is possible for eggs to hatch in three days rather than an even multiple of 2. However, more important problems are age effects that exist within stages. A newly hatched larva does not have the same chance of becoming a pupa as a larva that has been a larva for eight days.

TABLE 8.4. Summary Statistics for the Mediterranean fruit fly *Ceratitis capitata*

Stage	Days in stage	Survival per day	Fraction leaving	Fraction staying	g_x	s_x
Egg	3	1.0	0.5	0.5	0.5	0.5
Larva	9	0.90321	0.071449	0.928551	0.064532	0.838668
Pupa	12	0.981576	0.075077	0.924923	0.073694	0.907883
Adult	56	0.951695	0.0	1.0	0.0	0.951695

Note: fraction leaving is determined using Equation 8.20 and fraction staying = (1 - fraction leaving); g_x = fraction leaving × survival; s_x = fraction staying × survival.

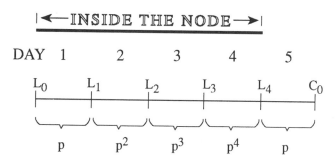

FIGURE 8.4 Diagram of transition from stage L to stage C. It takes 5 days from the beginning of stage L to the instant when the organism enters stage C. Four days would be spent in a node and one day in transit from stage L to stage C. The probability of getting from L_0 to L_4 is p^4.

It is possible to introduce an age-associated condition on being ready to transfer to the next stage by asking: What fraction of members of a particular stage have lived long enough to be ready to enter the next stage? The concept for analysis of the problem is shown in Figure 8.4. L_0 is 1.0, because it is the probability of being alive at the beginning of the stage for individuals that have just entered the stage; it is like l_0 that we have been using in previous chapters. Successive terms in Figure 8.4 are

$$L_0 + L_1 + L_2 + L_3 + L_4 = 1 + p + p^2 + p^3 + p^4$$

and the fraction, f_L, ready to move on to stage C is

$$f_L = \frac{p^4}{1 + p + p^2 + p^3 + p^4}. \tag{8.18}$$

Notice that if $p = 1$ or if there is no age effect and 5 days is the mean time in residence, f_L is 1/5. With an age effect, it is possible to calculate f_L by using the formula for determining the nth partial sum, s_n, of a geometric series

$$\sum_{n=1}^{\infty} ap^{n-1} = a + ap + ap^2 \ldots + ap^{n-1}$$

$$s_n = \frac{a(1-p^n)}{1-p} \qquad p \neq 1. \tag{8.19}$$

For example, with a = 1 and p=0.5 we can calculate the denominator of Equation 8.18 as

$$1 + 0.5 + 0.5^2 + 0.5^3 + 0.5^4 = \boxed{1.9375}$$

or, with Equation 8.19,

$$s_n = \frac{1-0.5^5}{1-0.5} = \boxed{1.9375}.$$

To complete the estimate of the fraction f_L that is ready to pass from stage L to stage C, we would complete the division in Equation 8.18 after determining the denominator either from Equation 8.18 or Equation 8.19

$$f_L = \frac{(1-p)p^{n-1}}{1-p^n}. \tag{8.20}$$

Using our example with p = 0.5, $p^4 = 0.0625$ and so Equation 8.18 is

$$f_L = \frac{0.0625}{1.9375} = \boxed{0.032258}.$$

Entering numbers directly into Equation 8.20 gives the same result

$$f_L = \frac{(1-0.5)0.5^{5-1}}{1-p^5} = \boxed{0.032258}.$$

It now should be clear that the fraction ready to leave a stage can change substantially if suitability of an individual to pass to the next stage depends on age. If there is no age effect and the average time interval is 5 days then the fraction ready on any day is 1/5 or 0.20. If there is an age effect then a smaller fraction of the total number in the stage are ready to move. If p = 0.5 then this fraction is 0.032 compared with 0.20.

Now we can reanalyze the Mediterranean fruit fly data with this additional correction to account for changes in the likelihood that an individual will be ready to move to the

TABLE 8.5 Summary Statistics for the Mediterranean Fruit Fly *Ceratitis capitata*

Stage	Days in stage	Survival per day	Fraction leaving	Fraction staying	g_x	s_x
Egg	3	1.0	0.472813	0.527187	0.472813	0.527187
Larva	9	0.90321	0.041437	0.958562	0.037426	0.865775
Pupa	12	0.981576	0.037597	0.962403	0.036904	0.944672
Adult	56	0.951695	0.0	1.0	0.0	0.951695

Note: fraction leaving is determined using Equation 8.22 and fraction staying is 1-fraction leaving; λ is taken as 1.115 from Carey (1982).

TABLE 8.6 Iteration to Improve λ_1 for the Mediterranean Fruit Fly Classified by Stages using Equation 8.22

Iteration	Old λ	Stage	New g_x	New s_x	New λ
0	1.115	egg	0.472813	0.527187	
		larva	0.037426	0.865775	
		pupa	0.036904	0.944672	
					1.1219010
1	1.1219010	egg	0.4712755	0.5287246	
		larva	0.03621493	0.866995	
		pupa	0.03534524	0.9462308	
					1.12068613
2	1.12068613	egg	0.4715455	0.5284545	
		larva	0.03642575	0.8667842	
		pupa	0.03561517	0.9459608	
					1.12088951
3	1.12088951	egg	0.4715003	0.5284997	
		larva	0.03639038	0.8668196	
		pupa	0.03556984	0.9460062	
					1.12122458
4	1.12122458	egg	0.4714258	0.5285742	
		larva	0.03633218	0.8668778	
		pupa	0.03549532	0.9460807	
					1.12104610
5	1.12104610	egg	0.4714654	0.5285346	
		larva	0.03636317	0.8668468	
		pupa	0.03553500	0.9460410	
					1.12114114
6	1.12114114	egg	0.4714444	0.5285556	
		larva	0.03634334	0.8668634	
		pupa	0.03551385	0.9460621	
					1.12109047
7	1.12109047	egg	0.4714556	0.5285444	
		larva	0.03635546	0.8668545	
		pupa	0.03552514	0.9460509	
					1.12111753
8	1.12111753	egg	0.4714497	0.5285503	
		larva	0.03635076	0.8668592	
		pupa	0.03551911	0.9460569	1.12110309

Note: transitions for adults do not change and the probability of remaining in the adult state is the survival probability, 0.951695.

TABLE 8.7 Roots of the Characteristic Equation Obtained from Transitions at Iteration 8 in Table 8.5

Real	Imaginary	Modulus
1.1211	0.0	
0.4771	0.0	
0.8475	±0.2114	0.8735

TABLE 8.8 Stable-Stage (c_x) and Reproductive-Value (v_x) Vectors for the Mediterranean Fruit Fly using $\lambda = 1.1211$.

Stage	c_x	v_x
Egg	1.0000 (30.12%)	1.0000
Larva	1.8543 (55.85%)	1.2569
Pupa	0.3851 (11.60%)	8.7908
Adult	0.0807 (2.43%)	43.3231

next stage. The new estimates of g_x and s_x are shown in Table 8.4. The transition matrix **A** for the data in Table 8.4 is formed in the usual way by remembering that a transition *from* a node *to* a node means from a *column* to a *row*

$$
\mathbf{A} = \begin{pmatrix} 0.5 & 0 & 0 & 7.3393 \\ 0.5 & 0.83867 & 0 & 0 \\ 0 & 0.06453 & 0.90788 & 0 \\ 0 & 0 & 0.07369 & 0.95170 \end{pmatrix}
$$

and the characteristic equation is

$$
\lambda^4 - 3.19825\lambda^3 + 3.77272\lambda^2 - 1.93643\lambda + 0.34487 = 0.
$$

The population growth rate λ_1 is 1.196 day^{-1}, which is lower than was estimated by assuming no age effects and it is closer to the growth rate determined using an age structured analysis, 1.115 day^{-1}.

A final complication is introduced by the fact that the fraction at a node that is ready to move to the next node is sensitive to population growth rate. If the population is growing then each day adds more individuals than were added the previous day and so the higher the growth rate the smaller the fraction of the terminal age group just before they pass to the next stage. This additional correction really is just completing the Z-transformation of Figure 8.2.

Equation 8.18 can be rewritten to include population growth and these changes (Equations 8.21 and 8.22) are the basis for transitions in Table 8.5

$$
f_L = \frac{(p/\lambda)^4}{1 + p/\lambda + (p/\lambda)^2 + (p/\lambda)^3 + (p/\lambda)^4} \tag{8.21}
$$

or

$$
f_L = \frac{(1-p/\lambda)(p/\lambda)^{n-1}}{1-(p/\lambda)^n}. \tag{8.22}
$$

The transition matrix **A** for the data in Table 8.5 is

$$
\mathbf{A} = \begin{pmatrix} 0.52719 & 0 & 0 & 7.3393 \\ 0.47281 & 0.86577 & 0 & 0 \\ 0 & 0.03743 & 0.94467 & 0 \\ 0 & 0 & 0.03690 & 0.95170 \end{pmatrix}
$$

and the characteristic equation is

$$
\lambda^4 - 3.28933\lambda^3 + 3.99703\lambda^2 - 2.11788\lambda + 0.40555 = 0
$$

which has a dominant root of 1.1219 day^{-1}.

The value of λ_1, 1.1219, is not the same as the value of λ that was used with Equation 8.38 ($\lambda_1 = 1.115$ from Carey 1982) to determine the transitions of g_x and s_x shown in Table 8.5. A better solution for λ_1 can be obtained by iteration. The procedure is to use an estimate of λ_1 to change values of g_x and s_x and iterate until the value of λ_1 from the characteristic equation of the matrix is similar to the value that was used in Equation 8.22. As with all iterative procedures, one can get as close as one wants but never attain an exact solution (Table 8.6).

The difference between the new λ and old λ in iteration 8 is only 1.45×10^{-5} compared with 6.90×10^{-3} at iteration 0. A value of $\lambda = 1.1211$ day^{-1} is a good estimate of the growth rate for the stage-structured analysis. The difference between this value and the growth rate obtained from an age-structured analysis ($\lambda = 1.115$) is only a 0.5% change and probably is due to the fact that all females do not produce the same number of eggs.

Several aspects of the analysis are worth noting. First, when all roots are determined (Table 8.7) it is apparent that the second largest modulus is for the conjugate complex pair of roots and so the damping ratio, ρ, is

$$
\rho = \frac{1.1211}{0.8735} = \boxed{1.28},
$$

which means that return to a stable-stage distribution following disturbance would be fairly slow. Given annual variation in life-table values, it probably would be reasonable to conclude that Medflies are non-equilibrium species and never attain a stable-stage distribution; they

TABLE 8.9 Elasticities of Matrix Elements, a_{ij}, for the Mediterranean Fruit Fly

Element	Name	Description	Value	$\frac{\partial \lambda}{\partial a_{ij}}$	e_{ij}
a_{11}	s_1	remaining as an egg	0.5286	0.0979	0.0416
a_{22}	s_2	remaining as a larva	0.8669	0.2282	0.1764
a_{33}	s_3	remaining as a pupa	0.9461	0.3314	0.2797
a_{44}	s_4	adult survival	0.9517	0.3425	0.2907
a_{21}	g_1	hatching	0.4715	0.1231	0.0518
a_{32}	g_2	pupating	0.0364	1.5960	0.0518
a_{43}	g_3	eclosing	0.0355	1.6334	0.0518
a_{44}	f_4	eggs/female × one day survival of eggs	7.3393	0.0079	0.0518

Note: $\lambda = 1.1211$; data from Carey (1982) and based on Shoukry and Hafez (1979)

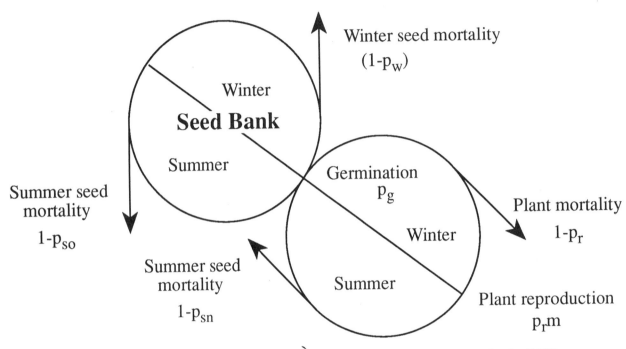

FIGURE 8.5 Life cycle of a plant species with a seed bank following the representation of Schmidt and Lawlor (1983).

always are aimed in that direction but probably never attain it before the environment changes around them.

The stable-stage distribution (Table 8.8) shows the typical distribution for an insect; namely, that the adult stage, which would be the one that usually would be observed, represents a very small fraction (2.43%) of the stable population and the cryptic stages dominate the population structure. It is particularly worth noting this because the adult stage is long-lived relative to the other stages so intuition might lead to the erroneous conclusion that most of the population should be adults.

A final important point emerges from the calculation of elasticities (Table 8.9). Contrary to what might be imagined, eggs/female (f_4) has a relatively small elasticity. A 10% increase in egg production from 7.339 to 8.073, that is adding about one more *female* egg or two eggs total to the daily clutch, would increase λ by only 0.5%. The largest elasticities are for survival of pupae and adults. A 10% change in either s_2 or s_3 would change λ by about 3%. A 10% change in larval survival (s_1) would change λ by about 2%.

Based on the elasticities it is possible to suggest where one should concentrate effort for population control. All things being equal, effort should focus on reducing survival of pupae and adults and, to a slightly lesser extent, survival of larvae. In order to achieve the same changes in λ with modifying egg production, it would be necessary to change clutch size by nearly 100%; however, the changes that are possible depend on the control techniques that are available. Large changes in production of fertile eggs with release of sterile males may be easier to attain than small changes in adult survival.

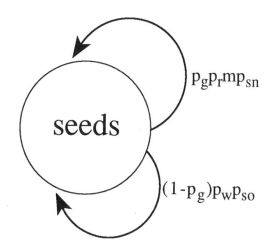

FIGURE 8.6 The model of Schmidt and Lawlor (1983) recast as a graph with one node for a seed bank; p_g is the probability of germination and $(1-p_g)$ is the probability of not germinating so the top arrow is for survival to flowering (p_r), reproduction (m) and seed survival to the next time when germination could occur.

PLANTS WITH SEED BANKS

Schmidt and Lawlor (1983) developed a model for plant species with seed banks (Figure 8.5). In their model, transitions are defined for seeds and for plants:

Seeds

p_w = probability of winter survival of seeds,

p_{so} = probability of summer survival of seeds in seed bank,

p_g = germination probability,

Plants

p_r = probability of winter survival of plants,

m = seeds per individual,

p_{sn} = probability of summer survival of seeds.

Schmidt and Lawlor's model can be rearranged so that it is in the same form as Figure 8.1 and one way of doing this is to use just one stage: namely, seeds (Figure 8.6).

Figure 8.6 can be treated like any other life-cycle graph in order to determine the characteristic equation. Each transition should be multiplied by λ^{-1}, and then the characteristic equation can be formed. There are just two loops and each loop has just one term so

$$\lambda - p_g p_r m p_{sn} - (1-p_g) p_w p_{so} = 0$$

or $\quad \lambda = p_g p_r m p_{sn} - p_w p_{so} + p_g p_w p_{so}$. (8.23)

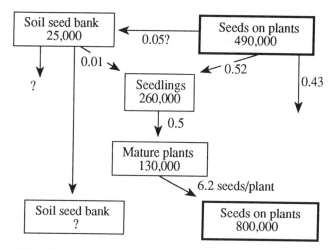

FIGURE 8.7 Flow chart for *Pogogyne abramsii* in vernal pools in San Diego; data from P. Zedler (unpublished).

Figure 8.6 and Equation 8.23 provide a framework for structuring data on a rare and endangered plant, *Pogogyne abramsii,* that lives in drying vernal pools following winter rains in restricted areas near San Diego, California. Figure 8.7 is a representation of population transitions from 1985-1986 in an average pool with an area of 100 square meters (P. Zedler, unpublished).

Transition to the seed bank is estimated as 0.05 so germination rate is 0.95. The transition from seeds to seedlings is 0.52 which includes summer survival and germination

$$0.95 p_{sn} = 0.52, (8.24)$$

or $\quad p_{sn} = 0.547.$ (8.25)

Survival of seedlings to mature plants is
$$p_r = 0.5, (8.26)$$

and $\quad m = 6.2$ seeds/plant. (8.27)

If 25,000 is the number of seeds in the seed bank when 490,000 seeds are produced, then

$$25,000 p_w p_s 0.95 = 5,200,$$

$$p_w p_{so} = 0.2189473,$$

and $\quad p_w = p_{so} = \sqrt{0.2189473} = \boxed{0.468}$. (8.28)

Values from Equations 8.24-8.28 can be substituted into Equation 8.23 in order to determine the population growth rate of the population of seeds. The characteristic equation is

$$\lambda = 0.95 \cdot 0.5 \cdot 6.2 \cdot 0.547 - (1-0.95) \cdot 0.468^2$$

TABLE 8.10 Sensitivities and Elasticities for the Life Cycle of *Pogogyne abramsii*

TABLE 8.10 Sensitivities and Elasticities for the Life Cycle of *Pogogyne abramsii*

Parameter	$\dfrac{\partial \lambda}{\partial a_i}$	Elasticity
p_g	1.4767	0.8649
p_r	3.2218	0.9932
m	0.2598	0.9932
p_{sn}	2.9450	0.9932
p_w	0.0234	0.0067
p_{so}	0.0234	0.0067

Note: data from P. Zedler (unpublished).

or $\qquad \lambda = \boxed{1.6219 \text{ yr}^{-1}}$.

Based on this estimate of λ, 1985-86 was a good year for *Pogogyne abramsii* in San Diego.

A sensitivity analysis can be done to explore how variation in parts of the life cycle would change λ. The characteristic equation is

$$f(\lambda, a_i) = \lambda - p_g p_r m p_{sn} - p_w p_{so} + p_g p_w p_{so}$$

and the partials are

$$\frac{\partial f(\lambda, a_i)}{\partial \lambda} = \boxed{1},$$

$$\frac{\partial f(\lambda, a_i)}{\partial p_g} = -p_r m p_{sn} + p_w p_{so} = \boxed{-1.476676},$$

$$\frac{\partial f(\lambda, a_i)}{\partial p_r} = -p_g m p_{sn} = \boxed{-3.22183},$$

$$\frac{\partial f(\lambda, a_i)}{\partial m} = -p_g p_r p_{sn} = \boxed{-0.259825},$$

$$\frac{\partial f(\lambda, a_i)}{\partial p_{sn}} = -p_g p_r m = \boxed{-2.954},$$

$$\frac{\partial f(\lambda, a_i)}{\partial p_{so}} = -p_w + p_g p_w = \boxed{-0.0234},$$

and $\qquad \dfrac{\partial f(\lambda, a_i)}{\partial p_w} = -p_{so} + p_g p_{so} = \boxed{-0.0234}$.

The calculation of $\dfrac{\partial \lambda}{\partial a_i}$ is easy because $\partial f(\lambda, a_i)/\partial \lambda = 1$ and so, *in this particular case*, $\partial \lambda / \partial a_i = -\partial f(\lambda, a_i)/\partial \lambda$. More generally, of course, Equation 7.7 (Chapter 7) would be used.

TABLE 8.11 Transition Probabilities for Loggerhead Sea Turtles

Stage	Class	Approx. age (yr)	Years in the stage	p_x
0	eggs, hatchlings	<1	1	0.6747
1	small juveniles	1-7	7	0.7857
2	large juveniles	8-15	8	0.6758
3	subadults	16-21	6	0.7425
4	adults	22	30+	0.8091

Note: data from Crouse *et al.* (1987).

Elasticities, $\dfrac{a_i}{\lambda} \dfrac{\partial \lambda}{\partial a_i}$ are given in Table 8.10. The conclusion is that, with respect to proportional change in λ (elasticity),

$$m = p_r = p_{sn} > p_g \gg p_{so} = p_w.$$

Over all, λ is most sensitive to changes in seed production (m), summer seed survival (p_{sn}) and survival of plants following germination in winter (p_r). Changes in the survival of seeds in the seed bank (p_w and p_{so}) would have a very limited effect on λ.

SEA TURTLE DEMOGRAPHY

Sea turtle demography has been analyzed by Crouse *et al.* (1987) with a goal of making recommendations concerning sea turtle conservation. The problems of constituting a life-cycle graph and transition matrix are much like those encountered with the Medfly example. Transitions from one class to the next include the fact that individuals that have just entered the class will remain in the stage for a number of years before they are ready to transfer to the next stage. First, an analysis is done without correcting for λ within a node and then the consequences of this assumption are explored.

The annual survival probabilities, p_x, for five stages are shown in Table 8.11 together with the estimates of the numbers of years in each stage (Frazer 1983a, 1983b). The numbers of eggs produced/year by breeding females is 120 per clutch with a mean of 2.99 clutches per season (Richardson 1982). Assuming a 50:50 sex ratio, the total female eggs, m_A, produced by a reproductive female would be

$$m_A = 120 \times 2.99 \times 0.5 = 179.4.$$

The elements in the matrix for transition from one stage to the next were calculated assuming an age effect. The

TABLE 8.12 Transitions for Loggerhead Sea Turtles

Stage	Years in stage	Survival	Fraction leaving	Fraction staying	g_x	s_x
Eggs, hatchlings	1	0.6747	1.0	0.0	0.6747	0.0
Small juveniles	7	0.7857	0.0618	0.9381	0.0486	0.7371
Large juveniles	8	0.6758	0.0218	0.9782	0.0147	0.6610
Subadults	6	0.7425	0.0698	0.9302	0.0518	0.6907

Note: fraction leaving was determined using Equation 8.29 and fraction staying = (1 - fraction leaving).

TABLE 8.13 Fraction of Current Breeding Females Whose Last Reproduction was 1 to 5 Years Ago

Fraction of breeders	Time since last breeding (years)
0.0358	1
0.4989	2
0.3221	3
0.1119	4
0.0313	5

Note: data from Crouse *et al.* (1987) based on Frazer (1984) for Little Cumberland Island, Georgia.

fraction ready to leave a node, f_L, was determined as was done for Medflies,

$$f_L = \frac{(1-p)p^{n-1}}{1-p^n}. \qquad (8.29)$$

Transition from one stage (node) to the next, g, is the fraction ready to leave, f_L, multiplied by the survival rate, p,

$$g = f_L p \text{ or } p\left(\frac{(1-p)p^{n-1}}{1-p^n}\right) \qquad (8.30)$$

The probability of staying in the stage or node, s, is

$$s = p\left(1 - \frac{(1-p)p^{n-1}}{1-p^n}\right) \qquad (8.31)$$

For example, for stage 2, the small juveniles, p = 0.7857 and n = 7 so

$$g = 0.7857\left(\frac{(1-0.7857)0.7857^{7-1}}{1-0.7857^7}\right) = \boxed{0.0486}.$$

The probability that a small juvenile stays as a small juvenile is

$$s = 0.7857\left(1 - \frac{(1-0.7857)0.7857^{7-1}}{1-0.7857^7}\right) = \boxed{0.7371}.$$

Calculations of retention, s, and transition to the next stage, g, are shown in Table 8.12.

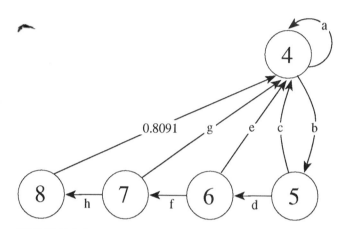

FIGURE 8.8 Transitions from breeding adult sea turtles (node 4) to breeding one year later (transition a), not breeding for 2 years (node 5), not breeding for 3 years (node 6), 4 years (node 7) and 5 years (node 8); annual adult survival is given as 0.8091 and all survival transitions out of a node must sum to 0.8091.

There is a complication in creating the life-cycle graph for loggerhead sea turtles because adult females do not breed every year and are unlikely to breed two years in a row. Frazer (1984) calculated the fraction of females in 5 reproductive classes based on the time since last reproduction (Table 8.13). It is more that just a passing challenge to come up with transitions for non-breeders such that Table 8.13 is satisfied and Figure 8.8 shows the nature of the problem to be solved. The first step in analysis is to create the loops that can be used to estimate the values in Table 8.13 (Table 8.14). The values in Table 8.13 sum to 1.0 and so to match values in Table 8.14, it is necessary to divide each loop by the sum of the loops, T

$$T = a + bc + bde + bdfg + bdfh(0.8091). \qquad (8.32)$$

Several restrictions on finding a solution for the transitions are based on the adult mortality rate of 0.8091. All survival transitions leaving a node must sum to 0.8091, which means that

$$a + b = 0.8091, \qquad (8.33)$$

$$c + d = 0.8091, \qquad (8.34)$$

$$e + f = 0.8091, \qquad (8.35)$$

and $\quad g + h = 0.8091. \qquad (8.36)$

TABLE 8.14 Loops and Transitions to Go from a Breeding Adult Back to a Breeding Adult

Loop	Transitions	Relative value	"Observed" value
1	a	a/T	0.0358
2	bc	bc/T	0.4989
3	bde	bde/T	0.3221
4	bdfg	bdfg/T	0.1119
5	bdfh(0.8091)	bdfh(0.8091)/T	0.0313

Note: time periods can be 1 year (loop 1), 2 years (loop 2), etc.; "observed" values are the estimates from Frazer (1984).

Finding the solution for all transitions can be done using nonlinear regression and the most difficult part is seeing how to combine all the information in Table 8.14 with the restrictions in Equations 8.32 to 8.33. Here is one way of doing it. First determine all of the elements that must enter the regression. These will be the separate values for a through h, because they are separate in Equations 8.31 to 8.35, and the 5 elements that are given as the "relative" values in Table 8.14: a/T, bc/T, bde/T, bdfg/T, and bdfh(0.8091)/T. I have created a single equation that includes all of these elements (Equation 8.36). The values x1 through x13 are dummy variables that are used either as 1 or 0 to show which parts of the equation are used for a particular value of y. For example, for Equation 8.32, y = 0.8091, x1 = 1 and x2 = 1 and all other values of xi from x3 through x13 are 0.

$$y = x1*a + x2*b + x3*c + x4*d + x5*e + x6*f + x7*g + x8*h +$$

$$x9*a/T +$$

$$x10*b*c/T +$$

$$x11*b*d*e/T +$$

$$x12*b*d*f*g/T +$$

$$x13*b*d*f*h*.8091/T \hspace{2cm} (8.37)$$

In order to have a single equation, T in Equation 8.37 must be replaced by a+b*c+b*d*e+b*d*f*g+b*d*f*h*.8091 (Equation 8.38).

$$y = x1*a + x2*b + x3*c + x4*d + x5*e + x6*f + x7*g + x8*h +$$

$$x9*a/(a+b*c+b*d*e+b*d*f*g+b*d*f*h*.8091) +$$

$$x10*b*c/(a+b*c+b*d*e+b*d*f*g+b*d*f*h*.8091) +$$

$$x11*b*d*e/(a+b*c+b*d*e+b*d*f*g+b*d*f*h*.8091) +$$

$$x12*b*d*f*g/(a+b*c+b*d*e+b*d*f*g+b*d*f*h*.8091) +$$

$$x13*b*d*f*h*.8091/(a+b*c+b*d*e+b*d*f*g+b*d*f*h*.8091)$$

$$(8.38)$$

The data set is shown in Table 8.15. I used NONLIN in SYSTAT (1992) to estimate transitions with the model specified by Equation 8.38 and the data in Table 8.15. It also would be possible to use the BASIC program **SIMPLEX. BAS**, which is given at the end of Chapter 12. Results are shown in Table 8.16.

Transitions provide estimates of fractions from the breeding and each non-breeding category that enter the breeding class (node 4 in Figure 8.8). These are the relative or "expected" values in Table 8.17. The values reported by Frazer (1984) for Little Cumberland Island, Georgia, are listed as "observed". The fit is perfect, which is not very surprising because there are 8 unknowns and just 9 equations.

TABLE 8.15 Data for Estimating Transitions a — h for Adult Sea Turtles (Figure 8.8)

y	x1	x2	x3	x4	x5	x6	x7	x8	x9	x10	x11	x12	x13
0.8091	1	1	0	0	0	0	0	0	0	0	0	0	0
0.8091	0	0	1	1	0	0	0	0	0	0	0	0	0
0.8091	0	0	0	0	1	1	0	0	0	0	0	0	0
0.8091	0	0	0	0	0	0	1	1	0	0	0	0	0
0.0358	0	0	0	0	0	0	0	0	1	0	0	0	0
0.4989	0	0	0	0	0	0	0	0	0	1	0	0	0
0.3221	0	0	0	0	0	0	0	0	0	0	1	0	0
0.1119	0	0	0	0	0	0	0	0	0	0	0	1	0
0.0313	0	0	0	0	0	0	0	0	0	0	0	0	1

Note: females do not breed every year and transitions are for paths that lead from breeding back to breeding with lags of 1 to 5 years.

$$A = \begin{pmatrix} 0.7371 & 0 & 0 & 121.041 & 0 & 0 & 0 & 0 \\ 0.0486 & 0.6610 & 0 & 0 & 0 & 0 & 0 & 0 \\ 0 & 0.0147 & 0.6907 & 0 & 0 & 0 & 0 & 0 \\ 0 & 0 & 0.0518 & 0.0203 & 0.3582 & 0.5128 & 0.6012 & 0.8091 \\ 0 & 0 & 0 & 0.7888 & 0 & 0 & 0 & 0 \\ 0 & 0 & 0 & 0 & 0.4509 & 0 & 0 & 0 \\ 0 & 0 & 0 & 0 & 0 & 0.2963 & 0 & 0 \\ 0 & 0 & 0 & 0 & 0 & 0 & 0.2079 & 0 \end{pmatrix} \qquad (8.39)$$

TABLE 8.16. Transitions Between Breeding and Non-Breeding Adult Loggerhead Sea Turtles

label	transition
a	0.020274
b	0.788826
c	0.358165
d	0.450935
e	0.512798
f	0.296302
g	0.601243
h	0.207857

Note: transition labels refer to Figure 8.8; a is the transition for females that breed two years in a row; transition b is for females that breed and do not breed the next year, etc.

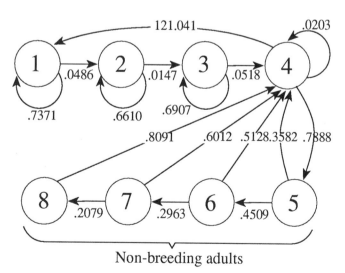

FIGURE 8.9 Life cycle graph for loggerhead sea turtles based on data provided by Frazer (1984) and Richardson (1982) and summarized by Crouse *et al.* (1987); $f_x = p_0 m_x$

Estimates of transitions for return breeders can be combined with reproduction and survival through the sub-adult stage to create the complete life-cycle graph (Figure 8.9), which can be converted to a matrix in the usual manner by establishing as many columns and rows as nodes in the life-cycle graph and then using the rule that a transition *from* a node *to* a node means *from* a column *to* a row. The matrix shown in Equation 8.39 is the same as Figure 8.9. The following data file for the matrix elements in Equation 8.39 can be analyzed using the program **MATRIX.BAS**

```
8, 8
1, 1, .7371
1, 4, 121.041
2, 1, .0486
2, 2, .6610
3, 2, .0147
3, 3, .6907
4, 3, .0518
4, 4, .0203
4, 5, .3582
4, 6, .5128
4, 7, .6012
4, 8, .8091
5, 4, .7888
6, 5, .4509
7, 6, .2963
8, 7, .2079
0, 0, 0
```

The dominant root for Equation 8.39, $\lambda_1 = 0.9406$ can be used to create a new life-cycle graph using $f_x = m_x \lambda$ (Figure 8.10). The appropriate matrix for Figure 8.10 is Equation 8.40.

TABLE 8.17. Transition Estimates for Breeding and Non-Breeding Sea Turtles

Transition	Estimate	Loop	Loop value	Relative = expected	"Observed"
a	0.020274	a	0.020274	0.0358	0.0358
b	0.788826	bc	0.282530	0.4989	0.4989
c	0.358165	bde	0.182407	0.3221	0.3221
d	0.450935	bdfg	0.063369	0.1119	0.1119
e	0.512798	bdfh.8091	0.017725	0.0313	0.0313
f	0.296302	total	0.566305		
g	0.601243				
h	0.207857				

Note: equation 8.36 was used as the model in NONLIN (SYSTAT 1992) together with data presented in Table 8.14; relative (=expected) values are the loop values/T; T = the sum of all of the loops, which is 0.566305.

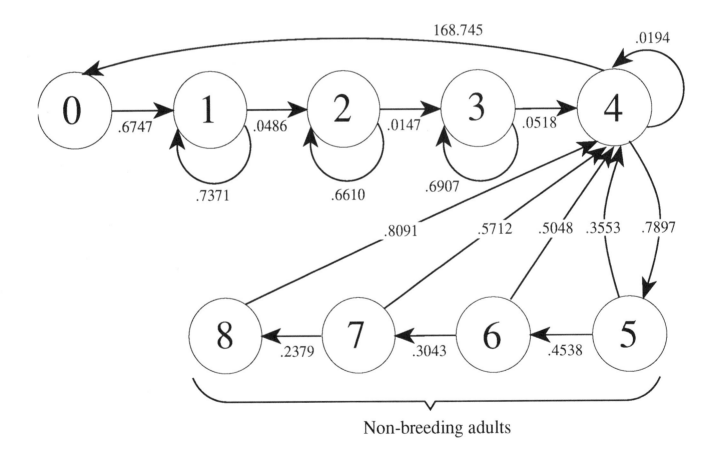

FIGURE 8.10 Life cycle graph for loggerhead sea turtles based on data provided by Frazer (1984) and Richardson (1982) and summarized by Crouse *et al.* (1987); $f_x = m_x\lambda$.

$$A = \begin{pmatrix} 0 & 0 & 0 & 0 & 168.745 & 0 & 0 & 0 & 0 \\ 0.6747 & 0.7371 & 0 & 0 & 0 & 0 & 0 & 0 & 0 \\ 0 & 0.0486 & 0.6610 & 0 & 0 & 0 & 0 & 0 & 0 \\ 0 & 0 & 0.0147 & 0.6907 & 0 & 0 & 0 & 0 & 0 \\ 0 & 0 & 0 & 0.0518 & 0.0203 & 0.3582 & 0.5128 & 0.6012 & 0.8091 \\ 0 & 0 & 0 & 0 & 0.7888 & 0 & 0 & 0 & 0 \\ 0 & 0 & 0 & 0 & 0 & 0.4509 & 0 & 0 & 0 \\ 0 & 0 & 0 & 0 & 0 & 0 & 0.2963 & 0 & 0 \\ 0 & 0 & 0 & 0 & 0 & 0 & 0 & 0.2079 & 0 \end{pmatrix} \quad (8.40)$$

The data file for Equation 8.40 must be formed by starting index values at 1 rather than at 0, which means that column and row numbers start at 1 and go to 9 rather than from 0 to 8. This is not difficult but does require some care in creating a data file using a word processor. The data file for Equation 8.40 to be used with **MATRIX.BAS** is

```
9, 9
1, 5, 168.745
2, 1, .6747
2, 2, .7371
3, 2, .0486
3, 3, .6610
4, 3, .0147
4, 4, .6907
5, 4, .0518
5, 5, .0203
5, 6, .3582
5, 7, .5128
5, 8, .6012
5, 9, .8091
6, 5, .7888
7, 6, .4509
8, 7, .2963
9, 8, .2079
0, 0, 0
```

The coefficients of the characteristic equation are similar for Figures 8.9 and 8.10 (Table 8.18). The coefficient, for λ^0, $-2.512487 \times 10^{-21}$, is sufficiently close to 0 that it can be removed and make the characteristic equation an 8th order polynomial rather than a 9th. Using $f_x = p_0 m_x$ or $f_x = m_x \lambda$ provided approximately the same roots (Table 8.19) and so

TABLE 8.18 Coefficients of the Characteristic Equations for Figure 8.9 ($f_x = p_0 m_x$) and Figure 8.10 ($f_x = m_x \lambda$)

Exponent of λ	Coefficients: $f_x = p_0 m_x$	Coefficients: $f_x = m_x \lambda$
8	1	1
7	-2.1091	-2.1091
6	1.212745	1.212745
5	0.04178	0.04178
4	-0.090546	-0.086067
3	-0.055291	-0.059504
2	0.006355	0.006355
1	-0.004434	-0.004434
0	0.005966	0.005966
delete		-2.5125×10^{-21}

Note: the two underlined values are for coefficients that differed with the two different ways of forming f_x.

interpretations of the results would not change. For example, the dominant root is 0.94061 for both methods and the second largest modulus is 0.733519 for $f_x = p_0 m_x$ and 0.750262 for $f_x = m_x \lambda$. The associated damping ratios, ρ, are 1.28 and 1.25 respectively. The length of time it would take for the dominant root to have 1000× the importance of the second largest root is

$$t = \frac{ln(1000)}{ln(\rho)}$$

or

TABLE 8.19 Comparison of Roots of the Characteristic Polynomial Using $f_x = p_0 m_x$ or $f_x = m_x \lambda$;

$f_x = p_0 m_x$		$f_x = m_x \lambda$	
Real	Imaginary	Real	Imaginary
0.940609	0	0.940608	0
0.708788	±0.188864	0.722359	±0.20271
0.539803	0	0.515688	0
-0.03965	±0.356993	-0.042242	±0.35636
-0.354793	±0.208272	-0.353716	±0.211069
		1.1655 × 10⁻¹⁸	0

Note: with $f_x = p_0 m_x$ there are 8 roots and with $f_x = m_x \lambda$ there are 9; the root 1.1655×10^{-18} is best considered to be 0 and can be ignored.

$$\frac{6.907755}{0.248674} = \boxed{28 \text{ years for } f_x = p_0 m_x}$$

and

$$\frac{6.907755}{0.226104} = \boxed{30 \text{ years for } f_x = m_x \lambda}$$

With either definition of f_x, the conclusion is that loggerhead turtles would take a long time to achieve a stable-stage distribution.

The stable-stage and reproductive value distributions (Figure 8.11) show that most individuals would be in the pre-reproductive classes. Only 0.3% are adults and, of these, only 42% breed during any year. The reproductive value, as expected, peaks at the breeding-adult stage, drops and then increases slightly as each year passes without breeding. The reason for this is clear in Figure 8.10 which shows that the probability of breeding increases from node 5 to node 8.

An error is introduced into the analysis because of the assumption in Equation 8.29 that $\lambda = 1$ when, in fact, it seems to be less than one. As with the Medfly data, it is possible to iterate to improve the estimate of λ by performing a Z-transformation within each node. The formula for fraction leaving a node, f_L, is

$$f_L = \frac{(1-p/\lambda)(p/\lambda)^{n-1}}{1-(p/\lambda)^n}. \tag{8.41}$$

Transfer to the next node, g_x, and retention in a node, s_x, are

$$g_x = p f_L$$

and $s_x = p(1-f_L)$

for small juveniles, large juveniles, and subadults. Calculating values for g_x and s_x is tedious when done by hand, but is made tolerable with a very simple BASIC program,

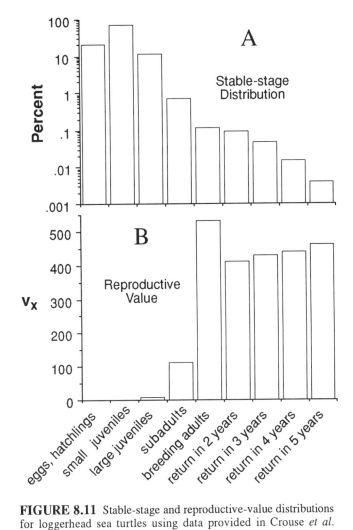

FIGURE 8.11 Stable-stage and reproductive-value distributions for loggerhead sea turtles using data provided in Crouse *et al.* (1987).

```
20 Input "lambda";L
30 Input "number of time periods in a
   node";N
40 Input "survival rate: ";p
50   fL=(1-p/L)*(p/L)^(N-1):fL=fL/(1-
     (p/L)^N)
60 g=fL*p:s=(1-fL)*p
70 print fL,1-fL,g,s
80 goto 30
```

which provides estimates that are needed to modify a file used with **MATRIX.BAS**. Table 8.20 shows the sequence followed in improving the estimate of λ for loggerhead sea turtles. The value of λ after six iterations was 0.962479, which indicates that λ is about 0.96yr⁻¹. The best estimate that can be attained is based on the limit of calculating λ_{i+1} given λ_i. Figure 8.12 plots the iterations shown in Table 8.20 and indicates that the iterations are converging on a value where $\lambda_{i+1} = \lambda_i$. This value can be approximated by

TABLE 8.20 Improving the Estimate of λ by Accounting for λ≠1 within Nodes that have Retentions of More than One Year

n	p	yrs	(1) λ=0.94061		(2) λ=0.975826		(3) λ=0.956128	
			s_X	g_X	s_X	g_X	s_X	g_X
1	0.7857	7	0.7243	0.0614	0.7323	0.0534	0.7280	0.0577
2	0.6758	8	0.6556	0.0204	0.6590	0.0168	0.6572	0.0186
3	0.7425	6	0.6793	0.0632	0.6863	0.0562	0.6825	0.0600

n	p	yrs	(4) λ=0.966489		(5) λ=0.960910		(6) λ=0.963861	
			s_X	g_X	s_X	g_X	s_X	g_X
1	0.7857	7	0.7303	0.0554	0.7290	0.0567	0.7297	0.0560
2	0.6758	8	0.6582	0.0176	0.6576	0.0181	0.6579	0.0179
3	0.7425	6	0.6845	0.0580	0.6834	0.0591	0.6840	0.0585

Note: ;the model used is $f_X = m_X\lambda$ and so in addition to changing values for g_X and s_X, f_X also is changed for each iteration; n = node in Figure 8.10; iterations number is in **bold**.

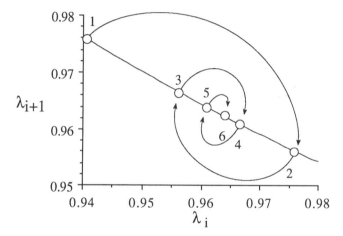

FIGURE 8.12 Convergence to a solution for λ when λ is included in the calculation of g and s for loggerhead sea turtles; convergence is at $\lambda = 0.963$ yr^{-1}.

first obtaining parameters of a function that fits the data points of λ_{i+1} as a function of λ_i. I chose a second-order polynomial,

$$\lambda_{i+1} = 3.2159 - 4.1393\lambda_i + 1.8688\lambda_i^2,$$

and the solution is where

$$\lambda_{i+1} = \lambda_i$$

or

$$1.8688\lambda_i^2 - 5.1393\lambda_i - 3.2159 = 0.$$

There are two roots but the one we want is $\lambda_1 = 0.96288$ yr^{-1}.

Elasticity values are shown in Table 8.21. Other than the elasticity for m_X, all elasticities are for survival rates and so can be summed to show the importance of survival of different stages with respect to changing λ (Figure 8.13).

TABLE 8.21 Elasticities for the Life-Cycle Graph of Loggerhead Sea Turtles Shown in Figure 8.10

Element	Elasticity
m_X	0.0619
Hatchling to small juvenile	0.0619
Remain small juvenile	0.1940
Small juvenile to large juvenile	0.0619
Remain large juvenile	0.1337
Large juvenile to subadult	0.0619
Remain subadult	0.1520
Subadult to adult	0.0619
Return breeder, 1yr	0.0035
Return breeder, 2yr	0.0505
Return breeder, 3yr	0.0339
Return breeder, 4yr	0.0122
Return breeder, 5yr	0.0036
Delay 4-5	0.1001
Delay 5-6	0.0496
Delay 6-7	0.0158
Delay 7-8	0.0036

Note: s_X are retention at a node and g_X is transmission to the next node.

GENERAL COMMENTS

Stage-structured analysis can be a useful technique in cases where age is less important in predicting fecundity or survival than some other measure of an organism such as "juvenile" or "reproductive *vs.* non-reproductive" adult. There are problems with the analysis. There is no general agreement on how to treat elasticities for g and s. Some authors (*e.g.* Silvertown *et al.* 1993) keep the elasticities of g and s separate and call g "growth" and s "survival." My

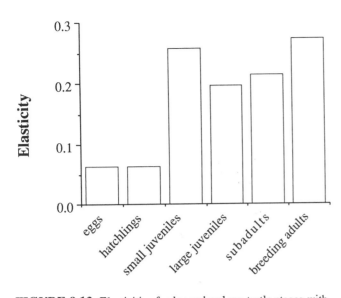

FIGURE 8.13 Elasticities for loggerhead sea turtle stages with all transitions for a stage combined.

TABLE 8.22 Survival and Fecundity Schedules for Blanding's Turtle *Emydoidea blandingii*

Age	$p_x \, yr^{-1}$	$m_x \, yr^{-1}$
0	0.2610	0
1-13	0.7826	0
14	0.9600	1
15	0.9600	2
16	0.9600	3
17+	0.9600	4

Note: data from Congdon *et al.* (1993).

a. Construct a life-cycle graph to show the transitions and then create the appropriate matrix.
b. Complete the analysis by determining r, the stable-stage and reproductive-value vectors, ρ, and elasticities.
c. Based on elasticities, where should conservation efforts be placed?

2. Snapping turtles (*Chelydra serpentina*) were studied by Congdon *et al.* (1994) from 1975 through 1992 in southeastern Michigan in the same reserve where Blanding's turtles were studied (Problem 1). The mean clutch size for fully mature females is 28 eggs. Females do not necessarily reproduce each year and the probability of reproduction is 0.85 (min-max = 0.77 to 0.88). If a 50:50 sex ratio is assumed, annual fecundity is 12 (Table 8.23).

TABLE 8.23 Survival and Fecundity Schedules for Snapping Turtles *Chelydra serpentina*

Age	$p_x \, yr^{-1}$	$m_x \, yr^{-1}$
0-1	0.23	0
1-2	0.47	0
2-11	0.777	0
11	0.82	4
12	0.82	6
13	0.93	8
14	0.93	10
15+	0.93	12

Note: data from Congdon *et al.* (1994).

a. Based on analysis of elasticities, how would you approach snapping turtle conservation? Would your approach be different from that used for Blanding's turtle in problem 1?
b. Construct a life-cycle graph that has a non-reproductive class for females so that you incorporate the 0.85 probability that a female reproduces each year. How does this change in the graph change stability characteristics (damping ratio and oscillatory behavior)?

sense is that both are just survival because Equation 8.31 for s ("survival") is the same as Equation 8.31 for g ("growth") except for the difference between the probability of leaving a node versus the probability of staying in a node. These probabilities are determined by how many age classes are contained in a node and so it is possible to change g and s just by changing the number of age-classes that are present with a limit where a stage is equal to one year and the fraction leaving is 1.0 and fraction remaining is 0. In such a case, there would be only g values and no s values and so the only elasticities would be for growth; there would be no elasticities for survival. This problem becomes even more complicated when size categories are used for analysis, which is the topic of the next chapter.

PROBLEMS

1. Congdon *et al.* (1993) provide a life table for Blanding's turtle (*Emydoidea blandingii*) with 110 age classes. It would be possible to do an analysis using age structure, but most matrix programs will flat refuse to deal with a matrix of order 110. The analysis could be done using the Euler equation (but you wouldn't get elasticities) OR, and I think much better, it could be done as a stage-structured analysis with a shortened life table (Table 8.22). The annual fecundity of 4 eggs is based on a mean clutch size of 10 eggs, a reproductive frequency of 0.85, and the assumption that the sex ratio is 1:1.

3. Table 8.24 shows data for the Antarctic krill *Euphausia superba* based on Astheimer *et al.* (1985) and Astheimer (1986). Some values presented by Astheimer *et al.* (1985) are hypothetical and based on Mauchline (1980); other estimates are based on Fraser (1936) and Kikuno 1982.

a. Construct a stage-structured life-cycle graph and transition matrix and calculate λ.

b. Determine the stable-stage and reproductive-value distributions.

c. Do a sensitivity analysis to determine which stages most influence λ; that is, calculate elasticities and interpret them.

d. Which parts of the life cycle require study in greater detail and which could vary substantially with little effect on λ? There are two parts to this. The first is your decision as to which estimates of survival and fecundity are most poorly know and, secondly, of those that are poorly known, which have substantial influence on λ?

TABLE 8.24 Growth of Krill *Euphausia superba*

Molt no.	Stage	L	Age	Molt no.	Stage	L	Age
0	egg	0.6	0.0	19	Adol.	22.6	204.8
1	N1	0.6	8.1	20	Adol.	24.6	219.3
2	N2	0.7	16.6	21	Adol.	26.7	234.5
3	Mn	0.9	25.4	22	Adol.	28.8	250.3
4	C1	1.7	34.5	23	Adol.	31.0	266.8
5	C2	2.7	43.8	24	Adol	33.2	284.1
6	C3	4.0	53.5	25	Adol.	35.5	302.2
7	f1	5.3	63.3	26	Adol.	37.7	321.2
8	f2	6.1	73.4	27	Adult	40.0	341.1
9	f3	7.3	83.8	28	Adult	42.3	361.9
10	f4	8.0	94.4	29	Adult	44.6	383.8
11	f5	9.5	105.3	30	Adult	46.8	406.7
12	f6	11.3	116.5	31	Adult	49.1	430.7
13	Adol.	12.5	128.0	32	Adult	51.3	455.8
14	Adol.	13.9	139.8	33	Adult	53.5	482.2
15	Adol.	15.5	151.9	34	Adult	55.7	509.8
16	Adol.	17.1	164.4	35	Adult	57.9	538.6
17	Adol.	18.9	177.4	36	Adult	60.0	568.8
18	Adol.	20.7	190.9	37	Adult	62.0	600.3

Note: larval stages are: N = nauplius; Mn = metanauplius; C = calyptopis, F = furcilia; Adol = pre-reproductive; lengths, L, are millimeters; ages are days; data from Astheimer *et al.* (1985).

TABLE 8.25 Stage-specific Mortality (d) and Survival (p) Rates for *Euphausia superba*

Stage	d day^{-1}	p day^{-1}
C and F	0.117 to 0.134	0.88958 to 0.87459
Subadults	0.007	0.99302
Adults	0.002	0.99800

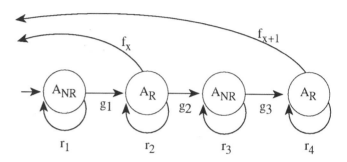

FIGURE 8.14 Suggestion for structuring a life-cycle graph for krill to separate reproductive and non-reproductive nodes; A_{NR} = non-reproductive adults and A_R = reproductive adults.

The *total* number of eggs (E) produced by a female L millimeters long during the spawning season is

$$E = 330.2L - 7714.0.$$

Stage-specific survival rates (Table 8.25) must be calculated from the instantaneous mortality coefficients (d)

$$N_t = N_0 e^{-dt}$$

and the survival rate, p, is

$$p = \frac{N_t}{N_0} = e^{-dt}$$

Mauchline (1965) sampled in such a manner that eggs and nauplii were under-represented in net catches and so there are no direct estimates of survival rates for these stages. Astheimer (1986) assumed that these stages had survival rates similar to the calyptopis and furcilia stages.

Hints and suggestions:

1. There is no particular reason why you should use a time-constant of one day. You might use 90 days to match the length of the breeding season or use one year to match the annual cycle. Do what seems right to you. If you are consistent, then it shouldn't make any difference what time-constant you choose.

2. First assume no age-effects and so all members of a stage can be treated as identical units and then redo the analysis assuming age-effect within a stage.

3. Because krill reproduce just during the breeding season you might want to construct the life-cycle graph so adults have nodes for both the reproductive and non-reproductive seasons (Figure 8.14).

4. Include λ within nodes by using Equation 8.41 to calculate the fraction ready to leave and fraction remaining.

4. Return to problem in Chapter 2 on the human louse *Pediculus humanus*. Analyze the data gathered by Evans and Smith as using a stage-structured life cycle

   a. First draw an appropriate life-cycle graph and calculate the transitions assuming structure within each node.

   b. Determine λ and then improve it using the approach presented in the Med fly example.

5. Conduct a stage-structured analysis using the data for Adélie penguins (Ainley *et al.* 1983) that were presented as problem 7 in Chapter 5.

6. *Phlox drummondii* is a winter annual plant that is a conspicuous element of the spring flora of south Texas. Leverich and Levin (1979) studied a population near the northern edge of Nixon, Texas (Table 8.26); m_x is seed production during the time interval and l_x, as usual, is survival to the beginning of the interval; ages 0 through 184 days are when seeds are dormant; ages and length of intervals are in days.

   a. Determine population growth rate using a single stage (i.e. one node); namely, seeds. [*Hint*] You will have to calculate 8 separate transits and then sum them. The first transit will be survival from time 0 to the midpoint of the interval 299-306 *times* m_x *times* survival to the end of the year. The second transit will be survival from 0 to the midpoint of the interval 306-313 *times* m_x *times* survival to the end of the year, etc.

   b. Write the characteristic equation and determine elasticities for m_x and survival rates.

TABLE 8.25 Life Table for *Phlox drummondii*

Age	Interval	l_x	m_x
0-63	63	1.0000	
63-124	61	0.6707	0
124-184	60	0.2962	0
184-215	31	0.1908	0
215-231	16	0.1767	0
231-247	16	0.1747	0
247-264	17	0.1737	0
264-271	7	0.1727	0
271-278	7	0.1707	0
278-285	7	0.1677	0
285-292	7	0.1657	0
292-299	7	0.1596	0
299-306	7	0.1586	0.3394
306-313	7	0.1546	0.7963
313-320	7	0.1516	2.3995
320-327	7	0.1476	3.1904
327-334	7	0.1365	2.5411
334-341	7	0.1054	3.1589
341-348	7	0.0743	8.6625
348-355	7	0.0221	4.3072
355-362	7	0.0000	0.0000

Note: times are days; data from Leverich and Levin (1979).

7. Here is a problem that in many ways is very similar to the problem with *Phlox drummondii*. *Gemma gemma* is a suspension-feeding bivalve and its life-cycle (Figure 8.15) was analyzed by (Weinberg *et al.* 1986). *Gemma gemma* broods its young and releases two broods during a year. Juvenile clams that survive the winter will, on average, produce their first clutch in July which show-up in node-1 in October with a survival rate p_0. A second clutch is released in October. For 1-year animals that survive the winter, the first clutch of the year is produced in April and the second, and last clutch, in July. According to results presented by Weinberg *et al.* (1986) none of these 1-year old clams survive to reproduce in October. Data for 1979 are shown in Table 8.27.

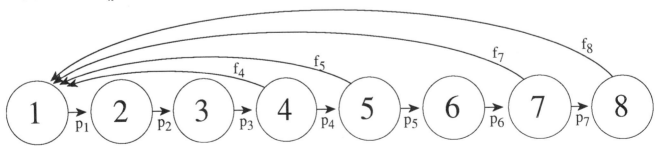

FIGURE 8.15 Life cycle of a suspension-feeding bivalve *Gemma gemma* in Connecticut based on Weinberg *et al.* (1986); time constant is 3 months starting with October as node 1.

TABLE 8.27 Survival and Fecundity Estimates for *Gemma gemma* **in Connecticut for 1979**

Age (months)	p_x	m_x
0	0.334	0
3	0.652	0
6	0.610	0
9	0.889	0
12	0.344	11.71
15	1.000	15.68
18	0.206	0
21	0.585	18.89
24	0	32.90

Note: data from Weinberg *et al.* (1986).

Analyze the data presented by Weinberg *et al.* (1986). *Hint:* it isn't quite as easy as it first appears. Start by drawing nodes that represent times of the year and make appropriate transitions. For example, 1-year old clams first reproduce in July and because the time constant is 3 months, young clams appear in the October node. When do young clams appear in the life-cycle graph that are the result of the fall reproductive event? If you don't quite see how to proceed, try the *Phlox* problem first and then come back to this problem.

8. The gray mangrove, Avicennia marina, was studied in southeastern Australia by Clarke (1992) and Clarke and Allaway (1993) and demographic data were synthesized by Clarke (1995). Table 8.28 shows Clarke's summary for two different levels of disturbance. Type a disturbance is small scale with a size of a few tree crowns in area whereas type b is larger (>0.1 hectare) and due to creation of new habitat by sedimentation or by storms, disease, clearing, etc.

a. Draw the life-cycle graphs for the two disturbance types and then create the two transition matrices.

b. Determine λ and elasticities for each disturbance type. Do elasticities shift with different disturbances so that different parts of the life cycle are more important for λ when the population is perturbed by small *vs.* large disturbances?

TABLE 8.28 Annual Transition Probabilities and Fecundities for the Gray Mangrove, *Avicennia marina,* **in Southeastern, Australia**

Stage	g_x	s_x	m_x
Type a disturbance (small)			
Propagules	0.200	0.000	0
Cotyledonary seedlings	0.083	0.666	0
Seedlings	0.010	0.825	0
Saplings	0.073	0.909	
Young tree	0.008	0.963	0
Tree	0.012	0.980	500
Older tree	0.000	0.999	1000
Type b disturbance (large)			
Propagules	0.200	0.000	0
Cotyledonary seedlings	0.083	0.666	0
Seedlings	0.010	0.825	0
Saplings	0.073	0.909	
Young tree	0.008	0.963	0
Tree	0.012	0.980	500
Older tree	0.000	0.999	1000

Note: data from Clarke (1995).

9

Size-Structured Demography

I don't keep the same size for ten minutes together.

Lewis Carroll,
Alice's Adventures in Wonderland, 1865

INTRODUCTION

A special case of stage-structured analysis is where organisms are classified by size so that stages are artificial categories rather than more natural ones such as egg, larva or pupa. There is no fundamental difference between the analysis of natural stages and artificial stages such as arbitrary size categories but size classification permits certain transitions that are unusual if not impossible where stages are natural. For example, organisms can shrink to smaller sizes and they also can leap-frog over size categories. Figure 9.1 and Equation 9.1 illustrate possible transitions for the same life cycle.

$$A = \begin{pmatrix} r_1 & f_2+s_{12} & 0 & f_4 \\ g_1 & r_2 & 0 & s_{13} \\ g_{31} & g_2 & r_3 & 0 \\ 0 & 0 & g_3 & r_4 \end{pmatrix} \qquad (9.1)$$

A complication in Equation 9.1 is that elements can have two or more parts. A single element, a_{ij}, can have sexual reproduction, f_x, shrinkage, s_x, and, possibly, retention, r_x. Where two or more transitions share the same location in the matrix, they are added together as shown for element a_{12} in

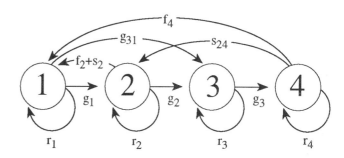

FIGURE 9.1 Some possible transitions in a size-classified life-cycle graph; r_x is retention within a size class; g_x is growth to another size class, which may skip one or more classes; s_x is shrinkage to a smaller size; f_x is size-specific fecundity ($p_0 m_x$ or $m_x \lambda$); all transitions are adjusted so it takes one time unit to traverse any arc.

Equation 9.1, where f_2 is a reproduction and s_{12} a shrinkage from size class 2 to size class 1.

In doing size-structured analysis, it is necessary to make decisions about how many size classes to have and what boundaries to choose for each class. The nature of the data will drive these decisions. The more size classes created, the finer the resolution of the elasticity analysis; a large number of size categories, however, also means fewer numbers in each class. The more individuals in each class, the more confidence one can place in the transition estimates and if the transitions are complex with growing and shrinking 0, 1, 2, 3 or more categories, then more individuals must be observed in order to have sufficient numbers in each category. Creating a size-based transition matrix is something of an art even with the rules given by

TABLE 9.1 Growth and Survival Transitions and Size-Specific Seed Production of *Araucaria cunninghamii* Near Bulolo, Papua New Guinea; A, Transfers up to a Height of 300 cm; B, Transfers Starting at 2 cm dbh.

A		Height (cm)		
	seeds	0-20	20-100	100-300
N/ha	1150595	22912	929	324
p_x	0.01646	0.0512	0.9511	0.9838
G_x	1.0	0.2381	0.0315	0.0184
R_x	0	0.7619	0.9685	0.9816
m_x	0	0	0	0

B	Diameter at breast height (cm)						
	2-10	10-30	30-50	50-70	70-90	90-110	>110
N/ha	222	48	13	11	21	6	5
p_x	0.9432	0.9822	0.9983	0.9948	0.9917	0.9954	0.9722
G_x	0.0313	0.0139	0.0167	0.0124	0.0079	0.0118	0
R_x	0.9687	0.9861	0.9833	0.9876	0.9921	0.9882	1
m_x	0	0	13356	30952	21111	19269	15510

Note: transfer out of the 100-300 height class in **A** is to the 2-10 cm dbh class in **B**; G_x and R_x are growth transitions for individuals that survived ($G_x + R_x = 1$ for each column) and p_x values are annual survival rates; in a transition matrix or life-cycle graph $g_x = G_x \times \pi_x$ and $r_x = R_x \times p_x$.; m_x is annual seed production; data from Enright and Ogden (1979).

$$
A = \begin{pmatrix}
.0390 & 0 & 0 & 0 & 0 & 219.8 & 509.5 & 347.5 & 317.2 & 255.3 \\
.0122 & .9211 & 0 & 0 & 0 & 0 & 0 & 0 & 0 & 0 \\
0 & .0300 & .9657 & 0 & 0 & 0 & 0 & 0 & 0 & 0 \\
0 & 0 & .0181 & .9137 & 0 & 0 & 0 & 0 & 0 & 0 \\
0 & 0 & 0 & .0181 & .9685 & 0 & 0 & 0 & 0 & 0 \\
0 & 0 & 0 & 0 & .0136 & .9816 & 0 & 0 & 0 & 0 \\
0 & 0 & 0 & 0 & 0 & .0167 & .9825 & 0 & 0 & 0 \\
0 & 0 & 0 & 0 & 0 & 0 & .0123 & .9839 & 0 & 0 \\
0 & 0 & 0 & 0 & 0 & 0 & 0 & .0078 & .9836 & 0 \\
0 & 0 & 0 & 0 & 0 & 0 & 0 & 0 & .0117 & .9722
\end{pmatrix} \tag{9.2}
$$

Vandermeer (1978) and Moloney (1986). The following examples show problems with analysis as well as some solutions to these difficulties.

ANALYSIS WITH ALL TRANSITIONS ESTIMATED

A Tropical Forest Tree, *Araucaria cunninghamii.*

Enright and Ogden (1979) provide data for a two-year study of a tropical conifer, *Araucaria cunninghamii*, in a 1.5 hectare plot near Bulolo, Papua New Guinea. All trees >2 cm dbh (*i.e. d*iameter at *b*reast *h*eight) were tagged in 1975 and, if still alive, again measured in 1976 and 1977. Also, two 20 × 20 m plots were used to measure transitions for plants <2 cm dbh. Estimates of reproduction were based on cone counts from other plots. The stratified approach used by Enright and Ogden (1979) is typical of size-based studies of long-lived organisms. It is necessary to obtain transitions from plots of different sizes or by making use of supplemental numbers of individuals of particular sizes. Table 9.1 shows the combined data for *A. cunninghamii* and Equation 9.2 shows an appropriate transition matrix **A** with f_x defined as $p_0 m_x$ and p_x and r_x determined by using $g_x = G_x \times p_x$ and $r_x = R_x \times p_x$.

The following data file is the correct representation of matrix **A** for the program **MATRIX.BAS**:

TABLE 9.2 All Roots of the Characteristic Equation for *Araucaria cunninghamii*

Root	Real	Imaginary	Modulus
1	1.0220	0	1.0220
2,3	0.9822	±0.0094	0.9822
4,5	0.9804	±0.0472	0.9815
6	0.9661	0	0.9661
7	0.9535	0	0.9535
8,9	0.9030	±0.0282	0.9034
10	0.0390	0	0.0390

Note: data gathered near Bulolo, Papua New Guinea, from 1975-77 by Enright and Ogden (1979).

10, 10
1, 1, .0390
1, 6, 219.8
1, 7, 509.5
1, 8, 347.5
1, 9, 317.2
1, 10, 255.3
2, 1, .0122
2, 2, .9211
3, 2, .0300
3, 3, .9657
4, 3, .0181
4, 4, .9137
5, 4, .0295
5, 5, .9685
6, 5, .0136
6, 6, .9816
7, 6, .0167
7, 7, .9825
8, 7, .0123
8, 8, .9839
9, 8, .0078
9, 9, .9836
10, 9, .0117
10, 10, .9722
0, 0, 0

The characteristic equation has 10 roots (Table 9.2) and the dominant root, λ_1, is 1.022 yr^{-1}, which means that the 1975-77 data indicate that the population of *Araucaria cunninghamii* near Bulolo had characteristics that would lead to approximately stationary structure.

The damping ratio, ρ, for the transition matrix Equation 9.2 is

$$\rho = \frac{1.022026}{0.982215} = \boxed{1.04053},$$

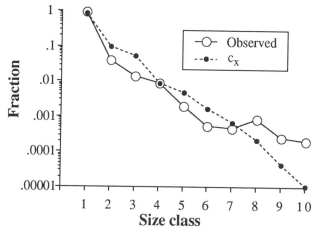

FIGURE 9.2 Observed fraction of the *Araucaria cunninghamii* population in each size class compared with the fraction of the population in the calculated stable-size distribution, c_x; data gathered near Bulolo, Papua New Guinea from 1975-77 by Enright and Ogden (1979).

which indicates that it would take a long time to arrive at a stable size distribution. The time, t, it would take for λ_1 to have 1000× the effect of λ_2 on population growth is

$$t = \frac{ln(1000)}{ln(r)} = \frac{6.907755}{0.039730} = \boxed{174 \text{ years}},$$

which means that *Araucaria cunninghamii* probably never attains a stable-size distribution.

The difference between the observed numbers of individuals in each size class and the number based on the stable-size distribution is a way of seeing whether *A. cunninghamii* is close to a stable-size distribution. Enright and Ogden (1979) estimate 24,491 individuals per hectare with about 94% being in the first size class, 0-20 cm. Figure 9.2 shows the distribution of observed (or estimated) frequency per hectare in each size class as well as the stable-size distribution, c_x, with $\lambda = 1.022$ yr^{-1}. The observed and expected (c_x) distributions in Figure 9.2 are similar; a better way, however, of comparing the observed size distribution and the stable-size distribution, c_x, is shown in Figure 9.3 in which observed minus expected is plotted and emphasizes the differences between the two distributions.

The comparison of observed and expected numbers of *A. cunninghamii* (Figure 9.3) shows that the major differences reside with the smallest size classes, which are the height classes 1-20, 20-100 and 100-300 cm. With year-to-year variation in reproductive success and early survival, the first size classes are exactly the ones that would be expected to deviate the most. Again, the indication is that the Araucaria population has not achieved the stable-size distribution that is expected based on the 1975-77 transitions; however, it seems to be developing in the direction that is predicted by the 1975-77 data.

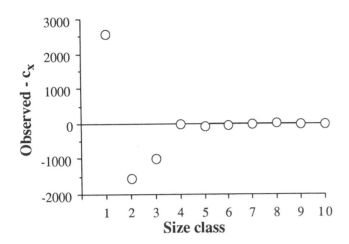

FIGURE 9.3 Differences between observed numbers of *Araucaria cunninghamii* per hectare and expected numbers based on the stable-size distribution, c_x; data gathered near Bulolo, Papua New Guinea from 1975-77 by Enright and Ogden (1979).

TABLE 9.3. Elasticities, e_{ij}, for Each Element of the Transition Matrix for *Araucaria cunninghamii*

Element	Name	e_{ij}
$a_{1,6}$	f_6	0.0044
$a_{1,7}$	f_7	0.0043
$a_{1,8}$	f_8	0.0009
$a_{1,9}$	f_9	0.0002
$a_{1,10}$	f_{10}	<0.0001
$a_{1,1}$	r_1	0.0004
$a_{2,1}$	g_1	0.0099
$a_{2,2}$	r_2	0.0902
$a_{3,2}$	g_2	0.0099
$a_{3,3}$	r_3	0.1694
$a_{4,3}$	g_3	0.0099
$a_{4,4}$	r_4	0.0833
$a_{5,4}$	g_4	0.0099
$a_{5,5}$	r_5	0.1788
$a_{6,5}$	g_5	0.0099
$a_{6,6}$	r_6	0.2399
$a_{7,6}$	g_6	0.0055
$a_{7,7}$	r_7	0.1361
$a_{8,7}$	g_7	0.0012
$a_{8,8}$	r_8	0.0299
$a_{9,8}$	g_8	0.0002
$a_{9,9}$	r_9	0.0054
$a_{10,9}$	g_9	<0.0001
$a_{10,10}$	r_{10}	0.0006

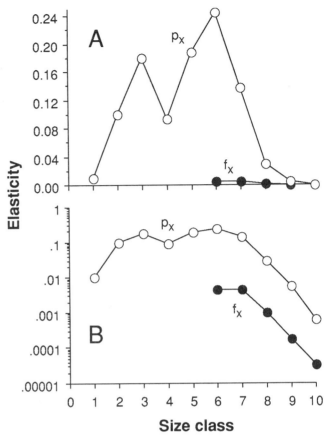

FIGURE 9.4 Elasticity values for size classes of *Araucaria cunninghamii*; p_x values are survival rates and include both survival while transferring to the next larger size class (g_x) as well as remaining in the same size class and surviving for one year (r_x); data from Enright and Ogden 1979.

An elasticity analysis was done for the *A. cunninghamii* data. Table 9.3 shows the elasticity estimates for each element and Figure 9.4 shows results of combining some elasticity estimates. Elasticity estimates for g_x and r_x transitions are combined for each size class because both are survival estimates for the size class. Accordingly, just two sets of elasticity measurements are shown in Figure 9.4: survival, p_x, and fecundity, f_x, where $f_x = p_0 m_x$.

The elasticity values can be summed to show how changes in an entire group of transitions can change population growth rate, λ, and elasticities for p_x and f_x are summarized in Table 9.4. The sum of all f_x elasticities is 0.0099, which also is the elasticity value for p_0; it is a very small value. A 10% change in seed production or first year survival, p_0, would change λ by only 0.1%, a value that would be nearly impossible to detect. Year to year seed production and early survival may vary by several orders of magnitude and a 1000% change in seed production could change λ by about 10%, assuming approximate linearity

TABLE 9.4 Combined g_x and r_x Elasticities for *Araucaria cunninghamii*.

Size class	p_x elasticity	f_x elasticity
1	0.0103	0
2	0.1000	0
3	0.1793	0
4	0.0932	0
5	0.1886	0
6	0.2454	0.0044
7	0.1372	0.0043
8	0.0301	0.0009
9	0.0054	0.0002
10	0.0006	<0.0001

over such a large change. A 1000% increase means taking a value, multiplying by 10 and adding this to the value, which is the same as multiplying by 11. For example, f_6 is 219.8 so a 1000% increase would be 2417.8. The estimate of p_0 provided by Enright and Ogden (1979) is 0.01646 yr^{-1}. A 1000% increase would make it 0.18106 yr^{-1}. An actual test of this showed that a 1000% increase in f_x values in fact increased λ by only about 3% rather than the predicted 10%, which is a problem with the linear extrapolation.

Small changes in survival of established trees possibly, though not certainly, are more likely to occur over many years. The summed elasticities of pre-reproductive trees (sizes 1 through 5) is 0.5714. The highest single elasticity for p_x is for size-6, the 10- 30 dbh class, and if this is added to the pre-reproductive classes the summed elasticity is 0.8168, which means that a 10% change in survival through the size-class of first reproduction would cause about an 8% increase in λ. Although details are somewhat different, the analysis presented above for *Araucaria cunninghamii* is the same as presented by Enright and Ogden (1979).

ANALYSIS WITH λ ASSUMED TO EQUAL 1.0

The Red Sea Urchin, *Strongylocentrotus franciscanus*

A sensitivity analysis was done for the red sea urchin *Strongylocentrotus franciscanus* with a goal of aiding in development of a fishery management plan (Ebert 1998). Animals were tagged with tetracycline or calcein at sites along the Pacific coast of the US (Ebert *et al.* ms) but the following data are primarily from six sites in Washington and Oregon together with four samples of very small animals gathered in California. All animals spent one year in the field and then were collected to determine growth transitions.

Table 9.5 shows growth transitions based on 1 cm size classes. Part A of Table 9.5 shows the actual numbers of tagged animals in a size class that were recaptured. The chemical tags, tetracycline or calcein, were used to determine a growth increment, which was then subtracted from the diameter to determine the original size at the time of tagging. Sorting a large data set of original and final sizes is prone to error if done by hand and so I have written a short BASIC program to create tables such as 9.5A. The program **TRANSITION.BAS** is provided at the end of this chapter.

The 1-centimeter intervals selected for red sea urchins are such that some individuals may grow sufficiently rapidly to skip one or more size classes. Examination of Table 9.5A shows as many as four possible transitions during a year for a single size class. The associated life-cycle graph (Figure 9.5) shows some of these possibilities for the first 5 size classes.

Table 9.5B and Figure 9.5 show the growth transitions as probabilities and these are identical with the G_x and R_x values in Table 9.6. Growth transitions must be multiplied by a survival rate, p_x, in order to produce transitions g_x and r_x. A single survival rate was assumed (Ebert 1998) and was determined from growth and size-frequency data, which is a topic that is covered in Chapter 13. The estimate of the common survival rate was 0.95 yr^{-1} and so all transitions in Table 9.5B were multiplied by this value to obtain transitions from each size class. For the size class 1.1-2.0 cm, the transitions are to four size classes (1) retention in size class 1.1-2.0, (2) growth to size class 2.1-3.0, (3) growth to size class 3.1-4.0, and (4) growth to size class 4.1-5.0. Growth transitions, R_x and G_x, and calculation of r_x and g_x for the transition matrix are shown in Table 9.6.

The fecundity values, m_x, for *S. franciscanus* were determined from changes in gonad mass before and after spawning and the relationship between egg number and spawn volume (Ebert 1998). The size-specific egg production is shown in Figure 9.6.

The final parameter that is required for completing the transition matrix is first-year survival, p_0, which for *S. franciscanus* includes fertilization rate, survival in the plankton, and early post-settlement survival. No field-based estimates of p_0 exist; however, the current population growth rate, λ, probably is close to 1.0 and so p_0 can be approximated by trial and error until λ is 1.0 or as close as one wishes to get. An analysis of this type means that population growth rate can not be estimated because it is being assumed to equal 1.0; the focus of the analysis is to examine the elasticities. The final version of the transition matrix is shown in Table 9.7. The dominant root is 0.9946, which for purposes of further analysis is sufficiently close to 1.0.

Elasticity values were calculated and, as was done in the analysis for *Araucaria cunninghamii*, survival transitions for

TABLE 9.5 Growth Transitions for Red Sea Urchins (*Strongylocentrotus franciscanus*)

A

to↓	0.1-1.0	1.1-2.0	2.1-3.0	3.1-4.0	4.1-5.0	5.1-6.0	6.1-7.0	7.1-8.0	8.1-9.0	9.1-10.0	10.1-11.0	11.1-12.0	12.1-13.0	13.1-14.0	14.1-15.0	15.1-16.0	≥16.1
0.1-1.0	0	0	0	0	0	0	0	0	0	0	0	0	0	0	0	0	0
1.1-2.0	58	45	0	0	0	0	0	0	0	0	0	0	0	0	0	0	0
2.1-3.0	31	201	0	0	0	0	0	0	0	0	0	0	0	0	0	0	0
3.1-4.0	7	103	10	0	0	0	0	0	0	0	0	0	0	0	0	0	0
4.1-5.0	0	37	22	10	0	0	0	0	0	0	0	0	0	0	0	0	0
5.1-6.0	0	0	2	59	15	0	0	0	0	0	0	0	0	0	0	0	0
6.1-7.0	0	0	0	18	54	18	5	0	0	0	0	0	0	0	0	0	0
7.1-8.0	0	0	0	0	9	16	19	13	0	0	0	0	0	0	0	0	0
8.1-9.0	0	0	0	0	0	3	6	45	119	0	0	0	0	0	0	0	0
9.1-10.0	0	0	0	0	0	0	0	7	76	101	0	0	0	0	0	0	0
10.1-11.0	0	0	0	0	0	0	0	0	0	26	34	0	0	0	0	0	0
11.1-12.0	0	0	0	0	0	0	0	0	0	0	9	12	0	0	0	0	0
12.1-13.0	0	0	0	0	0	0	0	0	0	0	0	6	66	0	0	0	0
13.1-14.0	0	0	0	0	0	0	0	0	0	0	0	0	12	193	0	0	0
14.1-15.0	0	0	0	0	0	0	0	0	0	0	0	0	0	6	242	0	0
15.1-16.0	0	0	0	0	0	0	0	0	0	0	0	0	0	0	3	95	0
≥16.1	0	0	0	0	0	0	0	0	0	0	0	0	0	0	0	1	14
total	96	386	34	87	78	37	30	65	195	127	43	18	78	199	245	96	14

Original size (cm)

B   Original size (cm)

to↓	0.1-1.0	1.1-2.0	2.1-3.0	3.1-4.0	4.1-5.0	5.1-6.0	6.1-7.0	7.1-8.0	8.1-9.0
0.1-1.0	0	0	0	0	0	0	0	0	0
1.1-2.0	0.6042	0.1166	0	0	0	0	0	0	0
2.1-3.0	0.3229	0.5207	0	0	0	0	0	0	0
3.1-4.0	0.0729	0.2668	0.2941	0	0	0	0	0	0
4.1-5.0	0	0.0959	0.6471	0.1149	0	0	0	0	0
5.1-6.0	0	0	0.0588	0.6782	0.1923	0	0	0	0
6.1-7.0	0	0	0	0.2069	0.6923	0.4865	0.1667	0	0
7.1-8.0	0	0	0	0	0.1154	0.4324	0.6333	0.2000	0
8.1-9.0	0	0	0	0	0	0.0811	0.2000	0.6923	0.6103
9.1-10.0	0	0	0	0	0	0	0	0.1077	0.3897

	9.1-10.0	10.1-11.0	11.1-12.0	12.1-13.0	13.1-14.0	14.1-15.0	15.1-16.0	≥16.1
9.1-10.0	0.7953	0	0	0	0	0	0	0
10.1-11.0	0.2047	0.7907	0	0	0	0	0	0
11.1-12.0	0	0.2093	0.6667	0	0	0	0	0
12.1-13.0	0	0	0.3333	0.8462	0	0	0	0
13.1-14.0	0	0	0	0.1538	0.9698	0	0	0
14.1-15.0	0	0	0	0	0.0302	0.9878	0	0
15.1-16.0	0	0	0	0	0	0.0122	0.9896	0
≥16.1	0	0	0	0	0	0	0.0104	1

Note: animals <2cm from California and >2 from Oregon and Washington; n=1828; columns are original diameters and rows (**to↓**) are sizes one year later; **A** shows numbers in each size class and **B** shows the fractions or probabilities of transfer.

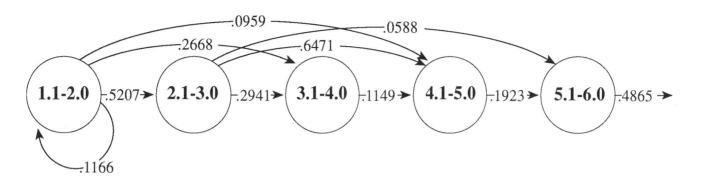

FIGURE 9.5 Graphical representation of growth transitions for small size classes of *Strongylocentrotus franciscanus* showing that some rapidly growing individuals pass through entire size classes during a year and enter the next+1 or next+2 size class; nodes are 1-cm size classes; analysis from Ebert (1998).

TABLE 9.6 Sample Calculations for Matrix Transitions for Red Sea Urchins

to↓	original size =1.1-2.0			
	R_x	G_x	r_x	g_x
1.1-2.0	0.1166	-	0.1107	-
2.1-3.0	-	0.5207	-	0.4946
3.1-4.0	-	0.2668	-	0.2535
4.1-5.0	-	0.0959	-	0.0910

Note: growth transitions are R_x and G_x and are multiplied by the annual survival rate, p, which is 0.9499 yr^{-1}, to obtain r_x and g_x the matrix transitions; example shows the initial size class 1.1-2.0 cm in Table 9.4; the sum of R_x and G_x is 1.0 and the sum of r_x and g_x is 0.9499.

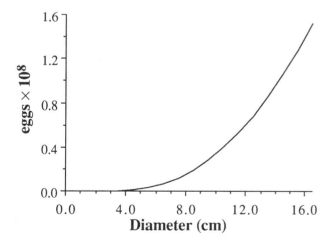

FIGURE 9.6 Egg production of *Strongylocentrotus franciscanus* based on spawn mass (Baker 1973, Ebert 1998) and the conversion of 7.764 x 10^5 eggs cm^{-3} (Levitan 1993).

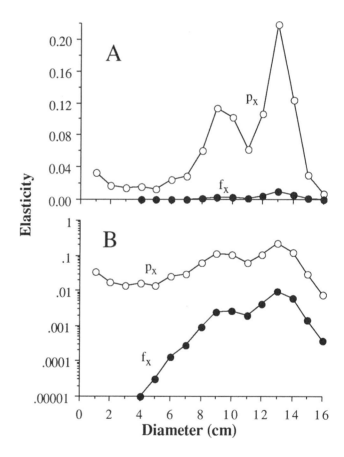

FIGURE 9.7 Elasticity values for survival (p_x) and f_x based on the transition matrix in Table 9.7; the drop in p_x elasticity at 11 cm is because raw transitions were used in the analysis: no data smoothing was done; no biological significance should be attached to this dip; **A,** linear elasticity scale; **B,** log-scale for elasticity values.

TABLE 9.7. Final Transition Matrix for Red Sea Urchins with Growth and Survival Transitions, r_X and g_X, in a Band Along the Diagonal and f_X Values in the First Row.

Original diameter (cm)

to↓	1.1-2	2.1-3	3.1-4	4.1-5	5.1-6	6.1-7	7.1-8	8.1-9	9.1-10	10.1-11	11.1-12	12.1-13	13.1-14	14.1-15	15.1-16	≥16.1
1.1-2	0.1107	0	0	0.0006	0.0027	0.0065	0.0120	0.0193	0.0286	0.0398	0.0531	0.0685	0.0860	0.1056	0.1276	0.1515
2.1-3	0.4946	0	0	0	0	0	0	0	0	0	0	0	0	0	0	0
3.1-4	0.2535	0.2794	0	0	0	0	0	0	0	0	0	0	0	0	0	0
4.1-5	0.0910	0.6146	0.1092	0	0	0	0	0	0	0	0	0	0	0	0	0
5.1-6	0	0.0559	0.6442	0.1827	0	0	0	0	0	0	0	0	0	0	0	0
6.1-7	0	0	0.1965	0.6576	0.4621	0.1583	0	0	0	0	0	0	0	0	0	0
7.1-8	0	0	0	0.1096	0.4107	0.6016	0.1900	0	0	0	0	0	0	0	0	0
8.1-9	0	0	0	0	0.0770	0.1900	0.6576	0.5797	0	0	0	0	0	0	0	0
9.1-10	0	0	0	0	0	0	0.1023	0.3702	0.7554	0	0	0	0	0	0	0
10.1-11	0	0	0	0	0	0	0	0	0.1945	0.7510	0	0	0	0	0	0
11.1-12	0	0	0	0	0	0	0	0	0	0.1988	0.6332	0	0	0	0	0
12.1-13	0	0	0	0	0	0	0	0	0	0	0.3166	0.8037	0	0	0	0
13.1-14	0	0	0	0	0	0	0	0	0	0	0	0.1461	0.9212	0	0	0
14.1-15	0	0	0	0	0	0	0	0	0	0	0	0	0.0286	0.9382	0	0
15.1-16	0	0	0	0	0	0	0	0	0	0	0	0	0	0.0116	0.9400	0
≥16.1	0	0	0	0	0	0	0	0	0	0	0	0	0	0	0.0099	0.9499

each size class were combined and include both the probability of staying in a size class and surviving for one year and the probability of leaving to another size class and surviving for one year. Each size class has only one elasticity value (Figure 9.7) and these are interpreted by choosing a size class and reading the elasticity for that size class either for survival p_X or for f_X, which represents the relative contribution of p_X or f_X to population growth. For example, the survival (p_X) elasticity for the size-class 9.1-10 cm, shown at 9 in Figure 9.7A, is about 0.12. If the survival rate of this size class would be changed by 10%, the population growth would be changed by 10% × 0.12 or 1.2%. The elasticity value for survival of animals in the 13.1-14 cm size class, shown as 13 in the graph, is about 0.22 so a 10% change in survival of 13.1-14 cm sea urchins would change population growth by 10% × 0.22 or 2.2%.

Elasticity values for f_X values are very small. Using a logarithmic scale (Figure 9.7B) shows that differences between elasticity values for p_X vs. f_X for a size class may exceed several orders of magnitude. The sum of the f_X elasticities is 0.03, which also is the elasticity of p_0, and the sum for p_X values is 0.97, which means that an overall change in all f_X values by 10% would change population growth by only 0.3% whereas an overall change in post-juvenile survival by 10% would change population growth by 9.7%. It also should be clear from Figure 9.7 that survival of large individuals is more important for population growth than survival of small sea urchins. The summed elasticity of p_X values for animals <9.0 cm is 0.21 whereas the summed elasticity of p_X values ≥9.0 cm is 0.76. A 10% change in survival of animals less than 9.0 cm would change population growth by 2.1% whereas a 10% change in survival of animals ≥9.0 cm would change population growth by 7.6%. Management of the red sea urchin fishery in Oregon and Washington would be best if individuals >9 or 10 cm would be conserved because the large animals contribute most to population maintenance.

ANALYSIS WITH FIXED RECRUITMENT

A coral, *Agaricia agaricites*

Hughes (1984) presents transition matrices for a Jamaican coral, *Agaricia agaricites* based on analysis of photographs taken of 1-m quadrats. Colonies were assigned in the first year to size classes: 0-10 cm^2, 10-50 cm^2, 50-200 cm^2 and >200 cm^2. During a storm-free year from 1977 to 1978 he obtained the transitions shown in Equation 9.3, which do not include any additions to the population from sexual reproduction. The top row as well as entries above the diagonal represent shrinkage including fragmentation of coral colonies.

$$A = \begin{pmatrix} 0.6020 & 0.1167 & 0.0217 & 0.0741 \\ 0.1681 & 0.6167 & 0.1087 & 0.0370 \\ 0 & 0.2000 & 0.8043 & 0.1481 \\ 0 & 0.0171 & 0.0217 & 0.9259 \end{pmatrix} \quad (9.3)$$

The roots of the characteristic equation (Table 9.8) have no imaginary parts and so the size-structured coral population would go to a stable-size distribution without any oscillatory behavior; however, it would do so fairly slowly because the damping ratio, ρ, is

$$\rho = \frac{0.9823101}{0.8522989} = \boxed{1.152}.$$

The dominant root, λ_1, would assume 1000× times the importance of the second root, λ_2, after t years,

$$t = \frac{ln(1000)}{ln(1.15254)} = \boxed{49 \text{ years}},$$

and would be 100× more important after about 30 years. The stable-size and reproductive-value distributions are shown in Table 9.9.

Figure 9.8 shows how the population assumes a stable-size distribution from starting values that are far from the stable structure. The figure was produced by using population projection with matrix multiplication. The program **SIMPOP.BAS** was used and is provided at the end of this chapter. An initial size-distribution vector, Equation 9.4, was multiplied by the transition matrix

TABLE 9.8 Roots of the Characteristic Equation for Size Transitions of the coral *Agaricia agaricites* During 1977-78

Root	Real	Imaginary
1	0.98231	0.0
2	0.85230	0.0
3	0.66933	0.0
4	0.44501	0.0

TABLE 9.9 Stable-Size, c_x, and Reproductive-Value, v_x, Vectors for *Agaricia agaricites* During 1977-78

Size cm^2	c_x	v_x
0-10	1.0000 (13.37%)	1.0000
10-50	1.5814 (21.14%)	9.1038
50-200	3.1916 (42.67%)	11.7807
>200	1.7071 (22.82%)	50.0364

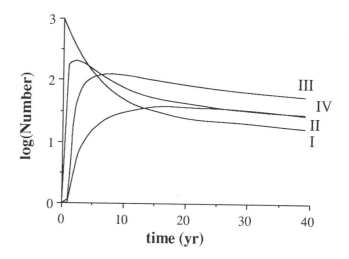

FIGURE 9.8 Convergence to a stable-size distribution of *Agaricia agaricites* starting with 1000 individuals 0.1-10cm^2 (size-class I) and 1 individual in each of the other size classes: 10-50cm^2 (size-class II), 50-200cm^2 (class III), and >200cm^2 (class IV); data from Hughes (1984).

(Equation 9.3) to produce a new size distribution vector which was again multiplied by the transition matrix and this projection was continued for 40 iterations (Figure 9.8).

$$\begin{pmatrix} 1000 \\ 1 \\ 1 \\ 1 \end{pmatrix} \quad (9.4)$$

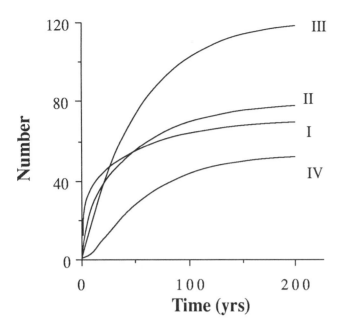

FIGURE 9.9 Simulation of size-structure and density changes of *Agaricia agaricites* starting with 20 small individuals and 1 colony in each of the other three size categories; each year, 20 new colonies were added to the smallest size class to simulate density independent recruitment; data from Hughes (1984).

TABLE 9.10 Stable-Size Distribution for *Agaricia agaricites* **with Fixed Recruitment of 20 Colonies Year**[-1]

Size cm^2	Number	Fraction
<10	70.28	21.55%
10-50	79.46	24.37
50-200	122.20	37.48
>200	54.11	16.60

In Figure 9.8, lines are very close to parallel at about 30 years, which is consistent with λ_1 being 100× more important with respect to population growth than λ_2 at that time. There are small adjustments following 30 years and again this is predicted by the length of time it would take for λ_1 to be 1000× more important with respect to population growth than λ_2.

For the coral population to persist ($\lambda \geq 1$), it appears that asexual reproduction is insufficient and so some new individuals must come from the plankton as the result of sexual reproduction. The problem of sexual reproduction is difficult for species with planktonic larvae because individuals that are present at a particular site generally are not the parents of individuals that settle from the plankton. This problem was handled for red sea urchins by assuming that size-specific m_x values were appropriate generally as well as for particular sites. This is another way of saying that the population or populations that are studied are representative samples of the larger spatially-structured populations.

Hughes (1984) took a different tack with *Agaricia agaricites* by assuming that recruitment was independent of adult density or size structure. He used the transition matrix to generate size distributions and each year he added 20 new colonies to the smallest size class. He then examined size structure of the coral populations with different storm frequencies. Figure 9.9 shows results of a simulation using **SIMPOP.BAS** similar to those done by Hughes. Initial conditions were

$$\begin{pmatrix} 20 \\ 1 \\ 1 \\ 1 \end{pmatrix}$$

With density-independent recruitment, the coral population eventually attains a stable size distribution that also is stationary, that is population density stops changing. The stable-size distribution for Figure 9.9 is shown in Table 9.10.

With density-independent recruitment, it takes much longer to achieve a stable size structure. In fact, it is so long that it is unlikely that the population ever would be stable and stationary. Very severe storms have been estimated to occur about every 20 years (Tunnicliffe 1980, cited by Hughes 1984) but a severe and destructive storm caused extensive damage in March 1979 and 17 months later in August 1980, Hurricane Allen hit Jamaica and caused extensive damage to the reef (Woodley *et al.* 1981). Non-equilibrium conditions probably are the rule on coral reefs.

GENERAL COMMENTS

Age-, stage-, and size-based analyses should not be thought of as three mutually exclusive approaches. It is possible to mix ages and stages or stages and sizes all in a single analysis. All that is necessary is to be very careful in keeping track of the meaning of nodes and transitions.

Size-based analysis is very appealing because size is something that can be measured without making any assumptions about the previous history of an individual; however, a problem with the analysis is reminiscent of one of the problems in stage-based analysis. In stage-based analysis it was obvious that structure within a node was important because, for example, a newly-hatched larva does not have the same probability of transfer to the pupal stage as a larva that has spent the average length of time as a larva. A related problem exists for size-based analyses where all individuals within a node are treated equally without respect to how long they have been in the node and how close their size is to the upper boundary. Each node of a size-structured analysis has an internal size structure that is important in determining how many individuals stay or grow out of the node. If the internal structure has most individuals close to the lower boundary then the fraction leaving will be very small but if most individuals are clustered near the upper boundary then a very large fraction of them will leave. Actual individual growth characteristics might be exactly the same in both cases but the appearances would be very different.

There are interesting problems of assigning paternity in some size-structured models. This problem was raised in the example with sea urchins and corals where larvae may travel great distances before settlement. One suggestion was to measure size-specific fecundity and use this to scale the contributions of size-classes to new recruits and this was the approach used with red sea urchins. Another approach was to make recruitment completely independent of size-structure at the study site, which was the approach used by Hughes (1984).

There are cases where it is not possible to establish size-specific fecundity and so some relationship simply must be assumed. In a study of stem dynamics of the shrub *Alnus incana*, Huenneke and Marks (1987) were faced with the problem of assigning new individuals, which were root sprouts, to adult size classes. It was not reasonable to excavate new shoots to see which stem should be given credit for reproduction. It was assumed that stem area at breast height was the best way of scaling so that twice the area meant twice the reproductive contribution. This may or may not be true but it probably is the best that can be done in the analysis. As shown in the analysis of *Araucaria cunninghamii*, size and reproductive output were not monotonically related; there was a rise in output from the size at first reproduction to the second reproductive size class but this class had the maximum fecundity. The last three size classes all showed a negative relationship between size and fecundity.

Size-based analysis is very general and can be extended from discrete units such as individuals (*e.g.* sea urchins) to colonies with discrete boundaries (*e.g.* corals) to units of a colony or colonies that have indistinct boundaries such as the alder shrubs studied by Huenneke and Marks (1987). One further step is to use just parts of plants or animals. Branches intersecting a transect line were defined as "individuals" and were used to study dynamics of dwarf birch *Betula nana* (Ebert and Ebert 1989). Branches intersecting a line really are much the same as root shoots emerging from the ground. In both cases, "individuals" are attached to other "individuals" but growth of a population means that all parts of the population are growing and so branch segments form a virtual population that, in some

respects, behaves in exactly the same way as the parent population. Populations can be defined in many different ways.

PROBLEMS

1. Åberg (1992a+b) studied the seaweed *Ascophyllum nodosum* in a size-based demographic context and determined values for a transition matrix **A** during a normal ice year (1986-87) at Göteborg on the Swedish coast.

$$\mathbf{A} = \begin{pmatrix} 0.38 & 0.10 & 0.17 & 0.04 & 0 \\ 0.11 & 0.18 & 0.10 & 0.07 & 0 \\ 0.01 & 0.10 & 0.23 & 0.26 & 0.10 \\ 0 & 0 & 0.10 & 0.15 & 0.60 \\ 0 & 0 & 0.02 & 0.04 & 0.10 \end{pmatrix}$$

The matrix presented above is just asexual and really represents just growth and shrinkage. Åberg includes a scaled fertility by adding the values in Table 9.11 to the first row of the matrix.

a. In order to get a better feeling for the meaning of the transition matrix, change it into a life-cycle graph. The size classes are 1 (<5g), 2 (5 - <15g), 3 (15 - <54g), 4 (54 - <190g) and 5 (≥190g).
b. Modify your life-cycle graph to include sexual reproduction.
c. Estimate λ and r.

TABLE 9.11 Adjustment of f_x Values for *Ascophyllum nodosum* to Include Sexual Reproduction

Element	Old value	m_x	New value
a_{11}	0.38	0	0.38
a_{12}	0.10	0.13	0.23
a_{13}	0.17	0.62	0.79
a_{14}	0.04	2.36	2.40
a_{15}	0	7.82	7.82

Note: data from Åberg (1992a,b).

2. The soft coral *Alcyonium* changes size by growing, shrinking and by undergoing fission; there also is sexual reproduction. An undescribed species of *Alcyonium* was studied by McFadden (1991) at Botanical Beach on Vancouver Island, British Columbia, and at Tatoosh Island in the Straits of Juan de Fuca, Washington. Here are two separate transition matrices for 6 month periods.

$$A = \begin{pmatrix} 0 & 0 & 0.06 & 0.13 & 0.77 \\ 0 & 0.36 & 0.14 & 0.02 & 0.05 \\ 0 & 0.14 & 0.46 & 0.35 & 0.08 \\ 0 & 0 & 0.24 & 0.50 & 0.36 \\ 0 & 0 & 0.08 & 0.17 & 0.56 \end{pmatrix} \quad \text{5/85 to 8/85}$$

$$A = \begin{pmatrix} 0 & 0 & 0 & 0 & 0 \\ .007 & 0.27 & 0.16 & 0.02 & 0.13 \\ 0 & 0.40 & 0.38 & 0.23 & 0.28 \\ 0 & 0.20 & 0.27 & 0.43 & 0.22 \\ 0 & 0.07 & 0.27 & 0.38 & 0.83 \end{pmatrix} \quad \text{8/85 to 5/86}$$

The transitions are 1 (planula), 2 (<0.14 cm^2), 3 (0.14 - 0.24 cm^2), 4 (0.25 - 0.36 cm^2), 5 (> 36 cm^2). The time constant for each matrix is 6 months.

a. Construct a life-cycle graph that combines both of these matrices into a single analysis. [*Hint:* start by labeling the columns for the May-August matrix 1-5 and the rows 6-10; label the columns of the September-May matrix 6-10 and the rows 1-5].

b. Determine the stable-size and reproductive-value vectors, damping ratio for the life-cycle, and do an elasticity analysis.

3. The genus *Calochortus* is in the Liliaceae and is represented by a large number of species along the west coast of North America, some of which are rare or threatened. Fiedler (1987) examined four species of this genus and did a size-based analysis. Here are two matrices for *C. pulchellus*, a rare species that occurs on Mount Diablo in Contra Costa Co, California

$$A = \begin{pmatrix} 0.49 & 0.01 & 13.07 & 10.04 \\ 0 & 0.84 & 0.14 & 0 \\ 0 & 0.15 & 0.63 & 0.29 \\ 0 & 0 & 0.23 & 0.71 \end{pmatrix} \quad \text{1981-82}$$

$$A = \begin{pmatrix} 0.50 & 0.01 & 1.99 & 7.03 \\ 0.07 & 0.85 & 0.02 & 0.12 \\ 0 & 0.11 & 0.89 & 0.04 \\ 0 & 0 & 0.07 & 0.37 \end{pmatrix} \quad \text{1982-83}$$

The size categories are based on the width of the basal leaf (cm): 1 (<0.25), 2 (0.25 - 1.34), 3 (1.35 - 2.04), 4 (≥ 2.05).

a. Estimate λ using each matrix separately and determine the mean (geometric mean?)

b. Estimate λ by determining the mean values for each element first.

c. Estimate λ by assembling a single matrix in which the two states alternate.

4. A population of the surf thistle, *Cirsium rhothophyllum*, was studied in the coastal sand dunes of Santa Barbara County, California, by Zedler *et al.* (1983). Several different transitions were measured. There are changes in the numbers of apices per plant, which can be treated as new vegetative individuals (Table 9.12). Rosettes were measured and there are size transitions (Table 9.13) which include transitions to a flowering stage that does not have a rosette. Seed production was estimated to be 319 achenes per flowering individual. Maximum emergence rate is 60% and plants grow to a rosette size of about 2 cm in the summer following seed production of the previous year. Values of f_x can be approximated from achenes per individual, number of flowering individuals, and number of new individuals that appear during a year. In July 1978 and January 1979 combined, there were 89 individuals that flowered. In July 1979, a total of 297 new individual were present in the plots so a reasonable estimate of f_x is:

$$f_x = \frac{\text{total new individuals}}{\text{number of flowering individuals}} = \frac{297}{89} = \boxed{3.337}$$

The size-frequency distribution in 1979 is shown in Table 9.14.

TABLE 9.13 Change in Apex Number for Non-flowering *Cirsium rhothophilum* that Survived as Vegetative Individuals from July 1978 to July 1979

Number in 1979↓	Original number of apices in 1978			
	1	2	>2	Total
1	257	8	1	266
2	44	11	4	59
>2	13	14	14	41
Total	314	33	19	366

Note: data from Zedler et al. (1983).

TABLE 9.13 Single Apex Rosette Size-Class Transitions for *Cirsium rhothophilum* from July 1978 to July 1979

to↓ (1979)	Original size in 1978				
	0-2	2-6	6-10	10-14	>14
0-2	39	3	1	1	0
2-6	58	21	3	0	2
6-10	9	19	4	3	0
10-14	7	12	5	5	3
>14	0	11	13	14	24
Flowering	2	4	2	3	31
Dead	96	61	18	10	33
Total	211	131	46	36	93

Note: data from Zedler *et al.* (1983).

Table 9.14 Size-Frequency Distribution of *Cirsium rhothophilum* Measured in 1979

Size class	Number
<2	223
2-6	171
6-10	53
10-14	38
>14	60

Note: data extracted from Figure 1 of Zedler *et al.* (1983)

a. Draw a life cycle graph for *Cirsium rhothophilum* that includes a separate node for flowering.
b. Transfer values from the graph to a matrix and determine and interpret λ and elasticities.
c. Compare the observed size-frequency distribution (Table 9.14) with the calculated stable-size distribution. If perturbed, how rapidly should the population return to a stable configuration?

BASIC PROGRAMS

TRANSITION.BAS

The program **TRANSITION.BAS** is used to take growth transitions and create a table showing transitions. The data are very simple. The first number in the file must be N, the number of data pairs. The rest of the file has data pairs with the first value being the size at time t and the second value being size at t+Δt. The data pairs are separated by commas. All time period, Δt, must be the same for a data set; you can not have some size pairs based on a time period of 2 months and others based on 1 year.. Here are a few diameter measurements of the red sea urchin at San Nicolas Island, CA, to show the structure of data files. The number of data pairs is 15.

```
15
2.6042, 4.1045
3.6932, 5.1278
1.5824, 3.1343
2.7064, 4.1732
3.0069, 5.1904
3.2899, 5.2108
2.4662, 4.2316
3.6277, 5.2654
2.4326, 4.5178
2.6908, 4.5212
1.8403, 3.5712
2.5526, 4.6469
2.4233, 4.6523
3.7005, 5.8661
3.1901, 4.9060
```

When the program is run, the first question is

Do you want to continue? (Y/N)?

Answer with a Y or an N followed by a <cr>. You may use either upper of lower case letters to answer.

The next question is

Filename with data:

Enter the filename. Be sure that the file is a *text* file. If it is not a text file, strings of ??????????????? may appear on the screen. If this happens, kill the program. Try Option-Control Esc.

If the file was read correctly, the value of N will appear on the screen and you will be asked another question

How many state categories? (i.e. the order of the matrix)

How many size classes do you want? It probably would have been a good idea to have sorted the data so you know both the smallest and largest individual. If you know smallest and largest you can try equal sized categories as a first pass. With the sample data given above, and for purposes of illustration, I decided to use 5 classes all with and equal size of 1 cm. So, in response to the question about the order of the matrix, I answered

5 <cr>

There now will be as many questions as the order of the matrix. Because there are going to be 5 size classes, there will be 5 questions

TABLE 9.12 Output from TRANSITION.BAS to Show the Meaning of Columns and Rows

to↓	1.0000-2.0000	2.0001-3.0000	Original size 3.0001-4.0000	4.0001-5.0000	5.0001-6.0000
1.0001-2.0000	0	0	0	0	0
2.0001-3.0000	0	0	0	0	0
3.0001-4.0000	2	0	0	0	0
4.0001-5.0000	0	7	1	0	0
5.0001-6.0000	0	0	5	0	0

Upper bound for state #1

I answered

2 <cr>

Upper bound for state #2

I answered 3 meaning that the second size class would start at 2.0001 and go to 3.0000. Notice that the boundaries are defined by the data. The size data given in the example have 4 decimal places. If there would be 3 decimal places the size class would be 2.001 - 3.000. If there would be 2 decimal places the interval would be 2.01-3.00 and with 1 decimal place the interval would be defined as 2.1-3.0.
There will be three more questions concerning upper boundaries

Upper bound for state #3

answer 4

Upper bound for state #4

answer 5

Upper bound for state #5

answer 6

The output will appear on the screen with numbers separated by spaces

```
0 0 0 0 0
0 0 0 0 0
2 0 0 0 0
0 7 1 0 0
0 0 5 0 0
```

The meaning of the output is best seen by adding labels to columns and rows (Table 9.12):
An examination of the sample data will reveal 2 individuals that initially were between 1.0001 and 2.0000 and in one year both grew to be in the size class 3.0001-4.0000.

You may same the output to a file with any name you choose. The same information presented on the screen appears in the output except values are separated by tabs and there are two tables produced. The first contains counts in each size class together with a final row that has the totals for each column; that is, the initial numbers in each size class. The second table has the relative values or growth transition rates.

```
0        0        0        0        0
0        0        0        0        0
2        0        0        0        0
0        7        1        0        0
0        0        5        0        0
2        7        6        0        0

0        0        0        0        0
0        0        0        0        0
1        0        0        0        0
0        1        0.166667 0        0
0        0        0.833333 0        0
```

TRANSITION.BAS

```
10 dim s1(3000),s2(3000),a(20,20),b(20),
   tc(20),p(500)
20 input "Want to continue? (Y/N): ";q$
30 if q$ = "N" or q$ = "n" then goto 760
40 for i = 1 to 20
50 b(i) = 0
60 for j = 1 to 20
70 a(i,j) = 0
80 next j
90 next i
100 input "Filename with data: ";f$
110 open f$ for input as #9
120 input #9,n
130 for i = 1 to n
140 input #9,s1(i),s2(i)
150 next i
160 close #9
170 print "N=";n
180 input "How many state categories?
    (i.e. the order of the matrix)";iz
```

```
190 for i = 1 to iz
200 print "Upper bound for state #";i;
210 input b(i)
220 next i
230 for i = 1 to n
240 for j = 1 to iz
250 if s1(i) <= b(j) then k1 = j : goto
    270
260 next j
270 for j = 1 to iz
280 if s2(i) <= b(j) then k2 = j : goto
    300
290 next j
300 a(k2,k1) = a(k2,k1)+1
310 next i
320 for i = 1 to iz
330 print
340 for j = 1 to iz
350 print a(i,j);
360 next j
370 next i
380 print
390 input "Want to save results? (Y/N)
    ";q$
400 if q$ = "N" or q$ = "n" then goto 20
410 input "Name of file for output: ";f$
420 open f$ for output as #9
430 print #9,f$,"n = ";n
440 for i = 1 to iz
450 print #9," "
460 for j = 1 to iz
470 print #9,a(i,j),
480 next j
490 next i
500 print #9," "
510 for j = 1 to iz
520 for i = 1 to iz
530 tc(j) = tc(j)+a(i,j)
540 next i
550 next j
560 for j = 1 to iz
570 print #9,tc(j),
580 next j
590 for j = 1 to iz
600 if tc(j) = 0 then goto 640
610 for i = 1 to iz
620 a(i,j) = a(i,j)/tc(j)
630 next i
640 next j
650 print #9," "
660 print #9," "
670 for i = 1 to iz
680 print #9," "
690 for j = 1 to iz
```

```
700 print #9,a(i,j),
710 next j
720 next i
730 print #9," "
740 close #9
750 goto 20
760 end
```

SIMPOP.BAS

Once you have a transition matrix, you can simulate population growth just by matrix multiplication starting with some initial conditions. In **SIMPOP.BAS**, the transition matrix is **A** and the initial vector of numbers in each category is **B**. As was true in **MATRIX.BAS**, you only have to enter non-zero values and you can do this at the keyboard or by creating a data file using a word processor. Remember to save the *data as a text file*. Here is the transition matrix for the coral *Agaricia agaricites* that was studied by Hughes.

$$\mathbf{A} = \begin{pmatrix} 0.6020 & 0.1167 & 0.0217 & 0.0741 \\ 0.1681 & 0.6167 & 0.1087 & 0.0370 \\ 0 & 0.2000 & 0.8043 & 0.1481 \\ 0 & 0.0171 & 0.0217 & 0.9259 \end{pmatrix}$$

with the following initial conditions

$$\mathbf{B} = \begin{pmatrix} 1000 \\ 0 \\ 0 \\ 0 \end{pmatrix}$$

The data file is

```
4,4
1,1,.6020
1,2,.1167
1,3,.2017
1,4,.0741
2,1,.1681
2,2,.6267
2,3,.1087
2,4,.0370
3,2,.2000
3,3,.8043
3,4,.1481
4,2,.0171
4,3,.0217
4,4,.9259
0,0,0
1,1000
0,0
```

Notice that matrix **A** is first and **B** is second.

In the program, you are permitted to have a value for constant recruitment or not and can specify the length of the simulation. Output can be saved to a file, which can be cleaned and used with a plotting program such as CricketGraph. "Cleaning" the file means eliminating E-formatting. All that is required is to change numbers such as 2.46867E-3 to 0.00246867. CricketGraph does not recognize E-formatting. Making the conversions can be done by hand or the output can be converted by using a spreadsheet such as Excel.

```
10 dim a(30,30),b(30),c(30)
20 print "Population growth simulation"
30 print "Another fine product from
   Cornered Rat Software©"
40 print "                    T. A.
   Ebert 1996"
50 input "Do you want to continue?
   (y/n)";f$
60 if f$ = "y" or f$ = "Y" then goto 90
70 if f$ <> "n" and f$ <> "N" then goto
   50
80 goto 1100
90 input "Are data from a file or from
   the keyboard? (F/K)";f$
100 if f$ = "F" or f$ = "f" then goto
    510
110 if f$ <> "K" and f$ <> "k" then goto
    90
120 input "Order of the transition
    matrix: ";n
130 m = n
140 for i = 1 to n
150 for j = 1 to n
160 a(i,j) = 0
170 next j
180 next i
190 print "Enter row #, column #, and
    element value. Separate numbers with
    commas" : print "Enter 0 to
    terminate"
200 input i,j,z
210 if i = 0 then goto 240
220 a(i,j) = z
230 goto 200
240 for i = 1 to n
250 b(i) = 0
260 next i
270 print "And now for the vector of
    initial numbers in each class"
280 print "Enter row # and element
    value. Separate numbers with a
    comma"
290 print "Enter 0 to terminate"
300 input i,z
310 if i = 0 then goto 340
320 b(i) = z
330 goto 300
340 input "File name for matrices: ";f$
350 open f$ for output as #9
360 print #9,n;",";n
370 for i = 1 to n
380 for j = 1 to n
390 if a(i,j) = 0 then goto 410
400 print #9,i;",";j ",";a(i,j)
410 next j
420 next i
430 print #9,"0, 0, 0"
440 for i = 1 to n
450 if b(i) = 0 then goto 470
460 print #9,i;",";b(i)
470 next i
480 print #9,"0, 0"
490 close #9
500 goto 690
510 input "Name of file with matrices:
    ";f$
520 open f$ for input as #9
530 input #9,n,n
540 for i = 1 to n
550 for j = 1 to n
560 a(i,j) = 0
570 next j
580 next i
590 input #9,i,j,a(i,j)
600 if i = 0 then goto 620
610 goto 590
620 for i = 1 to n
630 b(i) = 0
640 next i
650 input #9,i,b(i)
660 if i = 0 then goto 680
670 goto 650
680 close #9
690 input "Constant recruitment to be
    added (enter 0 if zero) ";rr
700 input "How many interations in the
    simulation? ";im
710 input "Would you like results saved
    to a file? (y/n) ";p$
720 if p$ = "N" or p$ = "n" then goto
    770
730 if p$ <> "Y" and p$ <> "y" then goto
    710
740 p$ = "y"
750 input "Name of file for output: ";f$
760 open f$ for output as #9
```

```
770 jq = 0
780 print jq,
790 for i = 1 to n-1
800 print b(i),
810 next i
820 print b(n)
830 if p$ = "N" or p$ = "n" then goto
    890
840 print #9,jq,
850   for i = 1 to n-1
860 print #9,b(i),
870 next i
880 print #9,b(n)
890 for jq = 1 to im
900 for i = 1 to n
910 c(i) = 0
920 for k = 1 to n
930 c(i) = c(i)+a(i,k)*b(k)
940 next k
950 next i
960 print jq, : for i = 1 to n-1 : print
    c(i), : next i : print c(n)
970 if p$ = "N" or p$ = "n" then goto
    1030
980 print #9,jq,
990 for i = 1 to n-1
1000 print #9,c(i),
1010 next i
1020 print #9,c(n)
1030 for i = 1 to n
1040 b(i) = c(i)
1050 next i
1060 b(i) = b(i)+rr
1070 next jq
1080 close #9
1090 goto 50
1100 end
```

10

Confidence Intervals for λ

INTRODUCTION

There are a number of reasons for wanting to estimate the variance of λ but the major reason is to obtain confidence limits. There is uncertainty in the estimate of the transitions in a matrix, some of which is due to the number of individuals that contribute to the calculations, termed demographic stochasticity (Goodman 1987, Menges 1986), and some due to the stochastic nature of environmental conditions (*e.g.* Bierzychudek 1982, Kalisz and McPeek 1992). Measurements may be made over several years and annual transitions may vary substantially. An average value of λ can be determined but one also needs an explicit statement concerning variability and the confidence that can be placed in an estimate.

In this chapter I present two different ways for estimating confidence limits. The first makes use of the variances of individual matrix elements, $V(a_{ij})$, and the second method is computer intensive and projects a population based on the random selection of transition matrices. These two methods should be viewed as introductions to a large and rich field that includes Monte Carlo methods (*e.g.* Alvarez-Buylla and Slatkin 1991, 1993) and discussions of computer-intensive resampling methods (*e.g.* Lenski and Service 1982, Meyer *et al.* 1986, McPeek and Kalisz 1993).

DEMOGRAPHIC STOCHASTICITY

The first method (Equation 10.1) for estimating confidence limits for λ utilizes not only the variance of matrix elements, $V(a_{ij})$, which were introduced in Chapter 3, but also the sensitivities of λ to changes in a_{ij} (Lande 1988, Caswell 1989). It makes sense that the contributions of the individual elements to the variance of λ should be adjusted so if λ is not very sensitive to changes in a_{ij}, the element a_{ij} should not contribute very much to $V(\lambda)$,

$$V(\lambda) \approx \sum \left(\frac{\partial \lambda}{\partial a_{ij}} \right)^2 V(a_{ij}). \tag{10.1}$$

The sensitivities, $\frac{\partial \lambda}{\partial a_{ij}}$, can be obtained from the program **MATRIX.BAS** and $V(a_{ij})$ for survival rates have a binomial sampling distribution

$$V(a_{ij}) = \frac{a_{ij}(1 - a_{ij})}{N} \tag{10.2}$$

where N is the sample size. For transitions that do not represent survival rates, that is, f_x values, some independent measure of variance is required.

The following example uses a data set assembled by John Rae for a rare and endangered cactus in Florida, *Cereus eriophorus*. Estimate of the variance of λ is in three parts. The first part focuses on size-specific mean and variation of fruit production; the second part assembles a transition matrix in order to determine $\partial \lambda / \partial a_{ij}$; and, the third part, assembles information from parts one and two to estimate variance using Equation 10.1.

TABLE 10.1 Fruit Numbers for *Cereus eriophorus*

Size (dm)	N	Mean size (dm) ±se	Mean # fruits±se	Var. of fruit #	Var. × 0.03448	$V(f_x)$
1-4	105	2.48±0.09	0.0286±0.0163	0.02800	9.662×10^{-4}	9.202×10^{-6}
4-8	98	5.94±0.12	0.1122±0.0320	0.10067	3.471×10^{-3}	3.542×10^{-5}
8-16	95	11.32±0.24	0.4000±0.0724	0.49787	1.717×10^{-2}	1.807×10^{-4}
16-32	146	23.03±0.37	0.8014±0.0970	1.37407	4.738×10^{-2}	3.245×10^{-4}
>32	42	42.90±2.04	2.1905±0.3014	3.8165	1.316×10^{-1}	3.133×10^{-3}

Note: sizes are in decimeters (0.1 m); 0.03448 is recruits per fruit based on the average annual recruitment of 1.8 plants and a total of 261 fruits, and a 5 year period (1.8 × 5/261); $V(f_x)$ is calculated using Equation 10.4; data from John Rae.

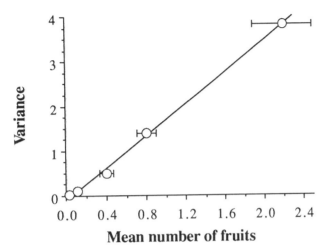

FIGURE 10.1 Variance, V, in fruit number in *Cereus eriophorus* as a function of mean number of fruits, M, per plant; error bars are 1 se; the regression equation is V = -0.097 + 1.784M; $r^2 = 0.998$; data from John Rae.

TABLE 10.2 Numbers of Individual *Cereus eriophorus* Transferring Among Size Classes During One Year

To next↓	Size class at the start of the year					
	≤1	1-4	4-8	8-16	16-32	>32
≤1	20	2				
1-4	5	60	10	6		
4-8		12	58	17	3	
8-16		1	14	50	12	
16-32			1	19	105	8
>32					20	32
Dead	11	30	15	3	6	2
Total	36	105	98	95	146	42

Note: units are 0.1 m; all data from John Rae combined from 1988 to 1993.

$$V(f_x) = \frac{V(f) \times 0.03448}{N}. \qquad (10.3)$$

Data on fruit numbers for *Cereus eriophorus* (Table 10.1), were analyzed to show how variance changed with the mean number in each size class. Analysis consisted of sorting the data file by size class and then determining the mean number of fruits per plant and the variance of this estimate, which is called "variance of fruit #" in Table 10.1. The relationship between mean and variance (Figure 10.1) is very close to linear with the variance 1.8× the mean.

The annual recruitment was about 1.8 new cacti per year during the period from 1988-93 and I will assume that recruitment is related to fruit production so when few fruits are produced, recruitment is reduced. The total number of fruits produced from 1988-1992 was 261 and so recruits/fruit is (1.8 × 5)/261 = 0.03448. Recruits per fruit, 0.03448, times fruit number in a size class, divided by the number of individuals in the size class, is f_x. The estimate is f_x rather than m_x because early survival was included in the original observation of numbers of recruits, 1.8 yr^{-1}.

The variance, V, of f_x for a size class (Equation 10.4) is based on how the variance of fruit number V(f) changes with size class and the conversion of fruits to recruits. N is the number of plants used to estimate V(f) in Table 10.1.

Data on growth and survival for *Cereus eriophorus* are shown in Table 10.2. For this analysis all years have been combined but data from individual years will be treated separately in a following section of this chapter to evaluate the effect of year to year variation of transitions on λ.

Because totals in Table 10.2 are known in each initial size class, growth, g_x, and retention, r_x, transitions are calculated as number in each size class divided by the column total. For example, in the ≤1 dm column (initial size), 20 remained in the ≤1 dm size class and 5 grew to the 1-4 size class. The initial number was 36 (the column total) and so r_x = 20/36 or 0.5555 yr^{-1} and g_x = 5/36 or 0.1389 yr^{-1}. Just to clarify this relationship, it is possible to separate growth or retention from survival as was presented in Chapter 9. Survival rate for the ≤1 dm size class, p_x, is 25/36 or 0.6944 yr^{-1}. A total of 25 plants survived for one year and these could be used to calculate growth, G_x, and retention, R_x, transitions that *do not* include survival. G_x is 5/25 or 0.2 and R_x is 20/25 or 0.8. G_x and R_x always sum to 1.0. As shown in Chapter 9, the g_x and r_x transitions are $G_x \times p_x$ and $R_x \times p_x$ or g_x = 0.2 × 0.6944 = 0.1389 and r_x = 0.8 × 0.6944 = 0.5555, which are the same as the calculations presented at the beginning of this paragraph.

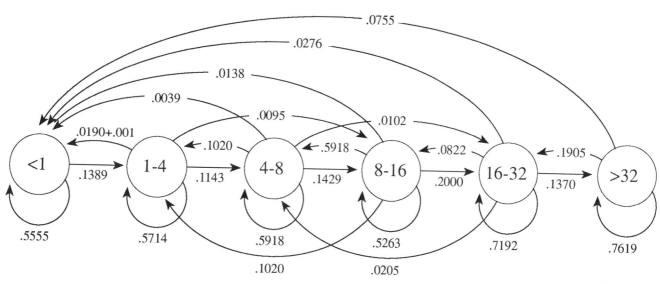

FIGURE 10.2 Life-cycle graph of *Cereus eriophorus*; transitions based on all data combined from 1988 to 1993; time for all transitions is 1 year; transition from the second node (1-4 dm) to the first node (<1 dm) is 0.0190 + 0.001 which is shrinkage (0.0190) plus f_x (0.001); data from John Rae.

TABLE 10.3 Calculation of Contributions of Matrix Elements g_x, r_x, and s_x to the Variance of λ for the Cactus *Cereus eriophorus*

Element	Number transferred	Total N	p	V(a_{ij})	Sensitivity $\partial\lambda/\partial a_{ij}$	Contribution to V(λ)
a_{11}	20	36	0.5555	0.00686	0.007601	3.9627×10^{-7}
a_{12}	2	105	0.0190	0.00018	0.005926	6.2339×10^{-9}
a_{21}	5	36	0.1389	0.00332	0.020824	1.4407×10^{-6}
a_{22}	60	105	0.5714	0.00233	0.016233	6.1461×10^{-7}
a_{23}	10	98	0.1020	0.00093	0.018329	3.1400×10^{-7}
a_{24}	6	95	0.0632	0.00062	0.018306	2.0885×10^{-7}
a_{32}	12	105	0.1143	0.00096	0.043714	1.8424×10^{-6}
a_{33}	58	98	0.5918	0.00247	0.049358	6.0053×10^{-6}
a_{34}	17	95	0.1789	0.00155	0.049295	3.7574×10^{-6}
a_{35}	3	146	0.0205	0.00014	0.154845	3.2976×10^{-6}
a_{42}	1	98	0.0095	0.00010	0.084623	6.8759×10^{-7}
a_{43}	14	98	0.1429	0.00125	0.095549	1.1410×10^{-5}
a_{44}	50	95	0.5263	0.00262	0.095427	2.3898×10^{-5}
a_{45}	12	146	0.0822	0.00052	0.299755	4.6430×10^{-5}
a_{53}	1	98	0.0102	0.00010	0.141182	2.0534×10^{-6}
a_{54}	19	95	0.2000	0.00168	0.141002	3.3485×10^{-5}
a_{55}	105	146	0.7192	0.00138	0.442917	2.7136×10^{-4}
a_{56}	8	42	0.1905	0.00367	0.348505	4.4594×10^{-4}
a_{65}	20	146	0.1370	0.00081	0.493701	1.9738×10^{-4}
a_{66}	32	42	0.7619	0.00432	0.388464	6.5179×10^{-4}

Note: probability, p, is # transferred/N; V(a_{ij}) = p(1-p)/N; contribution of a_{ij} to V(λ) is $(\partial\lambda/\partial a_{ij})^2$V($a_{ij}$) as shown in Equation 10.1; based on data from John Rae.

Values of f_x are determined from Table 10.1 using the mean number of recruits per fruit; that is, 0.03448, and the mean number of fruits/individual in each size class. The formula for f_x is

$$f_x = \frac{\text{mean number of fruits} \times 0.03448}{N}. \qquad (10.4)$$

For example, in the >32 dm size class, mean number of fruits/individual = 2.190 so $2.190 \times 0.03448 = 0.0755$, which is f_6 in the transition matrix **A** (Equation 10.5). Equation 10.5 combines r_x, g_x, and f_x transitions and these are shown as a life-cycle graph in Figure 10.2.

$$\mathbf{A} = \begin{pmatrix} 0.5555 & 0.0200 & 0.0039 & 0.0138 & 0.0276 & 0.0755 \\ 0.1389 & 0.5714 & 0.1020 & 0.0632 & 0 & 0 \\ 0 & 0.1143 & 0.5918 & 0.1789 & 0.0205 & 0 \\ 0 & 0.0095 & 0.1429 & 0.5263 & 0.0822 & 0 \\ 0 & 0 & 0.0102 & 0.2000 & 0.7192 & 0.1905 \\ 0 & 0 & 0 & 0 & 0.1370 & 0.7619 \end{pmatrix} \quad (10.5)$$

Analysis of the transition matrix (Equation 10.5) using **MATRIX.BAS** provides estimates of the sensitivities, $\partial\lambda/\partial a_{ij}$, that can be used to calculate the contribution of each element, a_{ij}, to the variance of λ. Contributions of variability of g_x, r_x, and s_x to variance in λ are shown in Table 10.3 and contributions of variability of fecundity to the variance of λ are shown in Table 10.4.

The summed contributions to $V(\lambda)$ are 1.70232×10^{-3} from survival and 1.0076×10^{-6} from fecundity for a total of

$$V(\lambda) = \boxed{1.7033 \times 10^{-3}};$$

the square root of $V(\lambda)$ gives the standard error, se,

$$se = \sqrt{1.7033 \times 10^{-3}} = \boxed{0.04127}.$$

The estimated value of the population growth rate, λ_1, for combined data from 1988-1992 is:

$$\lambda_1 = \boxed{0.936 \text{ yr}^{-1}}$$

and approximate 95% confidence limits are ± 1.96 se or ± 0.081 so λ probably lies between 0.855 and 1.017.

TABLE 10.4 Contribution of Fecundity, f_x, of *Cereus eriophorus* to Variance of λ

Element	Variance	Sensitivity	Contribution
a_{12}	9.2019×10^{-6}	0.005926	3.2315×10^{-10}
a_{13}	3.5420×10^{-5}	0.006691	1.5857×10^{-9}
a_{14}	1.8070×10^{-4}	0.006682	8.0681×10^{-9}
a_{15}	3.2451×10^{-4}	0.02099	1.4297×10^{-7}
a_{16}	3.1332×10^{-3}	0.016516	8.5467×10^{-7}

Note: calculation of $V(f_x)$ is given in Table 10.1; data from John Rae.

TEMPORAL STOCHASTICITY

A second approach to estimating confidence limits explicitly includes year to year variation in transition matrices and uses them randomly in a simulation. Each year the estimate of population growth rate per individual is $ln(N_{i+1}/N_i)$ and a mean r can be estimated from a series of iterations (Heyde and Cohen 1985). If the yearly estimates of r are sorted from high to low, then the top 2.5% and the lower 2.5% define the 95% confidence limits.

A first step is to create the appropriate fecundity values for each year. As with the total matrix, the adjustment is based on numbers of fruits produced and the assumption of 1.8 recruits/year (Table 10.5) and this is followed by determining the survival data for each year (Table 10.6). Fecundity and survival rates can be combined into transition matrices for each year (Equations 10.6-10.10)

88-89

$$\mathbf{A} = \begin{pmatrix} 0.4444 & 0 & 0.0034 & 0.0029 & 0.0312 & 0.0460 \\ 0.3333 & 0.7308 & 0.05 & 0.0417 & 0 & 0 \\ 0 & 0.1538 & 0.55 & 0.1667 & 0 & 0 \\ 0 & 0 & 0.25 & 0.4583 & 0.0625 & 0 \\ 0 & 0 & 0 & 0.2917 & 0.875 & 0 \\ 0 & 0 & 0 & 0 & 0.0315 & 1 \end{pmatrix}$$

$$(10.6)$$

89-90

$$\mathbf{A} = \begin{pmatrix} 0.6 & 0.0013 & 0.0034 & 0.0131 & 0.0239 & 0.1121 \\ 0.2 & 0.5185 & 0.0952 & 0.1429 & 0 & 0 \\ 0 & 0.0741 & 0.5238 & 0.1429 & 0.0278 & 0 \\ 0 & 0 & 0.0952 & 0.4762 & 0.1389 & 0 \\ 0 & 0 & 0 & 0.1429 & 0.6667 & 0.25 \\ 0 & 0 & 0 & 0 & 0.1111 & 0.5 \end{pmatrix}$$

$$(10.7)$$

TABLE 10.5 Number of Fruits Produced by Each Size Class of *Cereus eriophorus* from 1988-1989

	≤1	1-4	4-8	8-16	16-32	>32	Total
1988 N	9	26	20	24	32	3	
# fruits	0	0	2	2	29	4	37
f_x	0	0	0.0034	0.0029	0.0312	0.0460	
1989 N	5	27	20	21	36	4	
# fruits	0	1	2	8	25	13	49
f_x	0	0.0013	0.0034	0.0131	0.0239	0.1121	
1990 N	4	23	20	17	28	6	
# fruits	0	0	0	2	10	6	18
f_x	0	0	0	0.0041	0.0123	0.0345	
1991 N	8	17	19	17	25	14	
# fruits	0	1	3	7	25	30	66
f_x	0	0.0020	0.0054	0.0142	0.0345	0.0739	
1992 N	9	12	19	19	25	15	
# fruits	0	1	4	19	28	39	91
f_x	0	0.0029	0.0073	0.0345	0.0386	0.0896	

Note: recruits per fruit was 0.03448 and f_x is recruits per fruit x fruit number in a size class ÷ n in the size class; data from John Rae.

TABLE 10.6 Size Transitions and Survivors for *Cereus eriophorus*

'88-'89 n=114	≤1.0	1-4	4-8	8-16	16-32	>32
≤ 1.0	4	0	0	0	0	0
1 - 4	3	19	1	1	0	0
4 - 8	0	4	11	4	0	0
8 - 16	0	0	5	11	2	0
16 - 32	0	0	0	7	28	0
>32	0	0	0	0	1	3
Dead	2	3	3	1	1	0
Total	9	26	20	24	32	3

'91-'92 n=100	≤1.0	1-4	4-8	8-16	16-32	>32
≤ 1.0	7	2	0	0	0	0
1 - 4	1	6	2	0	0	0
4 - 8	0	1	13	3	1	0
8 - 16	0	1	3	11	1	0
16 - 32	0	0	1	2	18	4
>32	0	0	0	0	5	10
Dead	0	8	0	0	0	0
Total	8	18	19	16	25	14

'89-'90 n=114	≤1.0	1-4	4-8	8-16	16-32	>32
≤ 1.0	3	0	0	0	0	0
1 - 4	1	14	2	3	0	0
4 - 8	0	2	11	3	1	0
8 - 16	0	0	2	10	5	0
16 - 32	0	0	0	3	24	1
>32	0	0	0	0	4	2
Dead	1	11	6	2	2	1
Total	5	27	21	21	36	4

'92-'93 n=96	≤1.0	1-4	4-8	8-16	16-32	>32
≤ 1.0	4	0	0	0	0	0
1 - 4	0	10	3	1	0	0
4 - 8	0	0	12	5	1	0
8 - 16	0	0	0	9	3	0
16 - 32	0	0	0	1	17	3
>32	0	0	0	0	2	11
Dead	6	1	4	0	2	1
Total	10	11	19	16	25	15

'90-'91 n=98	≤1.0	1-4	4-8	8-16	16-32	>32
≤ 1.0	2	0	0	0	0	
1 - 4	0	11	2	1	0	
4 - 8	0	5	11	2	0	
8 - 16	0	0	4	9	1	
16 - 32	0	0	0	6	18	
>32	0	0	0	0	8	6
Dead	2	7	2	0	1	
Total	4	23	19	18	28	6

Note: data from John Rae.

90-91

$$A = \begin{pmatrix} 0.5 & 0 & 0 & 0.0041 & 0.0123 & 0.0345 \\ 0 & 0.4783 & 0.1053 & 0.0556 & 0 & 0 \\ 0 & 0.2174 & 0.5789 & 0.1111 & 0 & 0 \\ 0 & 0 & 0.2105 & 0.5 & 0.0357 & 0 \\ 0 & 0 & 0 & 0.3333 & 0.6429 & 0 \\ 0 & 0 & 0 & 0 & 0.2857 & 1 \end{pmatrix}$$

(10.8)

91-92

$$A = \begin{pmatrix} 0.875 & 0.1131 & 0.0054 & 0.0142 & 0.0345 & 0.0739 \\ 0.125 & 0.3333 & 0.1053 & 0 & 0 & 0 \\ 0 & 0.0556 & 0.6842 & 0.1875 & 0.04 & 0 \\ 0 & 0.0556 & 0.1579 & 0.6875 & 0.04 & 0 \\ 0 & 0 & 0.0526 & 0.125 & 0.72 & 0.2857 \\ 0 & 0 & 0 & 0 & 0.2 & 0.7143 \end{pmatrix}$$

(10.9)

and

92-93

$$A = \begin{pmatrix} 0.4 & 0.0029 & 0.0073 & 0.0345 & 0.0386 & 0.0896 \\ 0 & 0.9091 & 0.1579 & 0.0625 & 0 & 0 \\ 0 & 0 & 0.6316 & 0.3125 & 0.04 & 0 \\ 0 & 0 & 0 & 0.5625 & 0.12 & 0 \\ 0 & 0 & 0 & 0.0625 & 0.68 & 0.2 \\ 0 & 0 & 0 & 0 & 0.08 & 0.7333 \end{pmatrix}$$

(10.10)

A simulation using all 5 transition matrices selected randomly for 1000 iterations was used to estimate $ln(\lambda)$ and the 95% confidence interval of $ln(\lambda)$. The program is called **SIM_VAR.BAS** and is a modification of **SIMPOP.BAS**. A major modification is the addition of a subscript to the elements of the transition matrices so that each element is a_{kij}. The subscript k is the matrix number from 1 to 5, which is selected at random using two lines of code. A new seed value is selected for the random number generator each time the program is run. A line further in the program uses a random number to select a matrix. The two important lines for selecting a matrix based on a random number are

```
20 randomize timer
580 k=int(m*rnd+1)
```

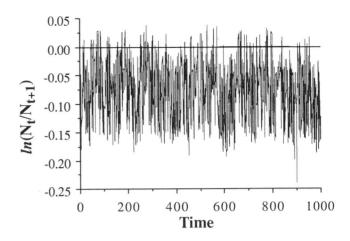

FIGURE 10.3 $ln(N_{i+1}/N_i)$ for 1000 iterations (i) using 5 randomly selected transition matrices for a cactus, *Cereus eriophorus* based on data from John Rae.

TABLE 10.7 Comparison of Estimates of λ and 95% Confidence Limits for Demographic Stochasticity (Equation 10.1) and Temporal Stochasticity (Simulation)

Stochasticity	λ	lower 95%	upper 95%
Demographic	0.936	0.854	1.016
Temporal	0.928	0.843	1.023

The data file for **SIM_VAR.BAS** must contain all five transition matrices (Equations 10.6-10.10) together with a vector of starting conditions. The appropriate file for the cactus data is provided at the end of this chapter.

Results of the simulation (Figure 10.3) gave a mean value of -0.0757 for $ln(N_{i+1}/N_i)$ with a median of -0.0746. The 1000 values of $ln(N_{i+1}/N_i)$ values of the output from the simulation were sorted and the 26th value from the top and from the bottom were used as estimates of the upper and lower 95% confidence limits, which were -0.17067 to 0.02285, which are appropriate for $ln(\lambda)$ and so the 95% confidence limits for λ are 0.843 to 1.023 yr^{-1}.

Estimation of λ and 95% confidence limits by simulation or by determining $V(\lambda)$ using Equation 10.1 with all data combined provide similar though not identical results (Table 10.7). The similarity suggests that most of the observed variability probably is associated with demographic stochasticity rather than temporal variation.

There are additional complications in estimating λ that were not included in the above analysis. The first is that Equation 10.1 includes an assumption that the covariance of a_{ij} terms is equal to zero. A second complication arises from an assumption of no correlation among years, which was the explicit assumption in the simulation that was done for *Cereus eriophorus*. An extended analysis should account for these complications and an approach is provided by Tuljapurkar (1982, 1990) for estimating population growth in stochastic environments

$$a \approx ln(\lambda) - \frac{\tau^2}{2\lambda^2} + \frac{\theta}{\lambda^2} \qquad (10.11)$$

where a is the stochastic analogue of r, τ^2 is a term similar to $V(a_{ij})$ in Equation 10.1 but includes covariance of the a_{ij} elements, and θ is the autocorrelation term for correlation of environmental states from year to year.

Patterns of autocorrelation may be important (Tuljapurkar 1982, 1990; Orzack 1985); however, Orzack's work is rather difficult to translate into predictions of what one should find in the field and, possibly, this is the most important lesson to be learned. It *may* be that everything is simpler than it first seems because his analysis of covariance of traits and the influence of environmental autocorrelation are both based on having nearly equal environmental conditions: 50% good years and 50% bad years. Negative autocorrelation, under these circumstances, means that good and bad years flip back and forth very frequently with the maximum negative autocorrelation being when good and bad years alternate. Positive autocorrelation with 1:1 good and bad would require a long string of good years followed by a long string of bad years. Neither of these seem reasonable for long-lived species primarily because good and bad years tend not to be 1:1. It is much more likely that species live under generally good conditions with occasional catastrophes or generally poor conditions with occasional super years when the population is rebuilt to a high level. For field conditions it *may* be that if environmental conditions are normally distributed they approach Orzack's requirement and so if there are as many super years as catastrophes and as many ok+ and ok- years then 50% good to 50% bad may still be reasonable. Because "good" and "bad" are defined by organisms rather than by weather data, it is unknown whether this is true or not. The fact that some long-term records indicate non-normality of recruitment (*e.g.* Shreve 1917, Bowman and Lewis 1977, Shepherd 1990, Grant and Grant 1992, Ebert *et al.* 1994, Noda and Nakao 1996) suggests that normality of environmental conditions should not be assumed but rather must be measured.

Few researchers have actually attempted to use Equation 10.11 with field data probably because of the demands placed on having long-term measurements of all of the appropriate parts. Benton *et al.* (1995) have analyzed a 21 year data set for red deer, *Cervus elephas*, on the island of Rum off the coast of Scotland. They conclude that "...there is little difference in the results calculated using stochastic and deterministic techniques, thus allowing the considerable convenience of the use of standard methodology (Caswell 1989)." For many species and many studies, "standard methodology" is sufficiently daunting that it would be

unfortunate if confidence limits for population growth were not estimated because Equation 10.11 was overwhelming.

PROBLEMS

1. Noda and Nakao (1996) studied a subtidal snail *Umbonium costatum* in Hakodate Bay, Japan. They simulated population growth using 8 transition matrices with substantial variation in annual recruitment. Age-specific survival and fecundity are shown in Table 10.8 and annual variation in recruitment, u_0, is presented in Table 10.9. Noda and Nakao calculate f_x as

$$f_x = u_0 m_x p_0.$$

The post-settlement survival rates were considered to be the same from year to year and so only annual recruitment varied.

a. First estimate λ by using a mean value for u_0 and discuss the meaning of the elasticity for $u_0 \times p0$ (*i.e.* the Σf_x)

b. Construct 8 transition matrices, one for each year, and estimate λ by simulation using random matrices. Determine the 95% confidence limits and discuss the importance of variation in recruitment in light of your original elasticity analysis.

TABLE 10.8 Age-Specific Survival (p_x) and Fecundity (m_x) for *Umbonium costatum* in Hakodate Bay, Japan

Age (x)	$p_x \pm se$	n	m_x
0	0.278±0.064	6	-
1	0.565±0.134	5	2100
2	0.770±0.097	4	17000
3	0.757±0.091	3	27000
4	0.722±0.023	2	32000
5	0.785±0.078	4	35000
6+	0.785±0.078		37000

Note: data from Noda and Nakao (1996).

TABLE 10.9 Recruitment (u_0) for a Subtidal Snail *Umbonium costatum* in Hakodate Bay, Japan

Year	$u_0 \times 10^{-4}$
1982	5.977
1983	0.069
1984	1.331
1985	0.071
1986	0.131
1987	0.038
1988	0.482
1992	0.147

Note: data from Noda and Nakao (1996).

TABLE 10.10 Stages and Sizes of *Calathea ovandensis*

	Stage	Leaf area (cm^2)
1	Seeds	
2	Seedlings	<1450
3	Juveniles	<1450
4	Pre-reproductive	<1450
5	Small reproductive	<1450
6	Medium reproductive	≥1450 and <2050
7	Large reproductive	≥2050 and <2950
8	Extra-large	≥2950

Note: data from Horvitz and Schemske (1995).

2. Horvitz and Schemske (1995) provide four years of transition data for *Calathea ovandensis,* a perennial monocot in a lowland secondary forest near San Andrés Tuxtla, Veracruz, Mexico. They used a mixed stage- and size-structured analysis (Table 10.10) and define $f_x = p_x m_{x+1}$; that is, individuals in a particular node survive and then reproduce. Because survival can be to various other stages, f_x is a summation of the contributions of all individuals that grew (and survived) into other stages. Use the transition matrices (Table 10.11) to find the long-term value for λ and the 95% confidence limits by simulation.

TABLE 10.11 Transitions for *C. ovandensis* in Plot #3

1982-83	stage	1	2	3	4	5	6	7	8
	1	0.4580	0	0	28.6	30.4	45	57.2	72.4
	2	0.1378	0	0	0	0	0	0	0
	3	0.0041	0.0508	0.1875	0	0.0385	0	0	0
	4	0	0	0.2500	0.1667	0.0385	0	0	0
	5	0	0	0	0.6667	0.3462	0.1429	0.0645	0
	6	0	0	0	0	0.1923	0.1905	0	0
	7	0	0	0	0	0.1923	0.3333	0.5161	0.1667
	8	0	0	0	0.1667	0.0769	0.2381	0.3871	0.8333
1983-84									
	1	0.5710	0	0.7	2.9	16.5	13.5	26.2	39.3
	2	0.0287	0	0	0	0	0	0	0
	3	0.0003	0.0097	0.1154	0.3333	0	0	0	0
	4	0	0	0.0769	0.5000	0.1111	0.3333	0.0345	0
	5	0	0	0.0385	0.1667	0.7222	0.5556	0.3793	0.1200
	6	0	0	0	0	0.1111	0.1111	0.3448	0.4000
	7	0	0	0	0	0	0	0.1724	0.2800
	8	0	0	0	0	0	0	0	0.1600
1984-85									
	1	0.5500	0	0.9	5.3	11.3	15.7	16.4	27.5
	2	0.0478	0.0148	0.0500	0	0	0	0	0
	3	0.0021	0.0148	0.1000	0.1429	0.1111	0.0417	0	0
	4	0	0	0.0500	0.1429	0.2222	0.2917	0.3636	0.0909
	5	0	0	0.0500	0.2857	0.4444	0.2500	0.3636	0.4545
	6	0	0	0	0	0.1111	0.3333	0.0909	0.2727
	7	0	0	0	0	0	0.0417	0.1818	0.0909
	8	0	0	0	0	0	0	0	0.0909
1985-86									
	1	0.5560	0	4.2	18.8	36.6	60.2	83.8	116.6
	2	0.0438	0.0351	0	0	0	0	0	0
	3	0.0002	0.0351	0.2500	0.0625	0	0	0	0
	4	0	0	0.0833	0.2500	0.0455	0	0	0
	5	0	0	0.1667	0.5000	0.5000	0.0769	0	0
	6	0	0	0	0.1250	0.4091	0.6923	0.2500	0
	7	0	0	0	0	0.0455	0.0769	0.5000	0
	8	0	0	0	0	0	0.1538	0.2500	1.0000

Note: data from Horvitz and Schemske (1995).

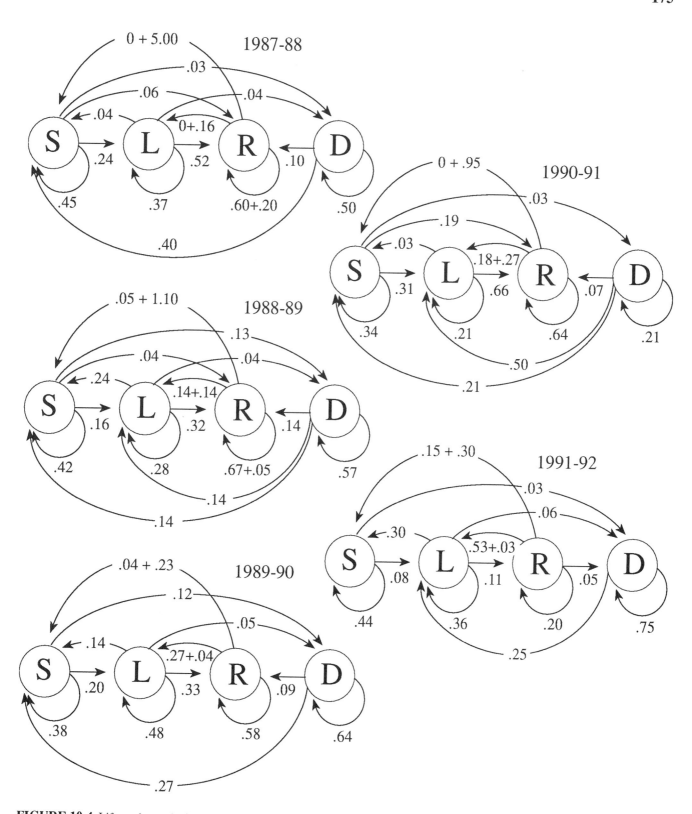

FIGURE 10.4 Life-cycle graphs for a perennial herb, *Astragalus scaphoides*; data from Lesica (1995).

3. Lesica (1995) presents data on annual transitions of stages for a perennial herb, *Astragalus scaphoides*, in Beaverhead County, Montana. The particular study site is called Haynes Creek. Stages are defined as

S   Small nonreproductive (1-3 leaves),

L   Large nonreproductive (≥4 leaves),

R   Reproductive,

D   Dormant (no above ground parts observed).

In life cycle graphs shown in Figure 10.4, some transitions from reproductive individuals (the 'R' node) have two numbers. The first number is the transfer of reproductive individuals to another stage and the second number is f_x. If only one number is given, it is a transfer and not f_x. When you constitute transition matrices, you will just add these two numbers together.

Use the life cycle graphs to develop matrices and then find the long-term value for λ and the 95% confidence limits by estimating variance (Equation 10.1) and also by simulation. For Equation 10.1 you will need an estimate of $V(f_x)$ which can be approximated using the f_x values for the five matrices. In order to estimate variance for survival terms, you need values of N (Table 10.12).

TABLE 10.12 Numbers of *Astragalus scaphoides* Sampled Each Year

Year	N
1987	75
1988	97
1989	108
1990	97
1991	112
1992	104

Note: data from Lesica (1995)

4. Croxall *et al.* (1990) provide demographic data for wandering albatrosses, *Diomedea exulans*, at Bird Island, South Georgia. Breeding females lay one egg by the end of January when a census was taken each year to determine the number of successful breeders; some nesting attempts failed. Fledging was based on chicks tagged (ringed) in late October (Table 10.13) and, assuming a 50:50 sex ratio, one half of this number is a good estimate of p_0m_x.

Birds return to South Georgia at about age 5 years as juveniles and survival estimates from time of banding as fledged chicks are shown in Table 10.14. Survival estimates were based on sightings of marked birds (Cormack 1964), which is a technique that was introduced in Chapter 3. Survival to first breeding, which is from 10 to 12 years, is shown in Table 10.15 and adult survival in Table 10.16.

TABLE 10.13 Breeding Population Size and Nesting Success of Wandering Albatrosses at Bird Island

Season	Pairs attempting	Pairs at 31 Jan.	Chicks hatched	Chicks fledged
1976	1433	1318	997	848
1977	1541	1418	1093	800
1978	1382	1271	1023	(939)
1980	1339	1173	906	719
1981	1415	1320	1047	886
1982	1404	1261	1028	914
1983	1453	1263	972	876
1984	1366	1302	1103	993
1985	1232	1103	890	783
1986	1491	1431	1217	1059
1987	1233	1176	948	859
1988	1366	1281	1045	959

Note: the difference between pairs attempting and pairs at 31 Jan. is the number of failed nests; successful nesting is determined at a census on 31 January and fledging is determined at the end of October; data from Croxall *et al.* 1990.

TABLE 10.14 Estimated Survival Rates of Juvenile Wandering Albatrosses from Fledging to Age 5 years

Cohort	Birds ringed	Survival rate
1972	368	0.453
1973	75	0.428
1975	854	0.472
1976	847	0.425
1977	806	0.525
1978	871	0.506

Note: survival is for both sexes combined; data from Croxall *et al.* (1990).

TABLE 10.15 Recruitment Probability from Fledging to First Breeding for Wandering Albatrosses

Cohort	Birds Ringed	10-11 years	12 years
1962/63	1400	0.129	0.140
1972	368	0.155	0.052
1973	75	0.227	0.040
1975	854	0.190	0.046
1976	847	0.170	—

Use the data on wandering albatrosses to

a. Construct an average life cycle and matrix and discuss the meaning of the elasticities. Be sure to include nodes for return breeders, both those that are successful and those that fail (Table 10.17).

b. Construct life cycle graphs for each year.

c. Develop transition matrices from the graphs.

d. Simulate the population using random matrices to determine λ and 95% confidence limits.

TABLE 10.16 Annual Survival of Adult Female Wandering Albatrosses Based on Cormack's Method

Year	A	B	C	Survival (se)
1976	96	0	13	0.935 (0.044)
1977	97	26	13	0.927 (0.031)
1978	43	80	8	0.963 (0.024)
1979	10	155	19	0.912 (0.027)
1980	8	110	8	0.943 (0.023)
1981	2	171	16	0.919 (0.022)
1982	3	149	11	0.941 (0.021)
1983	0	151	17	0.895 (0.026)
1984	0	135	9	0.945 (0.022)
1985	0	126	11	0.936 (0.027)
1986	1	123	29	0.873 (—)
1987	1	97	61	
1988	0	63	63	
Geometric mean survival 1976-84				0.931 (0.006)

Note: data are for three cohorts tinged as chicks in 1958, 1962, and 1963; A = number first captured; B = number of previously captured birds seen; C = number seen for last time; annual survival rates are not different at p>0.10; data from Croxall *et al.* (1990).

TABLE 10.17 Breeding Frequency of Wandering Albatrosses Based on Current Breeding Success

	Successful breeders				Failed breeders		
Year	N	+2	+3	+>4	N	+1	+2	+>3
1976	128	0.66	0.13	0.05	63	0.71	0.16	0.06
1977	103	0.71	0.11	0.06	118	0.38	0.44	0.09
1978	146	0.59	0.18	0.07	64	0.63	0.09	0.12
1979					47	0.71	0.15	0.06
1980	275	0.66	0.07	0.06	212	0.67	0.10	0.06
1981	394	0.76	0.00	0.05	200	0.78	0.02	0.05
1982	326	0.74	0.04	0.05	174	0.67	0.16	0.04
1983	355	0.77	0.05		258	0.61	0.15	0.07
1984	443	0.75			224	0.85	0.03	
1985					295	0.75		

BASIC PROGRAM

The program **SIM_VAR.BAS** will read up to 30 transition matrices and an initial vector of starting conditions. The program randomly selects a matrix for projection of one time unit and produced a new vector of numbers of individuals in each state such as age class, stage, or size class. Another matrix is then randomly selected to produce the next vector, and so on.

```
10 dim a(30,30,30),b(30),c(30)
20 randomize timer
30 print "Population growth simulation
   with variation"
40 print "Another fine product from
   Cornered Rat Software©"
50 print "                        T. A.
   Ebert 1996"
60 input "Do you want to continue?
   (y/n)";f$
70 if f$ = "y" or f$ = "Y" then goto 100
80 if f$ <> "n" and f$ <> "N" then goto
   60
90 goto 880
100 input "Name of file with matrices:
    ";f$
110 open f$ for input as #9
120 input #9,m
130 rem m is the number of matricies
    that are used in the analysis
140 print "m = ";m
150 for k = 1 to m
160 input #9,n,n
170 for i = 1 to n
180 for j = 1 to n
190 a(k,i,j) = 0
200 next j
210 next i
220 input #9,i,j,a(k,i,j)
230 if i = 0 then goto 250
240 goto 220
250 next k
260 for i = 1 to n
270 b(i) = 0
280 next i
290 input #9,i,b(i)
300 if i = 0 then goto 320
310 goto 290
320 close #9
330 input "How many interations in the
    simulation? ";im
340 input "Would you like results saved
    to a file? (y/n) ";p$
350 if p$ = "N" or p$ = "n" then goto
    400
360 if p$ <> "Y" and p$ <> "y" then goto
    340
370 p$ = "y"
380 input "Name of file for output: ";f$
390 open f$ for output as #9
400 t = 0
410 for i = 1 to n
420 t = t+b(i)
430 next i
```

```
440 jq = 0
450 print jq,
460 for i = 1 to n-1
470 print b(i)/t,
480 next i
490 print b(n)/t
500 if p$ = "N" or p$ = "n" then goto
    560
510 print #9,jq,
520 for i = 1 to n-1
530 print #9,b(i)/t,
540 next i
550 print #9,b(n)/t
560 ts = t
570 for jq = 1 to im
580 k = int(m*rnd +1)
590 k = k+1
600 t = 0
610 for i = 1 to n
620 c(i) = 0
630 for j = 1 to n
640 c(i) = c(i)+a(k,i,j)*b(j)
650 next j
660 t = t+c(i)
670 next i
680 r = log(t/ts)
690 ts = t
700 print jq,
710 for i = 1 to n
720 print c(i)/t,
730 next i
740 print r
750 if p$ = "N" or p$ = "n" then goto
    810
760 print #9,jq,
770 for i = 1 to n
780 print #9,c(i)/t,
790 next i
800 print #9,r
810 for i = 1 to n
820 b(i) = c(i)
830 next i
840 b(i) = b(i)
850 next jq
860 close #9
870 goto 60
880 end
```

The following data file contains all five transition matrices together with a final vector containing initial conditions. The format for matrices is exactly the same as used in other Cornered Rat programs.

```
5
6, 6
1, 1, 0.4444
1, 3, 0.0034
1, 4, 0.0029
1, 5, 0.0312
1, 6, 0.0460
2, 1, 0.3333
2, 2, 0.7308
2, 3, 0.05
2, 4, 0.0417
3, 2, 0.1538
3, 3, 0.55
3, 4, 0.1667
4, 3, 0.25
4, 4, 0.4583
4, 5, 0.0625
5, 4, 0.2917
5, 5, 0.875
6, 5, 0.0315
6, 6, 1
0, 0, 0
6, 6
1, 1, 0.6
1, 2, 0.0013
1, 3, 0.0034
1, 4, 0.0131
1, 5, 0.0239
1, 6, 0.1121
2, 1, 0.2
2, 2, 0.5185
2, 3, 0.0952
2, 4, 0.1429
3, 2, 0.0741
3, 3, 0.5238
3, 4, 0.1429
3, 5, 0.0278
4, 3, 0.0952
4, 4, 0.4762
4, 5, 0.1389
5, 4, 0.1429
5, 5, 0.6667
5, 6, 0.25
6, 5, 0.1111
6, 6, 0.5
0, 0, 0
6, 6
1, 1, 0.5
1, 4, 0.0041
1, 5, 0.0123
1, 6, 0.0345
2, 2, 0.4783
2, 3, 0.1053
2, 4, 0.0556
```

```
3,  2,  0.2174                      5,  6,  0.2857
3,  3,  0.5789                      6,  5,  0.2
3,  4,  0.1111                      6,6,0.7143
4,  3,  0.2105                      0,  0,  0
4,  4,  0.5                         6,  6
4,  5,  0.0357                      1,  1,  0.4
5,  4,  0.3333                      1,  2,  0.0029
5,  5,  0.6429                      1,  3,  0.0073
6,  5,  0.2857                      1,  4,  0.0345
6,  6,  1                           1,  5,  0.0386
0,  0,  0                           1,  6,  0.0896
6,  6                              2,  2,  0.9091
1,  1,  0.875                       2,  3,  0.1579
1,  2,  0.1131                      2,  4,  0.0625
1,  3,  0.0054                      3,  3,  0.6316
1,  4,  0.0142                      3,  4,  0.3125
1,  5,  0.0345                      3,  5,  0.04
1,  6,  0.0739                      4,  4,  0.5625
2,  1,  0.125                       4,  5,  0.12
2,  2,  0.3333                      5,  4,  0.0625
2,  3,  0.1053                      5,  5,  0.68
3,  2,  0.0556                      5,  6,  0.2
3,  3,  0.6842                      6,  5,  0.08
3,  4,  0.1875                      6,  6,  0.7333
3,  5,  0.04                        0,  0,  0
4,  2,  0.0556                      1,500000
4,  3,  0.1579                      2,1400000
4,  4,  0.6875                      3,1800000
4,  5,  0.04                        4,1200000
5,  3,  0.0526                      5,2100000
5,  4,  0.125                       6,1300000
5,  5,  0.72                        0,0
```

11

Growth Functions for Individuals

INTRODUCTION

Caswell(1982a) showed how to incorporate individual growth and survival functions into transition matrices that describe population growth. The functions can describe reproductive output and survival and so provide a means for smoothing values that are used in matrix projection (*e. g.* Ebert 1985). This approach tends to minimize apparent population behavior that arises solely from sampling problems that determine values in a transition matrix. The purpose of this chapter is to present a number of growth functions and show how to estimate parameters that can be used to describe not only overall size of individuals as a function of time but also the reproductive output with increasing age or size.

Several growth functions are commonly used and two are particularly popular: the von Bertalanffy or Brody-Bertalanffy curve (Brody 1927, 1945; von Bertalanffy 1934, 1938; Ricker 1975a) and the logistic equation which,

although used as a model of individual growth (*e. g.* Arambasic *et al.* 1987, Soukupová 1988), usually is found associated with population growth (see any introductory ecology text). A third model for growth, the Gompertz function, was developed for modeling survival (Gompertz 1825) and still is used for that purpose (*e. g.* Rosen *et al.* 1981, Libertini 1988, Finch 1990). It also was one of the earliest models of individual growth (Winsor 1932) and continues to be used for growth studies (*e. g.* Kruger 1978, Gage 1987, Watanabe *et al.* 1988, Jacobsen and Kushlan 1989, Kingsley 1989, Nakaoka and Matsui 1994) The Brody-Bertalanffy, logistic and Gompertz models all are part of a family of curves that can be written as a single function, now known as the Richards function after its developer (Richards 1959). It is presented in various forms but the one that I prefer is

$$S_t = S_\infty (1 - be^{-Kt})^{-n} \qquad (11.1)$$

in which

S_t = the size of an individual at time t,

S_∞ = asymptotic size,

K = the growth rate constant with units of time^{-1},

n = a shape parameter,

S_0 = size at t=0,

and

$$b = \frac{S_\infty^{-1/n} - S_0^{-1/n}}{S_\infty^{-1/n}}. \qquad (11.2)$$

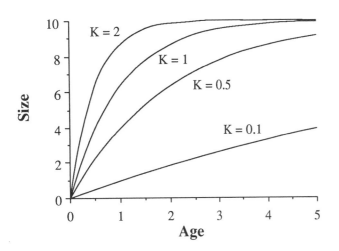

FIGURE 11.1 The effect of changes in the growth-rate constant K on changes in the shapes of growth curves all with the same shape parameter n=-1, the Brody-Bertalanffy model; maximum size S_∞, equal to 10, and size at time 0, S_0, equal to 0.01; the larger the value of K, the faster maximum size is approached.

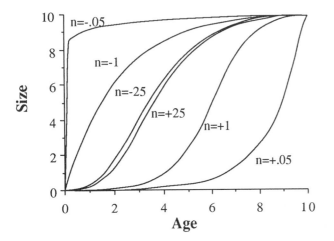

FIGURE 11.2 Examples of the Richards family of growth curves; the Brody-Bertalanffy curve has a shape parameter n=-1, the logistic is n=+1; note that the curves for n=+25 and n=-25 are very close together and they are converging on the Gompertz equation as |n| → ∞; the parameter K was selected so growth is 99% of S_∞ at age 10.

When n = -1, Equation 11.1 is the Brody-Bertalanffy equation. When n is +1 the function is the logistic and as |n| → ∞ the function approaches the Gompertz equation. The function has a discontinuity at n=0.

To provide a bit more intuitive feeling for the Richards family of curves, first consider the growth-rate constant K. The meaning of this parameter is shown in Figure 11.1 using n=-1, the Brody-Bertalanffy model. All curves were constructed with a asymptotic length of 10 and size at time 0, S_0, equal to 0.01. The larger the value of K, the faster the organism approaches its maximum size

TABLE 11.1 Values of K Used with Different Shape Parameters to Generate Figure 11.2 so that 99% of Maximum Size was Attained at Age 10

n	K	b
+25	0.667392	-0.31826
+1	1.150187	-999
+0.05	13.9657	-1×10^{60}
-0.05	0.1703237	1.0
-1.0	0.460517	0.999
-25	0.639802	0.24142

The consequences of changing the shape parameter n is less intuitive (Figure 11.2). The graph was generated by having all curves attain 99% of their maximum size at an age of 10 and maximum size was again equal to 10 with S_0 = 0.01. In order to achieve this, it was necessary to change the growth-rate constant K (Table 11.1).

The curves in Figure 11.2 cover a wide range of possibilities some of which have never been observed in actual organisms. The region of the function where the shape parameter n is near zero, approached either from the positive or negative direction is very strange and under normal circumstances organisms do not have growth curves this extreme. From the positive side, an organism shows exponential growth nearly up to maximum size and then has an inflection point and maximum size is attained. In Figure 11.2 the curve for n = +0.05 has an inflection at about age 11.5. At the other extreme, for n = -0.05, organisms achieve most of their size within only about 0.2 time units and then grow quite slowly towards maximum size. Most organisms seem to have absolute values of n no smaller than about 0.2.

FIXED TIME INTERVALS

A variety of procedures can be used to estimate parameters if data are *size* and *age* (Richards 1959, Nelder 1961, Causton 1969, Causton *et al.* 1978, Pienaar and Thomson 1973, Johnson *et al.* 1975, Schnute 1981) but data from field populations often must be gathered without knowing actual age. In many studies, natural growth lines are assumed to represent age but validation of natural lines is rare and without validation the translation of natural growth lines into ages must be viewed with suspicion. When age is not known, data frequently are size at time of tagging or mapping for sessile animals or plants, size some time later, and the time interval between the observations. Equation 11.1 is unsuitable for parameter estimation and what is needed is a difference equation.

The difference equation model is a regression with size at t+Δt as a function of size at time t; that is, size at the time of recapture as a function of size at the time of tagging. The first step is to rearrange the exponent n in Equation 11.1 so individual size measurements are transformed

$$S_t^{-1/n} = \text{size at tagging after transformation,}$$

$$S_{t+\Delta t}^{-1/n} = \text{size at recapture after transformation,}$$

and

$$S_\infty^{-1/n} = \text{asymptotic size after transformation;}$$

so Equation 11.1 becomes

$$S_t^{-1/n} = S_\infty^{-1/n}(1-be^{-Kt})$$

or $$S_t^{-1/n} = S_\infty^{-1/n} - S_\infty^{-1/n}be^{-Kt}, \qquad (11.3)$$

which can be rearranged to

$$be^{-Kt} = 1 - \frac{S_t^{-1/n}}{S_\infty^{-1/n}}. \qquad (11.4)$$

At time t+Δt, the size is

$$S_{t+\Delta t}^{-1/n} = S_\infty^{-1/n}(1-be^{-K(t+\Delta t)})$$

or

$$S_{t+\Delta t}^{-1/n} = S_\infty^{-1/n}(1-\mathbf{be^{-Kt}}e^{-K\Delta t}). \qquad (11.5)$$

Now, substitute

$$1 - \frac{S_t^{-1/n}}{S_\infty^{-1/n}}$$

from Equation 11.4 into Equation 11.5 for be^{-Kt}

$$S_{t+\Delta t}^{-1/n} = S_\infty^{-1/n}\left[1 - \left(1 - \frac{S_t^{-1/n}}{S_\infty^{-1/n}}\right)e^{-K\Delta t}\right],$$

which is

$$S_{t+\Delta t}^{-1/n} = S_\infty^{-1/n} - (S_\infty^{-1/n} - S_t^{-1/n})e^{-K\Delta t} \qquad (11.6)$$

or

$$S_{t+\Delta t}^{-1/n} = S_\infty^{-1/n} - S_\infty^{-1/n}e^{-K\Delta t} + S_t^{-1/n}e^{-K\Delta t}$$

and so finally

$$S_{t+\Delta t}^{-1/n} = S_\infty^{-1/n}(1-e^{-K\Delta t}) + S_t^{-1/n}e^{-K\Delta t}. \qquad (11.7)$$

If Δt is a constant, that is, all measurements were made over the same time period such as one year, then Equation 11.7 is a linear equation with

$$\text{slope} = e^{-K\Delta t} \qquad (11.8)$$

and

$$\text{intercept} = S_\infty^{-1/n}(1-e^{-K\Delta t})$$

or

$$\text{intercept} = S_\infty^{-1/n}(1-\text{slope}). \qquad (11.9)$$

If the Brody-Bertalanffy model is the one of choice, then no transformation is needed (remember n = -1 and -1/n = 1); if the logistic is the model of choice (n = +1), however, then an inverse transformation of Equation 11.1 is appropriate (*cf.* Brewer 1976)

$$S_t = S_\infty(1-be^{-Kt})^{-1},$$

which is

$$S_t = \frac{S_\infty}{1-be^{-Kt}}$$

or

$$\frac{1}{S_t} = \frac{1}{S_\infty}(1-be^{-Kt})$$

so Equation 11.7 is

$$\frac{1}{S_{t+\Delta t}} = \frac{1}{S_\infty}(1-e^{-K\Delta t}) + \frac{1}{S_{t+\Delta t}}e^{-K\Delta t}.$$

For the Gompertz equation, Equation 11.1 could be written

$$S_t = S_\infty(1-be^{-Kt})^{\pm 25}$$

or some exponent greater than 25 could be used; however, as |n| → ∞ the equation converges on

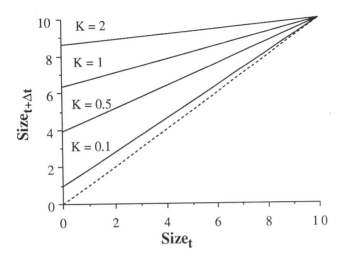

FIGURE 11.3 Walford graphs of Brody-Bertalanffy growth curves shown in Figure 11.1; the dotted line is the 45° line of 0-growth; $S_\infty = 10$ and so all lines intersect the 45° line at 10.

$$S_t = S_\infty e^{(1-be^{-Kt})}$$

and so, as a difference equation, the Gompertz is

$$ln(S_{t+\Delta t}) = ln(S_\infty)(1 - e^{-K\Delta t}) + ln(S_{t+\Delta t}) e^{-K\Delta t}.$$

Obviously, other values for n could be used for the transformation and thereby turn the problem into one of linear regression. Two points need to be emphasized: (1) the above procedure works ONLY when Δt is a constant so there is a single slope; and, (2) you can't estimate n by this method because n enters as a nonlinear parameter. We come back to this in a few pages. And finally, a minor point: K has units of 1/time so it is possible to change K in years to K in days by dividing by 365, just as we did with r in previous chapters.

Graphs of size at $t+\Delta t$ *vs.* size at t, with fixed Δt, have been used for a long time to estimate parameters of the Brody-Bertalanffy growth equation. This derivation was developed by Ford (1933) and by Walford (1946) and has been used to estimate growth parameters graphically as well as by linear regression using least-squares.

Growth curves shown in Figure 11.1 have been re-done as difference equations (Figure 11.3) and show the relationships indicated by Equation 11.7, namely, increased values of K flatten the slope and increase the Y-intercept. Also, the growth curves shown in Figure 11.2 have been recalculated in difference-equation form (Equation 11.7) in order to show how different shape parameters give rise to different shapes in the Walford plot (Figure 11.4). Plotting has been done in Figure 11.4 without transformation. If data were transformed, then all would be linear.

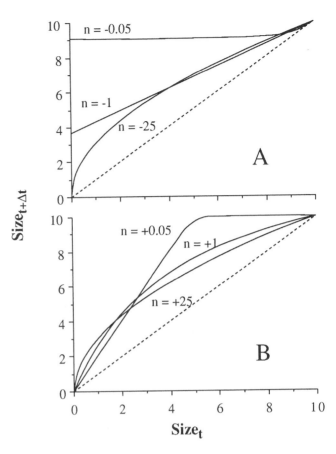

FIGURE 11.4 Walford graphs for growth curves shown in Figure 11.2 with negative values of the shape parameter n in A and positive values in B; all graphs are for a fixed value of Δt, the time interval between measurements; the dotted line is the 45° line of 0 growth; $S_\infty = 10$ and so all lines intersect the 45° line at 10.

TABLE 11.2 Size-Pair Data for *Macoma baltica*

Size 1	Size 2
6.632	10.011
10.011	12.496
12.496	14.201
14.201	15.528
15.528	16.189

Note: size pairs are separated by one year and were determined using natural growth lines; data from Bachelet (1980).

A clam, *Macoma baltica*

Data for *Macoma baltica* (Bachelet 1980) (Table 11.2) are an example of growth data separated by a fixed time interval (Figure 11.5). The data are based on natural growth lines that are *assumed* to be separated by one year. The best model to use for analysis is indicated by the distribution of data points in Figure 11.5. Because the data points have a

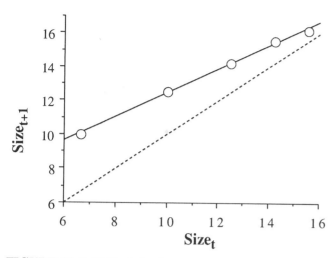

FIGURE 11.5 Walford plot for growth of a common mud-flat clam, *Macoma baltica* in the Gironde estuary of France (Bachelet 1980) based on growth lines that are assumed to be annual; the 45° line is the line of zero growth.

linear distribution in a Walford plot, the Brody-Bertalanffy model is the best one to use. Had the scatter of data points not been linear, a model other than the Brody-Bertalanffy would have been better. Figure 11.4 would have provided some possibilities.

The regression equation for Figure 11.5, with standard errors, is

$$S_{t+1} = 0.704(\pm0.017)S_t + 5.307(\pm0.206)$$

and the correlation coefficient is 0.998. Slope and intercept can be used to estimate K and S_∞ using Equation 11.8 and 11.9

$$K = -ln(0.704) = \boxed{0.351 \text{ yr}^{-1}}$$

and

$$S_\infty = \frac{5.397}{(1 - 0.704)} = \boxed{18.23\text{mm}}.$$

Because the correlation coefficient is nearly 1.0 for the *Macoma* data, it is possible to ignore one important problem that usually must be addressed. A major assumption of predictive linear regression has been violated: namely, that the covariate, x, is error free, which is not true in a Walford plot. Size at t is measured with the same error as size at t+Δt and there also is inherent natural variability. There are methods for addressing this problem called ordinary major axis and reduced major axis regressions. Reduced major axis regression also is called Model II regression or geometric mean regression. These methods have been described by a variety of investigators (*e. g.* Tessier 1948, Kermack 1954, Ricker 1973, Lande 1979, Seim and Sæther

1983, Smith 1980, Laws and Archie 1981, Rayner 1985, Ebert 1988, McArdale 1988, reviewed by LaBarbera 1989, Jolicoeur 1990). There is not universal agreement on what should be done. Ricker (1973, 1975a, 1975b, 1982) has advocated use of Model II or geometric mean regression in which the functional regression lies between lines drawn using predictive regressions of y=f(x) and x=g(y).

The relationship between GM functional and ordinary least-squares or Model I regressions is quite simple and the two can be related by the correlation coefficient, r. The functional slope, v, is related to the predictive slope, b

$$v = \frac{b}{r}.$$

Both slopes pass through the center of the data cloud, $\overline{x}, \overline{y}$, and so once the slope has been determined, the intercept is easily calculated and hence functional estimates of K and S_∞. In terms of sums of squares,

$$b = \frac{\sum xy}{\sum x^2}$$

and

$$v = \sqrt{\frac{\sum y^2}{\sum x^2}} = \frac{b}{r} \qquad (11.10)$$

There also are opponents of the *ad hoc* use of functional regression (Jolicoeur 1975, Sprent and Dolby 1980, 1982). A major criticism leveled by these authors is that there is no good reason to assume that the errors in x and y are *equal*. Prairie *et al.* (1995) argue that alternatives of the ordinary least-squares (Model I) regression "...should be tempered with the realization that, without an estimate of the natural error variability component (not just the measurement errors), these estimators are equally likely to be worse than they are to be better...." On the other hand, if it is reasonable to assume equal errors in x and y, then Model II (geometric mean) regression would be appropriate. If S_t and $S_{t+\Delta t}$ are measured in the same way, then it is reasonable to assume equal errors, otherwise errors should be measured.

Another clam, *Protothaca staminea*

Here is an example where the regression model influences parameter estimates. A death assemblage of the clam *Protothaca staminea* was collected from South Slough, Coos Bay, Oregon. Natural growth lines were used to measure growth increments and the time interval between measurements was assumed to be one year (Figure 11.6). Natural growth lines for this species have been validated in southern California (Smith 1974) and so, probably, can be applied to samples in Oregon.

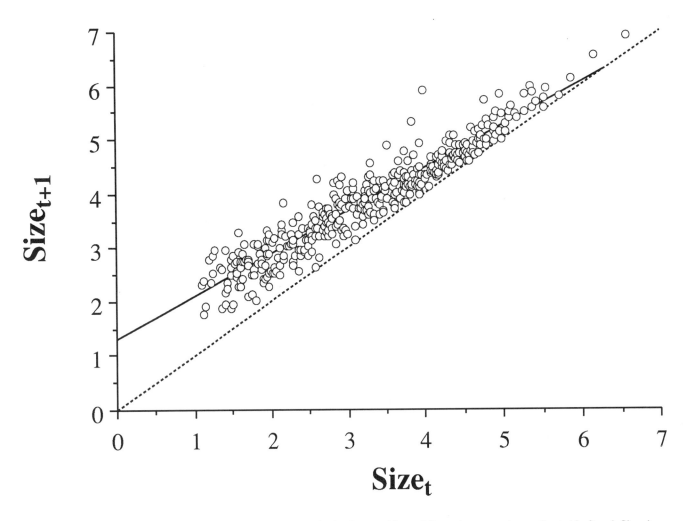

FIGURE 11.6 Walford graph based on check lines on shells of a death assemblage of *Prototheca staminea* collected in Sough Slough, Coos Bay, Oregon; N = 422; (Population Ecology class, Oregon Institute of Marine Biology, August 1994, unpublished).

Using linear regression with N = 422,

$$r^2 = 0.905,$$

$$slope = 0.7942 \pm 0.0125(se),$$

and

$$intercept = 1.3074 \pm 0.0427(se).$$

Accordingly,

$$K = 0.2304 \text{ yr}^{-1}$$

with 95% limits of

$$0.2000 < K < 0.2463 \text{ yr}^{-1}$$

and

$$S_\infty = 6.3533 \text{ cm}$$

with 95% limits of

$$5.9474 < S_\infty < 6.7591 \text{ cm}.$$

The functional regression has a slope that is determined from the predictive slope divided by the correlation coefficient, which is 0.9513.

The functional slope, Equation 11.10, is

$$\frac{0.79422}{0.9513} = 0.83488$$

so the functional estimate of K is

$$\boxed{K = 0.1805 \text{ yr}^{-1}}.$$

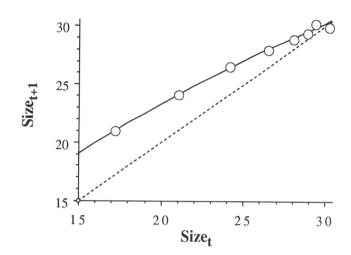

FIGURE 11.7 Walford graph for length (cm) of ciscoes (*Coregonus artedii*) from Vermilion Lake, Minnesota; time, Δt, is one year (Carlander 1950); fitted line is for the logistic equation determined using linear regression of $1/S_{t+\Delta t}$ *vs.* $1/S_t$; $r^2 = 0.996$.

TABLE 11.3 Size Pairs Separated by One Year for Ciscoes (*Coregonus artedii*) from Vermilion Lake, Minnesota

Size$_t$	Size$_{t+1}$	Size$_t$	Size$_{t+1}$
17.2	21.0	28.9	29.4
21.0	24.1	29.4	30.2
24.1	26.5	30.2	29.9
26.5	28.0	29.9	30.6
28.0	28.9		

Note: data from Carlander (1950).

The intercept for the functional regression requires knowing the means for X and Y because

$$\overline{Y} = C + v\,\overline{X}$$

The mean of $S_t = 3.2226$ cm and the mean of $S_{t+1} = 3.8669$ cm; so,

$$3.8669 = C + 0.83488 \times 3.2226$$

or

$$C = 1.1764$$

and so

$$S_\infty = \frac{1.1764}{1 - 0.83488} \text{ or } \boxed{S_\infty = 7.125 \text{ cm}},$$

TABLE 11.4 Regression Analysis of Inverse-Transformed Data for Ciscoes (*Coregonus artedii*) from Vermilion Lake, Minnesota

Parameter	Value	se
Slope	0.5865	0.0130
Intercept	0.0135	0.0005
K	0.5335	
$1/S_\infty$	0.0327	
S_∞	30.5813	

Note: $r^2 = 0.996$; N = 9; K = -*ln*(slope), $1/S_\infty$ = intercept/(1-slope); data from Carlander (1950).

which is outside the 95% confidence limits set by using Model I regression where x is a fixed effect. Which is the better way? There certainly are errors in both x and y and in a Walford graph equality of errors seems reasonable and so Model II regression probably should be the preferred way.

A fish, cisco, *Coregonus artedii*

Sometimes data in a Walford plot are curved rather than linear (Figure 11.7). If the curve is real then the logistic or Gompertz functions may be better models than the Brody-Bertalanffy. A curve may very well be real in the case of the ciscoes in Figure 11.7 because means are plotted and total N in the sample was 533 (Carlander 1950).

To estimate parameters of the logistic equation for data in Figure 11.7, sizes (Table 11.3) were first inverse-transformed followed by linear regression (Table 11.4).

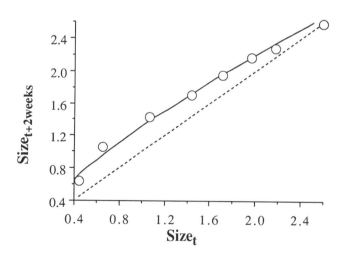

FIGURE 11.8 Walford graph of change in size over 2-week periods of a cattail shoot (*Typha glauca*) measured in meters (Waters and Shay 1991; fitted line is based on the Gompertz function.

TABLE 11.5 Growth of a Cattail (*Typha glauca*) Shoot

L_1	L_2	L_1	L_2
0.440	0.650	1.710	1.960
0.650	1.060	1.960	2.170
1.060	1.430	2.170	2.290
1.430	1.710	2.590	2.590

Note: L_1 is length at time t in meters and L_2 is length two weeks later; data from Waters and Shay (1991).

TABLE 11.6 Regression Analysis of Log-Transformed Data for a Cattail (*Typha glauca*) Shoot

Parameter	Value	se
Slope	0.7397	0.0376
Intercept	0.2745	0.0239
K	0.3015	
$ln(S_\infty)$	1.0546	
S_∞	2.87m	

Note: data from Waters and Shay (1991); N = 8; K = -ln(slope), $ln(S_\infty)$ = intercept/(1-slope); $r^2 = 0.982$.

Cattails, *Typha glauca*

Another example of a Walford plot where data show a curve is provided in Figure 11.8 for growth of a cattail (*Typha glauca*) shoot (Waters and Shay 1991); data are shown in Table 11.5.

To estimate parameters of the Gompertz equation for data in Figure 11.8, sizes (Table 11.5) were first subjected to a logarithmic transformation followed by linear regression (Table 11.6). The problem of measurement error in x obviously is not a problem for either the ciscoes or cattails given the high values of r^2. Given the gentle curves in both examples, it would be possible to use either the logistic or the Gompertz or, for some purposes, even the Brody-Bertalanffy model, without serious changes in conclusions concerning general growth trends. This is not always true, and as more data points are added over a wider range of sizes, different models provide very different degrees of goodness-of-fit.

A tropical sea urchin, *Diadema setosum*

Figure 11.9 shows a plot that obviously is not linear and hence not suitable for the Brody-Bertalanffy model nor does it bend in such a manner that the logistic or Gompertz models would be reasonable. Comparison with curves in Figure 11.4 suggests that some value of n between -0.05 and -1 probably would be best. The degree of scatter also suggests that estimation probably would be best using functional regression.

The data for the sea urchin *Diadema setosum* are given in Table 11.7 and because n can not be assumed to be any

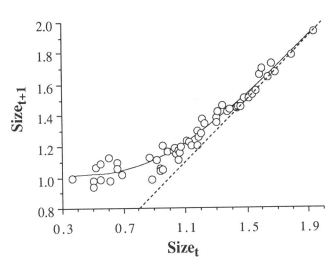

FIGURE 11.9 Growth of demi-pyramids (jaw elements) of a tropical sea urchin, *Diadema setosum* from Zanzibar (Ebert 1980); measurements are in centimeters and based on marks produced by tetracycline tagging; time interval is one year; the dotted line is the line of 0 growth; fitted line has 3 parameters of the Richards function.

particular value, n must be made a parameter to be estimated. One procedure to do this is to transform data by raising both $S_{t+\Delta t}$ and S_t to the -1/n power, to use a linear regression equation,

$$S_{t+\Delta t}^{-1/n} = S_\infty^{-1/n}(1 - be^{-K(t+\Delta t)}),$$

and to minimize the residual sum of squares of the function:

$$SSE = \sum_{i=1}^{N}\left[Y_i - (mX_i + C)^{-n}\right]^2, \qquad (11.11)$$

where

$$Y = S_{t+\Delta t} \text{ and } X = S_t^{-1/n}$$

with the restriction that m and C are the functional slope and intercept.

A minimization procedure (**RICH.BAS**) is given in Ebert (1980) where data are transformed by raising each value to the -1/n power, doing a *functional* linear regression and then using Equation 11.11 to obtain SSE. The discontinuity at n=0 means that the pattern of search requires two runs in order to explore both negative and positive values of n (Figure 11.10); C and m are used to estimate K and S_∞

$$K = -ln(\text{m}),$$

and

$$S_\infty = \left(\frac{C}{1-m}\right)^{-n}.$$

TABLE 11.7 Demi-Pyramids (Jaws) of *Diadema setosum* Tagged with Tetracycline in June 1976 at the East African Marine Fisheries Research Organization Laboratory in Zanzibar and Collected in June 1977

S_t	S_{t+1}	S_t	S_{t+1}	S_t	S_{t+1}	S_t	S_{t+1}	S_t	S_{t+1}
0.36	1.00	0.87	1.13	1.05	1.11	1.21	1.38	1.48	1.51
0.50	0.94	0.88	0.99	1.05	1.18	1.22	1.35	1.51	1.51
0.50	0.98	0.91	1.11	1.06	1.16	1.30	1.36	1.535	1.54
0.52	1.07	0.935	1.06	1.07	1.20	1.30	1.39	1.555	1.56
0.55	0.99	0.94	1.04	1.11	1.24	1.31	1.43	1.58	1.66
0.55	1.09	0.95	1.05	1.125	1.23	1.34	1.47	1.60	1.70
0.60	1.13	0.95	1.21	1.14	1.21	1.37	1.43	1.635	1.65
0.61	0.98	0.98	1.17	1.18	1.21	1.385	1.44	1.66	1.73
0.66	1.06	1.02	1.18	1.18	1.30	1.433	1.45	1.68	1.68
0.66	1.10	1.03	1.19	1.183	1.26	1.438	1.46	1.79	1.79
0.69	1.02	1.04	1.15	1.20	1.29	1.46	1.46	1.94	1.94

Note: S_t is the length of the glowing image under UV light; S_{t+1} is the length of the jaw; measurements in centimeters (Ebert 1980).

TABLE 11.8 Comparison of Different Models for Growth of Demi-pyramids of *Diadema setosum* from Zanzibar

	Logistic	Brody-Bertalanffy	Gompertz	Richards
Shape parameter n	+1.0	-1.0	±25	-0.213
SSE	1.12079	0.35858	0.60249	0.1234
Walford slope	0.304074	0.64342	0.48044	0.9808
Walford intercept	0.488402	0.56667	0.52809	1.0134
K yr^{-1}	1.19048	0.44096	0.73305	0.0194
S_∞ cm	1.42491	1.58918	1.50272	2.3286
Scaling parameter b	-13.24906	0.93707	0.10273	1.0000

Note: data from Ebert (1980).

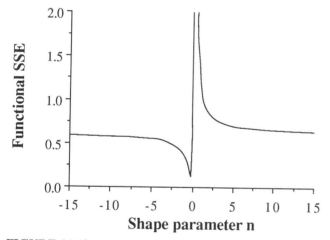

FIGURE 11.10 Sum of squares of the residuals for data in Table 11.7. The minimum is for the best value for n in a functional regression where both size$_t$ and size$_{t+1}$ are subject to equal errors; because exponents are -1/n, there is a discontinuity at n=0.

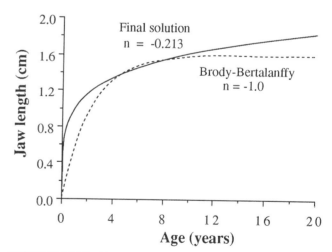

FIGURE 11.11 Size of demi-pyramids (jaws) as a function of time for *Diadema setosum* with a fit for the Brody-Bertalanffy equation (n= -1.0) and the best fit solution for the Richards function (n = -0.213).

Growth of jaws of *Diadema setosum* from Zanzibar illustrate how changes in the shape parameter change SSE; the value of n that provides the smallest SSE is the best estimate. Each time n is changed, both S_t and S_{t+1} are transformed by raising each to the -1/n power and linear regression parameters are estimated. Functional regression parameters are then estimated. An estimated value of $(S_{t+1})^{-1/n}$ is first transformed back to S_{t+1}, before subtracting it from the observed value of S_{t+1}, squaring the difference and summing to determine the SSE.

Table 11.8 shows the changes in SSE with different growth models. The difference between the poorest fit, the logistic, and the best fit using a shape parameter of -0.213, is nearly an order of magnitude change in SSE. For the second best fit, the Brody-Bertalanffy model, the difference is about 200%, which is reasonable considering that the scatter of data (Figure 11.9) is far away from linear. A comparison of the Brody-Bertalanffy (n=-1.0) and the Richards function with n = -0.213 (Figure 11.11) illustrates the difference between the two models. The small, negative shape parameter translates into rapid initial growth followed by a prolonged period of slow growth towards maximum size.

VARIABLE TIME INTERVALS

So far, all examples have used some fixed value for the time interval between measurements. There are times when this is not possible even though it really is a good idea to do so particularly if work is being done under field conditions. Seasonal changes can result in substantial changes in growth increments during different seasons and so if some size pairs are for a 6-month period from spring through summer and other size pairs are for 3 months during winter, combining all data into a single analysis confounds seasonal effects with size effects on growth. However, some studies can not be conducted to accommodate restrictions of having time intervals exactly equal to one year and so analysis must be done with mixtures of intervals or, in terms of the equations we have been using, Δt no longer is a constant but rather is a variable.

Variable time intervals were brought into growth analysis by Fabens (1965) for the Brody-Bertalanffy model; however, if data are transformed before applying his method, then the approach works very well for the logistic or Gompertz or, with known n, for other Richards function transformations.

A good starting point is Equation 11.6

$$S_{t+\Delta t}^{-1/n} = S_{\infty}^{-1/n} - (S_{\infty}^{-1/n} - S_{t}^{-1/n})e^{-K\Delta t}.$$

Now add and subtract $S_{t}^{-1/n}$ *to the right size of the equation* and rearrange

$$S_{t+\Delta t}^{-1/n} = S_{t}^{-1/n} + (S_{\infty}^{-1/n} - S_{t}^{-1/n}) - (S_{\infty}^{-1/n} - S_{t}^{-1/n})e^{-K\Delta t}.$$

The result is the model presented by Fabens (1965)

$$S_{t+\Delta t}^{-1/n} = S_{t}^{-1/n} + (S_{\infty}^{-1/n} - S_{t}^{-1/n})(1 - e^{-K\Delta t}), \quad (11.12)$$

which is rather pleasing because it has two parts on the right size of the equation: original size,

$$S_{t}^{-1/n},$$

and a growth increment,

$$S_{\infty}^{-1/n} - S_{t}^{-1/n})(1 - e^{-K\Delta t}).$$

Given $S_{t}^{-1/n}$ and Δt_i one can estimate $S_{t+t}^{-1/n}$. The procedure is to choose S_{∞} and K to minimize the sum of squared residuals, which is a nonlinear regression problem that can be solved with any nonlinear regression module in statistics software. It also can be solved using the program **FABENS.BAS** at the end of this chapter.

The final parameter that may be determined is b, which can not be estimated from tagging data. To determine b, it is necessary to know the age of at least one individual of a particular size. Selection of some arbitrary value will not influence estimates of K or S_{∞} but will determine the position of the growth curve on the time axis; that is, pinning a particular size to a particular age. Typical data that might be known are sizes at birth or settlement or hatching so data are one or more sizes at time = 0. The parameter b is

$$b = \frac{\sum e^{-Kt}(S_{\infty}^{-1/n} - S_{t}^{-1/n})}{S_{\infty}^{-1/n}\sum\left[e^{-Kt}\right]^2}, \quad (11.13)$$

where t is known age at a particular size S_t. Note that with a single value at t=0, Equation 11.13 is that same as Equation 11.2. Now, here is an example.

TABLE 11.9 Growth Data for Seven *Nautilus belauensis* Marked and Recaptured in Palau

Date of tagging	Date of recapture	Δ Time (days)	Δ Time (years)	Original diameter	Final diameter
25 May 1982	14 Aug. 1982	82	0.225	19.22	19.65
5 July 1981	5 Aug. 1982	330	0.904	20.82	21.28
30 May 1982	15 July 1982	46	0.126	17.70	17.84
29 May 1982	17 July 1982	45	0.123	20.23	20.30
5 July 1981	12 June 1982	342	0.937	17.90	18.98
20 June 1981	8 June 1982	353	0.967	19.23	20.41
8 June 1981	29 May 1982	355	0.973	20.30	20.48

Note: measurements are in centimeters; data from Saunders (1984).

TABLE 11.10 Growth Analysis for *Nautilus belauensis* Using the Brody-Bertalanffy Model

Parameter	Estimate	Asymptotic se
S_∞	21.99 cm	±0.98
K	0.38 yr^{-1}	±0.17

Note: N = 7; computer code for standard errors adapted from the FORTRAN code of Sims (1985); total SS = 7.79489; residual SS = 0.30739; r^2 = 0.96; data from Saunders (1984).

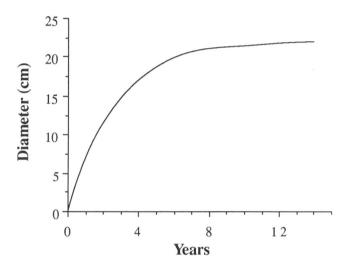

FIGURE 11.12 Brody-Bertalanffy growth curve estimated for *Nautilus belauensis* Saunders from Palau using data published by Saunders (1984); asymptotic diameter, S_∞, is estimated to be 21.99 cm, the growth rate constant, K, is 0.381 yr^{-1}, and b = 0.998.

Nautilus belauensis in Palau

The following analysis uses tag and recapture data gathered by Saunders (1984) for a cephalopod *Nautilus belauensis* in Palau (Table 11.9). The BASIC program, **FABENS.BAS**, based on Fabens' FORTRAN I program was used to estimate K and S_∞ (Table 11.10). Nothing is known about actual age and so in order to draw a reasonable growth curve, I pinned the curve at 0.05 cm for age 0.

Graphically, with the curve pinned to 0.05 cm at t=0 (Figure 11.12), a reasonable relationship can be seen between age and size. Now, is it the biologically correct one? Or even the best model? With available data, it is impossible to tell. A problem with the analysis is that the smallest individuals that were tagged had diameters of nearly 18 cm, which in Figure 11.12 would have been close to 5 years old. What if very teeny *N. belauensis* do not understand the Brody-Bertalanffy equation and have some other model in mind? The point is that there is a very large gap in size from the very smallest individuals to the sizes that could be captured and tagged. The growth curve probably is very good for the region covered by the data and is biologically plausible for the extrapolated region but the age-size relations might be off and by an unknown number of years.

OTHER GROWTH MODELS

Sometimes the Richards function is not adequate. For example, a Walford graph for red sea urchins (*Strongylocentrotus franciscanus*) from San Nicolas Island, California (Ebert and Russell 1993) is curved rather than following a straight line (Figure 11.13). Animals were collected, tagged with tetracycline and returned to a site and again collected after one year. After cleaning in sodium hypochlorite bleach, the tetracycline marks can be seen by fluorescence under ultraviolet light and growth increments measured. The Richards-function family of growth models is unable to accommodate an inflection point in a graph such as Figure 11.13 and so, assuming that the inflection is biologically real, some different growth model is needed. The Tanaka function (Tanaka 1982, 1988) is a four parameter equation that accommodates an early lag and exponential phase followed by a declining growth rate. The function does not always have an asymptotic size and animals may continue to grow for as long as they live

$$S_t = \frac{1}{\sqrt{f}} ln \left| 2f(t-c) + 2\sqrt{f^2 (t-c)^2 + fa} \right| + d. \quad (11.14)$$

In Equation 11.14, "||" means that the absolute value of the enclosed expression should be used and "*ln*" means natural or base-e logarithms. Tanaka (1988) provides biological meanings for the four parameters

a = a measure of the maximum growth rate, which is at $\frac{1}{\sqrt{a}}$,

c = age at which growth rate is maximum,

d = a parameter that shifts the body size at which growth is maximum, and

f = a measure of the rate of change of the growth rate.

Tanaka (1982, 1988) wanted to formulate a growth curve in which initial growth rate was low, increased to a maximum growth rate, and then decreased but with potentially some growth throughout life; that is, no asymptotic size. Tanaka started with an equation for growth rate and then integrated this to produce Equation 11.14. The instantaneous growth rate, dS/dt, with respect to time that was selected by Tanaka (1982) is

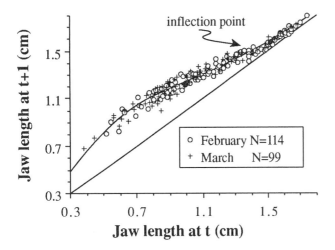

FIGURE 11.13 Growth of demi-pyramids (jaws) of the red sea urchin (*Strongylocentrotus franciscanus*) at San Nicolas Island, CA. (Ebert and Russell 1993); inflection point indicates the place where the curve changes shape from descent to the 45° line to running nearly parallel with the 45° line of 0 growth.

$$\frac{dS}{dt} = \frac{1}{\sqrt{f(t-c)^2 + a}}.$$ (11.15)

Parameters are the same as in Equation 11.14 and are defined mathematically

$$c = \frac{a}{E} - \frac{E}{4f}$$

and

$$E = \exp(\sqrt{f}\,(S_0 - d)),$$

where S_0 is size at time 0. The size at which the growth rate is maximum, S_a is

$$S_a = \frac{1}{\sqrt{f}} ln\left|2\sqrt{fa}\right| + d,$$

which shows that with fixed values of f and a, changes in the parameter d change the size at which maximum growth occurs.

The difference equation for the Tanaka function is

$$S_{t+1} = \frac{1}{\sqrt{f}} ln\left|2G + 2\sqrt{G^2 + fa}\right| + d$$ (11.16)

where

$$G = E/4 - fa/E + f$$

and

$$E = \exp(\sqrt{f}\,(S_t - d)).$$

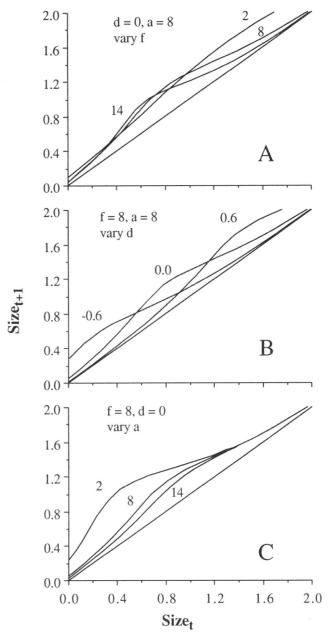

FIGURE 11.14 Simulation of curves for different parameters of the Tanaka function; **A**: parameters d and a held constant at 0 and 8 and f varied (2, 8 and 14); **B**: parameters f and a held constant at 8 and d varied (-0.6, 0, and +0.6); **C**: parameters f and d held constant at 8 and 0 and a varied (2, 8, 14); all three graphs have a common curve based on f = 8, d = 0, and a = 8.

Equation 11.16 is a bell-shaped curve that is rotated so that it is asymptotic to a 45° line in a plot of S_{t+1} *vs.* S_t. How parameters f, d and a influence the growth curve is shown in Figure 11.14. Starting values for showing effects of changing parameters were f = 8, d = 0 and a = 8. Each graph in Figure 11.14 contains a plot with these starting values together with parameter changes above and below

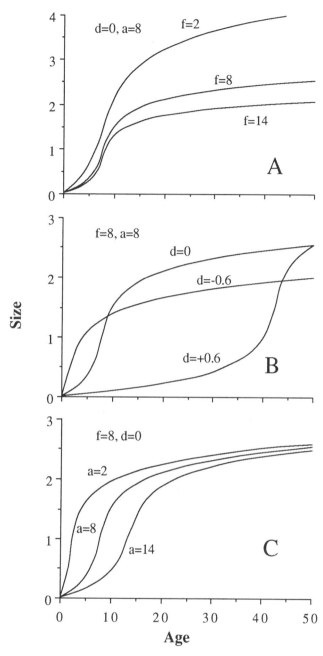

FIGURE 11.15 Size as a function of age using the Tanaka function; parameters are the same as in Figure 11.12 and so it is useful to examine the figures together; base growth curve is with f=8, d=0, and a=8; A: f is changed holding d=0 and a=8; B: d is changed holding f=8 and a=8; C: a is changed holding f=8 and d=0.

TABLE 11.11 Tanaka Parameters Estimated Using SIMPLEX.BAS for *Strongylocentrotus franciscanus* at San Nicolas, Is., CA.

Parameter	Value
f	9.192
d	-0.112
a	7.396
SSE	0.3450

Note: data from Ebert and Russell (1993).

phase. The long lag phase is followed by a very rapid rise to the maximum growth rate and this in turn is followed by a rapid decline in growth to a long period of nearly constant growth. Small values of f would produce a growth curve with a more gentle rise to maximum growth and gentle decline. The size at which growth rate is maximum also is changed and is at a smaller size for large values of f, but the effect is modest over a realistic range of values.

The influence of changes in the parameter d (Figure 11.14B) is to shift the size of maximum growth. Negative values of d show maximum growth at a small size whereas increases in d shift maximum growth rate to ever larger sizes. Changes in the parameter a (Figure 11.14C) show an inverse relationship with maximum growth rate. As the parameter a decreases, the maximum growth rate increases, but the size at which maximum growth is attained becomes smaller. With the parameters that were used, growth rate was maximum with a = 2.

The size-based growth relations shown in Figure 11.14 also can be shown as size as a function of time (Figure 11.15). Figure 11.15 illustrates how change in a parameter changes the shape of the growth curve. In each of the parts of Figure 11.15, A through C, a growth curve with parameters f = 8, d = 0, and a = 8 is included to provide a reference for changes in one parameter. In Figure 11.15A, the parameter f is changed and shows that with small f, growth is rapid and continues at a higher rate than with larger values of f. The slowest growth is with f=14. None of the lines cross so the line with f=2 always is higher. This figure should be compared with 11.14A where the same parameters are used showing S_{t+1} *vs.* S_t.

Figure 11.15B shows how changes in the parameter d change the shape of the growth curve. Results are more complex than with changes in f. The base for comparison is d=0. With d=-0.6, growth is more rapid initially but then crosses the line for d=0 and sizes are smaller after about age 8. With d=+0.6 a very long lag phase is followed by very rapid growth so at age 50 size at age is the same as it is with d=0. The line for d = +0.6 crosses the line of d=0 and for ages greater than 50, individuals are larger than with d=0. In Figure 11.15C, the parameter a is changed. Curves do

these values. The parameter f was changed in Figure 11.14A by increasing and decreasing the starting value, 8, by 6 so one line is for f=2, d = 0, and a = 8 and another line is drawn using parameter values of f = 14, d = 0 and a = 8. The plots show that increasing f makes plots more leptokurtic, which means that increasing f prolongs the lag

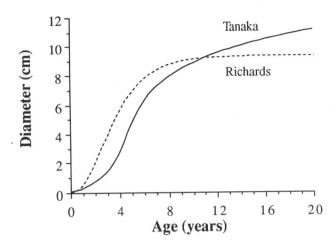

FIGURE 11.16 Comparison of Tanaka and Richards function growth curves for red sea urchins (*Strongylocentrotus franciscanus*) at San Nicolas Island, California.

The parameters of the Tanaka function lack clear biological meaning. This is a definite problem both in developing non-mathematical definitions of parameters but also in understanding differences among data sets. The Richards family of curves, which includes the Brody-Bertalanffy model, also suffers in the sense that none of the parameters (K, S_∞, n) has units of *growth* associated with it. This problem was addressed by Schnute (1981) who developed a growth model in which two parameters represent growth at two specified sizes; Schnute's 1981 presentation, however, requires knowing the ages of individuals and so is not appropriate for mark-recapture studies. This problem was addressed by Baker *et al.* (1991) and most recently by Francis (1995) who presents a mark-recapture analogue of Schnute's model.

not cross at any point and at age 50 are approaching the same size. Initial growth is different with fastest growth for a=2 and slowest for a=14.

Parameters a, d, and f for the Tanaka function can be estimated by nonlinear regression using the simplex program at the end of Chapter 12 or with nonlinear regression modules that are provided by good statistical software (*e. g.* SYSTAT, SAS). Obviously, if you have access to such software then it should be used in preference to **SIMPLEX.BAS** because **SIMPLEX.BAS** is *very* slow. Parameter estimates for the red sea urchin data are shown in Table 11.11.

The red sea urchin data were analyzed using the Richards function as the growth model and Figure 11.16 shows that the initial lag with the Richards function is much shorter than with the Tanaka function and, possibly the most important difference, the Tanaka function shows the continuing growth of animals whereas the Richards function provides an asymptotic size.

Francis' mark-recapture analogue of Schnute's growth model

Francis (1995) presents his model using change in size, ΔS, as a function of original size, S, rather than size at t+Δt as a function of size at time, t. There are five parameters in the model but two of them are fixed by the investigator at the beginning of analysis so only 3 parameters are actually estimated. The two that are fixed are y_1 and y_2, which are two sizes that are selected to be close to the lower, y_1, and upper, y_2, bounds of the data set. Selection of y_1 and y_2 is arbitrary and should be done just by inspection of the data set. Two of the fitted parameters, g_1 and g_2, are the growth rates of individuals with sizes y_1 and y_2 and so would be very useful for comparing data sets. The final parameter, b, "may be thought of as describing curvature" (Francis 1995). It is like n in the Richards function. The model is complex and Francis defines four additional parameters, λ_1, λ_2, a and c

$$\Delta S = \begin{cases} -S + \left[S^b e^{-a\Delta t} + c\left(1 - e^{-a\Delta t}\right) \right]^{1/b} & \text{if } a\neq0 \text{ and } b\neq0 \\[2em] -S + S^{\exp(-a\Delta t)} \exp\left[c\left(1 - e^{-a\Delta t}\right) \right] & \text{if } a\neq0 \text{ and } b=0 \\[2em] -S + \left[S^b + (\lambda_1^b - y_1^b)\Delta t \right]^{1/b} & \text{if } a=0 \text{ and } b\neq0 \\[2em] -S + S(\lambda_1/y_1)^{\Delta t} & \text{if } a=0 \text{ and } b=0 \end{cases}$$

$$(11.17)$$

TABLE 11.12 Analysis of Red Sea Urchin Growth Using the Model of Francis (1995)

Parameter	Estimate	se.	Lower<95%>Upper	
g_1 cm yr^{-1}	0.3636	0.0163	0.3314	0.3958
g_2 cm yr^{-1}	-0.0032	0.0074	-0.0177	0.0113
b	0.2324	0.1770	-0.1164	0.5813

Note: $r^2 = 0.862$; fixed jaw lengths are $y_1 = 0.4$ cm and $y_2 = 1.7$cm; g_1 is the annual growth-rate at y_1 and g_2 is the rate at y_2; data from Ebert and Russell (1993).

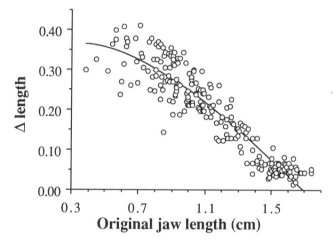

FIGURE 11.17 Growth increment of demi-pyramids of the red sea urchin *Strongylocentrotus franciscanus* at San Nicolas Island, CA fitted using the model of Francis (1995); data from Ebert and Russell (1993).

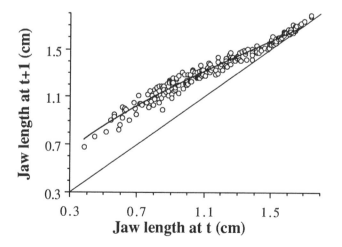

FIGURE 11.18 Growth of demi-pyramids of the red sea urchin *Strongylocentrotus franciscanus* at San Nicolas Island, CA fitted using the model of Francis (1995) and then changing from Δsize to size at t+1 for the dependent variable; data from Ebert and Russell (1993).

$$\lambda_1 = y_1 + g_1,$$

$$\lambda_2 = y_2 + g_2,$$

$$a = \begin{cases} ln\left[\dfrac{y_2^b - y_1^b}{\lambda_2^b - \lambda_1^b}\right] & \text{if } b \neq 0 \\[2em] ln\left[\dfrac{ln(y_2/y_1)}{ln(\lambda_2/\lambda_1)}\right] & \text{if } b = 0 \end{cases},$$

and

$$c = \begin{cases} \dfrac{y_2^b \lambda_1^b - y_1^b \lambda_2^b}{\lambda_1^b - y_1^b + y_2^b - \lambda_2^b} & \text{if } b \neq 0 \\[2em] \dfrac{ln(y_2)\,ln(\lambda_1) - ln(y_1)\,ln(\lambda_2)}{ln(\lambda_1 y_2) - ln(\lambda_2 y_1)} & \text{if } b = 0 \end{cases}$$

It is best to think of the model (Equation 11.17) as just one equation with $a \neq 0$ and $b \neq 0$ and to change the model only if a fit, or lack of it, indicates that a or b are tending towards 0.

Assembling all of the pieces into a single model is necessary for certain statistical programs such as SYSTAT. For the SYSTAT model each line ends with a comma and carriage return, ¶. The comma and carriage return mean that the next line is a continuation and so the model actually is a single line. The values of y_1 and y_2 are set equal to 0.4 and 1.7 respectively and ΔS is DS.

```
MODEL DS=-S+(S^B*EXP(-LOG((1.7^B-.4^B)/((1.7+,¶
G2)^B-(.4+G1)^B)))+(1.7^B*(.4+G1)^B-.4^B*(1.7+,¶
G2)^B)/((.4+G1)^B-.4^B+1.7^B-(1.7+G2)^B)*(1-EXP,¶
(-LOG((1.7^B-.4^B)/((1.7+G2)^B-(.4+G1)^B)))))^(1/B)¶
```

Red sea urchin data from San Nicolas Is., CA, were analyzed using the Francis analogue of Schnute's model (Table 11.12) and the data, together with the model, are plotted in Figures 11.17 and 11.18. The fits seem very good; in both figures, however, it is clear that, as was true with the Richards function, large individuals are not modeled correctly so that the continuing growth of large animals would be miss judged. Mean growth of a 1.7cm jaw is estimated to be -0.003 cm yr^{-1} with 95% confidence limits of -0.018 to +0.011. In both figures, the fitted line falls ever more below the data cloud for large jaws.

The various analyses of growth for red sea urchins point to the Tanaka function as biologically the most reasonable for these animals because it includes both the slow growth of small animals, a size of maximum growth followed by a decline in growth but never with cessation. Sea urchins, like trees, continue to grow throughout life. The Francis size-growth version of Schnute's model is attractive because, unlike the Tanaka function, has two out of three parameters that actually have "growth" as part of the definition whereas none of the Tanaka parameters has simple biological meaning. There is a need to combine the virtues of both of these models. In cases where there is an asymptotic size, the approach proposed by Francis would be superior to all others but if there is no asymptotic size then the best model seems to be the Tanaka function.

MODELS WITH SEASONAL COMPONENTS

Additional complications of studying growth are seasonal effects. A number of authors have tried to incorporate seasonal parameters into the Brody-Bertalanffy model (*e. g.* Cloern and Nichols 1978, Pauly 1981). A seasonal model also was developed by Sager (1982) and Sager and Gosselck (1986)

$$S_t = S_\infty \left(1 - be^{-Kt}\right)\left[1 + \frac{(1-\varepsilon)K}{2\pi} \sin 2\pi(t - t_A)\right]. \quad (11.18)$$

Sager's model is attractive because its second element operates to modify the first (the basic growth model, Equation 11.1) with a component that varies with season. When the parameter $\varepsilon = 1.0$, the model has no seasonal component but seasonality becomes more pronounced as $\varepsilon \to 0$.

Equation 11.18 can be recast as a difference equation using size pairs (S_t and $S_{t+\Delta t}$), a time interval (Δt), and Julian day/365 (t)

$$S_{t+\Delta t} = S_\infty - (S_\infty - S_t)e^{-K\Delta t}\left[1 + \frac{(1-\varepsilon)K}{2\pi} \sin 2\pi(t + \Delta t - t_A)\right]. \quad (11.19)$$

The parameters are

S_∞ = asymptotic size,
K = the growth rate constant,
ε = parameter measuring strength of the seasonal effect; equal to 1 with no effect, and
t_A = parameter that adjusts the time of minimum growth (yr).

An example of analysis of growth with a seasonal component is provided by the predatory snail *Nucella lamellosa*

TABLE 11.13 Growth Parameters with Seasonality for the Predatory Snail *Nucella lamellosa* at Hogg Bay, Prince William Sound, Alaska

	Number of parameters			
	4	3	2	2
SSE	1352.22	1353.05	1357.61	1484.72
S_∞	29.49mm	29.49	29.55	29.059
K	1.389	1.389	1.287	1.841
t_a	0.093	0.033	0 (fixed	0 (fixed)
	3 Feb	12 Jan.	1 Jan	1 Jan
ε	0.00001	0 (fixed)	0 (fixed)	1 (fixed)

Note: N = 420 size pairs; data from Ebert and Lees (1996).

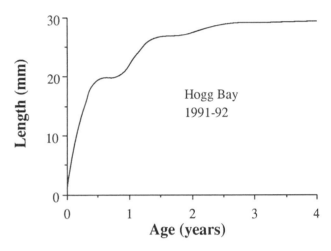

FIGURE 11.19 Growth with seasonal adjustment; *Nucella lamellosa* at Hogg Bay, prince William Sound, Alaska (Ebert and Lees 1996) .

that was tagged at Hogg Bay in Prince William Sound, Alaska (Ebert and Lees 1996). Tagging was done during April/May and July 1991. Samples of animals tagged in May were measured in July 1991; samples from most sites were again measured in September 1991 and July 1992. Time periods of up to 14 months were used to evaluate growth.

Table 11.13 shows the results of using four different models. The first model used four parameters and indicated that ε was sufficiently small that it might be fixed at 0, which was used in the second model. The small change in SSE from model 1 to model 2 was from 1352.22 to 1353.05 and suggests that fixing ε rather than estimating it is a good idea and furthermore makes biological sense because $\varepsilon = 0$ is for a maximum seasonal effect. In the third model, time of the growth minimum was fixed at 1 January, *i. e.* $t_a = 0$. The change in SSE to 1357.61 again is a very small increase and would seem to justify fixing t_a rather than estimating it.

The fourth model is the Brody-Bertalanffy model without seasonality; that is, ε is fixed at 1. The change in SSE is substantial and indicates that a seasonal model probably is best for *Nucella lamellosa* (Figure 11.19), which makes biological sense for species in Alaskan waters.

GROWTH MODELS AND THE ALLOMETRY OF REPRODUCTION

You boil it in sawdust: you salt it in glue:
You condense it with locusts and tape:
Still keeping one principal object in view –
To preserve its symmetrical shape.

L. Carroll ,
The Hunting of the Snark, 1876

Within the context of demographic analysis using matrix or life-cycle methods, one reason for describing growth of individuals is to provide a general description of reproductive output of particular ages or sizes. To do this it is necessary to combine information on individual growth with the allometric relationship between size and reproduction.

The allometric equation describes the change in the size of one body part relative to the size of another

$$y = \alpha x^{\beta}. \tag{11.20}$$

If β is equal to 1.0 then y is a fixed proportion of x and y is said to grow isometrically. Although β always should be estimated, it is assumed to be equal to 1.0 in all cases where so-called body indices are calculated such as the ever popular 'gonad index', GI, in various studies mostly of fish and marine invertebrates. The gonad index, GI, has various definitions but one is

$$GI = \frac{gonad\ wt}{total\ body\ wt} \times 100\%.$$

In terms of Equation 11.20, the gonad index is α and so

$$GI = \frac{y}{x^{\beta}} \times 100\%$$

and unless β is estimated, it must be assumed to be some value and the assumed value always seems to be 1.0. If $\beta \neq 1.0$, then the gonad index changes purely as a function of size. Allometric relationships tend to be ignored in studies that use body indices and so β is never mentioned. Body indices, such as the gonad index, with assumed $\beta=1.0$ should not be used for anything important. If comparisons

are to be made across seasons or among areas, an analysis of covariance with size as a covariate should be used.

A common approach to finding α and β is to transform x and y using logarithms and then do a simple Model I regression

$$ln(y) = ln(\alpha) + \beta ln(x). \tag{11.21}$$

Because both x and y are subject to natural variation and errors of measurement, some workers (*e. g.* Ricker 1973) suggest using a geometric mean functional regression to obtain parameter estimates; however, unlike the problem with S_t and $S_{t+\Delta t}$ in Walford plots, it is not a good idea to just assume equality of errors in x and y. The problem is not at all simple. The goal is to estimate the true relationship between x and y, which means determining the true slope. In order to do this, it is necessary to estimate the error variance of x, var(ε), and this is no mean feat. An interesting approach is presented by Prairie *et al.* (1995), which they call the slope-range method. For purposes of allometry, the approach would use a third variable, z, which is correlated with the true value of x but not its errors. The variable z is used to create subsets of x and y by selecting ranges of z and using this range to select the x-y pairs. Within each of these subsets, ordinary least-squares regressions are done and a slope obtained. The variance of x within each subset also is calculated and a graph is constructed of slopes *vs.* 1/var(x). The y-axis intercept is the true slope of the entire data set. What is clear from their presentation is that the slope-range method is a data-hungry procedure because they suggest that as a rough practical guide, about 10 groups be used with, ideally, more than 20 observations in each group. Data sets of this size seldom are available for describing size-related reproduction although if adequate data sets are available, then the slope-range method of Prairie *et al.* (1995) should be tried.

For gonad development or related reproductive processes (*e. g.* Emlet 1989, Russell and Huelsenbeck 1989, Ebert and Russell 1994), another problem arises; namely, that gonad growth does not begin until the organism has attained some minimum size. Consequently, Equation 11.19 must be modified to accommodate this additional biological complexity

$$y = \alpha(x - \gamma)^{\beta} \tag{11.22}$$

or

$$y = \alpha x^{\beta} + c. \tag{11.23}$$

Modification of the basic allometry equation (Equation 11.19) by adding a parameter seems to have been introduced by Robb (1929) and viewed as a more inclusive or general

TABLE 11.14 Gonad Weights and Test Lengths of a Tropical Sea Urchin, *Heterocentrotus mammillatus*

Wet weight (g)	Length (cm)	Gonad wet (g)
1.63	1.51	0.0
7.11	2.25	0.0
23.70	2.00	0.0
53.96	3.33	0.85
57.98	3.73	1.74
64.28	3.54	1.65
71.64	3.65	1.00
78.09	3.74	0.64
109.67	4.54	2.21
114.03	4.33	2.53
126.38	4.67	2.35
130.78	4.46	5.08
152.93	5.01	6.05
172.02	4.95	4.12
203.03	5.41	5.00
213.00	5.66	5.87
226.19	5.83	5.90
257.06	5.44	4.60
262.22	6.13	10.39
397.03	6.49	7.04

Note: animals collected at Honaunau Bay, (Big Island), Hawaii, 22 August 1975 (Ebert 1987a).

allometry equation (Huxley 1932, Reeve and Huxley 1945). In the following presentation, I refer to Equation 11.23 as the "general" allometry equation and Equation 11.22 as the "adjusted" allometry equation because γ adjusts x so when x-γ = 0, y = 0. In this form, the additional parameter (γ) is the x-intercept and represents the point in ontogeny where the development of y begins relative to x.

Reeve and Huxley (1945) viewed equations more complex than Equation 11.20 as "...useless in practice, since the ... constants could not be estimated even very roughly without quite prohibitive labour in computation." The ubiquity of nonlinear regression software has removed the problems of laborious computation for models such as Equation 11.22 or 11.23 or Lumer's (1937) four parameter equation

$$y = \alpha(x - \gamma)^\beta + \delta. \qquad (11.24)$$

The slate-pencil sea urchin *Heterocentrotus mammillatus*

Data in Table 11.14 can be used to illustrate one approach to estimating the parameters in Equation 11.22. The data are for a tropical sea urchin, *Heterocentrotus mammillatus*, in Hawaii and are from dissections of animals of a range of

TABLE 11.15 Analysis of Gonad Size *vs.* Length with a Correction for Size When Gonads Begin to Develop (γ) in the Sea Urchin *Heterocentrotus mammillatus*

Parameter	Estimate	se	Lower <95%> Upper	
α	1.6166	2.0134	-2.7017	5.9348
γ	2.8037	1.0788	0.4899	5.1174
β	1.2692	0.7278	-0.2917	2.8301

Note: animals collected at Honaunau Bay, (Big Island), Hawaii, 22 August 1975 (Ebert 1987a); $r^2 = 0.80$.

TABLE 11.16 Analysis of Gonad Size as a Function of Length without a Correction for Size When Gonads Begin to Develop in *Heterocentrotus mammillatus*

Parameter	Estimate	se	Lower <95%> Upper	
α	0.0374	0.0298	-0.0262	0.1010
β	2.9253	0.4620	1.9406	3.9099

Note: Honaunau Bay, (Big Island), Hawaii, 22 August 1975 (Ebert 1987a); $r^2 = 0.78$.

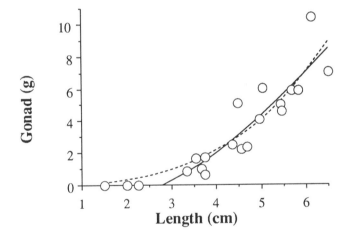

FIGURE 11.20 Relationship between gonad weight, G, and body length, L, in a tropical sea urchin *Heterocentrotus mammillatus* (Ebert 1987a); solid line is for $G = 1.62(L-2.80)^{1.27}$ and the dotted line is for $G = 0.0374L^{2.925}$.

sizes. Analysis is done by excluding all individuals ≤2.0 cm that have no gonads. Two functions are appropriate for the data

$$y = 0 \text{ for } x \le \gamma$$

and

$$y = \alpha(x - \gamma)^\beta \text{ for } \gamma > 0$$

and we want to estimate parameters for the second function. Length is used because in any further analysis, either size-structured or age-structured, length would be used rather than weight.

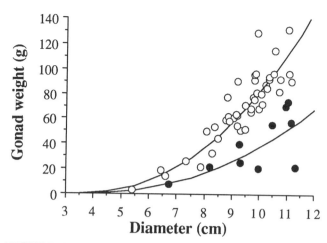

FIGURE 11.21 Size-specific gonad weights for red sea urchins (*Strongylocentrotus franciscanus*) based on data gathered by Baker (1973); open circles are for October 1972 through January 1973 and solid circles are for April 1972; the difference between the two lines is the spawn mass; analysis in Table 11.14.

Results of a nonlinear regression (NONLIN in SYSTAT) are shown in Table 11.15. With small numbers of individuals, the standard errors are very large and indicate that dissection of considerably more individuals should be done.

For comparison, analysis also was done using the basic allometry equation (Equation 11.20), which lacks the correction for a size when gonads begin to develop. Results are shown in Table 11.16 and lines are compared in Figure 11.20. The basic allometry equation overestimates reproduction in small *H. mammillatus*.

Gonad size in many organisms is an indication of relative reproductive contribution but a better estimate would include an estimate of spawn mass together with conversion from mass to numbers of eggs. The following illustration, again for a sea urchin, shows how this can be done.

Size-specific egg production in the red sea urchin *Strongylocentrotus franciscanus*

In order to develop the size-based model for red sea urchins that was presented in Chapter 9, an estimate of size-specific egg production was needed. No estimates of size-specific egg production for *Strongylocentrotus franciscanus* are in the literature and to I used an indirect approach. Data from Baker (1973) were gathered in southern California and include size-specific gonad weights. Data were gathered throughout the year and showed maximum gonad sizes from fall through the winter with a sudden drop in spring. The difference between maximum and minimum was taken as the mass of gametes released. Size-specific spawn mass was converted to egg number using the relationship between volume and egg number given by Levitan (1993).

TABLE 11.17 Analysis of Size-Specific Gonad Weights

A. Homogeneity of Slopes; N = 54

Source	SS	df	MS	F-ratio	p
time period, T	0.0157	1	0.0157	0.229	0.63
ln(D-3.45)	9.7818	1	9.7818	142.423	0.00
T x *ln*(D-3.45)	0.0378	1	0.0378	0.550	0.46
error	3.4341	50	0.0687		

B. Analysis for Significance of Adjusted Means

Source	SS	df	MS	F-ratio	p
time period	4.8231	1	4.823	70.850	<0.001
ln(D-3.45)	17.5962	1	17.596	258.483	<0.001
error	3.4718	51	0.0681		

C. *ln*(Gonad) Weight, *ln*(G), Adjusted for a Common *ln*(D-3.45) Mean with a Common Slope of β = 2.16956

Time period	*ln*(G)	se	N	α
Oct. through March	4.13548	0.03938	44	1.3758
April	3.36134	0.08292	10	0.6345

Note: G is gonad weight in grams; test diameters, D (cm), were adjusted by subtracting 3.45 cm, which is the diameter when gonads were first measurable (Baker 1973); all data were transformed using natural logarithms and so a test for homogeneity of slopes is a test of equality of the allometric exponent, β; common slope, β, = 2.1696; α determined using $ln(G) = ln(\alpha) + (2.16956)(1.758961)$; April is the month of minimum and March through October were pooled as maximum gonad size.

Size-specific gonad measurements for *Strongylocentrotus franciscanus* are shown in Figure 11.21, using data from Baker (1973). Variability in the data was too great to estimate γ in Equation 11.22 and so a reasonable value had to be selected. Baker dissected a red sea urchin with a diameter of 3.45 cm that did not have developed gonads and so I selected 3.45 as γ and subtracted it from all diameter measurements, D, before *ln*-transformation and estimation of allometric parameters for gonad weight *vs.* diameter. The next step in analysis was an ANCOVA to determine whether a single allometric exponent was appropriate (Table 11.17). A natural logarithm transformation was applied and the interaction term of *ln*(D-3.45) × season had an associated p=0.46 so slopes (allometric exponents, β) can be considered to be the same for pre and post spawning. A second ANCOVA without the interaction term showed that adjusted means were different at p < 0.00001. The common slope is β of the allometry equation and adjusted means can be used to estimate the two different values of α because the regression lines must pass through the common mean of *ln*(D-3.45) and the estimate of *ln*(gonad weight) at this common mean. The final equation for spawn mass, S, is the

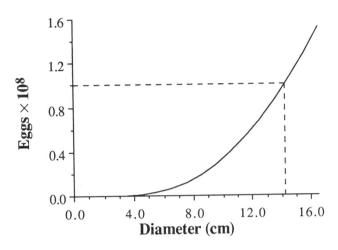

FIGURE 11.22 Egg production based on spawn mass of red sea urchins (*Strongylocentrotus franciscanus*) (Figure 11.21) and the conversion of 7.764×10^5 eggs cm^{-1} (Levitan 1993).

difference between the pre and post spawning curves of gonad mass, G, as a function of test diameter, D

$$S = \alpha_1(D - 3.45)^\beta - \alpha_2(D - 3.45)^\beta,$$

which is

$$S = (\alpha_1 - \alpha_2)(D - 3.45)^\beta$$

or

$$S = 0.741267(D - 3.45)^{2.1696}.$$

Spawn mass was converted to egg number using the conversion provided by Levitan (1993) for *S. franciscanus*: 7.764×10^5 eggs cm^{-3}. Assuming that 1 cm^3 = 1 gram, the number of eggs spawned by an animal of diameter D can be estimated (Figure 11.22). For example, a 9.25 cm *S. franciscanus* would have a spawn mass of 38.3g and so would release about 29.7×10^6 eggs (Figure 11.22).

GENERAL COMMENTS

... the classical approach to growth analysis might well be buried in the sands of time.

F. I. Woodward,
Review of *Plant Growth Analysis*, 1980

More growth models are available than I have been presented here. Hunt (1982) provides a wide array of models that have been used for plants, all of which could also be applied to animals, and Sager has published a dazzling array of modifications of asymptotic functions (*e. g.* Sager 1978, 1979a-d, 1980a-c, 1982, 1984a+b, 1986, Sager and Sammler 1984). There also are models with more parameters. A six parameter model is presented by Kanefuji and Shohoji (1990), two seven-parameter models by Jolicoeur *et al.* (1988, 1992) and a nine parameter "triple logistic" model by Brock and Thissen (1976). Also, polynomial models have been suggested for cases where there may be a decrease in length past a certain age (*e. g.* Brown 1988, Chen *et al.* 1992, Hearn and Leigh 1994). Searching for the model that will provide the "best fit" can become a search for the Grail with all of the fun being in the search because there may be no end. Does this mean that you should go back to the beginning of this chapter and then skip it? Well — maybe not quite yet.

Two points must be considered. The first is that an accurate modeling of growth is a summary of physiological processes that include plasticity, trade-offs and constraints. As a consequence, parameters of a growth curve represent a synthesis of features that contribute to survival and reproduction and hence are a link between physiology and m_x and l_x vectors (*e. g.* Charnov 1993). A growth curve is an emergent property of physiological processes just as l_x and m_x are emergent properties. It seems very unlikely that the physiology of most organisms ever will be studied in sufficient detail so energy and nutrient-based models could be used to predict demographic events although from a theoretical standpoint this would be very reasonable and desirable (*e. g.* Metz and Diekmann 1986, Łomnicki 1988, DeAngelis and Gross 1992). If growth models are to be used, however, then data must be gathered so that life-time activities can be discussed without having to base much of the life of the organisms on extrapolation. This is a plea for gathering data over the entire range of sizes of the organisms being studied. Also, attention should be paid to independence of measurements in data sets; 3 fish with 15 measurements taken on otoliths of each fish would not be as good as 45 individual tagged fish. On the other hand, if the fish were exceedingly rare, coelacanths for example, I would be quite happy with 3 fish.

And now the second point: Growth curves can be used as intermediate steps in obtaining some other metric of interest. For example, when combined with size structure of a population, growth-curve parameters can be used to estimate survival rates (Chapter 13). For this purpose it is necessary only to have a growth function that holds for the range of individuals in a sample. It is not necessary to worry about which model is "right" if the Gompertz is used instead of the Brody-Bertalanffy equation and the curves are very similar over the range of animals studied (*e. g.* sizes 4 to 6 in Figure 11.6).

The summary is that what you want to do with a growth function should dictate both the data that you gather and how much you should worry about which model to use.

PROBLEMS

1. Estimate growth parameters for male African buffalo, *Syncerus caffer* (Sinclair 1974) using data in Table 11.18. The weight at birth of a normal calf is 38-45 kg (Sinclair 1970 cited in Sinclair 1974). Do the analysis two ways. First, use size pairs with Δtime and, secondly, use age and size. Enter data so observations are correctly weighted (*i. e.* if n = 4 for a particular age, enter the age and size 4 times). Try the Brody-Bertalanffy, logistic and Gompertz models.

TABLE 11.18 Size at Age Observations for Male African Buffalo

N	Age	Mean wt.	N	Age	Mean wt.
1	2 months	60	3	5.5 years	505
5	6 months	133	3	6.5	587
1	9 months	163	2	7.5	614
3	1.5 years	269	1	8.5	557
2	2.5	343	2	9.5	648
7	3.5	398	2	10.5	690
4	4.5	503	2	11.5	690

Note: weights are in kilograms; data from Sinclair (1974).

2. *Fouquieria* (=*Idria*) *columnaris*, the cirio or boojum tree of Baja California, Mexico, was studied by Humphrey and Humphrey (1990) from 1969 to 1986 at several sites. Heights were measured no more frequently than once per year and often a year would be skipped. Here are the data (Table 11.19) from one site. Estimate growth parameters and draw a growth curve. Examine the goodness of fit by examining a plot of residuals. Time in years for each date are given in Table 11.20.

TABLE 11.20. Time in Years Between First Measurement on 26 July 1969 and Subsequent Measurements of *Fouquieria* (=*Idria*) *columnaris*.

Date	Δ years	Date	Δ years
26 Jul. 69	0.000	25 Oct. 77	8.255
9 Mar 71	1.619	20 Oct. 78	9.241
9 Jun. 72	2.874	29 May 80	10.849
6 Jun. 73	3.866	17 Oct. 83	14.236
20 Jun 74	4.904	14 Oct. 85	16.230
11 Jun 75	5.879	18 Jun 86	16.907

Note: data from Humphrey and Humphrey (1990)

3. Find Richards function parameters for growth of male singing voles, *Microtus miurus muriei*, from Alaska (Morrison, *et al.* 1977). Data are shown in Table 11.21.

TABLE 11.21 Weight at Age Data for Male Singing Voles

Age	Mean wt.	Age	Mean wt.
0	3.0	24	38.6
1	6.7	28	41.5
2	10.7	32	43.4
3	14.4	36	45.2
4	19.2	40	46.9
8	25.4	44	50.3
12	29.0	48	51.7
16	32.4	52	53.5
20	35.7	56	55.1

Note: ages are in weeks and weights are in grams; data from Morrison *et al.* 1977.

TABLE 11.19 Height in meters of Twelve *Fouquieria* (=*Idria*) *columnaris* Trees at Idria Island, BC, Mexico

| Date | Tree number | | | | | | | | | | | |
	1	5	6	7	8	9	12	13	14	15	16	18
26 Jul. 69	6.43	1.14	1.04	0.66	6.25	4.75	2.52	2.69	4.98	1.58	4.04	0.86
9 Mar 71	6.58	1.19	1.22	0.74	6.35	4.88	2.64	2.74	4.98	1.65	4.14	0.99
9 Jun. 72	6.55	1.22	1.25	0.74	6.33	4.88	2.67	2.72	4.98	1.65	4.14	0.91
6 Jun. 73	6.60	1.22	1.25	0.76	6.35	4.93	2.67	2.72	4.98	1.68	4.14	-
20 Jun 74	6.63	1.22	1.27	0.74	6.40	4.93	2.69	2.72	4.98	1.68	4.14	-
11 Jun 75	6.58	1.25	1.27	0.76	6.40	4.90	2.67	2.69	5.00	1.68	4.17	-
25 Oct. 77	6.63	1.35	1.27	0.79	6.40	4.90	-	2.72	5.00	1.68	4.17	-
20 Oct. 78	6.68	1.27	1.37	0.81	6.40	4.90	2.72	2.74	5.06	1.70	4.22	-
29 May 80	6.71	1.37	1.32	0.91	6.50	4.98	2.72	2.82	5.28	1.96	4.37	1.17
17 Oct. 83	6.90	1.53	2.14	1.10	6.68	4.99	2.88	2.94	5.40	2.38	4.54	1.40
14 Oct. 85	7.01	1.70	2.36	1.14	6.73	5.08	3.02	3.00	5.49	2.53	4.63	1.45
18 Jun 86	6.94	1.75	2.42	1.15	6.78	5.07	3.02	3.04	5.50	2.57	4.62	1.44

Note: data from Humphrey and Humphrey (1990).

TABLE 11.23 Growth of Dairy Cows

Age (months)	Holstein	Jersey	Guernsey	Ayrshire
birth	29.1	25.7	26.6	27.6
1	30.6	27.0	28.2	28.6
2	32.3	28.9	29.8	30.2
3	34.3	30.6	31.6	31.9
4	36.2	32.6	33.5	34.0
5	37.7	34.5	35.3	35.5
6	39.7	36.2	36.9	37.2
7	41.1	37.7	38.4	38.5
8	42.3	39.0	39.9	39.9
9	43.5	40.1	40.9	40.9
10	44.4	40.9	41.7	41.7
11	45.3	41.7	42.6	42.5
12	46.0	42.2	43.3	43.2
13	46.7	42.8	43.9	44.0
14	47.3	43.3	44.6	44.8
15	47.9	43.9	45.0	45.1
16	48.5	44.4	45.3	45.7
17	48.9	44.7	45.9	46.2
18	49.3	45.2	46.4	46.5
19	49.8	45.5	46.7	46.8
20	50.2	45.9	47.0	47.4
21	50.6	46.2	47.3	47.6
22	51.0	46.4	47.7	47.8
23	51.3	46.7	47.9	48.1
24	51.7	46.9	48.0	48.3
27	52.2	47.7	48.9	48.1
30	52.5	47.9	49.3	48.3
33	52.7	48.0	49.7	48.9
36	53.0	48.2	49.9	48.7
39	53.1	48.6	50.0	49.1
42	53.2	48.6	49.9	49.9
45	53.2	48.5	50.1	50.0
48	53.3	48.5	50.4	50.2
51	53.5	48.5	50.6	49.4
54	53.6	48.6	50.5	50.3
57	53.7	48.6	50.5	50.3
60	53.6	49.0	50.6	50.4
63	53.5	49.0	50.4	49.2
66	53.7	48.7	50.0	49.2
69	53.7	48.6	49.8	49.3
72	53.7	48.4	49.7	49.1
75	53.9	48.5	49.3	48.3
78	54.0	48.6	49.4	48.9
81	53.8	48.4	49.2	49.1
84	53.7	48.0	49.3	48.7
87	53.6	48.3	49.4	48.8
90	53.7	48.2	49.4	48.7
93	53.5	48.4	48.9	48.5
96	53.2	47.7	49.6	49.2

Note: sizes are heights at the withers in inches; data from Ragsdale (cited in Brody 1945).

4. Waters and Shay (1991) studied the effects of water depth and other factors on growth of cattail (*Typha glauca*) shoots. Measurements of shoots were made once per week and an estimate of initial age was made. Calculate growth parameters for the following data set (Table 11.22).

TABLE 11.22 Length Measurements of a Cattail Shoot

Length$_t$	Length$_{t+2wks}$
0.43m	0.64m
0.64	0.92
0.92	1.23
1.23	1.47
1.47	1.72
1.72	1.85
1.85	2.05
2.30	2.30

Note: initial age is 2 weeks; data from Waters and Shay (1991).

5. Analysis of growth of dairy cows should be fun. Data are from Ragsdale *et al.* (1934) in the Missouri Station Bulletin #336 (cited in Brody 1945) and are heights at the withers in inches (Table 11.23). Although maximum sizes obviously are different for the four breeds, do they approach their maximum sizes at different rates? What is the *maximum* growth rate for each breed?

6. Shirakihara *et al.* (1993) studied growth in the finless porpoise, *Neophocaena phocaenoides* off Kyushu, Japan. A summary of male and female lengths in centimeters is shown in Table 11.24. Ages are in years. Try several different growth models. Which seems best to describe growth of this marine mammal?

TABLE 11.24 Length at Age Data for the Finless Porpoise, *Neophocaena phocaenoides*

Age	Body length (male)	n	Body length (female)	n
0	77.2	8	80.3	4
0.5	103.3	15	104.6	17
1	117.2	7	117.7	6
2	125.2	5	126.2	3
3	131.3	3	128.5	4
4			134.0	1
5				
6	145.7	3		
7	143.0	2	145.0	2
8	152.5	1		
9			148.5	2
10	151.0	1	165.0	1
11-23	161.1	6	151.4	6

Note: data from Shirakihara *et al.* (1993).

7. Fletcher (1984) followed cohorts of the limpet *Cellana tramoserica* at the Cape Banks Scientific Marine Research Area at Botany Bay, NSW, Australia. Data for mean sizes (mm) at his subtidal site are provided in Table 11.25. *Note:* there is a missing month when sampling wasn't done for some reason (November of the second year). For this particular example, ignore seasonal effects.

a. Compare Brody-Bertalanffy, the logistic, and the Gompertz functions for modeling growth of this species. Express K in terms of years.
b. Does the Richards function provide a better fit?

8. The following data (Table 11.26) are for the black turban snail, *Tegula funebralis*, that was studied in Sunset Bay, Oregon during 1963-65 by Frank (1965). Select the most appropriate growth model and estimate the parameters of the function. Justify the selection of a particular model by showing a plot of residuals.

9. The pumpkinseed, *Lepomis gibbosus*, an introduced species in the Camargue, was studied by Crivelli and Mestre (1988). Estimates of size at age are given in Table 11.27. Select the most appropriate growth model and estimate the parameters. Justify your selection of a growth model by an analysis of residuals.

TABLE 11.25 Mean Lengths for Cohorts (A - H) of the Limpet *Cellana tramoserica*

Cohort	Nov	Jan	Mar	May	July	Sept	Dec	Feb	May
A	41.0	41.5	42.5						
B	31.8	35.0	37.9	42.4	43.3				
C	23.8	29.7	33.2	37.0	38.5	43.0	43.2	44.6	46.0
D		12.7	20.8	30.1	35.3	36.2	37.3	37.0	39.1
E		5.0	17.0	22.0	29.0				
F			12.0	15.7	22.6	31.2	33.0		
G				12.0	17.0	22.6	28.0	32.5	
H							6.2	12.8	23.0

Note: data from Fletcher (1984).

TABLE 11.26 Size-Specific Annual Growth Increments of *Tegula funebralis*

Initial size	N	Ave annual growth	SE	Initial size	N	Ave annual growth	SE
6.1-6.5 mm	3	4.67 mm	0.38	18.1-18.5 mm	39	1.33 mm	0.10
6.6-7.0	7	4.33	0.32	18.6-19.0	34	1.43	0.088
7.1-7.5	3	5.07	0.30	19.1-19.5	26	1.09	0.11
7.6-8.0	20	4.20	0.11	19.6-20.0	13	1.28	0.15
8.1-8.5	20	4.36	0.091	20.1-20.5	13	1.04	0.14
8.6-9.0	32	4.26	0.10	20.6-21.0	7	1.60	0.20
9.1-9.5	26	3.94	0.19	21.1-21.5	16	0.98	0.095
9.6-10.0	39	4.03	0.087	21.6-22.0	9	0.63	0.13
10.1-10.5	27	4.00	0.12	22.1-22.5	8	0.70	0.11
10.6-11.0	21	3.43	0.12	22.6-23.0	13	0.70	0.16
11.1-11.5	14	3.44	0.13	23.1-23.5	14	0.58	0.15
11.6-12.0	26	3.33	0.13	23.6-24.0	17	0.65	0.20
12.1-12.5	20	3.08	0.18	24.1-24.5	20	0.49	0.14
12.6-13.0	28	2.80	0.10	24.6-25.0	12	0.37	0.13
13.1-13.5	39	2.70	0.095	25.1-25.5	16	0.42	0.14
13.6-14.0	47	2.47	0.10	25.6-26.0	26	0.54	0.10
14.1-14.5	44	2.37	0.091	26.1-26.5	18	0.47	0.13
14.6-15.0	28	2.06	0.12	26.6-27.0	25	0.22	0.11
15.1-15.5	39	2.22	0.10	27.1-27.5	14	0.53	0.091
15.6-16.0	33	1.86	0.078	27.6-28.0	13	0.24	0.12
16.1-16.5	57	1.49	0.077	28.1-28.5	8	0.09	0.21
16.6-17.0	43	1.64	0.076	28.6-29.0	4	0.58	0.30
17.1-17.5	46	1.67	0.085	29.1-29.5	4	-0.10	0.27
17.6-18.0	52	1.53	0.073	29.6-30.0	1	-0.3	----

Note: growth in millimeters; data from Frank (1965).

**TABLE 11.27 Back Calculated Total Length (TL) at the End of Each Year (1977-1983)
of 130 Pumpkinseeds, *Lepomis gibbosus***

Year class	N	TL at capture (mm)	Age at total length (mm)						
			1	2	3	4	5	6	7
1983	9	44.4 (41-49)	31.6						
1982	17	57.1 (50-63)	31.8	51.5					
1981	30	78.9 (65-91)	31.9	51.1	65.4				
1980	14	94.6 (91-97)	32.6	51.6	67.6	83.5			
1979	32	105.8 (98-114)	33.6	51.3	68.1	84.3	92.2		
1978	24	123.0 (115-136)	34.3	51.8	69.6	85.1	97.0	105.5	
1977	4	139.5 (138-140)	37.0	53.3	71.9	88.2	107.6	120.4	126.0

Note: data from Crivelli and Mestre (1988).

TABLE 11.28 Height (cm) of the Giant Saguaro Cactus, *Cereus giganteus*, at Saguaro National Monument

#	1969	1970	1971	1972	Year 1973	1974	1975	1976	1977
G6808	2.7	3.1	3.4	5.0	8.4	10.7	11.5	13.5	18.5
G6813	3.1	4.6	5.8	8.9	12.5	15.7	19.4	22.6	26.3
G6812	3.7	5.0	5.7	8.2	11.4	14.9	16.5	20.0	23.7
G6805	4.6	6.7	8.8	12.8	17.7	21.8	25.4	29.7	34.3
G6806	5.9	8.6	10.6	13.6	17.3	21.5	24.9	29.2	33.4
G6807	10.2	14.1	17.1	21.9	28.4	34.7	39.3	45.9	52.2
G6811	17.6	20.3	23.9	27.8	31.6	35.5	39.1	44.2	50.8
G6801	17.9	20.0	23.9	27.8	33.6	40.9	47.4	55.8	66.0
G6814	20.0	23.5	26.8	31.3	37.3	44.4	52.0	60.9	70.0
G6816	22.3	25.5	30.5	34.5	40.2	48.0	54.2	64.1	74.3
G6804	23.8	27.8	31.9	35.4	41.2	48.4	54.2	61.5	67.3
G6802	29.4	33.0	37.9	42.9	49.6	57.2	64.3	73.3	84.5
G6810	38.3	43.3	48.2	53.1	59.2	64.7	71.3	78.1	85.6
G6803	59.2	62.5	68.4	74.4	84.0	94.3	99.6	109.5	119.9
G6809	62.7	72.3	81.3	89.5	99.7	108.1	116.4	125.2	135.6
G6815	68.8	79.5	91.5	101.0	109.0	116.5	123.6	129.9	136.5
G6817	92.6	104.9	118.8	130.0	143.0	155.8	168.4	180.1	194.1
7139				145.7	156.9	168.0	181.8	195.7	211.5
7133				169.9	188.4	203.6	216.7	235.2	245.4
7138				174.3	184.0	194.9	200.6	214.0	223.4
7132				184.3	194.6	203.5	212.5	224.8	234.2
7134				228.6	247.9	265.1	279.3	296.5	313.7
7140				353.4	373.3	393.2	407.6	424.3	444.3
7135				395.0	416.5	436.0	452.7	473.0	491.8
69B	428.3	440.7	451.6	465.8	485.5	495.4	506.6	518.5	529.7
69D	449.8	461.0	473.5	485.1	504.4	514.7	531.2	544.7	558.0
7131				491.8	505.8	516.1	527.5	539.0	550.1
7507							581.0	595.0	609.3
7502							667.0	674.0	681.6
7503							732.0	742.6	755.6
7511							744.6	750.0	757.3
7505							789.5	796.8	806.0
7508							865.9	874.2	879.9
7510							900.5	905.9	912.6
7517							986.6	992.0	999.2
7521							1051.6	1059.1	1069.2
7513							1110.4	1117.1	1127.0
7518							1193.5	1201.9	1209.5
7520							1276.2	1283.2	1292.7

Note: # is the plant code number; data are for the East Site and are taken from Appendix 2 of Steenbergh and Lowe (1983).

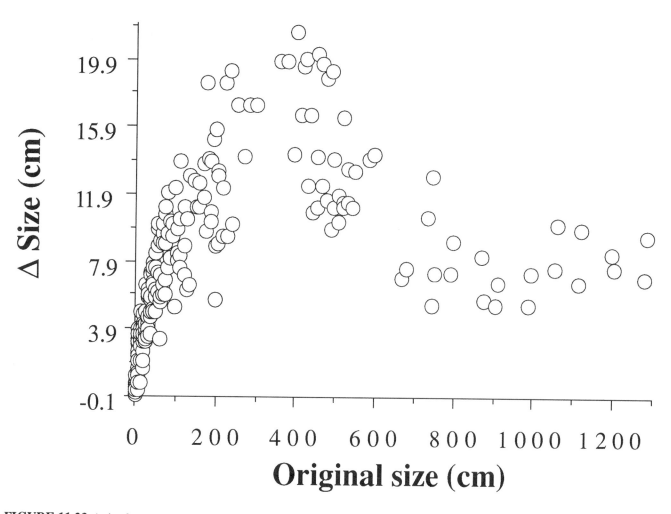

FIGURE 11.23 Δ size for one year as a function of size at the beginning of the year for the giant saguaro cactus, *Cereus giganteus*, at Saguaro National Monument; data from Steenbergh and Lowe (1983) and presented in Table 11.28.

TABLE 11.29 Growth Data for Saguaro Seedlings

#	Time in years							
	1	2	3	4	5	6	7	8
1	0.36	0.51	0.74	1.92	2.08	2.40	2.88	3.50
2	0.20	0.41	0.50	0.53	1.20	2.40		
3	0.22	0.45	1.23	0.65		2.57		
4	0.25	0.47	1.30	0.84				
5	0.40	0.48		1.00				
6	0.42	0.51		1.14				
7	0.46	0.55						
8	0.46	0.56						
9	0.53							

Note: measurements are above-ground stem heights; data taken from Table 2.2 of Steenbergh and Lowe (1983).

10. Growth data for the giant saguaro cactus, *Cereus giganteus* at Saguaro National Monument have been published by Steenbergh and Lowe (1983). Mapped individuals were measured and growth data for one site are shown in Table 11.28. Growth of seedlings is shown in Table 11.29. As an aid in selecting a growth model, a graph of Δsize during a year as a function of size at the beginning of the year is shown in Figure 11.23.

11. Lichenometry is an approach to dating substrates by knowing growth rates of lichens. The first step is to determine growth curves of lichens on structures of known age. McCarthy and Smith (1995) provide measurements of the largest thalli of *Xanthoria elegans* on tree-ring dated moraines and historically dated structures (walls, mausolea, etc.) in the vicinity of Peter Lougheed and Elk Lakes Provincial Parks in Alberta and British Columbia. Table 11.30 shows sizes and ages.

a. First graph the data and use this as the basis for selecting an appropriate growth model. [*Hint:* if there seems to be an asymptotic size, then the Richards function would be reasonable. If assuming an asymptotic size does not seem reasonable, then the Tanaka function probably would be better.]

b. Use either a statistical package or SIMPLEX.BAS for the next chapter to estimate growth parameters. [*Hint:* try estimating all parameters and then try fixing the size at t=0 at something reasonable such as 0.1mm.]

c. Discuss the graph your growth curve with the data as well as a plot of residuals.

TABLE 11.30 Growth Data for *Xanthoria elegans*.

Size	Age	Size	Age
15	40	40	82
17	54	45	71
17	50	52	121
21	64	54	121
22	62	54	116
25	51	54	89
28	67	55	98
28	38	62	108
29	38	62	123
30	59	85	196
30	55	85	236
35	74	85	260

Note: ages in years; data based on several authors and summarized by McCarthy and Smith (1995).

The SYSTAT functions for NONLIN are similar to the functions that would be written for SIMPLEX.BAS except for SIMPLEX.BAS it is necessary to write the function to be minimized, which is SSE.

Richards function:

SIZE =A*(1-(A^(-1/n)-S0^(-1/n))/A^(-1/n)*exp(-K*AGE))^(-n)

The value of S0, the size at t=0, could be estimated or replaced with some fixed value such as 0.1mm.

Tanaka function

SIZE =1/SQR(f)*LOG(ABS(2*f*(AGE-c)+2*SQR(f*f*((AGE-c)^2)+f*a)))+d

The parameter c includes size at t=0,

c = (a/EXP(SQR(f)*(S0-d))-EXP(SQR(f)*(S0-d)))/(4*f))

and so if S0 is to be fixed, then the following code would be appropriate for the entire function,

SIZE =1/SQR(f)*LOG(ABS(2*f*(AGE-(a/EXP(SQR(f)*(.1-d))-EXP(SQR(f)*(.1-d))/(4*f)))+2*SQR(f*f*((AGE-(a/EXP(SQR(f)*(.1-d))-EXP(SQR(f)*(.1-d))/(4*f)))^2)+f*a)))+d

BASIC PROGRAMS

FABENS.BAS

The program **FABENS.BAS** has been modified a number of times since Fabens wrote it in FORTRAN I in 1964. Estimating K and S_∞ of the Brody-Bertalanffy equation by this program has been called "Fabens' Method" although the actual method is the Newton-Raphson algorithm. Using the program is very easy. After launching the application you will be asked a few simple questions.

Do you want to continue? (Y/N)

enter

Y <cr> or y <cr>   The program understands both upper and lower cases.

Are the data from a file (F), entered on the keyboard (K) or from the last run (R)? (F/K/R)

To start with, you probably will want to enter data on the keyboard although for large data sets you should create a data file using a word processor. If you want to enter data at the keyboard enter...
K <cr>

Number of observations?

 Observations are the number of size pairs, S_t and $S_{t+\Delta t}$ together with Δt. Enter the correct number, say 5 for example.

5 <cr>

Enter size at tagging, size at recapture, and time interval. Separate data values with commas

1.03,4.52,1.35
2.44,5.00,1.00
etc.

Number of size values with a known age:

1<cr>

Enter size and age separated by a comma

.05,0

Name of file where data are to be stored?

NAUT.DAT <cr>   [or whatever. Also, you don't have to use all caps.]
After all of the estimates are printed out...

Do you want to continue? (Y/N)

Y <cr>

Do you want to use the Brody-Bertalanffy (B), the logistic (L), Gompertz (G) or Richards (R) model? (B/L/G/R)

enter a letter, say B, and then <cr>
Eventually estimates will appear on the screen and you will be asked whether you want to save the results to a file:
Do you want results stored in a file? (Y/N)

If you enter Y<cr> you will get another prompt

Filename for output:

enter an appropriate name such as

Nautilus.out<cr>

Parameter estimates as well as residuals will be stored in a text file that can be opened using a text editor or word processor.

The data file for the *Nautilus* example (Table 11.9)

```
7
19.22,19.65,0.225
20.82,21.28,0.904
17.70,17.84,0.126
20.23,20.30,0.123
17.90,18.98,0.937
19.23,20.41,0.967
20.30,20.48,0.973
1
.05,0
```

There are 7 size pairs and one estimate of size at known age (0.05 at age 0). A data file could be created using an editor or word processor. If a word processor is used then the file *must be saved as a text file.*

FABENS.BAS

```
10 dim x(1000),y(1000),d(1000),xs(1000),
   ys(1000),yh(1000)
20 dim v(50),w(50),vs(50)
```

```
30 print "Fitting the Brody-Bertalanffy,
   logistic and Gompertz growth
   equations"
40 print
50 print "This program is based on
   Fabens'1964 FORTRAN program."
60 print "It was translated into
   APPLESOFT BASIC by T. A. Ebert in
   1981,"
70 print "updated to MICROSOFT BASIC for
   the Macintosh, APRIL 1988 and"
80 print "modified in 1993, 1994 and
   1995 for Cornered Rat Software©"
90 print
100 print
110 print
120 input "Do you want to continue?
    (Y/N)";f$
130 if f$ = "N" or f$ = "n" then goto
    2770
140 if f$ <> "Y" and f$ <> "y" then goto
    120
150 print "Are data from a file (F),
    entered on the keyboard (K), or from
    the"
160 input "           last run (R)?
    (F/K/R)";f$
170 if f$ = "F" or f$ = "f" then goto
    530
180 if f$ = "K" or f$ = "k" then goto
    290
190 if f$ <> "R" and f$ <> "r" then goto
    150
200 n = ns
210 for i = 1 to n
220 x(i) = xs(i)
230 y(i) = ys(i)
240 next i
250 for i = 1 to n1
260 v(i) = vs(i)
270 next i
280 goto 640
290 input "Number of observations: ";n
300 if n <= 2 then print "You need more
    than 2 values to continue" : goto 90
310 print "Enter size at tagging, size
    at recapture, and time interval."
320 print "separate data values with
    commas."
330 for i = 1 to n
340 input x(i),y(i),d(i)
350 next i
360 input "Number of size value with a
    known age: ",n1
```

```
370 print "Enter size and age separated
    by a comma."
380 for i = 1 to n1
390 input v(i),w(i)
400 next i
410 input "Name of file for data: ";f$
420 open f$ for output as #1
430 print #1,n
440 for i = 1 to n
450 write #1,x(i),y(i),d(i)
460 next i
470 print #1,n1
480 for i = 1 to n1
490 write #1,v(i),w(i)
500 next i
510 close #1
520 goto 640
530 input "What file has the size pairs?
    ";f$
540 open f$ for input as #1
550 input #1,n
560 for i = 1 to n
570 input #1,x(i),y(i),d(i)
580 next i
590 input #1,n1
600 for i = 1 to n1
610  input #1,v(i),w(i)
620 next i
630 close #1
640 ns = n
650 f = 0
660 g = 0
670 h = 0
680 m = 0
690 p = 0
700 print
710 print
720 for i = 1 to n
730 xs(i) = x(i)
740 ys(i) = y(i)
750 next i
760 for i = 1 to n1
770 vs(i) = v(i)
780 next i
790 print "Do you want to use the Brody-
    Bertalanffy (B), the logistic (L),"
800 input " Gompertz (G) or Richards (R)
    model? (B/L/G/R)  ",m$
810 if m$ = "L" or m$ = "l" then goto
    2490
820 if m$ = "G" or m$ = "g" then goto
    2580
830 if m$ = "R" or m$ = "r" then goto
    2670
840 if m$ <> "B" and m$ <> "b" then goto
    790
850 for i = 1 to n
860 r = (y(i)-x(i))/d(i)
870 s = (x(i)+y(i))/2
880 f = f+r
890 g = g+s
900 h = h+r*s
910 p = p+s*s
920 next i
930 c = (f*g-n*h)/(n*p-g*g)
940 ap = 1
950 rem  Iterate to improve K and Smax
960 f = 0
970 g = 0
980 h = 0
990 m = m+1
1000 for i = 1 to n
1010 p = 1/exp(c*d(i))
1020 f = f+(1-p)*(y(i)-x(i)*p)
1030 g = g+(1-p)^2
1040 next i
1050 a = f/g
1060 for i = 1 to n
1070 p = 1/exp(c*d(i))
1080 h = h+d(i)*p*(y(i)+x(i)*(1-2*p)-
     2*a*(1-p))
1090 next i
1100 da = h/g
1110 f = 0 : g = 0
1120 for i = 1 to n
1130 p = 1/exp(c*d(i))
1140 r = a-x(i)
1150 s = a-y(i)
1160 f = f+d(i)*p*r*(s-r*p)
1170 g = g+d(i)*p*(da*(s+r*(1-
     2*p))+d(i)*r*(2*p*r-s))
1180  next i
1190 cp = c-f/g
1200 f = (cp-c)/c
1210 c = cp
1220 g = (a-ap)/ap
1230 ap = a
1240 if abs(f) > 2.000000E-06 then goto
     1270
1250 if abs(g) > 2.0E-06 then goto 1270
1260 goto 1330
1270 if m < 15 then goto 960
1280 print "After ";m;"   iterations, K=
     ";c;" S(max) = ";a
1290 print
1300 if m < 20 then goto 960
1310 print "DiffK = ";f;"DiffS(max) =
     ";g
```

```
1320 goto 1340
1330 print "Final values based on ";m;"
     iterations"
1340 s1 = 0
1350 s2 = 0
1360 s3 = 0
1370 rs = 0
1380 for i = 1 to n
1390 ed = exp(-c*d(i))
1400 s1 = s1+(1-ed)^2
1410 ax = a-x(i)
1420 s2 = s2+d(i)*ed*ax*(1-ed)-(y(i)-
     x(i)-ax*(1-ed))*d(i)*ed
1430 s3 = s3+(ax*d(i)*ed)^2+(y(i)-x(i)-
     ax*(1-ed))*ax*d(i)^2*ed
1440 next i
1450 va = s3/(s1*s3-s2^2)
1460 vk = s1/(s1*s3-s2^2)
1470 rs = 0
1480 sy = 0
1490 y2 = 0
1500 for i = 1 to n
1510 yh(i) = x(i)+(a-x(i))*(1-exp(-
     c*d(i)))
1520 rs = rs+(y(i)-yh(i))^2
1530 sy = sy+ys(i)
1540 y2 = y2+ys(i)^2
1550 next i
1560 va = va*rs/(n-2)
1570 vk = vk*rs/(n-2)
1580 va = sqr(va)
1590 vk = sqr(vk)
1600 f = 0
1610 g = 0
1620 for i = 1 to n1
1630 p = 1/exp(c*w(i))
1640 f = f+(a-v(i))*p
1650 g = g+p*p
1660 next i
1670 b = f/(a*g)
1680 if m$ = "B" or m$ = "b" then goto
     1870
1690 rs = 0
1700 if m$ <> "L" and m$ <> "l" then
     goto 1760
1710 for i = 1 to n
1720 yh(i) = (x(i)+(a-x(i))*(1-exp(-
     c*d(i))))^(-1)
1730 rs = rs+(ys(i)-yh(i))^2
1740 next i
1750 goto 1870
1760 if m$ <> "G" and m$ <> "g" then
     goto 1820
1770 for i = 1 to n
1780 yh(i) = exp(x(i)+(a-x(i))*(1-exp(-
     c*d(i))))
1790 rs = rs+(ys(i)-yh(i))^2
1800 next i
1810 goto 1870
1820 if m$ <> "R" and m$ <> "r" then
     print "error in instructions" : goto
     120
1830 for i = 1 to n
1840 yh(i) = (x(i)+(a-x(i))*(1-exp(-
     c*d(i))))^(-sm)
1850 rs = rs+(ys(i)-yh(i))^2
1860 next i
1870 ts = y2-sy^2/n
1880 r2 = (ts-rs)/ts
1890 print "Total SS: ";ts
1900 print "Residual SS: ";rs
1910 print "     R-squared:  ";r2 :
     print
1920 print "     K = ";c,
1930 print "     Standard Error = ";vk
     : print
1940 print "     S(max) = ";a,
1950 print "     Standard Error = ";va
1960 print "          b = ";b
1970 print " observed        expected
     residual"
1980 for i = 1 to n
1990 print ys(i),yh(i),
2000 print ys(i)-yh(i)
2010 next i
2020 rem  PART C
2030 input "Do you want results stored
     in a file? (Y/N) ";f$
2040 if f$ = "N" or f$ = "n" then goto
     120
2050 if f$ = "Y" or f$ = "y" then goto
     2070
2060 goto 2030
2070 input "Filename for output: ",f$
2080 open f$ for output as #1
2090 print #1,"Total SS: ";ts
2100 print #1,"Residual SS: ";rs
2110 print #1,"     R-squared:  ";r2 :
     print
2120 print #1,"     K = ";c,
2130 print #1,"     Standard Error =
     ";vk : print
2140 print #1,"     S(max) = ";a,
2150 print #1,"     Standard Error = ";va
     : print
2160 print #1,"          b = ";b
2170 print #1," observed
     expected         residual"
```

```
2180 for i = 1 to n
2190 print #1,ys(i),yh(i),ys(i)-yh(i)
2200 next i
2210 print #1,"Age and size values for
     plotting"
2220 p = 1/c
2230 i = int(0.434294*log(p)-1)
2240 n = int(p*0.1^i)
2250 p = n
2260 p = p*50^i
2270 r = 0
2280 t = r*p
2290 if m$ = "B" or m$ = "b" then let s
     = a*(1-b*(exp(-c*t))) : goto 2340
2300 if m$ = "L" or m$ = "l" then let s
     = (a*(1-b*(exp(-c*t))))^(-1) : goto
     2340
2310 if m$ = "G" or m$ = "g" then let s
     = exp(a*(1-b*(1/exp(c*t)))) : goto
     2340
2320 if m$ <> "R" and m$ <> "r" then
     print "Error in transformation" :
     goto 120
2330 s = (a*(1-b*(1/exp(c*t))))^(-sm)
2340 print #1,t,s
2350 if (r-5) >= 0 then goto 2380
2360 r = r+0.5
2370 goto 2280
2380 if (r-14) >= 0 then goto 2410
2390 r = r+1
2400 goto 2280
2410 if (r-30) >= 0 then goto 2440
2420 r = r+2
2430 goto 2280
2440 if (r-70) >= 0 then goto 2470
2450 r = r+5
2460 goto 2280
2470 close #1
2480 goto 120
2490 rem This is the subroutine that
     transforms data for the logistic
     model
2500 for i = 1 to n
2510 x(i) = 1/x(i)
2520 y(i) = 1/y(i)
2530 next i
2540 for i = 1 to n1
2550 v(i) = 1/v(i)
2560 next i
2570 goto 850
2580 rem This is the subroutine that
     transforms data for the Gompertz
     model
2590 for i = 1 to n
```

```
2600 x(i) = log(x(i))
2610 y(i) = log(y(i))
2620 next i
2630 for i = 1 to n1
2640 v(i) = log(v(i))
2650 next i
2660 goto 850
2670 rem Transformation of data for
     Richards function -- but you must
     know n
2680 input "Enter a value for the shape
     parameter, n:  ";sm
2690 sn = -(1/sm)
2700 for i = 1 to n
2710 x(i) = x(i)^sn : y(i) = y(i)^sn
2720 next i
2730 for i = 1 to n1
2740 v(i) = v(i)^sn
2750 next i
2760 goto 850
2770 end
```

RICHARDS.BAS

Using **RICHARDS.BAS** is very simple once a data file has been created. As currently constituted, the program does not provide for data entry; you must create a file using a word processor. *Be certain to save the data as a text file.*

The data file has four parts: (1) N1, the number of size pairs and T, the time interval; (2) the size pairs, S_t and $S_{t+\Delta t}$; (3) N2, the number of observations of size at known age; and, (4) size and age pairs. For example, here is a portion of the data file Dia1.dat for *Diadema setosum* from Zanzibar.

55, 1	① N, the number of size pairs; T, the time interval
.36, 1.00	② S_t and $S_{t+\Delta t}$
.50, .94	
.50, .98	
.52, 1.07	
.55, .99	
⋮	
1.68, 1.68	
1.79, 1.79	
1.94, 1.94	
1	③ N, the number of observations of size at age
.01,0	④ size and age.

The meaning of values in the file are that there are 55 data pairs and the value for Δt is 1 year. The first size measurements are 0.36 and 1.00 and these mean that the original size, S_t, was 0.36 and size at $t+\Delta t$, $S_{t+\Delta t}$, was 1.00. After 55 data pairs are read, the number of observations with size and known age are entered. In this case there is just one observation of size at known age: size is 0.01 and age is 0. Note: you must enter something for ③ and ④; however, these values are used to calculate the parameter b and will not change the estimates of n, K, and S_∞. The parameter b pins the growth curve to the time axis so there is a size at time=0.

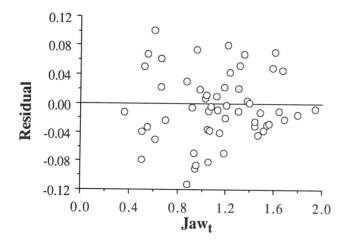

FIGURE 11.24 Residual plot of growth analysis of *Diadema setosum* (Ebert 1980) showing the scatter around the 0-line of perfect fit; there is no obvious trend which is a good result.

Running the program

Launch the program and respond to the questions.

Name of file with data:

Dia1.dat <cr> [or whatever name you have given your file]

Enter a control value to determine which values will be estimated by the program and which will be entered at the terminal. Enter a:

1 if all parameters are to be estimated.
2 if you wish to enter the shape parameter n but estimate other parameters
3 if all parameters except b are to be entered at the terminal

1 <cr>

Most of the time you will want to estimate all parameters but there also are times when you may want to force the fitting of the Brody-Bertalanffy or Gompertz models or you may want to use specific parameters in order to obtain residuals or the likelihood value for the fit.]

Do you want results sent to a file? (Y/N)

Y <cr>

The results of each iteration will be shown on the screen and, eventually, the results with the smallest SSE will be printed out.

Press <RETURN> to continue

Do you want to continue? (Y/N) [you are now back at the top of the program]

If results are saved to a file, you will have residuals, the observed value of $S_{t+\Delta t}$ minus the estimate of $S_{t+\Delta t}$. It is a good idea to plot the residuals *vs.* S_t in order to see how well the estimates fit the data. A plot of the residuals for the data for *Diadema setosum* is shown in Figure 11.24. The size pairs are plotted in Figure 11.9

RICHARDS.BAS

```
10 dim s1(1000),s2(1000),tb(20),sb(20)
20 print "Estimation of Richards
      Function parameters using the
      program "
30 print "developed by T. A. Ebert 1980.
      Can. J. Fish. Aquat. Sci. 37: 687-
      692"
40 print : print "Another fine product
      from Cornered Rat Software©"
50 print ;tab (15);"T. A. Ebert 1991" :
      print
60 input "Do you want to continue?
      (Y/N)";f$
70 if f$ = "n" or f$ = "N" then goto
      1690
80 print
90 print
100 input "Name of file with data: ";r$
110 open r$ for input as #1
120 input #1,n,t
130 for i = 1 to n
140 input #1,s1(i),s2(i)
150 next i
160 input #1,nb
170 for i = 1 to nb
180 input #1,sb(i),tb(i)
190 next i
200 close #1
210 print "Enter a control value to
      determine which values will be
      estimated by"
```

```
220 print "the program and which will be
    entered at the keyboard.  Enter a: "
230 print : print ;tab (3);"1 if all
    parameters are to be estimated."
240 print ;tab (3);"2 if you wish to
    enter the shape parameter n but
    estimate other parameters"
250 print ;tab (3);"3 if all parameters
    except b are to be entered at the
    keyboard"
260 input it
270 if it = 1 then goto 310
280 input "Enter an estimate of the
    shape parameter n  ";xn
290 if it = 2 then goto 310
300 input "Enter K and Smax: ";fk,sm
310 input "Do you want results sent to a
    file? (Y/N) ";p$
320 if p$ = "Y" or p$ = "y" then input
    "File for output: ";f$ : open f$ for
    output as #1
330 if p$ = "Y" or p$ = "y" then print
    #1,"n             SSE            K
    S∞"
340 print "n             SSE        K
    S∞"
350 if n <= 0 then goto 740
360 in = n
370 if it = 2 then goto 470
380 if it = 3 then goto 740
390 f = -1
400 s1 = 100000
410 xn = 25
420 dc = -5*f
430 xn = xn*f
440 k = 5
450 for jj = 1 to 8
460 for j = 1 to k
470 d = -1/xn
480 sx = 0
490 sy = 0
500 for i = 1 to n
510 a = s1(i)^d
520 b = s2(i)^d
530 sx = sx+a
540 sy = sy+b
550 next i
560 b1 = sx/n
570 c2 = sy/n
580 s2 = 0
590 s3 = 0
600 s4 = 0
610 for i = 1 to n
620 a = s1(i)^d
630 b = s2(i)^d
640 a = b1-a
650 b = c2-b
660 s2 = s2+a*a
670 s3 = s3+b*b
680 s4 = s4+a*b
690 next i
700 b = s4/s2
710 c = c2-b*b1
720 r = s4/sqr(s2*s3)
730 goto 800
740 b = exp(-fk*t)
750 c = sm^(-1/xn)*(1-b)
760 fb = b
770 fc = c
780 if it = 3 then goto 850
790 goto 850
800 fb = b/abs(r)
810 fc = c2-fb*b1
820 fk = -log(fb)/t
830 if (1-fb) < 0 then sm = 0 : goto 850
840 sm = (fc/(1-fb))^(-xn)
850 se = 0
860 n = in
870 for i = 1 to n
880 se = (s2(i)-(b*s1(i)^(-1/xn)+c)^(-
    xn))^2+se
890 next i
900 if it = 2 or it = 3 then goto 1150
910 print xn,se,fk,sm
920 if p$ = "Y" or p$ = "y" then print
    #1,xn,se,fk,sm
930 if se > s1 then goto 950
940 s1 = se : sn = xn
950 xn = xn+dc
960 if abs(xn) <= 0.02 then goto 980
970 next j
980 rem
990 if abs(sn) >= 25 then goto 1040
1000 xn = sn-dc
1010 dc = dc/5
1020 k = 10
1030 next jj
1040 f = 1
1050 if it <= 0 then goto 1110
1060 s5 = s1
1070 s6 = sn
1080 s1 = 100000
1090 it = 0
1100 goto 410
1110 it = 2
1120 if s1 > s5 then sn = s6
1130 xn = sn
1140 goto 470
```

```
1150 if it = 3 then goto 1170
1160 sm = (fc/(1-fb))^(-xn)
1170 c1 = 0
1180 b1 = 0
1190 d = -1/xn
1200 for i = 1 to nb
1210 b2 = exp(-fk*tb(i))
1220 c1 = c1+b2*b2
1230 b1 = b1+b2*(sm^d-sb(i)^d)
1240 next i
1250 b1 = b1/(c1*sm^d)
1260 print ;tab (5);"Final values for
     parameters" : print : print
1270 print "Shape parameter, n = ";xn
1280 print
1290 print "Error sums of squares: ";se
1300 print
1310 print "Walford slope: ";fb
1320 print
1330 print "Walford intercept: ";fc
1340 print
1350 print "Growth rate constant K: ";fk
1360 print
1370 print "Asymptotic size: ";sm
1380 print
1390 print "Scaling parameter b: ";b1
1400 l = se^(-n/2)
1410 print "Likelihood function: ";l
1430 input "Press <RETURN> to continue:
     ";r$
1440 if p$ = "N" or p$ = "n" then goto
     1670
1450 print #1,"Shape parameter n: ";xn
1460 print #1,"Error sums of squares:
     ";se
1470 print #1,"Walford slope: ";fb
1480 print #1,"Walford intercept: ";fc
1490 print #1,"Growth rate constant K:
     ";fk
1500 print #1,"Asymptotic size: ";sm
1510 print #1,"Likelihood function: ";l
1520 print #1,"Scaling parameter b: ";b1
1530 print #1,"Residuals"
1540 print
     #1,"Size(t)","Size(t+Δt)","estimated
     ","residual"
1550 for i = 1 to n
1560 c2 = (b*s1(i)^(-1/xn)+c)^(-xn)
1570 rs = s2(i)-c2
1580 print #1,s1(i),s2(i),c2,rs
1590 next i
1600 print #1,"Age","Size"
1610 for i = 0 to 49
1620 s = 1-b1*exp(-fk*i)
1630 if s < 0 then goto 1660
1640 s = sm*s^(-xn)
1650 print #1,i,s
1660 next i
1670 if p$ = "Y" or p$ = "y" then close
     #1
1680 goto 60
1690 end
```

12

Functions Describing Survival

INTRODUCTION

The survival of individuals can be modeled using general functions with one or more parameters. The usual way this is done is to use age as a covariate and numbers of individuals in a cohort as the dependent variable. A decaying exponential equation with one parameter is the simplest function

$$n_t = n_0 e^{-Zx}. \tag{12.1}$$

The parameter Z is the mortality coefficient and x is time. The larger the value of Z, the more rapidly a cohort of individuals disappears (Figure 12.1). Two rates are determined from Equation 12.1 and can be illustrated most easily by fixing time in some familiar units such as years. The annual survival rate is

$$p_x = e^{-Z} \tag{12.2}$$

and the annual mortality rate is

$$q_x = 1 - e^{-Z}.$$

If time, x, is in days then Equations 12.2 and 12.3 would be daily survival and mortality rates. Also note that it is possible to change from one set of time units to another by

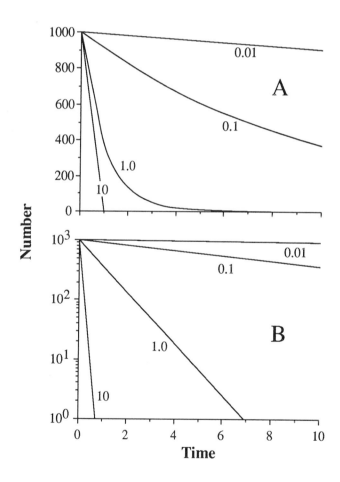

FIGURE 12.1 Survival of cohorts with different mortality coefficients; numbers on lines are values of Z, the mortality coefficient; A: linear scale for number; B: logarithmic scale for number.

adjusting Z, the mortality coefficient. If Z is appropriate for time in years (Z_y) then Z for time in days (Z_d) is

$$Z_d = Z_y/365.$$

Estimating Z from data is straightforward. The usual way of proceeding is to start with Equation 12.1, perform a log-transformation,

$$ln(n_t) = ln(n_0) - Zt,$$

and then use a linear regression with $ln(n_t)$ as a function of t. The slope of the line is -Z and the Y-intercept is $ln(N_0)$. In many cases this will be adequate; two problems arise, however, with this approach: (1) as with the determination of allometric parameters, if a log-transformation is used then small values are weighted relatively more than large values which may or may not be what is desired; (2) n_0 frequently is known and so it should not be viewed as a parameter to estimated. If a linear regression is used, it should be done with a fixed intercept. Estimating Z in Equation 12.1 can be done by non-linear regression to eliminate problems associated with weighting and prior knowledge of n_0.

A basic assumption in using Equation 12.1 is a *constant* survival probability; age does not influence survival. If one selects appropriate sections of a life table, many organisms seem to follow Equation 12.1. Once established, many trees and shrubs are virtually immortal. Many marine invertebrates and many fish appear not to age and so survival rate is no different for a 5 or a 50 year old individual. Use of Equation 12.1 in describing survivorship is a statement that organisms do not show senescence. Furthermore, nearly all organisms have low survival rates when very small or young: larval fish, planktonic stages, seeds and seedlings. *Complete* survivorship curves for all creatures probably show a changing survival rate with age.

CURVES WITH SIMPLE SHAPES

In a classic paper on life tables (Deevey 1947), shapes of survivorship curves were discussed and his shape classification is still used. Some organisms show good survival for a substantial time and then a rapid and accelerating increase in mortality. Such a curve is called 'Deevey Type I' and is how senescence manifests itself. Some species show constant survival; that is, they follow Equation 12.1. These are called 'Deevey Type II'. Some species have high mortality when young and then high survival for a long time and these are called 'Deevey Type III' (Figure 12.2). The designation I, II and III actually are from Pearl and Miner (1935) even though today they are associated with Deevey. Deevey, by the way, gave Pearl and Miner proper credit.

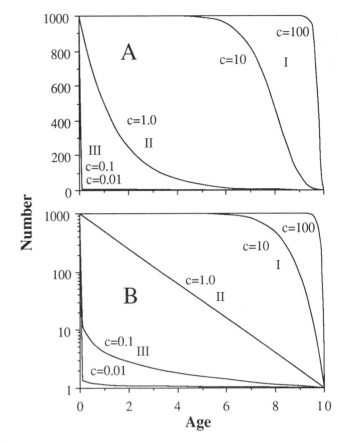

FIGURE 12.2 Survivorship curves showing types I, II, and II of Deevey (1947); numbers by curves are shape parameters, c, of the Weibull distribution.

Two commonly used functions can model all of the Pearl and Miner (Deevey) curves. One is the Weibull function (Weibull 1951) and the other is the Gompertz function (Gompertz 1825). The Weibull function has been used for ecological studies (Pinder *et al.* 1978, Caswell 1982a, Ebert 1985) and is

$$n_x = n_0 e^{-(x/b)^c}, \qquad (12.3)$$

where n_x is the number alive at time x. Recall that

$$p_x = n_{x+1}/n_x$$

and so

$$p_x = e^{[(x/b)^c - ((x+1)/b))^c]}. \qquad (12.4)$$

There are two parameters in the Weibull distribution, b and c. The parameter b is a function of maximum age, ω, at which the last individual is alive in a cohort, and also is a function of the shape parameter, c

TABLE 12.1 Relationship Between Gompertz and Weibull Parameters

Weibull	Gompertz		Deevey Type
c	a	g	
0.01	241.5252	-35.6031	III
0.1	80.0910	-14.1689	III
1	0.6900	1.071×10^{-8}	II
10	3.327×10^{-5}	1.2782	I
100	3.558×10^{34}	8.059	I

Note: Weibull curves all have $\omega = 1$ and $n_0 = 1000$; all curve shapes are shown in Figure 12.2.

$$b = \omega e^{-v/c}. \qquad (12.5)$$

If a cohort starts with 1000 individuals, the last individual is alive when $l_x = 0.001$ and so

$$v = ln(-ln(0.001)). \qquad (12.6)$$

If a cohort started at some other size, v would have a different value or one could keep v as defined in Equation 12.6 and scale survival so the starting value was 1000 individuals. The value of ω is the age at which $l_x = 0.001$ or whatever value is used to define v. Use of 1000 individuals as a starting number is arbitrary. Deevey (1947) used 1000 for n_0 but he wasn't the first; the tradition goes back at least to Halley (1693a,b) who used 1000 as the initial value in describing male mortality rates in Breslau for the years 1687-1691.

The parameter c defines the shape of the curve. Its effect can be seen in Figure 12.2A but is more easily understood by examining a semi-logarithmic plot (Figure 12.2B). When c = 1.0, the Weibull produces a constant survival rate, which is the straight line in Figure 12.2B (Deevey Type II). When c>1.0 the Weibull becomes a Type I curve of Deevey and when c<1.0 the curve is Type III.

The second general function that can model all of the curves in Figure 12.2 is the Gompertz function

$$n_x = n_0 e^{[a/g\,(1-e^{gx})]} \qquad (12.7)$$

or

$$n_x = n_0 * exp(a/g * (1 - exp(gx))).$$

The two parameters are a and g that, in suitable combinations, can generate the same curves as the Weibull function. A comparison is shown in Table 12.1 where the curves shown in Figure 12.2 have been fit using Gompertz parameters.

TABLE 12.2 Life Table of *Floscularia conifera*, a Sessile Rotifer

Age (days)	Number
0	1000
1	980
2	780
3	720
4	720
5	420
6	280
7	220
8	80
9	40
10	20

Note: all data were adjusted to a starting number of 1000; data from Edmondson (1945).

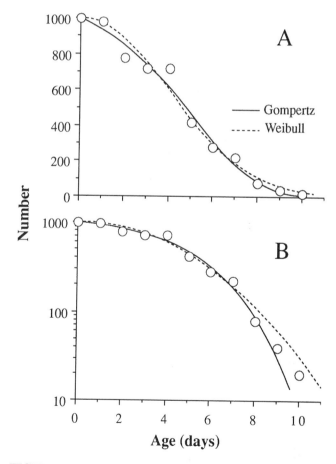

FIGURE 12.3 Survivorship of *Floscularia conifera* with fitted Gompertz and Weibull curves; Gompertz: a=0.0574, g = 0.3423; Weibull ω = 13.920 days and c = 2.082.; **A**, linear scale; **B**, logarithmic scale showing a downward bend of the curve which indicates increased mortality with age; data from Edmondson (1945).

TABLE 12.3 Analysis of *Floscularia conifera* Using the Weibull and Gompertz Functions

	Weibull $r^2 = 0.978$					Gompertz $r^2 = 0.983$			
Parameter	Estimate	se	Lower <95%>Upper		Parameter	Estimate	se	Lower <95%>Upper	
ω	13.9203	1.5087	10.5074	17.3332	a	0.0574	0.0119	0.0304	0.0844
c	2.0818	0.2342	1.5521	2.6115	g	0.3423	0.0554	0.2171	0.4676

Note: data from Edmondson 1945.

In order to cover all the basic shapes of the Weibull, the Gompertz requires both very large and very small values for parameters with Deevey Types I and II; most organisms, however, do not show such extreme Type I curves and so both the Weibull and Gompertz should be tried with data sets. Estimating parameters is done by nonlinear regression, which can be done using **SIMPLEX.BAS** at the end of this chapter or any nonlinear module from statistics software such as NONLIN in SYSTAT.

A sessile rotifer *Floscularia conifera*

Data for *Floscularia conifera* were gathered by Edmondson (1945) and were used as an example by Deevey (1947). The data were adjusted to $n_0 = 1000$ and are shown in Table 12.2 and analysis using the Weibull and Gompertz models is provided in Table 12.3 and is shown in Figure 12.3. The fit is reasonably good with either model and the parameters would be adequate for describing the shape of the survival curve. The shape parameter c of the Weibull is greater than 1 and so the survival curve is tending towards Deevey Type I as can be seen by comparing shapes of Figure 12.3 with shapes in Figure 12.2.

TABLE 12.4. Human Survival in the US Based on the 1964 Census

Age	Male	Female	Age	Male	Female
0.0	1000.00	1000.00	47.5	894.86	933.85
1.0	972.85	978.93	52.5	862.04	914.50
7.5	968.79	975.53	57.5	811.36	885.66
12.5	966.31	973.72	62.5	738.26	845.71
17.5	963.76	972.22	67.5	641.32	787.86
22.5	957.37	969.55	72.5	517.74	704.17
27.5	948.54	966.07	77.5	383.78	591.67
32.5	940.02	961.83	82.5	251.28	444.95
37.5	929.91	955.73	87.5	132.39	272.14
42.5	916.01	946.85			

Note: data from Keyfitz and Flieger (1968).

Human survival in the US based on the 1964 census.

Survival data for men and women in the US based on the 1964 census are shown in Table 12.4. Analysis is shown in Table 12.5 and results are plotted in Figure 12.4 together with lines for the Gompertz and Weibull functions. The fit is fairly good using either of these two functions; neither function, however, correctly captures the substantial drop in numbers due to infant mortality. The fits are best for older ages and for some analyses this would be sufficient because the over-all shape of the curve is reasonable.

TABLE 12.5. Human Survival for the US Based on the 1964 Census

Sex	Model	r^2	Parameter	Estimate	se	Lower <95%>Upper	
Males	Weibull	0.987	ω	114.2833	3.1566	107.6235	120.9431
			c	5.0469	0.3102	4.3925	5.7013
	Gompertz	0.995	a	0.00027			
			g	0.07150			
Females	Weibull	0.981	ω	118.8443	3.5520	111.3503	126.3383
			c	5.8244	0.4108	4.9577	6.6910
	Gompertz	0.991	a	0.00010			
			g	0.07805			

Note: standard errors were not computable for Gompertz parameters; data (Keyfitz and Flieger 1967) shown in Table 12.4.

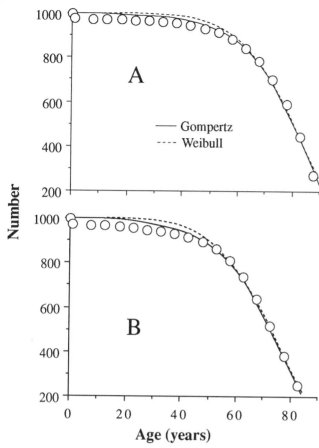

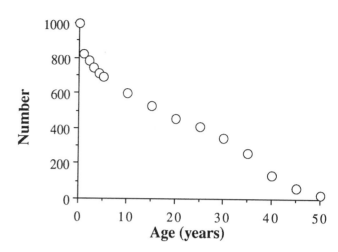

FIGURE 12.5 Survivorship curve for humans at the Libben site in Ohio; survival does not fit any of the patterns shown in Figure 12.2; data from Lovejoy *et al.* (1977) .

FIGURE 12.4 Human survivorship curves for the US based on the 1964 census with fitted lines for the Gompertz and Weibull functions; **A**, females; **B**, males; data from Keyfitz and Flieger (1967).

CURVES WITH INFLECTIONS:
THE RICHARDS FUNCTION

Human survival at the Libben site in Ohio

As long as shapes of survival curves fall within patterns shown in Figure 12.2, the Weibull or Gompertz functions work very well; there are survival curves, however, that do not follow the patterns of Figure 12.2. The data for human remains at the Libben site in Ohio (Lovejoy *et al.* 1977) that were introduced in Chapter 3 are of this sort (Figure 12.5).

The problem with Figure 12.5 is the very rapid initial drop followed by much improved survival but with senescence. It is like the Richards function for individual growth (Chapter 11) turned on its side with age, a, as a function of number, n, rather than the other way around

$$a_n = a_\infty(1-be^{-Kn})^{-m}. \qquad (12.8)$$

Of course what we want is number, n, as a function of age, a, and so Equation 12.8 must be rearranged. In order to simplify notation, I will define a new parameter, c

$$c = -\frac{1}{n}$$

and, with this new parameter, recall the definition of b from Chapter 11

$$b = \frac{a_\infty^c - a_0^c}{a_\infty^c} \qquad (12.9)$$

where a_0 is a slight adjustment in fitting and means the actual age when n=1000. Age when n = 1000 is given in data as age 0 and so a_0 can be set equal to 0 and so reduce estimation by one parameter. This eliminates b from Equation 12.8 but works only when c>0. If c<0, as it would for a logistic model, a_0 would be estimated. Solving for n in Equation 12.8 yields

$$n = -Gln\left(\frac{a_\infty^c - a_n^c}{a_\infty^c - a_0^c}\right) \qquad (12.10)$$

where

n = number of individuals,

a_∞ = asymptotic age, which is similar to ω in the Weibull function,

a_n = age at number in the survival curve,

a_0 = age at the beginning of the curve; can be fixed or estimated,

c = shape parameter equal to -1 if the shape is logistic curve lying on its side, and

G = decay-rate constant; the larger the value the *faster* the decay; it is 1/K in Equation 12.8.

TABLE 12.6 Analysis of Human Survival at the Libben Site, Ohio, using the Transposed Richards function (Equation 12.11)

Parameter	Estimate	se	lower<95%>upper	
G	0.4315	0.0457	0.3319	0.5310
A_∞	75.2731	6.0343	62.1254	88.4208
c	0.2533	0.0234	0.2024	0.3042

Note: $r^2 = 0.997$; data from Lovejoy *et al.* (1977).

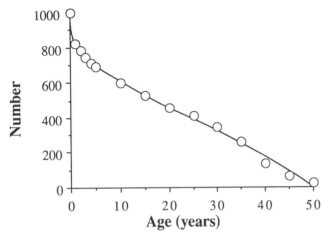

FIGURE 12.6 Human survival at the Libben Site, Ohio; fitted line has parameters shown in Table 12.4; data from Lovejoy *et al.* 1977) .

Equation 12.10 is a Richards curve on its side and would start with n = 0 and increase to a maximum number. The scale must be turned about so that n starts at a maximum and goes to a minimum, which is easy to do. With n_0 being the initial number

$$n_a = n_0 \left[1 + G ln \left(\frac{a_\infty^c - a_n^c}{a_\infty^c - a_0^c} \right) \right].$$ (12.11)

Here is the model for SYSTAT with n_0 equal to 1000 and AGE being the observed age with n individuals

```
N=1000*(1+G*LOG((A^C-AGE^C)/(A^C-A0^C)))
```
(12.12)

Rather than SYSTAT, **SIMPLEX.BAS** could be used to estimate parameters and the appropriate code is given at the end of this chapter. If the data have some other value for n_0, then Equation 12.12 would have to be changed accordingly. Results of analysis of the Libben Site data are shown in Table 12.6 and graphed in Figure 12.6. Analysis was first done including estimation of a_0. The results were c>0 and $a_0 < 10^{-43}$ so a_0 was fixed at 0 and the analysis redone.

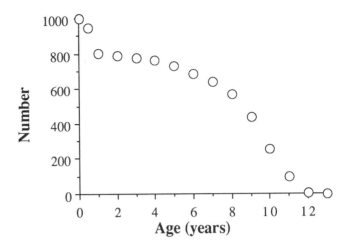

FIGURE 12.7 Survival of Dall sheep (*Ovus dalli dalli*) on Mt. McKinley; data from Murie (1944) presented by Deevey (1947).

The shape of the human survival curve at the Libben Site, shows a gentle sigmoid shape with the maximum age below the x-axis. The estimated maximum age, a_∞, of 75 years has a 95% confidence interval of 62 to 88 years.

CURVES WITH INFLECTIONS: THE SILER MODEL

Dall sheep

Sometimes the models that have been proposed are not up to the task of describing survival. The now famous Dall sheep data from Murie (1944) were used as an example by Deevey (1947). The data are 608 skulls of both sexes combined that were placed into age classes based on horn characteristics and, to a lesser extent, on teeth. Survival of these sheep can not satisfactorily be described with the functions that have been provided in this chapter so far (Figure 12.7).

A model that works quite well has been proposed by Siler (1979). It is a 5 parameter function for components of the mortality rate (hazard function), h_x

$$h_x = a_1 e^{-b_1 x} + a_2 + a_3 e^{b_3 x}.$$ (12.13)

The a_i terms are the initial hazard values and the b_i terms are the adjustment coefficients that determine how rapidly each hazard term changes with age. The a_2 term is a constant representing the mortality rate of adults after the initial drop in mortality and before senescence. There are three survival (l_x) terms associated with each of the three parts of the hazard function

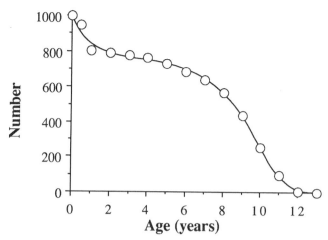

FIGURE 12.8 Survival of Dall sheep with line fitted using the method of Siler (1979) and parameters shown in Table 12.7; data from Murie (1944).

TABLE 12.7 Estimation of Parameters for Dall Sheep using the model of Siler (1979)

Parameter	estimate	se	lower<95%>upper	
a_1	0.2594	0.0534	0.1404	0.3783
b_1	1.0863	0.3767	0.2469	1.9257
a_2	0.0090	0.0093	-0.0117	0.0297
a_3	0.0007	0.0000	0.0007	0.0007
b_3	0.6868	0.0092	0.6662	0.7074

Note: $r^2 = 0.997$; data from Murie (1944).

$$l_x = l_1 l_2 l_3 \tag{12.14}$$

with

$$l_1(x) = \exp\{(-a_1/b_1)[1 - \exp(-b_1 x)]\}, \tag{12.15}$$

$$l_2(x) = \exp(-a_2 x), \tag{12.16}$$

and

$$l_3(x) = \exp\{(a_3/b_3)[1 - \exp(b_3 x)]\}. \tag{12.17}$$

The senescent adult component, l_3, is the Gompertz equation. Results of using Equations 12.14 - 12.17 in a nonlinear regression are shown in Table 12.7 and Figure 12.8. The model for SYSTAT is

```
N =1000*EXP((-A1/B1)*(1-EXP(-B1*AGE)))*
EXP(-A2*AGE)*EXP((A3/B3)*(1-EXP(B3*AGE)))
```

and appropriate code for **SIMPLEX.BAS** is presented at the end of this chapter.

The line in Figure 12.8 shows that the fit is very good. The pattern of survival for Dall sheep is typical for

mammals and Siler's model has been applied to primates including human populations (Gage 1988, Gage and Dyke 1988, Dyke *et al.* 1995). The high infant mortality, high adult survival, and then senescence make the three parts of Siler's model biologically reasonable and worth exploring in greater detail.

GENERAL COMMENTS

There are more survival models including 8-parameter and 10-parameter functions (Mode and Busby 1982, Mode and Jacobson 1984) but I think that Equation 12.1, the Weibull and Gompertz, Equation 12.11 (the transposed Richards), and Siler's model (Equation 12.13) are sufficient for purposes of describing survival in most organisms in order to create smoothed curves for forming matrix elements. In Chapter 14, models will be used to form transition matrices and elasticities for parameters, such as ω and c in the Weibull distribution, will be explored with respect to changes in λ.

PROBLEMS

1. White pine (*Pinus strobus*) was studied by Holla and Knowles (1988) in Ontario, Canada. Ages were determined by counting "growth lines" in cores (Table 12.8). Plot the data and then estimate parameters for the most appropriate survival model.

TABLE 12.8 White Pine (*Pinus strobus*)

Age interval	l_x	Age interval	l_x
0-10	1.000	100-110	0.141
10-20	0.365	110-120	0.106
20-30	0.267	120-130	0.082
30-40	0.224	130-140	0.075
40-50	0.200	140-150	0.055
50-60	0.188	150-160	0.039
60-70	0.184	160-170	0.024
70-80	0.176	170-180	0.008
80-90	0.173	180-190	0.004
90-100	0.153	190-200	0

Note: data from Holla and Knowles (1988).

2. Spinage (1972) collected skulls of large mammals in Akagera National Park, Rwanda. Compare the survivorship of these mammals with each other (Tables 12.9-12.11. Select the most appropriate model and examine patterns of the parameters.

TABLE 12.9. Water Buffalo *Syncerus caffer* (1967-69)

Females				Males			
Age	N	Age	N	Age	N	Age	N
0-1	1000	10-11	118	0-1	1000	12-13	193
1-2	500	11-12	91	1-2	500	13-14	143
2-3	465	12-13	46	2-3	469	14-15	107
3-4	428	13-14	27	3-4	458	15-16	93
4-5	392	14-15	18	4-5	458	16-17	71
5-6	355	15-16	18	5-6	458	17-18	43
6-7	255	16-17	9	6-7	135	18-19	36
7-8	219	17-18	9	7-8	430	19-20	29
8-9	182	18+	9	8-9	386	20-21	21
9-10	145			9-10	364	21-22	14
				10-11	321	22-23	7
				11-12	264		

Note: data from Spinage (1972).

TABLE 12.10 Zebra *Equus burchelli boehmi* (1967-69)

Female				Male			
Age	N	Age	N	Age	N	Age	N
0-1	1000	11-12	345	0-1	1000	12-13	358
1-2	805	12-13	278	1-2	822	13-14	328
2-3	738	13-14	278	2-3	760	14-15	254
3-4	672	14-15	278	3-4	730	15-16	164
4-5	656	15-16	180	4-5	371	16-17	134
5-6	625	16-17	131	5-6	641	17-18	119
6-7	525	17-18	66	6-7	581	18-19	60
7-8	493	18-19	33	7-8	569	19-20	30
8-9	443	19-20	16	8-9	551	20-21	15
9-10	443	20-21	16	9-10	508	21-22	15
10-11	443			10-11	492	22-23	15
				11-12	433	23-24	15

Note: data from Spinage (1972).

TABLE 12.11 Wart hog *Phacochoreus aethiopicus* sexes combined (1967-69)

Age	N	Age	N
0-1	1000	9-10	31
1-2	500	10-11	26
2-3	394	11-12	22
3-4	270	12-13	16
4-5	148	13-14	13
5-6	56	14-15	11
6-7	43	15-16	9
7-8	37	16-17	8
8-9	35		

Note: data from Spinage (1972).

3. Data for the 608 Dall sheep skulls from Mt. McKinley, Alaska (Murie 1944), presented in Figure 12.7 are shown in Table 12.12 together with a smaller set of 221 skulls. The 608 skulls were for sheep that were judged to have died before about 1937 and the smaller set of 221 were judged to have died between about 1937 and 1941 and so represented more recent conditions for the study. Based on confidence intervals for the parameters, would you feel justified in combining these two samples or does it seem as though conditions had changed? Note that for comparison you should use the actual numbers rather than adjusting to $n_0 = 1000$.

TABLE 12.12 Skulls of Dall Sheep Placed in Age Categories Based on Horns and Teeth

A. 608 skulls of sheep that died before about 1937.

Sex	0	1	2	3	4	5	6	7	8	9	10	11	12	13	14	?	Total
								Age category									
Both	33	88															121
Ewes			3	2	4	12	20	17	18	20	28	5	2			82	213
Rams			4	6	3	6	8	12	24	27	39	51	30	1	1	62	274
Total	33	88	7	8	7	18	28	29	42	47	67	56	32	1	1	144	608

B. 221 skulls of sheep judged to have died between about 1937 and 1941

Sex	0	1	2	3	4	5	6	7	8	9	10	11	12	13	14	?	Total
								Age category									
Both	8	29															37
Ewes			2	1	1	3	6	3	4	22	12	8	2			27	91
Rams			1		1	2		5	2	10	13	24	30		2	3	93
Total	8	29	3	1	2	5	6	8	6	32	25	32	32		2	30	221

Note: the column with the ? heading are animals that could not be aged exactly because of absence of horns, but teeth showed they were ≥9 years old; data from Murie (1947).

TABLE 12.13 Survival of the Eastern Fence Lizard, *Sceloporus undulatus*

S Carolina		Ohio		Texas		Colorado	
age	l_x	age	l_x	age	l_x	age	l_x
0.00	1.00	0.00	1.00	0.00	1.00	0.0	1.00
0.25	0.31	0.25	0.49	0.25	0.80	1.0	0.14
0.83	0.11	0.75	0.08	0.75	0.06	1.8	0.11
1.83	0.04	1.75	0.03	1.75	0.01	2.8	0.03
2.83	0.02	2.75	0.01			3.8	0.01
3.83	0.01	3.75	0.01				

Note: data from Tinkle and Ballinger (1972).

4. The eastern fence lizard, *Sceloporus undulatus* was studied by Tinkle and Ballinger (1972). Table 12.13 gives their estimates of survival at four sites.

a. Use the data from all sites in a single analysis of survival. Choose the most appropriate model for your analysis and estimate parameters.

b. Examine the residuals to see whether the data show a pattern of deviation from your estimated values.

BASIC PROGRAMS

The simplex algorithm

The simplex algorithm is a geometric approach to finding the minimum of a function (Nelder and Mead 1963, O'Neill 1985). A simplex is a figure that is drawn in N dimensions where the number of dimensions is equal to the number of parameters you are trying to find and it has N+1 points.

With 2 unknowns a simplex is 2-dimensional and is a triangle. A reasonable way of visualizing how a solution is found is to picture a function that you want to minimize as being shaped like a bowl. The solution is when you arrive at the very lowest point in the bowl. You create a simplex with some guesses for starting values of the parameters so that you have created a point on the solution bowl equal to f(a,b) where a and b are the two parameters you a trying to find. The simplex algorithm takes your two guesses for a and b and adds a small amount first to a, keeping b at its initial guess, and then to b, keeping a at its initial guess and so there are three points: f(a,b), f(a',b), and f(a,b'), where the primes indicate the parameter values that have been increased by a small amount. The "small amount" is a value that must be supplied and will be different for different problems. It is called the "step size".

The simplex is moved by following just four mechanisms that are shown in Figure 12.9: (1) reflection, (2) extension and (3) contraction of vertices, and (4) contraction of the entire simplex around one of the points. By following these rules the simplex moves down the inside of the N-dimensional bowl until it arrives at the lowest point — or as close as you want to get. In effect, it extends the vertices of its figure and tests to see which way to crawl in order to move down. Given this charming behavior, it should be no surprise that one implementation of the algorithm is called AMOEBA (Press *et al.* 1986).

In Figure 12.9, the first move would be reflection of the highest point f(a,b), and then extension. Point f(a,b') now is the highest point and it will be reflected and extended. And so it will go, creeping and oozing down towards the minimum.

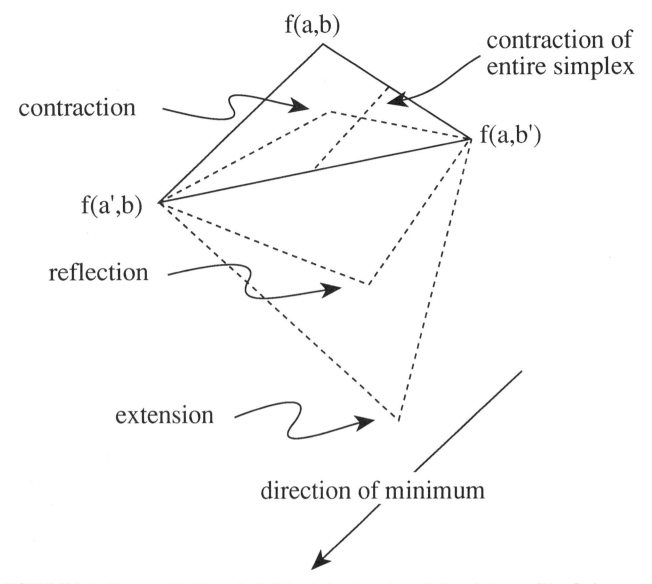

FIGURE 12.9 Possible moves of the highest point, f(a,b), in a simplex; the actual move in this particular case will be reflection.

For many purposes, the function to be minimized is the sum of squares of residuals, SS. There is an observed value for a dependent variable, y, and an estimate of y, ŷ, that is based on the model used to relate y with one or more independent variables. For example, for a straight line the model is

$$y = bx + c.$$

For every value of x, there is an observed value of y and an expected value of y, ŷ, where the expected value is based on multiplying x by some value of b and adding some value for c. The best values for b and c are determined by minimizing the sum of squares of the differences between the observed value of y and its estimate ŷ

$$SS = \Sigma(y - \hat{y})^2$$

or

$$SS = \Sigma(y - (bx+c))^2.$$

The better the match between y and ŷ, the smaller the value of SS.

If you used a simplex to do a linear regression, you would have two unknown parameters and so the simplex would have three points and look like Figure 12.8. The solution bowl would have the minimum SS at the bottom. It should be clear that *any* function could be minimized; it does not have to be the sum of squares.

SIMPLEX.BAS is based on O'Neill (1985), who modified the original program by Nelder and Mead (1965) to include remarks on the original program. I have tried to keep variable names as similar to O'Neill's as possible so that descriptions in his paper apply to **SIMPLEX.BAS**. There are three parts to **SIMPLEX.BAS**: (1) input-output, 2) the function to be minimized, and 3) the simplex algorithm.

Input-output

Input-output will tend to be fairly similar across uses; however, as written, just two variables are entered and these are called $x(i)$ and $y(i)$. If there are more variables to be entered, the input statements would have to be changed as well as the dimension statement. I would suggest $w(i)$ and $z(i)$ as two additional names. Check the dimension statement and any undimensioned name would work.

The function to be minimized

Lines 690 up to 800 can be used to define the function to be minimized although often no more than 10 lines will be needed. There are certain requirements:

1. Parameters are called $p(i)$. If there are two parameters to be estimated, call one $p(1)$ and the other $p(2)$.
2. The *value* of the function that is being minimized must be called ss. This stands for "sum of squares" in most of the things I do; however, it must be called ss even when some other function is to be minimized. The value of ss is passed back to the simplex algorithm where it will be used to branch in the program and then the function will be called again.

3. Index values called i and j are used in the simplex part of the program. *Do not use these index values in writing your function.* Choose something safe like iz or iq. For example *do not* write a loop that starts with For j=1 to 10. Remember that the simplex algorithm will call your function many times and because j is used in the algorithm, if you use j in the function it ends up being reset and will not be correct when the program returns to the simplex algorithm.

4. Just to be safe in writing the function to be minimized, use just two letters for names and use w as the second letter such as aw, bw, cw, etc. There are no two letter names with w as the second letter in other parts of the program so this will make your function safe.

The example that is given in the listing of **SIMPLEX.BAS** is for estimating ω and c for the Weibull function. The parameter ω is $p(1)$ and c is $p(2)$. As written, ω is defined as the age when the last individual of a cohort is alive. This means that you can not start with $n_0=1$; n_0

must be 1000 or 10000. In the code, the values of n are called $y(iw)$ and so $y(1)$ would be 1000 or 10000 or some such value. This is very important because of the way v is calculated in line 700. If $y(1)=1000$ then $1/y(1)$ would be 0.001 or if $y(1)$ would be 10000 then $1/y(1)$ would be 0.0001. If you always want to estimate ω so it is 0.001th of the starting number, then line 700 should be changed to $v = ln(-ln0.001)$.

```
680 rem  Weibull function
690 ss = 0
700 v = log(-(log(1/y(1))))
710 b = p(1)*exp(-v/p(2))
720 for iw = 1 to n
730 a = (x(iw)/b)^p(2)
740 q(iw) = y(1)*exp(-a)
750 s = y(iw)-q(iw)
760 ss = ss+s*s
770 next iw
780 print p(1),p(2),ss
790 return
800 end
```

An important point to notice is that all that happens in the function is that values of $p(1)$ and $p(2)$ are used to estimate a number, $q(iw)$ for each observed age, $y(iw)$. The difference between observed, $y(iw)$ and expected, $q(iw)$ is the residual, which is squared and added to the sum of squares (line 910). When all data values have been used, that is when iw is equal to n, the two parameters and the sum of squares are printed and the program branches back to the simplex algorithm where new values of $p(1)$ and $p(2)$ will be calculated.

SIMPLEX.BAS

```
10 dim x(300),y(300),t(300),q(300),
   pp(5,6),s(12),st(12),pb(12),ps(12),
   p2(12),p(12),xm(12),yp(12)
20 print " The SIMPLEX method. Based on
   Nelder and Mead (1965) and O'Neill
   (1977)"
30 print "         Yet another fine
   product from Cornered Rat
   Software©"
40 print "                    T. A.
   Ebert    April 1996"
50 print : print
60 input "Do you want to continue? (Y/N)
   ";f$
70 if f$ = "Y" or f$ = "y" then goto 100
80 if f$ <> "N" and f$ <> "n" then goto
   60
```

```
90 goto 800
100 input "Name of data file: ";f$
110 open f$ for input as #1
120 input #1,n
130 print n
140 for i = 1 to n
150 input #1,x(i),y(i)
160 print x(i),y(i)
170 next i
180 close #1
190 print : input " How many parameters
    do you want to estimate? ";np
200 print " Enter initial guesses for
    the parameters: "
210 for i = 1 to np : print "Parameter
    (";i;")"; : input s(i) : next i
220 print " Step size for parameters: "
230 for i = 1 to np : print "Step for
    parameter (";i;")", : input st(i) :
    next i
240 input " Convergence should be tested
    every __function evaluations.";kv
250 input " What is the maximum number
    of function evaluations? ";kc
260 print " What is the variance of
    function values that can be used to
    "
270 input "terminate the program (e.g.
    .000001)? ";rq
280 for i = 1 to np
290 pb(i) = 0
300 ps(i) = 0
310 p2(i) = 0
320 p(i) = 0
330 xm(i) = 0
340 yp(i) = 0
350 for j = 1 to np+1
360 pp(i,j) = 0
370 next j
380 next i
390 gosub 850
400 print
410 print "  Iterations = ";ic
420 print "SSE = ";ss
430 for i = 1 to np
440 print "Parameter (";i;")";xm(i)
450 next i
460 input "Would you like to save
    results and residuals in a file?
    (Y/N)";f$
470 if f$ <> "Y" and f$ <> "y" then goto
    60
480 input "File name for results: ";f$
490 open f$ for output as #1
500 for i = 1 to np
510 p(i) = xm(i)
520 next i
530 gosub 690
540 print #1," Iterations = ";ic
550 print #1,"SSE = ";ss
560 for i = 1 to np
570 print #1,"Parameter (";i;")",p(i)
580 next i
590 print #1,""
600 print #1,"Observed         Expected
    Residual"
610 for i = 1 to n
620 print #1,y(i),q(i),y(i)-q(i)
630 next i
640 close #1
650 goto 60
660 rem
670 rem Here is the function that is
    minimized, from 680-800
680 rem
690 ss = 0
700 v = log(-(log(1/y(1))))
710 b = p(1)*exp(-v/p(2))
720 for iw = 1 to n
730 a = (x(iw)/b)^p(2)
740 q(iw) = y(1)*exp(-a)
750 s = y(iw)-q(iw)
760 ss = ss+s*s
770 next iw
780 print p(1),p(2),ss
790 return
800 end
810 rem
820 rem Here is the SIMPLEX method based
    in Nelder, J. A. and R. Mead (1965)
830 rem Computer J 7: 308-313
840 rem
850 rc = 1
860 ec = 2
870 cc = 0.5
880 ep = 1.000000E-03
890 il = 1
900 if (rq < 0 or np < 1 or np > 20 or
    kv < 1) then return
910 il = 2
920 ic = 0
930 nr = 0
940 jc = kv
950 dn = np
960 nn = np+1
970 dp = nn
980 dl = 1
990 rn = rq*dn
```

```
1000 rem
1010 rem Construction of the initial
     SIMPLEX
1020 rem
1030 for i = 1 to np
1040 pp(i,nn) = s(i)
1050 p(i) = s(i)
1060 next i
1070 gosub 690
1080 yp(nn) = ss
1090 for j = 1 to np
1100 x = s(j)
1110 s(j) = s(j)+st(j)*dl
1120 for k = 1 to np
1130 pp(k,j) = s(k)
1140 p(k) = s(k)
1150 next k
1160 gosub 690
1170 yp(j) = ss
1180 s(j) = x
1190 next j
1200 ic = ic+nn
1210 rem
1220 rem SIMPLEX construction complete
1230 rem
1240 yl = yp(1)
1250 io = 1
1260 for i = 2 to nn
1270 if yp(i) > yl then goto 1300
1280 yl = yp(i)
1290 io = i
1300 next i
1310 yn = yp(1)
1320 ih = 1
1330 for i = 2 to nn
1340 if yp(i) < yn then goto 1370
1350 yn = yp(i)
1360 ih = i
1370 next i
1380 rem
1390 rem Calculate PB, the centroid of
     the SIMPLEX
1400 rem vertices excepting that with YP
     value YN
1410 rem
1420 for i = 1 to np
1430 z = 0
1440 for j = 1 to nn
1450 z = z+pp(i,j)
1460 next j
1470 z = z-pp(i,ih)
1480 pb(i) = z/dn
1490 next i
1500 rem
1510 rem Reflection through the centroid
1520 rem
1530 for i = 1 to np
1540 ps(i) = pb(i)+rc*(pb(i)-pp(i,ih))
1550 p(i) = ps(i)
1560 next i
1570 gosub 690
1580 ys = ss
1590 ic = ic+1
1600 if ys > yl then goto 1910
1610 for i = 1 to np
1620 p2(i) = pb(i)+ec*(ps(i)-pb(i))
1630 p(i) = p2(i)
1640 next i
1650 gosub 690
1660 y2 = ss
1670 ic = ic+1
1680 rem
1690 rem Check extension
1700 rem
1710 if y2 > ys then goto 1830
1720 rem
1730 rem Retain extension
1740 rem
1750 for i = 1 to np
1760 pp(i,ih) = p2(i)
1770 next i
1780 yp(ih) = y2
1790 goto 2530
1800 rem
1810 rem Retain reflection
1820 rem
1830 for i = 1 to np
1840 pp(i,ih) = ps(i)
1850 next i
1860 yp(ih) = ys
1870 goto 2530
1880 rem
1890 rem No extension
1900 rem
1910 l = 0
1920 for i = 1 to nn
1930 if yp(i) > ys then let l = l+1
1940 next i
1950 if l > 1 then goto 1830
1960 if l = 0 then goto 2190
1970 rem
1980 rem Contraction on the reflection
     size of the centroid
1990 rem
2000 for i = 1 to np
2010 p2(i) = pb(i)+cc*(ps(i)-pb(i))
2020 p(i) = p2(i)
2030 next i
```

```
2040 gosub 690                           2570 if jc <> 0 then goto 1310
2050 y2 = ss                             2580 rem
2060 ic = ic+1                           2590 rem Check to see if minimum has
2070 if y2 < ys then goto 2300                been reached
2080 rem                                 2600 rem
2090 rem Retain reflection               2610 if ic > kc then goto 2760
2100 rem                                 2620 jc = kv
2110 for i = 1 to np                     2630 z = 0
2120 pp(i,ih) = ps(i)                    2640 for i = 1 to nn
2130 next i                              2650 z = z+yp(i)
2140 yp(ih) = ys                         2660 next i
2150 goto 2530                           2670 x = z/dp
2160 rem                                 2680 z = 0
2170 rem Contraction on the YP(IH) side  2690 for i = 1 to nn
     of the centroid                     2700 z = z+(yp(i)-x)^2
2180 rem                                 2710 next i
2190 for i = 1 to np                     2720 if z > rn then goto 1310
2200 p2(i) = pb(i)+cc*(pp(i,ih)-pb(i))   2730 rem
2210 p(i) = p2(i)                        2740 rem Factorial tests to check that
2220 next i                                   YN is local minimum
2230 gosub 690                           2750 rem
2240 y2 = ss                             2760 for i = 1 to np
2250 ic = ic+1                           2770 xm(i) = pp(i,io)
2260 if y2 > yp(ih) then goto 2380       2780 p(i) = xm(i)
2270 rem                                 2790 next i
2280 rem Retain contraction              2800 yn = yp(io)
2290 rem                                 2810 if ic > kc then return
2300 for i = 1 to np                     2820 for k = 1 to np
2310 pp(i,ih) = p2(i)                    2830 dl = st(k)*ep
2320 next i                              2840 xm(k) = xm(k)+dl
2330 yp(ih) = y2                         2850 p(k) = xm(k)
2340 goto 2530                           2860 gosub 690
2350 rem                                 2870 z = ss
2360 rem Contract whole SIMPLEX          2880 ic = ic+1
2370 rem                                 2890 if z < yn then goto 3040
2380 for j = 1 to nn                     2900 xm(k) = xm(k)-dl-dl
2390 for i = 1 to np                     2910 p(k) = xm(k)
2400 pp(i,j) = (pp(i,j)+pp(i,io))*0.5    2920 gosub 690
2410 xm(i) = pp(i,j)                     2930 z = ss
2420 p(i) = xm(i)                        2940 ic = ic+1
2430 next i                              2950 if z < yn then goto 3040
2440 gosub 690                           2960 xm(k) = xm(k)+dl
2450 yp(j) = ss                          2970 p(k) = xm(k)
2460 next j                              2980 next k
2470 ic = ic+nn                          2990 il = 0
2480 if ic > kc then goto 2760           3000 return
2490 goto 1240                           3010 rem
2500 rem                                 3020 rem  Restart procedure
2510 rem check if YL has improved        3030 rem
2520 rem                                 3040 for i = 1 to np
2530 if yp(ih) > yl then goto 2560       3050 s(i) = xm(i)
2540 yl = yp(ih)                         3060 next i
2550 io = ih                             3070 dl = ep
2560 jc = jc-1                           3080 nr = nr+1
```

```
3090 goto 1030
3100 end
```

In order to use **SIMPLEX.BAS** to estimate parameters for the Gompertz equation, lines 680 to 800 would have to be replaced. This is the replacement

```
680 rem  Gompertz function; p(1) is a
    and p(2) is g
690 ss = 0
700 v = p(1)/p(2)
710 for iw = 1 to n
720 q(iw) = y(1)*exp(v*(1-
    exp(p(2)*y(iw))))
730 s = y(iw)-q(iw)
740 ss = ss+s*s
760 next iw
780 print p(1),p(2),ss
790 return
800 end
```

Replacement of lines 680 to 800 for the 3-parameter transposed Richards function shown in Equation 12.11 is as follows

```
680 rem  transposed Richards. p(1) is g,
    p(2) is A and p(3) is c
690 ss = 0
700 v = p(2)/p(3)
710 for iw = 1 to n
720 q(iw) = y(1)*(1+p(1)*log((v-
    y(iw)^p(3))/v))
730 s = y(iw)-q(iw)
740 ss = ss+s*s
760 next iw
770 print p(1),p(2),p(3),ss
790 return
800 end
```

Finally, here is the appropriate code of Siler's 5-parameter model

```
680 rem  Siler's model. l1, l2, and l3
    are Equations 12.15-12.17
655 rem q(iw) is eq. 12.14
690 ss = 0
695 v1 = p(1)/p(2)
700 v2 = p(4)/p(5)
710 for iw = 1 to n
720 l1 = exp((-v1*(1-exp(-p(2)*y(iw))))
730 l2 = exp(-p(3)*y(iw))
740 l3 = exp((v2*(1-exp(p(5)*y(iw))))
```

```
750 q(iw) = y(1)*l1*l2*l3
755 s = y(iw)-q(iw)
760 ss = ss+s*s
770 next iw
780 print p(1),p(2),p(3),p(4),p(5), ss
790 return
800 end
```

SIMPLEX.BAS has been referred to in other chapters. Here are two growth functions and the function for finding transitions for breeding and nonbreeding sea turtles.

Francis' analogue of Schnute's growth model.

There are 5 parameters in the Schnute model but two of them, ya and yb are arbitrary values that are at about the minimum (ya) and maximum (yb) of the scatter of data values. In the following code, I have assigned ya=.4 and yb=1.7, which are appropriate of the red sea urchin data. For other data sets, you would have to choose different values.

```
680 rem Francis' analogue of Schnute's
    growth model
685 ss = 0
687 ya=.4
689 yb=1.7
690 d1 = ya+p(1)
697 d2 = yb+p(2)
695 a = yb^p(3)-ya^p(3)
697 a1=d2^p(3)-d1^p(3)
699 a=log(a/a1)
700 c = yb^p(3)*d1^p(3)-ya^p(3)*d2^p(3)
710 c1 = d1^p(3)-ya^p(3)+yb^p(3)-d2^p(3)
720 c=c/c1
730 for iw = 1 to nf
740 q = x(iw)^p(3):q=q*exp(-a):q=q+c*(1-
    exp(-a))
745 q(iw)=q^(1/p(3))-x(iw)
750 s = y(iw)-q(iw)
755 ss = ss+s*s
760 next iw
770 print p(1),p(2),p(3),ss
790 return
800 end
```

The Brody-Bertalanffy model

Data are original size, x(i), and a final size, y(i). There are just two parameters, K and S_∞. I will call K p(1) and S_∞ p(2). With a fixed time interval of 1 year, the model is

$$y = x + (S_\infty - x)(1-e^{-K})$$

which, for our program is

$$y(i) = x(i) + (p(2) - x(i))*(1 - \exp(-p(1)))$$

```
690 ss = 0
695 for iw = 1 to n
710 q(iw) = x(iw)+(p(2) - x(iw))*(1-
    exp(-p(1)))
730 s = y(iw)-q(iw)
740 ss = ss+s*s
750 next iw
770 print p(1),p(2),ss
790 return
800 end
```

With variable time intervals, two more changes are needed. The first change has to do with data input. Line 280 would have to be changed so that 3 values would be read for each record, original size, x(i), final size, y(i), and the time interval, t(i). Note: **SIMPLEX.BAS** is written with a variable t(i) dimensioned as t(300). If you want to use some other name, then you must change line 10 by adding a new dimensioned variable.

```
150 input #1,x(i),y(i),t(i)
```

The function is very much like the function with fixed time

```
690 ss = 0
695 for iw = 1 to nf
710 q(iw) = x(iw)+(p(2) - x(iw))*(1-
    exp(-p(1)*t(iw)))
730 s = y(iw)-q(iw)
750 ss = ss+s*s
760 next iw
770 print p(1),p(2),p(3),ss
790 return
800 end
```

In Chapter 8 (Figure 8.8) the problem was presented of finding transitions from breeding adult sea turtles to breeding one year later and not breeding for 2-5 years. The input and function parts for **SIMPLEX.BAS** can be modified as follows to solve for 8 transitions.

```
10 dim x(500),y(500),t(500),q(300),
   pp(12,13),s(12),st(12),pb(12)
12 dim ps(12),p2(12),p(12),xm(12),
   yp(12), x1(15),x2(15),x3(15),x4(15)
13 dim x5(15),x6(15),x7(15),x8(15),
   x9(15),x10(15),x11(15)
14 dim x12(15),x13(15)
⋮
120 input #1,n
130 print n
```

```
140 for i = 1 to n
150 input #1,y(i),x1(i),x2(i),x3(i),
    x4(i),x5(i),x6(i),x7(i),x8(i),x9(i),
    x10(i),x11(i),x12(i),x13(i)
170 next i
175 nf=n
⋮
680 rem Here is the function that is
    minimized, from 680 to 800
685 rem
690 ss = 0
695 for iw=1 to nf
700 tw = p(1)+p(2)*p(3)+p(2)*p(4)*p(5)+
    p(2)*p(4)*p(6)*p(7)
705 tw = tw+p(2)*p(4)*p(6)*p(8)*.8091
710 q(iw)=x1(iw)*p(1)+x2(iw)*p(2)+
    x3(iw)*p(3)+x4(iw) *p(4)+x5(iw)*p(5)
715 q(iw)=q(iw)+ x6(iw)*p(6) +
    x7(iw)*p(7) + x8(iw)*p(8)
720 q(iw)=q(iw)+x9(iw)*p(1)/tw +
    x10(iw)*p(2)*p(3)/tw
725 q(iw)=q(iw)+x11(iw)*p(2)*p(4)*
    p(5)/tw
730 q(iw)=q(iw)+
    x12(iw)*p(2)*p(4)*p(6)*p(7)/tw
735 q(iw)=q(iw)+
    x13(iw)*p(2)*p(4)*p(6)*p(8)*.8091/tw
740 s = y(iw)-q(iw)
745 ss = ss+s*s
750 next iw
755 print ss
780 return
800 end
```

As shown in Chapter 8, the data set for estimating transitions for adult sea turtles is

```
9
.8091, 1, 1, 0, 0, 0, 0, 0, 0, 0, 0, 0, 0, 0
.8091, 0, 0, 1, 1, 0, 0, 0, 0, 0, 0, 0, 0, 0
.8091, 0, 0, 0, 0, 1, 1, 0, 0, 0, 0, 0, 0, 0
.8091, 0, 0, 0, 0, 0, 0, 1, 1, 0, 0, 0, 0, 0
.0358, 0, 0, 0, 0, 0, 0, 0, 0, 1, 0, 0, 0, 0
.4989, 0, 0, 0, 0, 0, 0, 0, 0, 0, 1, 0, 0, 0
.3221, 0, 0, 0, 0, 0, 0, 0, 0, 0, 0, 1, 0, 0
.1119, 0, 0, 0, 0, 0, 0, 0, 0, 0, 0, 0, 1, 0
0.038685, 0, 0, 0, 0, 0, 0, 0, 0, 0, 0, 0, 0, 1
```

Starting values for finding transitions should be between 0.001 and 0.8. Starting all transitions at 0.5 would not be bad for a first run with an initial step size of 0.5. The program will take a *very* long to time.

A final example will show the relationship between the model and the function to be minimized. The *models* were for growth: Francis' version of Schnute's model and the size-growth version of the Brody-Bertalanffy growth model. The *function to minimize* was the same in both cases: namely, the sum of squares of the residuals. O'Neill (1985) provides several test functions for the simplex algorithm. One of these is called Rosenbrock's parabolic valley (Rosenbrock 1960)

$$y = 100\left(x_2 - x_1^2\right)^2 + \left(1 - x_1\right)^2 \qquad (12.18)$$

Equation 12.18 is the function to be minimized and x_1 and x_2 are the parameters that will be given starting values and then changed in order to minimize the function. The function would be rewritten

ss = 100*((p(1) - p(1)^2))^2 + (1-p(1))^2

```
690 ss = 100*((p(2) - p(1)^2))^2 + (1-
    p(1))^2
750 print p(1),p(2),ss
780 return
800 end
```

Starting values for this test are p(2) = -1.2 and p(1) = 1 and other parameters are set as follows by answering prompts on the screen

Step for parameter p(1)

answer 1.0 <cr>

Step for parameter p(2)

answer 1.0 <cr>

Convergence should be tested every __function evaluations

answer 5 <cr>

What is the maximum number of function evaluations?

answer 1000 <cr>

What is the variance of function values that can be used to terminate the program (e.g. .000001)?

answer 10E-16 <cr>

The solution is for $x_1 = x_2 = 1$ and the minimum should be 0. In fact what the simplex will do is get fairly close with a value for the minimum of about 2 x 10-9 and both p(1) and

p(2) about equal to 1. The total number of interations is 177, exactly the number given by O'Neill.

The simplex method is very good, in the sense that it will crawl down to a minimum; however, there is a price in that the algorithm can be *very* slow. For example, getting a solution for Francis' version of Schnute's model with several hundred data pairs may take thousands of iterations and many hours of computer time on a typical desk-top computer. To arrive at a minimum for even moderate sized data sets probably would best be done by setting the number of iterations to 10,000 and leaving the computer to run overnight. The program will terminate based on the variance of function values, which should, in general, not be set smaller than 0.00005 or 0.00001. The program stores an ss value for each parameter as soon as the initial simplex is created. Convergence is tested by calculating the variance of the ss values and comparing it with the required minimum that was entered. For example, if there are three parameters, there will be three ss values stored, one for each parameter as it is changed during the movement of the simplex. Every 4 or 5 interations, depending on the answer to the question

Convergence should be tested every __function evaluations,

the variance will be calculated for the three ss values and compared with whatever answer was given to the question

What is the variance of function values that can be used to terminate the program (e.g. .000001)?

Here are a few hints for using the simplex program:

1. If at all possible, graph data so you have some feel for possible parameter values. Some of this can be done only after you have analyzed a number of data sets. For example, if you are using the Brody-Bertalanffy model you should always be able to guess S_∞ fairly accurately just from a graph of $S_{t+\Delta t}$ *vs.* S_t but it may take a bit of experience to make a good guess for K.
2. For step sizes, don't choose values that are larger than your guess for the parameter. In fact, as you get better at all of this you might want to choose step sizes that are 1/2 to 3/4 the size of your guesses. Pick a step appropriate for each parameter.
3. There always is a problem with local minima capturing the simplex and sucking it in. Sometimes the simplex can escape but sometimes, even with kicking and screaming, it can't. To improve your chances of locating the global minimum, try the program with somewhat different starting values and step sizes.
4. Nonlinear regression is something of an art. You will get better at it with practice. It also requires a fair amount of patience.

13

The Size Structure of Populations

*T*he construction of size-frequency distributions is a mush
used technique....

F. Surlyk,
Morphological Adaptations and Population Structures of
the Danish Chalk Brachiopods (Maastrichtiam, Upper
Cretaceous), 1972

INTRODUCTION

For many organisms, size data are easy to gather and so
size-frequency distributions are common in the literature. In
many cases, they provide the only clues to the underlying
dynamics of growth, survival, and recruitment and so it is
understandable that an extensive literature exists concerning
their analysis. One general research approach has focused on
the separation of size distributions into components
(Harding 1949, Cassie 1950, Bhattacharya 1967, Young and
Skillman 1975, Macdonald and Pitcher 1979). A second
approach has used size data to estimate either one or more
demographic parameters such as mortality with known
growth parameters (Beverton and Holt 1956; Smith 1972;
Van Sickle 1977a,b; Ebert 1981, 1987; Sainsbury 1982) or
both growth and mortality parameters (Green 1970; Ebert
1973, 1987b; Saila and Lough 1981; Fournier and Breen
1983; Pauly 1987). A third approach has used modeling of
size distributions to gain insight into the underlying
processes that give rise to observed distributions (Craig and
Oertel 1966; DeAngelis and Coutant 1982; Barry and
Tegner 1990; Hartnoll and Bryant 1990; Ebert *et al.* 1993).

All of these approaches can be useful and so will be treated
in this chapter; I start with modeling distributions, move on
to several methods for separation of components, and finish
with some techniques for estimating survival, and in some
cases growth, parameters from size data.

SIMULATION OF
SIZE-FREQUENCY DISTRIBUTIONS

Growth functions that describe size (S) as a function of
time (t) can be combined with survival functions that
describe number (N) as a function of time (t) to model size
frequency distributions that express number (N) as a
function of size (S).

The first simulation, which is taken from Barry and
Tegner (1990), is presented as a baseline for subsequent
simulations. Important assumptions for this simulation are:

1. Constant and continuous recruitment,

2. λ, population growth rate, = 1,

3. Brody-Bertalanffy growth model,

4. A constant survival rate,

5. Deterministic growth so σ=0 for size at an age,

6. Deterministic survival so σ=0 for numbers at an age,

and

7. A stable size distribution, which is equivalent to a stable age distribution.

The Brody-Bertalanffy model from Chapter 11 is

$$S_t = S_\infty(1 - be^{-Kt}) \tag{13.1}$$

and the exponential survivorship function from Chapter 12 is

$$N_t = N_0 e^{-Zt}. \tag{13.2}$$

Model #1

For purposes of our first simulation, b is fixed at 1.0 and so it drops from Equation 13.1. This is the same as saying that size is 0 at age 0.

A size distribution is the number of individuals, N_s, at a particular size and this can be related to the number, N_t, at a particular time, t. The number of individuals alive in the *age* interval t_1 to t_2 must be equal to the number in the corresponding *size* interval S_1 to S_2. Such an equivalence is clearer (possibly) if viewed as equating two definite integrals

$$\int_{t1}^{t2} N_t dt = \int_{S_{t1}}^{S_{t2}} N_s ds. \tag{13.3}$$

The change in size in the right integral is over a time interval as given in the left integral. Another way of viewing this is

$$N_s = N_t \frac{dt}{dS} \tag{13.4}$$

The problem is how to use Equations 13.1 and 13.2 with Equation 13.4 in order to model a size-frequency distribution.

The change in age with change in size, $\frac{dt}{dS}$, can be obtained from the Brody-Bertalanffy equation (Equation 13.1) with b=1,

$$S_t = S_\infty - S_\infty e^{-Kt} \tag{13.5}$$

or

$$S_\infty - S_t = S_\infty e^{-Kt}. \tag{13.6}$$

Now differentiate Equation 13.5 with respect to t

$$\frac{dS}{dt} = -KS_\infty e^{-Kt} \tag{13.7}$$

and substitute the left side of Equation 13.6 into Equation 13.7 for $S_\infty e^{-Kt}$. The result is

$$\frac{dS}{dt} = K(S_\infty - S_t)$$

or

$$\frac{dt}{dS} = \frac{1}{K(S_\infty - S_t)}. \tag{13.8}$$

Equation 1 can be rearranged to solve for age, t,

$$t = -\frac{1}{K} ln\left(1 - \frac{S_t}{S_\infty}\right) \tag{13.9}$$

and now substitute 13.9 into Equation 13.2 and multiply by Equation 13.8, which completes all of the substitutions needed in Equation 13.4

$$N_S = N_0 e^{Z/K \, ln(1 - S_t/(S_\infty))} \times \frac{1}{K(S_\infty - S_t)},$$

or

$$N_S = \frac{N_0}{KS_\infty}\left(1 - \frac{S_t}{S_\infty}\right)^{(Z/K - 1)}. \tag{13.10}$$

What does a graph of N *vs.* S look like? Here is a BASIC program to examine Equation 13.10:

```
10 dim n(200)
20 input "file for output: ";f$
30 open f$ for output as #1
40 n0 = 1
50 input "K = ";k
60 input "Z = ";z
70 input "maximum size: ";sm
80 g = n0/(k*sm) : h = z/k-1
100 if h < 0 then goto 160
110 for i = 0 to sm step .1
120 n(i) = g*(1-i/sm)^h
130 print #1,i,n(i)
140 next i
150 goto 210
160 h = abs(h)
170 for i = 0 to sm step .1
180 n(i) = g*(1/((1-i/sm)^h))
190 print #1,i,n(i)
200 next i
210 close #1
220 end
```

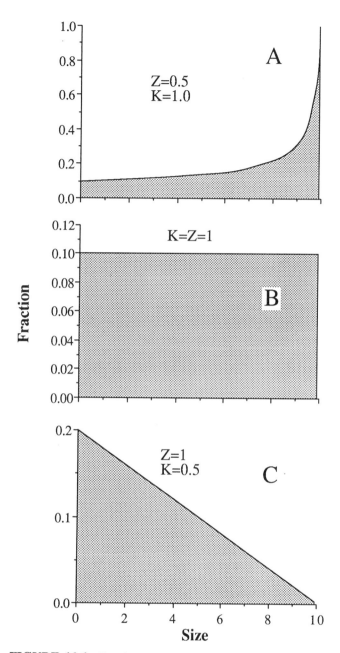

FIGURE 13.1 Size-frequency distributions with constant and continuous recruitment; **A**, K>Z; **B**, K=Z; **C**, K<Z.

Figure 13.1 shows the results of simulations with three different combinations of Z and K. Possibly the most important point of the simulation of size distributions is that the results really don't look very much like size distributions from the field. The clear message is that the models must be made more complex in order to obtain results that look like the size structure of actual organisms.

Model #2

The next model eliminates constant and continuous recruitment (Assumption 1) but preserves determinism

(Assumptions 5 and 6). The approach is to use Equations 13.1 and 13.2 to generate number-density distributions and then integrate over segments of arbitrary size to produce a size-frequency distribution (Ebert *et al.* 1993).

The first step is to calculate sizes at particular ages and then to estimate numbers in a cohort that survive to each age. The following program produces number-density distributions with the Richards function for growth (Chapter 11) and the Weibull function for survival (Chapter 12). Some of the parameters must be entered at the keyboard but two are calculated in the program. The parameter b is calculated in line 90 using S_∞ and S_0, where S_0 is size at time 0. Maximum age, ω, is determined by the growth parameters and is the age at which an individual has attained some specified fraction of its asymptotic size (line 120). The parameter ω is the age of the last survivor out of a cohort of 1000, hence the value 0.001 in line 130. The output from the following program may seem somewhat strange; however, it is in the correct format to draw Figure 13.2 when used with Cricket Graph. To eliminate this format, just remove lines 170 and 190.

```
10 dim n(10000),s(10000)
20 input "File for output: ";f$
30 open f$ for output as #9
40 input "c of the Weibull = ";c
50 input "  K = ";k
60 input "Smax = ";a : input "n = ";nr
70 input "Size at recruitment";sr
80 input "Number of recruitment episodes
   per year = ";stp : stp = 1/stp
90 b = (a^(-1/nr)-sr^(-1/nr))/(a^(-
   1/nr)) : print "B = ";b
100 input "Fraction of maximum size
    attained by a cohort before all are
    dead: ";flb
110 flb = 1-(flb^(-1/nr))
120 w = -(log(flb/b)/k) : print "Omega =
    ";w
130 v = log(-log(.001)) : b1 = w*exp(-
    v/c)
140 for i = 0 to 10000
150 n(i) = 1000*exp(-(i*stp/b1)^c)
160 s(i) = a*(1-b*exp(-k*i*stp))^(-nr)
170 print #9,0,s(i)
180 print #9,n(i),s(i)
190 print #9,0,s(i)
200 if n(i) < 1 then goto 220
210 next i
220 close #9
230 end
```

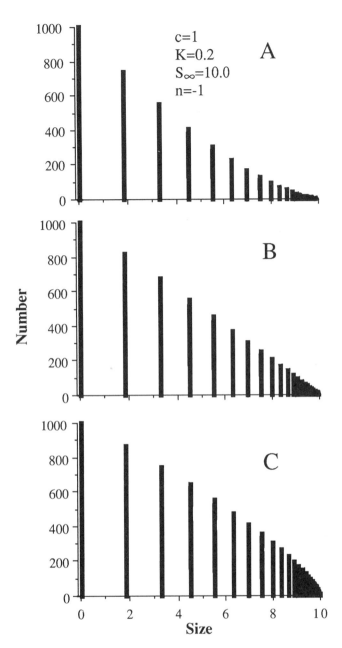

FIGURE 13.2 Number density with ω defined by the time required to attain a specified fraction of S_∞; in **A** the fraction is 0.99 and so ω = 23.00, which makes Z = 0.30 (K<Z); in **B** the fraction is 0.999 and so ω = 34.51, which makes Z = 0.20 (K=Z); in **C** the fraction is 0.9999 and so ω = 46.03 so Z = 0.15 or (K>Z); S_R = 0.05.

Model #3

Figure 13.2 shows shapes for size distributions that are not correct for several reasons. First, each number at size is

a single line that represents a point, which means that all organisms at a particular age have exactly the same size: no variation. Second, if a size-frequency distribution actually was being constructed, there would be size intervals such as 1 cm or 0.5 cm and these intervals would be uniform across the distribution. Figures 13.2A - 13.2C show a decreasing interval between vertical lines and so if a 1 cm interval were used some lines would be summed together. The following program sums all individuals within 1-unit intervals. For example, there is just one line in the interval 0.05-1 but two lines in the interval 6.0001-7.0000 and two in the interval 7.0001-8.0000. Lines get closer and closer together and so more of them would be summed together to provide the number in a 1-centimeter size class. The following program uses 1-cm intervals and creates size-frequency distributions with the same assumptions as in Figure 13.2.

```
10 dim n(10000),s(10000),f(20)
20 input "File for output: ";f$
30 open f$ for output as #9
40 input "c, the Weibull shape
      parameter: ";c
50 input "K, Smax, n, size at
      recruitment:    ";k,a,nr,sr
60 input "Recruitment episodes per year
      = ";stp : stp = 1/stp
70 input "Fraction of maximum size
      attained by a cohort before all are
      dead: ";flb
80 flb = 1-(flb^(-1/nr))
90 tot = 0 : tfi = 0 : b = (a^(-1/nr)-
      sr^(-1/nr))/(a^(-1/nr))
100 w = -(log(flb/b)/k) : print w
110 v = log(-log(.001)) : b1 = w*exp(-
      v/c)
120 for i = 0 to 10000
130 n(i) = 1000*exp(-(i*stp/b1)^c)
140 s(i) = a*(1-b*exp(-k*i*stp))
150 j = int(s(i)) : f(j) = f(j)+n(i)
160 tot = tot+n(i)
170 if n(i) < 1 then goto 190
180 next i
190 j = int(a+1)
200 for i = 0 to j
210 print #9,i,f(i)/tot
220 print #9,i+1,f(i)/tot
230 next i
240 close #9
250 end
```

Parameter values in the following simulation are

TABLE 13.1 Output from Size Distribution Simulation that can be Plotted to Create the Frequency Distribution Shown in Figure 13.3

Size x	y-coordinate
0	0.259566
1	0.259566
1	0.192229
2	0.192229
2	0
3	0
3	0.14236
4	0.14236
4	0.105428
5	0.105428
5	0.078078
6	0.078078
6	0.057822
7	0.057822
7	0.098021
8	0.098021
8	0.039813
9	0.039813
9	0.026683
10	0.026683
10	0

c of the Weibull = 1,
K = 0.20,
S_∞ = 10,
n = -1,

and

size at recruitment, S_R = 0.05.

The fraction of maximum size attained by the last individual out of a cohort of 1000 was set at 0.99 and there is just one recruitment episode per year. Output (Table 13.1), with the exception of x = 0, has two y-values for each x, which are the heights of two adjacent rectangles in the graph. For example, the two values for size 1 in Table 13.1 represent the height of the right-hand height of the 0-1 size class and the left-hand height of the 1-2 size class. The two y-values for each x-value are needed to draw Figure 13.3 using Cricket Graph or similar software. The values in Table 13.1 also are suitable for plotting on graph paper and connecting the dots. The resulting size-frequency histogram is much more satisfactory than size-frequency histograms that are produced as bar graphs by commercial software. Figure 13.3 is starting to look like a real size-frequency distribution with a major mode at the smallest size, a gap, and then and smaller mode at size 7-8; however, one final aspect of size distributions must be addressed, variation in size at age.

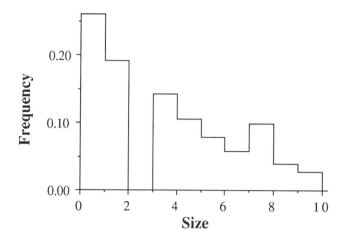

FIGURE 13.3 Integrated number-density distribution using a category width of 1.0 units of size; 1 recruitment episode per year; deterministic growth and survival with c = 1, K = 0.20, S_∞ = 10.0, S_R=0.05; ω defined by the time required to attain 0.99 of S_∞ and so ω = 23.

Model #4

The final model provides for variation in growth so size at an age is a mean and is surrounded by a normal curve. Two new steps are needed in this simulation. First it is necessary to establish a reasonable standard deviation surrounding each mean size. For many creatures, the standard deviation changes with size and often increases with size but it also can decrease. For the present simulation, the standard deviation (σ) increases with mean size

$$\sigma = 0.1\mu. \qquad (13.11)$$

In the accompanying program, a normal curve is calculated to ±4σ on each side of each mean. The procedure is first to calculate size at an age, then to calculate σ, then to determine the area under segments of the normal curve in units of σ/10 and finally to reduce each area by the survival rate. All segments are added together to form the final size distribution.

Areas under segments of the normal curve are determined by successive subtraction of terms obtained from a polynomial approximation of the area under the normal curve based on a program for the normal distribution given by Poole and Borchers (1979) who used an algorithm from Hastings (1955). This algorithm also is Function 26.2.16 by Abramowitz and Stegun (1972). P(x) is the area under the normal curve from the mean, μ, to a size, s, given a standard deviation of σ

$$P(x) = \frac{1}{2} - r(a_1t + a_2t^2 + a_3t^3) + \epsilon(x), \qquad (13.12)$$

$$x = \frac{s-\mu}{\sigma},$$

$$r = \frac{e^{-x^2/2}}{\sqrt{2\pi}},$$

$$a_1 = 0.4361836,$$

$$a_2 = -0.1201676,$$

$$a_3 = 0.9372980,$$

$$t = \frac{1}{1+px},$$

and

$$p = 0.33267$$

with an error of

$$\epsilon(x) < 10^{-5}.$$

The area under the normal curve, A, from s to s+Δs is

$$A = P(x)_{s+\Delta s} - P(x)_s. \qquad (13.13)$$

The following program simulates a size distribution with variation surrounding mean size at age.

```
10 dim vl(20),t(1200),n(10000),
   a(1200),m(10000),f(1200)
20 input "File for output: ";f$
25 open f$ for output as #9
30 input "c, the Weibull shape
   parameter: ";c
40 input "K, Smax, n, size at
   recruitment:    ";k,sm,nr,sr
50 input "Recruitment episodes per year
   = ";stp
55 stp = 1/stp
60 tot = 0
62 tfi = 0
64 input "Coefficient of variation: ";cv
70 b = (sm^(-1/nr)-sr^(-1/nr))/(sm^(-
   1/nr))
80 print "Fraction of maximum size
   attained when"
```

```
90 input " last individual is alive:
   ";flb
100 flb = 1-(flb^(-1/nr))
105 w = -(log(flb/b)/k)
108 print "omega = ";w
110 v = log(-log(.001))
115 b1 = w*exp(-v/c)
120 for i = 0 to 10000
130 n(i) = 1000*exp(-(i*stp/b1)^c)
140 m(i) = sm*(1-b*exp(-k*i*stp))^(-nr)
150 kk = i : if n(i) < 1 then goto 170
160 next i
170 for l = 0 to kk
180 for i = 0 to 20
190 vl(i) = 0
192 a(i) = 0
194 t(i) = 0
200 next i
210 s = m(l)*cv
215 dev = s*4/10
220 for i = 10 to 20
230 vl(i) = m(l)+(i-10)*dev
240 vl(20-i) = m(l)-(i-10)*dev
250 next i
260 for i = 10 to 20
270 y = vl(i)
280 z = y-m(l)
290 y = abs((y-m(l))/s)
300 r = exp(-(y^2)/2)/2.5066282746
310 y = 1/(1+0.33267*abs(y))
320 t = 0.5-r*(0.4361836*y-
    0.1201676*y^2+0.937298*y^3)
330 if z >= 0 then goto 350
340 t = 0.5-t
350 t(i) = t
360 next i
370 for i = 10 to 19
380 a(i) = t(i+1)-t(i)
390 a(19-i) = a(i)
400 next i
410 for i = 0 to 20
420 j = int(vl(i)+dev/2)
430 f(j) = f(j)+a(i)*n(l)
440 next i
450 tot = tot+n(l)
460 next l
470 j = int(sm+sm*cv*4+1)
480 for i = 0 to j
490 print #9,i,f(i)/tot
500 print #9,i+1,f(i)/tot
510 next i
520 close #9
530 end
```

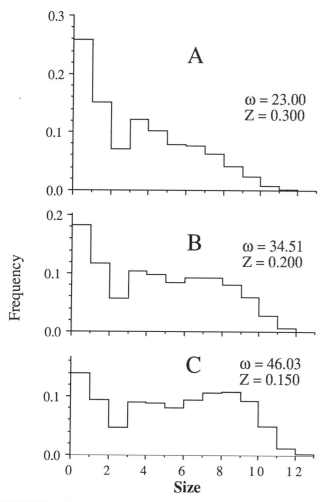

Figure 13.4 shows size distributions that represent sampling at the time of recruitment of new individuals and, because survivorship improves from 13.4A to 13.4C, the *relative proportion* of the smallest individuals *decreases* from 13.4A to 13.4C. The position of the second mode in the size distributions also increases from Figure 13.4A to 13.4C and the right-hand slope from the second mode becomes steeper as more and more individuals get closer to maximum size.

Different growth models

A different growth model changes the shape of the size distribution. For example, Figure 13.5 shows the outcome using four different shape parameters, n, of the Richards function: (A) logistic (n = +1), (B) Gompertz (n = -25), (C) Brody-Bertalanffy (n = -1), and (D) a shape parameter of n = -0.2. For all simulations, c of the Weibull was 1.0 and ω was 69 years so Z = 0.10. In order to keep mortality the same for all simulations, the growth parameter K had to be changed

$$K = -\frac{1}{\omega} ln\left[\frac{1 - (1-F)^{-1/n}}{b}\right], \qquad (13.14)$$

where F is the fraction of asymptotic size attained when the last survivor is alive out of a cohort of 1000. For all simulations in Figure 13.5, F was set at .999999, which means that 1 individual out of 1000 lived sufficiently long to get very close to the asymptotic size of 10.

A change in the growth model changes the shape of the distributions in Figure 13.5 profoundly. The reason for this is the length of time spent in the smallest sizes. The lag of the logistic is such that there is a piling-up of animals in the smallest size, 0.05-1.0, and because the lag is not as pronounced with the Gompertz model, the probability is greater that individuals will grow through the first size class. For the Richards function with n = -0.2, which is very much like the example of *Diadema setosum* in Chapter 11, growth through the small size classes is very rapid and sizes pile-up close to maximum size. Even though K was adjusted for each simulation using Equation 13.14, the real changes are due to the shape of the growth function; that is, n of the Richards function. Modifications of other parameters would have other sorts of interesting effects on the shapes of the size-distributions.

The important point of this section is that simulations can help to build intuition and so are useful when data are gathered. The simulations that have been shown also should suggest that interpretation of size-frequency distributions may be rather difficult because one has to worry about both growth and survival functions. If you know what functions

FIGURE 13.4 Size-frequency distribution simulation with different survival rates; survival determined by the age necessary to attain specified fractions of maximum size: time to reach 0.99 of S_∞ in **A**; time to reach 0.999 of S_∞ in **B** and 0.9999 in **C**.

Results of three simulations (Figure 13.4) show the effect of changing length of life but with growth parameters held fixed. The fraction of maximum size that is attained when the last individual out of 1000 is alive is used to determine ω in the Weibull distribution. Parameters for growth in Figure 13.4 are

$$K = 0.2 \text{ yr}^{-1},$$

$$S_\infty = 10,$$

$$S_R = 0.05,$$

$$n = -1,$$

$$c = 1,$$

and

$$CV = 0.1.$$

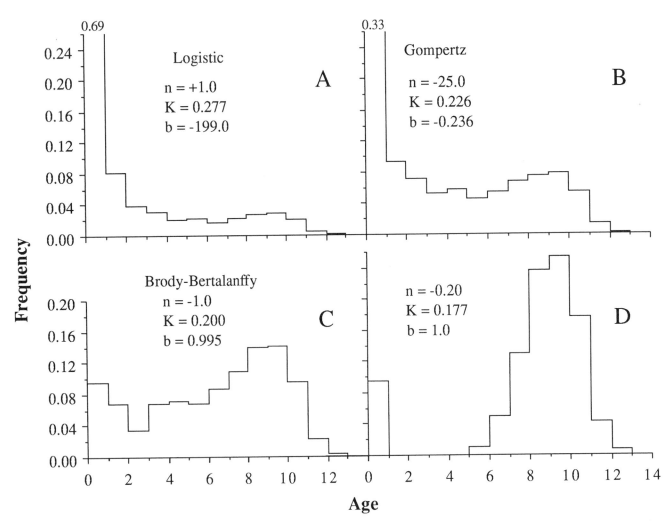

FIGURE 13.5 Simulation of size-frequency distributions with fixed survival, $Z = 0.10$, $S_\infty = 10$, $S_R = 0.05$; c of the Weibull = 1; Richards growth function shape parameters: (A) n=+1, logistic; (B) n=-25, Gompertz; (C) n=-1, Brody-Bertalanffy; (D) n=-0.20.

are the correct ones, then interpretation of size-frequency distributions can be fairly straightforward. On the other hand, if you know nothing about the underlying shapes of growth and survival functions then it may be necessary to assume that certain forms are the correct ones. Frequently, Brody-Bertalanffy growth and constant survival (c=1 of the Weibull or Equation 12.1) are assumed and this may be perfectly all right as long as it is stated up front that these are assumptions; however, based on models presented in Chapters 11 and 12, what should the default assumptions be? I don't know the answer, but my sense is there may be patterns across broad groups of organisms for shapes of growth and survival functions but we don't know as yet what these patterns are. For example, for mammals, the Siler survival model may be the best one (Siler 1979) and growth in echinoderms and molluscs may best be modeled using the Tanaka function (Tanaka 1982,1988). This is a fertile area for research.

SEPARATION OF DISTRIBUTIONS INTO COMPONENT PARTS

Although mark-recapture methods and natural growth lines of demonstrated periodic nature can be used to determine growth, decomposition of size-frequency distributions also has been used and remains popular, particularly in fisheries where some species (e.g. shrimp) have no hard parts with periodic lines or in tropical regions where here is no reliable environmental signal that is clear in hard parts of fish. The idea behind the use of size-frequency distributions is quite simple. If there is an annual pulse of recruits to a population, growth parameters can be extracted by following peaks in size-frequency histograms.

Separation of a polymodal size-frequency distribution into component normal distributions can be achieved in a number of ways. Use of probability paper is one method in which a cumulative size-frequency distribution is plotted and inflexion points used to separate component parts (Harding 1949; Cassie 1950,1954). Other approaches of this type include Hald (1952), Tanaka (1962), Taylor (1965), Hasselblad (1966), Bhattacharya (1967), Harris (1968), Shepherd (1987), Sparre (1987). There also are approaches that do not require that components be normal (McNew and Summerfelt (1978). Computer programs that separate normal components of polymodal distributions have been written, for example, by Young and Skillman (1975), Macdonald and Pitcher (1979) and Schnute and Fournier (1980). The oldest objective method for separating components is the use of probability paper that has one axis labeled as cumulative percent from 0.01 to 99.99 and scaled so that equal linear distances along the axis represent equal areas under a normal curve. If a size-frequency distribution is normally distributed, then if size is plotted *vs.* cumulative percentages on a probability scale, a straight line will be formed. It is worth spending a few pages examining the basis of the probability paper approach to decomposition of frequency distributions.

Obtaining the appropriate values for the probability scale can be done by using a table of proportions of a normal curve that lie beyond a given *normal deviate*, or x value, where

$$x = \frac{X - \mu}{\sigma}. \tag{13.15}$$

Usually tables of this sort are used to obtain a probability, p, of obtaining an observation equal to or more extreme (+ or -) than some value X, when μ (the mean) and σ (the standard deviation) are known. All that we are doing is working the table *backwards*: given p what is x; that is, $(X - \mu)/\sigma$?

It is possible to use a table from any statistics text or to use a computer program to calculate the normal deviate x to be associated with a tail probability p. We will start with the normal curve.

The probability-density, Z, for a normal curve based on the standardized normal deviate (*i.e.* $\overline{x} = 0$, $\sigma = 1$) is

$$Z = \frac{1}{\sqrt{2\pi}} e^{-x^2/2} \tag{13.16}$$

If you wanted to create a normal curve with x values on the abscissa and frequency values along the ordinate, Equation 13.16 would be the equation of choice. All you need are a mean and standard deviation

$$Z = \frac{1}{\sigma\sqrt{2\pi}} e^{-(x - \mu)^2/2\sigma^2}. \tag{13.17}$$

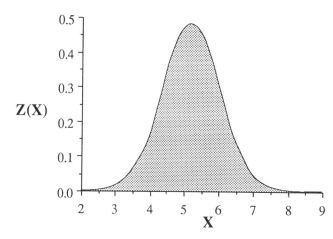

FIGURE 13.6 Normal curve with μ = 5.209 and σ = 0.829; Z(X) calculated using Equation 13.17.

For example, if μ = 5.209 and σ = 0.829, the following program could be used to produce Figure 13.6 by starting with x = 1 and incrementing x by 0.1

```
15 input "mean: ";m:input "sd: ";sd
20 input "file for output: ";f$
30 open f$ for output as #1
40 x0=1
42 rem x0 is the smallest size to be
   included.
45 c=1/(sd*sqr(2*pi))
47 rem The following line must be
   adjusted for each data set.
50 for i= 0 to 9 step .1
60 x=(x0+i-m)/sd
70 x=exp(-(x*x/2))
80 z=c*x
90 print #1, x0+i,z
100 next i
110 close #1: end
```

Tail probability values for a normal curve are obtained by integrating the probability-density function, Z(x) either from $-\infty$ to x (Equation 13.18) or from x to $+\infty$ (Equation 13.19)

$$p(x) = \int_{-\infty}^{x} Z(t)dt \tag{13.18}$$

or

$$q(x) = \int_{x}^{+\infty} Z(t)dt. \tag{13.19}$$

A rational approximation for x given $0<p\leq0.5$ (Equation 13.20) is given as Equation 26.2.23 in Abramowitz and Stegun (1964).

TABLE 13.2 Diameter Measurements in Centimeters for *Tripneustes gratilla* at Pupukea Beach Park, Oahu, Hawaii, 28 October 1975

Size	Freq	Size	Freq	Size	Freq	Size	Freq	Size	Freq	Size	Freq	Size	Freq
2.29	1	4.10	2	4.58	4	4.97	2	5.37	4	5.76	2	6.28	5
2.77	1	4.11	1	4.59	1	4.98	1	5.38	1	5.77	4	6.34	1
2.95	1	4.13	1	4.60	2	5.00	3	5.39	1	5.78	3	6.35	1
2.98	1	4.14	1	4.61	2	5.01	4	5.40	2	5.80	2	6.36	2
3.13	1	4.15	1	4.62	2	5.02	5	5.41	3	5.83	2	6.38	1
3.15	1	4.18	1	4.63	1	5.03	4	5.42	2	5.85	2	6.39	2
3.18	1	4.19	2	4.64	2	5.04	6	5.43	6	5.86	4	6.41	1
3.26	1	4.20	4	4.65	3	5.05	5	5.44	2	5.87	4	6.42	4
3.30	1	4.21	1	4.66	1	5.06	2	5.45	2	5.88	3	6.44	1
3.33	1	4.22	1	4.67	5	5.07	4	5.46	5	5.90	2	6.46	4
3.35	1	4.24	1	4.68	1	5.08	2	5.47	3	5.91	4	6.49	2
3.36	1	4.25	2	4.69	4	5.09	1	5.48	4	5.93	4	6.50	1
3.44	1	4.26	2	4.70	2	5.10	4	5.49	1	5.94	3	6.52	2
3.46	3	4.28	1	4.71	3	5.11	3	5.50	2	5.95	2	6.56	2
3.51	2	4.29	1	4.72	5	5.12	5	5.51	1	5.96	5	6.57	1
3.57	1	4.30	2	4.73	3	5.13	3	5.52	4	5.97	3	6.59	2
3.63	1	4.31	3	4.74	5	5.14	1	5.53	6	5.98	3	6.61	1
3.66	1	4.32	1	4.75	3	5.15	6	5.54	1	5.99	4	6.65	2
3.69	1	4.33	2	4.76	1	5.16	3	5.55	2	6.00	1	6.70	1
3.73	1	4.34	4	4.77	3	5.17	3	5.56	1	6.01	2	6.75	2
3.75	3	4.36	4	4.78	3	5.18	6	5.57	1	6.03	2	6.79	2
3.78	1	4.38	3	4.79	3	5.19	3	5.58	3	6.04	3	6.83	1
3.79	1	4.39	1	4.80	5	5.20	5	5.59	2	6.05	1	6.84	1
3.83	1	4.40	1	4.81	6	5.21	6	5.60	1	6.07	1	6.85	1
3.85	1	4.42	3	4.82	4	5.22	2	5.61	4	6.08	1	6.91	1
3.86	1	4.43	3	4.83	1	5.23	2	5.62	2	6.10	6	6.92	1
3.92	2	4.44	2	4.84	3	5.24	6	5.63	2	6.11	1	6.94	3
3.93	1	4.45	2	4.85	3	5.25	3	5.64	3	6.13	3	7.01	1
3.95	2	4.46	4	4.86	3	5.26	3	5.65	1	6.14	1	7.03	1
3.96	1	4.47	2	4.87	4	5.27	3	5.66	3	6.16	2	7.06	1
3.97	2	4.48	2	4.88	5	5.28	3	5.67	3	6.18	1	7.12	1
3.98	1	4.49	2	4.89	5	5.29	1	5.68	2	6.19	2	7.29	1
3.99	2	4.50	2	4.90	4	5.30	1	5.69	3	6.20	3	7.42	1
4.01	2	4.51	1	4.91	2	5.31	1	5.70	2	6.21	1	7.57	2
4.02	3	4.52	1	4.92	4	5.32	2	5.71	4	6.22	1	7.80	1
4.04	1	4.54	2	4.93	6	5.33	2	5.72	6	6.23	3	8.15	1
4.06	2	4.55	3	4.94	2	5.34	3	5.73	1	6.25	3	8.37	1
4.08	1	4.56	4	4.95	2	5.35	2	5.74	8	6.26	1		
4.09	1	4.57	1	4.96	6	5.36	2	5.75	3	6.27	1		

Note: data from Ebert (1982).

$$x = t - \frac{c_0 + c_1 t + c_2 t^2}{1 + d_1 t + d_2 t^2 + d_3 t^3} + \epsilon(p), \quad t = \sqrt{ln\frac{1}{p^2}} \quad (13.20)$$

$c_0 = 2.515517$	$d_1 = 1.432788$
$c_1 = 0.802853$	$d_2 = 0.189269$
$c_3 = 0.010328$	$d_3 = 0.001308$

$|\epsilon(p)| < 4.5 \times 10^{-4}$

with

The BASIC program, **X.BAS**, makes these calculations using a cumulative frequency distribution as input and provides x, the normal deviate, as output.

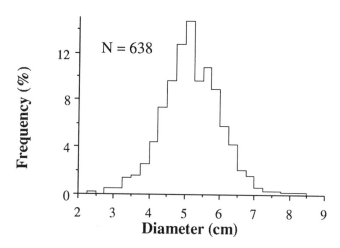

FIGURE 13.7 Size-frequency distribution of data in Table 13.2 for *Tripneustes gratilla* from Pupukea Beach Park, Oahu, Hawaii, 28 October 1975; data grouped using intervals of 0.25 cm. (Ebert 1982).

X.BAS

```
2 rem Program x.bas
5 dim p(100)
10 c0=2.515517:c1=.802853:c2=.010328
20 d1=1.432788:d2=.189269:d3=.001308
50 input "File with p-values: ";f$
55 open f$ for input as #1
57 input #1,n
60 for i=1 to n
62 input #1, p(i)
64 next i
66 close #1
68 input "file for output: ";f$
70 open f$ for output as #1
100 for i=1 to n
102 p = p(i)
105 if p(i)>.5 then let p=1- p(i)
107 if p = 0 then let p=1E-7
110 t=sqr(log(1/p^2))
120 x=c0+c1*t+c2*t*t
130 x=t-x/(1+d1*t+d2*t*t+d3*t*t*t)
135 if p(i)<=.5 then let x=-x
140 print #1, p(i),x
150 next i
155 close #1
160 end
```

A probit scale is the same as a scale in units of the normal deviate except that the probit is defined as the normal deviate+5. This simple addition of 5 eliminates negative values because the normal deviate is 0 at the mean, positive for observations greater than the mean and negative for values smaller than the mean. Keeping the normal deviate scale, as will be shown, has certain advantages.

TABLE 13.3 Summary Statistics for *Tripneustes gratilla*

N	638
Mean	5.2091
Median	5.1700
Variance	0.6872
Standard deviation (sd)	0.8290
Standard error (se)	0.0328
Skewness (g_1)	0.0983
Kurtosis (g_2)	0.6155

Note: size data gathered at Pupukea Beach Park, Oahu, Hawaii, 28 October 1975 (Ebert 1982).

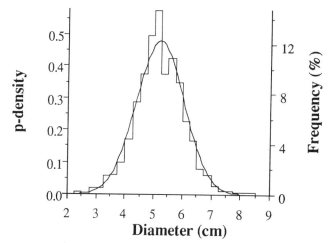

FIGURE 13.8 Superposition of Figures 13.6 and 13.7 for *Tripneustes gratilla* from Pupukea Beach Park, Oahu, Hawaii (Ebert 1982); N = 638.

Most size-frequency distributions are poly-modal but, occasionally, one occurs that is close to normal. The example in Table 13.2 for a tropical sea urchin, *Tripneustes gratilla*, at Pupukea Beach Park on Oahu, Hawaii, is very close to normal (Figure 13.7). Summary statistics are shown in Table 13.3 and include not only the mean and variance but also skewness and kurtosis. Skewness is a measure of asymmetry and if significant means that one tail of the curve is drawn out more than the other. Kurtosis is a measure of peakedness. A leptokurtic curve has values bunched around the mean as well as long symmetrical tails. A platykurtic curve is too flat or plate-like. Curves can be tested for skewness and kurtosis and some software packages include such tests. Skewness, g_1, of the *T. gratilla* data set is small and not statistically significant (Bliss 1967); however, the measure of kurtosis, g_2, is significantly different from 0, and indicates that the size distribution, though symmetrical, is leptokurtic; that is, too peaked. Figure 13.8 compares the actual data (Figure 13.7) to the theoretical normal distribution using the mean and standard deviation generated with the *Tripneustes gratilla* data. The

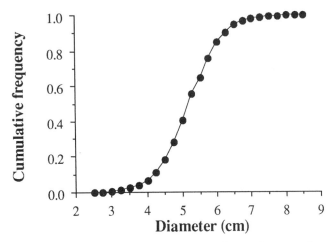

FIGURE 13.9 Cumulative frequency of test diameters for *Tripneustes gratilla* from Pupukea Beach Park, Oahu, Hawaii, 28 October 1975; N = 638; Ebert (1982).

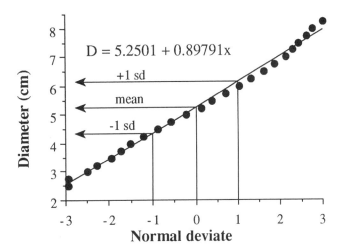

FIGURE 13.10 Diameter of *Tripneustes gratilla*, Pupukea, Hawaii, as a function of the normal deviate that was estimated from the cumulative frequency distribution; the mean is at a normal deviate of 0 and the standard deviation is ±1 normal deviate; the y-intercept at x=0 is the mean of the distribution and the slope is the standard deviation.

theoretical distribution (solid line) was drawn with no skewness or kurtosis. As indicated by the kurtosis, g_2, in Table 13.2, the size-frequency distribution for *Tripneustes gratilla* has a narrower shape and reaches a peak that is taller than the generated normal curve. Figure 13.7 is close to normal and for most purposes could be treated as a normal curve.

The data in Table 13.2 for *Tripneustes gratilla* can be replotted as a cumulative frequency distribution (Figure 13.9), which shows a typical sigmoidal or S-shape. The

mean of the distribution is at a cumulative fraction of 0.5 and plus or minus one standard deviation is at frequencies of 0.1587 and 0.8413. The cumulative frequency distribution is easier to interpret if it is expressed in units of the normal deviate, which is done by taking the values of the cumulative frequency distribution as data values in the program **X.BAS**. Results of this transformation (Figure 13.10) show a scatter of points that is approximately linear and with

1. an *intercept that is equal to the mean* (5.2501 cm),
2. a *slope equal to the standard deviation* (0.8979 cm),

which are close to the values shown in Table 13.3.

The scatter of points in Figure 13.10 does not follow a straight line but rather has a very gentle undulation; however, points are sufficiently close to a straight line that, for many purposes, they can be treated as straight and the entire distribution considered to be a single normal distribution. Treating a scatter of points on a plot of size *vs.* the normal deviate as straight even when it may have a small bend or two is the approach used in separating a distribution consisting of two or more modes into component parts. It also is the basis for separating components using probability paper.

Diadema setosum from Zanzibar

A cumulative frequency distribution can be used in conjunction with the program **X.BAS** to tease apart a frequency distribution that has more than one normal component and this approach is illustrated using data for a tropical sea urchin, *Diadema setosum*, that were collected at Yange Sand Bank off Zanzibar (Table 13.4). Diameters were measured and then grouped into size classes with a width of 0.25 cm. Animals were measured to the nearest 0.01 cm and so a 0.25 cm class means, for example, from 0.01-0.25 cm, 0.26-0.50 cm, 0.51-0.75 cm, etc.

The size data can be used to produce a frequency histogram (Figure 13.11) by selecting a suitable size interval that gives rise neither to too many nor too few categories. Grouping can be done by hand or by use of the program **FREQ_GENERAL.BAS** that is listed at the end of this chapter. A reasonable view is that one should have at least 15 size classes and preferably 20 or more. In order to do this it is necessary to have reasonably large samples. For calculating the normal deviate, data must be converted to cumulative frequency. For probability paper, data must be expressed as cumulative percent (cumulative frequency × 100%). Results of these calculations are shown in Table 13.5.

TABLE 13.4 Diameter Measurements in Centimeters of the Sea Urchin *Diadema setosum*

0.71	0.88	1.22	1.29	1.31	1.32	1.40	1.41	1.42	1.43	1.45	1.46	1.48
1.48	1.50	1.51	1.52	1.52	1.52	1.53	1.56	1.58	1.58	1.60	1.64	1.64
1.67	1.68	1.69	1.70	1.71	1.74	1.76	1.79	1.79	1.82	1.88	1.88	1.90
1.94	1.97	1.97	1.99	1.99	1.99	2.00	2.03	2.05	2.05	2.07	2.07	2.09
2.10	2.10	2.11	2.16	2.18	2.19	2.20	2.21	2.22	2.30	2.34	2.34	2.37
2.39	2.48	2.60	2.67	3.31	3.40	3.45	3.48	3.63	3.73	3.75	3.76	3.76
3.76	3.78	3.81	3.83	3.86	3.87	3.87	3.88	3.89	3.91	3.92	3.93	3.94
3.95	3.97	3.98	4.01	4.02	4.02	4.02	4.03	4.11	4.11	4.11	4.12	4.12
4.12	4.12	4.14	4.15	4.16	4.19	4.19	4.22	4.22	4.24	4.28	4.30	4.32
4.36	4.42	4.44	4.48	4.49	4.50	4.50	4.55	4.60	4.62	4.65	4.65	4.66
4.69	4.74	4.76	4.88	4.91	5.07	5.12	5.15	5.16	5.32	5.34	5.35	5.38
5.56	5.60	5.66	5.72	5.81	5.86	6.16	6.29	6.30	6.51			

Note: data from Ebert (1982) gathered at Yange Sand Bank off Zanzibar Harbor, 16 June 1976; N =153.

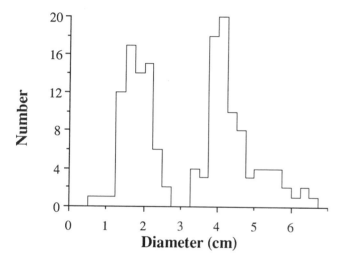

FIGURE 13.11 Size-frequency distribution for *Diadema setosum* from Yange Sand Bank, Zanzibar (Ebert 1982); 16 June 1976; N = 153.

Size is plotted *vs.* the normal deviates in Figure 13.12 together with the probability scale on the top of the graph. Unlike the plot for a unimodal curve, such as the data for *Tripneustes gratilla* from Pupukea (Figure 13.8), distributions with more than one mode produce lines that are far from linear using a normal deviate scale and this relationship is the basis for using probability paper to separate components of size-frequency distributions. The same curve shown in Figure 13.12 can be created using normal probability paper (Figure 13.13).

The first step in using probability paper to separate components is to plot size *vs.* cumulative percent. Because cumulative percent is used, the appropriate value for size is the *right* boundary of the size interval. Why the right boundary? Because this was how the area under the curve is calculated using Equation 13.18. For example, if the interval is from 1.26-1.50, 1.50 should be used. Points should be connected and then examined for inflection points

(Figure 13.13). An inflection point indicates an approximate point where two components can be separated and also the proportion of the total distribution contributed by that component. For example, the inflection is at about 45% and so the first component contributes about 45% to the total distribution and the second 55%.

TABLE 13.5 Size Data for *Diadema setosum*

Size	Number	Frequency	Cumulative %	Deviate
0.75	1	0.0065	0.6536	-2.4822
1.00	1	0.0065	1.3072	-2.2245
1.25	1	0.0065	1.9608	-2.0624
1.50	12	0.0784	9.8039	-1.2930
1.75	17	0.1111	20.9150	-0.8092
2.00	14	0.0915	30.0654	-0.5221
2.25	15	0.0980	39.8693	-0.2563
2.50	6	0.0392	43.7908	-0.1560
2.75	2	0.0131	45.0980	-0.1229
3.00	0	0.0000	45.0980	-0.1229
3.25	0	0.0000	45.0980	-0.1229
3.50	4	0.0261	47.7124	-0.0572
3.75	3	0.0196	49.6732	-0.0082
4.00	18	0.1176	61.4379	0.2903
4.25	20	0.1307	74.5098	0.6588
4.50	10	0.0654	81.0458	0.8795
4.75	8	0.0523	86.2745	1.0928
5.00	3	0.0196	88.2353	1.1869
5.25	4	0.0261	90.8497	1.3318
5.50	4	0.0261	93.4641	1.5116
5.75	4	0.0261	96.0784	1.7603
6.00	2	0.0131	97.3856	1.9412
6.25	1	0.0065	98.0392	2.0624
6.50	2	0.0131	99.3464	2.4822
6.75	1	0.0065		

Note: data collected at Yange Sand Bank, Zanzibar (Ebert 1982), with frequency, cumulative % and the normal deviate calculated using the program **X.BAS**.

TABLE 13.6 Analysis of *Diadema setosum* with Two Components

Size	N	Frequency	Cumulative %	N	Frequency A	Cumulative %	N	Frequency B	Cumulative %
0.75	1	0.0065	0.7	1	0.0145	1.4			
1.00	1	0.0065	1.3	1	0.0145	2.9			
1.25	1	0.0065	2.0	1	0.0145	4.38			
1.50	12	0.0784	9.8	12	0.1739	21.7			
1.75	17	0.1111	20.9	17	0.2464	46.4			
2.00	14	0.0915	30.1	14	0.2029	66.7			
2.25	15	0.0980	39.9	15	0.2174	88.4			
2.50	6	0.0392	43.8	6	0.0870	97.1			
2.75	2	0.0131	45.1	2	0.0290	100.0			
3.00	0	0.0000	45.1	0					
3.25	0	0.0000	45.1						
3.50	4	0.0261	47.7				4	0.0476	4.8
3.75	3	0.0196	49.7				3	0.0357	8.3
4.00	18	0.1176	61.4				18	0.2143	29.8
4.25	20	0.1307	74.5				20	0.2381	53.6
4.50	10	0.0654	81.0				10	0.1190	65.5
4.75	8	0.0523	86.3				8	0.0952	75.0
5.00	3	0.0196	88.2				3	0.0357	78.6
5.25	4	0.0261	90.8				4	0.0476	83.3
5.50	4	0.0261	93.5				4	0.0476	88.1
5.75	4	0.0261	96.1				4	0.0476	92.8
6.00	2	0.0131	97.4				2	0.0238	95.2
6.25	1	0.0065	98.0				1	0.0119	96.4
6.50	2	0.0131	99.3				2	0.0238	98.81
6.75	1	0.0065	100.0				1	0.0119	100.0

Note: size is the upper bound of the interval; relative frequency A is for the first component and frequency B is for the second component.

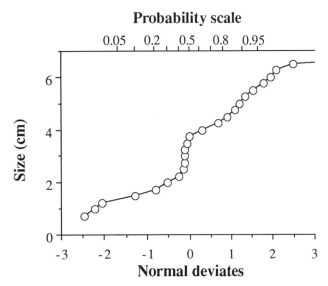

FIGURE 13.12 Size *vs.* the normal deviate (bottom scale) or cumulative probability (top scale) for *Diadema setosum* from Yange Sand Bank, Zanzibar (Ebert 1982).

The next step is to separate the relative frequencies in Table 13.5 for the two components and recalculate cumulative percent so that each component sums to 100% (Table 13.6). The cumulative percentages in each component can be calculated as shown in Table 13.6 or the cumulative % for the entire distribution up to the size where the inflection was, that is 3.25, can be divided by 0.451, the cumulative fraction up to that point. Now plot this new set of points (Figure 13.14). The same scatter of data values is shown in Figure 13.15 using the normal deviate scale.

Separation of three components makes the probability-paper technique clearer. Examination of Figure 13.13 or 13.14 indicates there probably is at least one more component that should be separated from the mixture. In addition to the inflection at 45.09% there is another inflection at about 87.5%. Two inflections define three components. The first component still is 45.10% of the population and the proportion of the second is 87.5% minus 45.10% or 42.41%. The third component is 100 - 87.5 or 12.5%. The procedure for separation consists of starting with the first component and then using the fitted line to aid in separation of the second component. With the fitted lines for the first and second components, it is possible to fit the third. The calculations for doing this are shown in Table 13.7. The steps are:

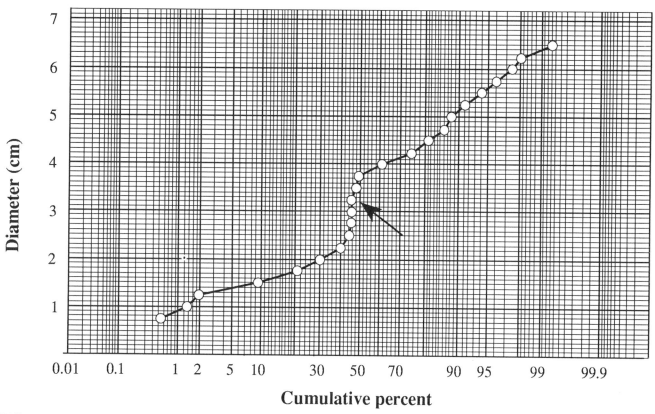

FIGURE 13.13 Diameter (cm) *vs.* cumulative percent of the size-frequency distribution for *Diadema setosum* (Ebert 1982) plotted on normal probability paper; the arrow indicates an inflection point where curvature changes and indicates where components should be separated.

1. Create the cumulative frequency distribution for the entire distribution. This is the 4th column from the left labeled "cum %". The points should be plotted on probability paper as was done in Figure 13.13.

2. Break the cumulative frequency distribution at the point of the first inflection: namely, 45.10%. Divide all of the entries up to this point by 0.451. This is column I (A) and changes the cumulative % values so they reach 100% at the point when the total cumulative % distribution is at 45.10%.

3. Plot these points on probability paper (Figure 13.16) and draw a line that seems like the best fit. Because of potential overlap of components, using linear regression probably is not a good idea. When overlap occurs, the ends the component lines will curve slightly and so will bias a regression. Fitting by eye, in this case, is better than a least-squares regression

4. Read percentage values from the best-fit line for component I shown in column I(B) in Table 13.7

5. Convert values in column I(B) back to appropriate values for the total cumulative frequency distribution by multiplying each value by 0.451 to produce column I(C). Notice that component I reaches a maximum of 45.1 and this fixed total contribution is continued down to the bottom of the table.

6. Begin extraction of the second component by starting at a test diameter of 3.50 cm. Subtract 45.1% from the cum % column up to the second inflection point at 87.5%. These values are shown in column II.

7. Convert column II into cumulative percent values appropriate for a single normal curve by dividing values in column II by 0.424 (remember the second component is 42.4% of the entire distribution). This division creates column II(A).

8. Plot values from column II(A) on probability paper (Figure 13.16) and draw the best line possible.

9. Read percent values from the line and enter them as column II(B).

10. Multiply column II(B) by 0.424 to produce column II(C) and add columns I(C) and II(C).

11. Repeat steps 6 through 10 for component III. The completed expected cumulative frequency distribution is shown as column I+II+III at the far right in Table 13.7.

Table 13.8 shows proportions, means and standard deviations. Means are at the 50% mark in Figure 13.16 and standard deviations are the size at 84.13% minus the size at 15.87% divided by 2. For example, for component I, size at 84.13% is 2.16cm and at 15.87% it is 1.36 cm. so (2.16-1.36)/2 = 0.4, the standard deviation.

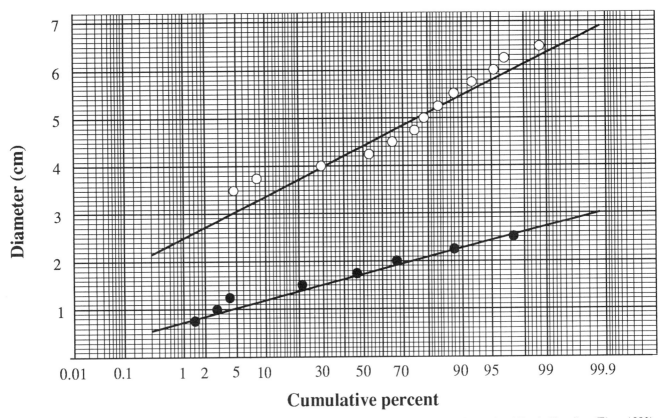

FIGURE 13.14 Two components of the size-frequency distribution of *Diadema setosum* from Yange Sand Bank, Zanzibar (Ebert 1982); intercepts are estimates of means of each component and slopes are standard deviations.

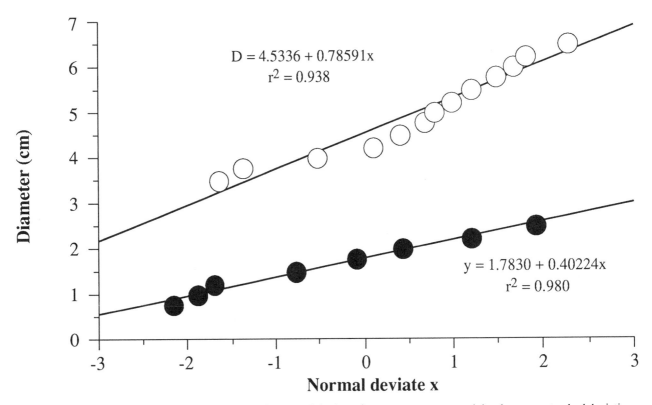

FIGURE 13.15 Data for *Diadema setosum* plotted using normal deviates; intercepts are means and the slopes are standard deviations.

TABLE 13.7 Analysis with 3 Components in a Size-Frequency Distribution of *Diadema setosum* from Zanzibar

Size	N	%	cum %	I (A)	I (B)	I (C)	II	II (A)	II (B)	II (C)	I+II %	III	III (A)	III (B)	III (C)	I+II+III
0.50	0				0.1	0.05										0.05
0.75	1	0.65	0.65	1.45	0.6	0.29										0.29
1.00	1	0.65	1.31	2.90	3.0	1.35										1.35
1.25	1	0.65	1.96	4.35	10.0	4.51										4.51
1.50	12	7.84	9.80	21.74	26.0	11.72										11.72
1.75	17	11.11	20.91	46.38	49.0	22.09										22.09
2.00	14	9.15	30.07	66.67	72.0	32.46										32.46
2.25	15	9.80	39.87	88.41	88.4	39.86										39.86
2.50	6	3.92	43.79	97.10	96.8	43.65										43.65
2.75	2	1.31	45.09	100.0	99.3	44.77										44.77
3.00	0	0	45.09		99.9	45.05										45.05
3.25	0	0	45.09		100.0	45.10			0.30	0.13	45.22					45.22
3.50	4	2.61	47.71			45.10	2.61	6.16	2.50	1.06	46.15					46.15
3.75	3	1.96	49.67			45.10	4.58	10.79	13.00	5.51	50.60					50.60
4.00	18	11.76	61.44			45.10	16.34	38.53	35.00	14.84	59.93					59.93
4.25	20	13.07	74.51			45.10	29.41	69.35	54.00	22.90	67.99			0.10	0.01	68.00
4.50	10	6.54	81.05			45.10	35.95	84.76	89.00	37.75	82.84			0.60	0.07	82.91
4.75	8	5.23	86.27			45.10	41.18	97.09	97.60	41.39	86.48			2.50	0.31	86.79
5.00	3	1.96	88.23			45.10			99.75	42.30	87.39	0.73	5.88	7.50	0.94	88.33
5.25	4	2.61	90.85			45.10			99.99	42.40	87.49	3.35	26.80	18.0	2.25	89.74
5.50	4	2.61	93.46			45.10				42.41	87.5	5.50	44.00	35.0	4.38	91.88
5.75	4	2.61	96.08			45.10				42.41	87.5	8.58	68.63	57.0	7.12	94.65
6.00	2	1.31	97.39			45.10				42.41	87.5	9.50	76.00	75.0	9.38	96.87
6.25	1	0.65	98.04			45.10				42.41	87.5	10.54	84.31	89.0	11.12	98.62
6.50	2	1.31	99.35			45.10				42.41	87.5	11.85	94.77	96.0	12.00	99.50
6.75	1	0.65	100.0			45.10				42.41	87.5			89.9	12.36	99.86
7.00	0					45.10				42.41	87.5			99.75	12.47	99.97
7.25	0					45.10				42.41	87.5			99.98	12.47	100.0

Note: inflections at 45.09 and 87.5 so proportions are 45.1% for I, 42.4% for II (87.5 - 45.1%) and 12.5% for III (100 - 87.5%); I (A) is cumulative %/0.4509 up to a size of 3.25 cm; I (B) are values taken from the graph; I (C) is the contribution of I to the total cumulative curve; II is cum% - 45.09 up to 4.75; II (A) is II/0.4241; II (B) values are taken from the graph; III is cum% - (I+II%).

Figure 13.17 shows the comparison of the observed and expected cumulative frequencies. Although the fit appears quite good, it is somewhat deceiving. If expected numbers in size classes are compared with observed numbers, results appear less good (Table 13.9). There is a problem at sizes 4.25 and 4.50 (Figure 13.18A) and so the proper approach would be to try to readjust lines to obtain a better fit. It should be fairly obvious that all of this is a great deal of work. Probability paper can be very useful but it requires a substantial investment of time in order to get results.

TABLE 13.8 Analysis of the Size-Frequency Distribution of *Diadema setosum* from Yange Sand Bank, Zanzibar, 16 June 1976

Component	Proportion	Mean cm	sd
I	0.451	1.75	0.40
II	0.424	4.11	0.33
III	0.125	5.68	0.47

Note: data from (Ebert 1982).

TABLE 13.9 Calculation of χ^2 for Goodness of Fit for Components of the Size-Frequency Distribution of *Diadema setosum* Separated by Using Probability Paper

Size	N	Freq.	Observed cum %	Expected cum %	%	Expected N
0.25	0		0	0.004	0.004	
0.50	0		0	0.045	0.041	
0.75	1	0.0065	0.6536	0.293	0.248	0.4482
1.00	1	0.0065	1.3072	1.353	1.06	1.6218
1.25	1	0.0065	1.9608	4.509	3.156	4.8287
1.50	12	0.0784	9.8039	11.72	7.211	11.033
1.75	17	0.1111	20.915	22.09	10.37	15.866
2.00	14	0.0915	30.065	32.46	10.37	15.866
2.25	15	0.0980	39.869	39.86	7.4	11.322
2.50	6	0.0392	43.791	43.65	3.79	5.7987
2.75	2	0.0131	45.098	44.77	1.12	1.7136
3.00	0	0.0000	45.098	45.05	0.28	0.4284
3.25	0	0.0000	45.098	45.22	0.17	0.2601
3.50	4	0.0261	47.712	46.15	0.93	1.4229
3.75	3	0.0196	49.673	50.60	4.45	6.8085
4.00	18	0.1176	61.438	59.93	9.33	14.275
4.25	20	0.1307	74.510	68.00	8.07	12.347
4.50	10	0.0654	81.046	82.91	14.91	22.812
4.75	8	0.0523	86.274	86.79	3.88	5.9364
5.00	3	0.0196	88.235	88.33	1.54	2.3562
5.25	4	0.0261	90.850	89.74	1.41	2.1573
5.50	4	0.0261	93.464	91.88	2.14	3.2742
5.75	4	0.0261	96.078	94.65	2.77	4.2381
6.00	2	0.0131	97.386	96.87	2.22	3.3966
6.25	1	0.0065	98.039	98.62	1.75	2.6775
6.50	2	0.0131	99.346	99.5	0.88	1.3464
6.75	1	0.0065	100.0	99.86	0.36	0.7650
7.00	0		100	99.97	0.11	
7.25	0		100	100.0	0.03	

Note: degrees of freedom are based on the number of size categories minus 3 times the number of components that are fitted; in this case there are 25 size classes starting with the first size category with an observation and ending with the last category with an observation and 3 components so df = 25 - 9 or 16; $\chi^2 = 30.64$, p<0.025; conclude that the fit is not as good as one might like because the smaller the χ^2 (the larger the value of p) the better the fit.

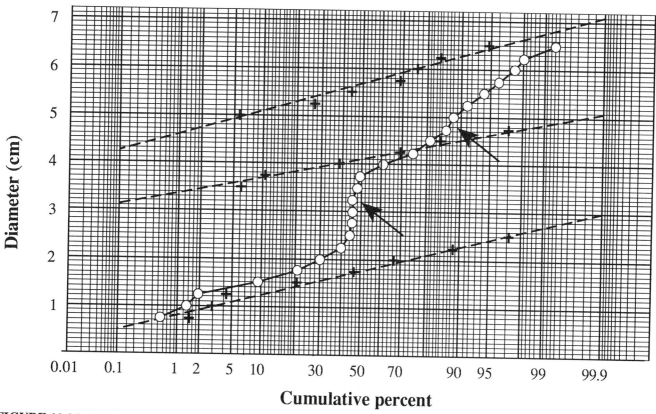

FIGURE 13.16 Size-frequency distribution of *Diadema setosum* from Zanzibar (Ebert 1982) separated into 3 components; arrows indicate inflection points; +'s are points from columns I(A), II(A) and III(A) in Table 13.7.

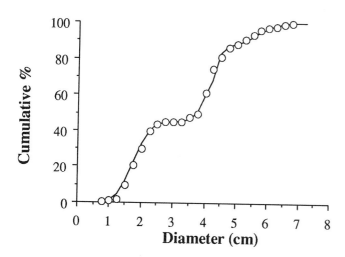

FIGURE 13.17 Comparison of observed (circles) and expected (line) cumulative percent for *Diadema setosum*; line is based on three components extracted using probability paper.

There are computer methods for extracting normal components from poly-modal distributions and they are real time savers. The BASIC program **SIZE_ANALYSIS.BAS** at the end of this chapter is based on a program developed by Macdonald and Pitcher (1979). The program does everything that the method with probability paper does, including the χ^2 test, plus standard errors are provided. Raw data first must be grouped into size classes to form a suitable data file for **SIZE_ANALYSIS.BAS** and this can be done using the program **FREQ_GENERAL.BAS**. Details of using both **FREQ_GENERAL.BAS** and **SIZE_ANALYSIS.BAS** are given at the end of this chapter.

SIZE_ANALYSIS.BAS can be used to extract up to 12 components but it would be very unusual that such a dissection would be meaningful. Remember that you lose 3 degrees of freedom for every component extracted if both mean and standard deviation are estimated. In general, you probably should not extract more than five or six components and the fewer the better. Results of extracting three components from the *Diadema setosum* data are shown in Table 13.10. The results with **SIZE_ANALYSIS.BAS** (Figure 13.18B) are similar to the results using probability paper but the fit is better based on χ^2 for goodness of fit.

With a good series of size-frequency distributions, it is possible to estimate means that can be used to determine parameters of a growth function. Several software packages are available that handle many distributions simultaneously and use the samples to obtain the best estimates of growth parameters. All of these applications emphasize the Brody-

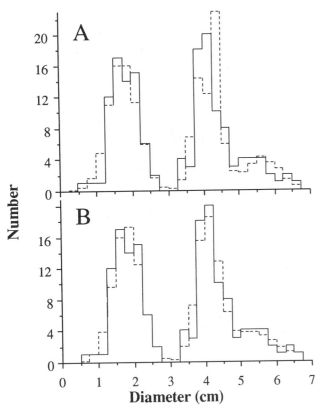

Number (vertical axis label)

Diameter (cm) (horizontal axis label)

FIGURE 13.18 *Diadema setosum*, Zanzibar (Ebert 1982); observed (solid line) and expected (dashed line) based on three extracted components; **A**, probability paper, $\chi^2=30.64$, df=16, $p<0.025$; **B**, **SIZE_ANALYSIS.BAS** based on Macdonald and Pitcher (1979), $\chi^2=16.06$, df=16, $p>0.25$.

Bertalanffy growth model. Packages include ELEFAN (Pauly 1987), MULTIFAN (Fournier *et al.* 1990, 1991), and MIX (Macdonald and Green 1988).

SURVIVAL ESTIMATION FROM SIZE STRUCTURE

The focus of this section is to explore some of the ways that size structure can be used to estimate survival rates. Most of these methods require knowledge concerning growth parameters but some can estimate both growth and survival from size structure.

Populations can be viewed as being members of two extreme types: those with continuous recruitment; and, those with pulsed recruitment concentrated at one instant in time. Existing analyses require that a population be at one of these extremes; most methods, however, are sufficiently robust, and demands of analysis frequently tend not to be severe, so useful survival rates for many populations can be estimated even though very few species show recruitment that is truly continuous or narrowly pulsed at an instant in time.

TABLE 13.10 Analysis of *Diadema setosum* Using the Method of Macdonald and Pitcher (1979)

Component	Proportion	Mean	sd
I	0.451±.040	1.802±.047	0.383±.035
II	0.375±.084	4.064±.063	0.313±.055
III	0.174±.047	5.288+.462	0.697±.274

Note: analysis was done with **SIZE_ANALYSIS.BAS**; values are given with standard errors; $\chi^2 = 16.056$, df=16, $0.50>p>0.25$

Techniques for estimating the instantaneous mortality, Z, from growth parameters and a size distribution are of three general types: (1) regression models that make use of the right descending slope of a size distribution to estimate Z; (2) models that use moment statistics, which currently use the mean of a size distribution; and (3) matrix methods that use size transitions together with initial and final size distributions.

There are a number of regression techniques all of which require constant and continuous recruitment. The "constant and continuous recruitment" restriction may not be fatal to these methods because the size at recruitment always is defined by the study and if recruitment size is sufficiently large then the actual ages at recruitment may be smeared out to the degree that even though birth or hatching or settlement was strongly pulsed, recruitment may be continuous. As a general rule, the greater the difference between sizes at birth and recruitment, the closer one gets to approximating continuous recruitment.

Regression Methods

Smith (1972) used the right descending slope of a size distribution with the Brody-Bertalanffy growth equation to estimate survival and his model was modified by Van Sickle (1977b). The regression equation is

$$ln(N_s) = m\,ln(X) + C \qquad (13.21)$$

where X is relative size

$$X = \frac{S_\infty^{-1/n} - S^{-1/n}}{S_\infty^{-1/n} - S_R^{-1/n}} \qquad (13.22)$$

Equation 13.22 is a modification of Van Sickle's formula and uses the shape parameter n of the Richards function as well as the parameter S_R, which is the size at recruitment to the population that was measured. With K as the Richards function growth-rate constant and m the slope of the regression, the mortality coefficient, Z is

$$Z = K(m + 1). \qquad (13.23)$$

One must assume constant and continuous recruitment and, also, size of animals in a size distribution may not equal or exceed S_∞. If size measurements exceed S_∞, the argument in the *ln*-function in Equation 13.21 becomes undefined.

Van Sickle (1977a) presented another regression model

$$Z = -\hat{g}b - \hat{g}' \qquad (13.24)$$

where $\hat{g}$ is the average growth rate over time period T and is equal to

$$\hat{g} = \frac{S_{t+T}^{-1/n} - S_t^{-1/n}}{T},$$

b is the slope of the regression of $ln(N)$ *vs.* $S^{-1/n}$; and

$$\hat{g}' = \frac{2nK - K(n+1)S_\infty^{-1/n}(S_t^{-1/n} + S_{t+T}^{-1/n})}{2}$$

is the average change in growth rate from S_t to S_{t+T}. For Brody-Bertalanffy growth $\hat{g}' = -K$ (Van Sickle 1977a). With the Richards function parameters (Ebert 1981)

$$\hat{g}' = \frac{nK[1 - (n + 1)(S_\infty S_t)^{-1/n}]}{n}.$$

Once again, it is necessary to assume constant and continuous recruitment; however, if S_R is sufficiently far from size at birth so that age at recruitment varies by several years, then continuous recruitment may be a reasonable assumption even if reproduction is highly seasonal.

Methods Based on Means: Continuous Recruitment

Beverton and Holt (1956) appear to be the first to have used moment statistics of size distributions to estimate survivorship. Their method makes use of the mean of a size distribution

$$Z = \frac{K(S_\infty^{-1/n} - \overline{X})}{\overline{X} - S_R^{-1/n}}. \qquad (13.25)$$

$\overline{X}$ is the mean of transformed sizes in the frequency distribution using $S_t^{-1/n}$ and it is necessary to assume constant and continuous recruitment. Here is an example of how the Beverton-Holt equation (Equation 13.25) can be applied.

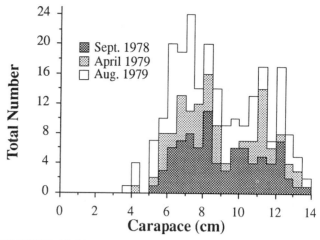

FIGURE 13.19 Size structure of male lobsters (*Panulirus penicillatus*) at Enewetak Atoll; N = 85 (Sept. 1978), N = 55 (April 1979), and N = 90 (Aug. 1979); Ebert and Ford (1986).

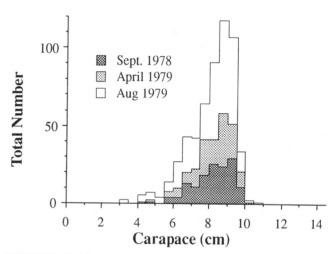

FIGURE 13.20 Size structure of female lobsters (*Panulirus penicillatus*) at Enewetak Atoll; N = 141 (Sept. 1978), N = 135 (April 1979), and N = 285 (Aug. 1979); Ebert and Ford (1986).

TABLE 13.11 Growth and Size Parameters for the Double-Spined Spiny Lobster *Panulirus penicillatus* at Enewetak Atoll, 1978-79

Parameter	Males	Females
K	0.211yr⁻¹	0.580yr⁻¹
S_∞	146.5mm	96.54mm
b	0.966	0.948
$\overline{X}$	88.25mm (N = 230)	81.29mm (N = 561)
S_R	45mm	45mm

Note: $\overline{X}$ is for all size distributions combined for each sex (Figures 13.19 and 13.20); data from Ebert and Ford (1986).

Spiny lobsters gathered at Enewetak Atoll (Ebert and Ford 1986) appear to satisfy the requirement of continuous recruitment because size structure did not change during a year (Figures 13.19 and 13.20)

Lobsters were tagged using a binary code of punched holes in the telson segments in September 1978 and April 1979 and carapace length was measured with vernier calipers. Recaptured animals were measured and the change in carapace length was used with the program **FABENS.BAS** (Chapter 11) to estimate growth parameters using $S_R = 5$ mm (Table 13.11). The Brody-Bertalanffy growth model was assumed.

Using the Beverton and Holt model (Equation 13.25), the mortality coefficient Z for males is

$$Z = \frac{0.211(146.5 - 88.25)}{88.25 - 45} = \boxed{0.284 \text{ yr}^{-1}}$$

and for females

$$Z = \frac{0.580(96.54 - 81.29)}{81.29 - 45} = \boxed{0.244 \text{ yr}^{-1}}.$$

The annual mortality rates are

Males $1 - e^{-Z} = 0.25$ or 25% per year

and

Females $1 - e^{-Z} = 0.22$ or 22% per year.

Males attain a larger size than females and approach this asymptotic carapace length slower than do females (K for males < K for females). Mean sizes are larger for males; however, when all data are assembled, the mortality rates are about the same for both sexes.

Methods based on means: Pulsed recruitment

If recruitment is periodic, a quite different approach must be used (Green 1970. Ebert 1973, 1981, 1987b). Average size in a population at time T following recruitment is

$$\overline{S_T} = \frac{\sum\limits_{t=0}^{\omega} N_t S_{T+t}}{\sum\limits_{t=0}^{\omega} N_{T+t}} \tag{13.26}$$

The index t represents age in years. For example, S_{T+0} is the average size of individuals in year-class 0 at time T

following annual recruitment. S_{T+5} is the average size of individuals in age-class 5 at time T following annual recruitment. The value of T ranges from 0 to 1 and represents time within a single year starting at the time of recruitment, T=0. For example, if recruitment was in March (T=0) and sampling was in July, four months later, T = 4/12 or 0.333. Actual sizes for each age class are important in determining mean size, which is why T must to added to ages for size.

The survival function is

$$N_t = N_T e^{-Zt} \tag{13.27}$$

with an initial number of N_T. Numbers of survivors are relative to N_T not N_0 and so T does not have to be added to t for numbers. Because just *relative* numbers in age classes are important in determining average size, N_T can be set equal to 1 in Equation 13.27.

If the Richards function is used to describe size, average size $\overline{S_T}$ at T time units following recruitment, is

$$\overline{S_T} = \frac{S_\infty \sum\limits_{t=0}^{\omega} e^{-Zt} \left[1 - b \sum\limits_{t=0}^{\omega} e^{-K(T+t)} \right]^{-n}}{\sum\limits_{t=0}^{\omega} e^{-Zt}}. \tag{13.28}$$

Now,

$$\sum\limits_{t=0}^{\infty} e^{-Zt} = \frac{e^Z}{e^Z - 1}$$

so

$$\frac{1}{\sum\limits_{t=0}^{} e^{-Zt}} = \frac{e^Z - 1}{e^Z} = e^{-Z}(e^Z - 1) = \boxed{1 - e^{-Z}}.$$

With the substitution of $1 - e^{-Z}$ for $1/\sum e^{-Zt}$, Equation 13.28 can be written:

$$\overline{S_T} = S_\infty (1 - e^{-Z}) \sum\limits_{t=0}^{\omega} e^{-Zt}(1 - b \cdot e^{-K(T+t)})^{-n} \tag{13.29}$$

Time since annual recruitment, T, and size at recruitment, S_R, are related and frequently neither can be estimated from existing data or extrapolated with confidence from growth parameters based on tagged individuals. A solution to this problem is to use the mean of the first mode of a size-frequency distribution as an estimate of S_R, which, as a consequence, makes T=0; animals recruit to the population at size S_R on the day of collection, T=0 (Ebert 1987b).

TABLE 13.12 Mean lengths of *Meganyctiphanes norvegica*

Sample date	Mean (mm)	N	Time	Max (mm)	Min (mm)
July 1973	21.5373	174	0.1667	29.67	11.17
Sept. 1973	26.4146	41	0.3333	35.67	18.67
Nov. 1973	27.9873	92	0.5	36.17	20.67
March 1975	31.9119	159	0.8333	43.67	18.17
May 1975	17.6349	21	0.0	34.33	10.00
Nov. 1975	29.1009	152	0.5	40.67	19.00
Feb. 1976	31.6436	123	0.75	41.00	18.50

Note: May was selected as the month of maximum recruitment because it is where mean size was smallest; data from Mauchline

Because S_R is an average size, animals smaller then S_R must be included when average size for the entire distribution is calculated; all data are used. Because the Richards function is used, b in Equation 13.29 has the following definition

$$b = \frac{S_\infty^{-1/n} - S_R^{-1/n}}{S_\infty^{-1/n}}. \qquad (13.30)$$

If average size is determined twice during a year then there are two equations for mean size (Equation 13.28) and so it is possible to estimate two unknowns, such as Z and K. Green (1970) used a graphical technique to solve for Z and K but with just two age classes. With three equations, three unknowns such as Z, K, and S_∞ can be found. The technique was extended to an infinite number of age classes and Newton's method was used to find parameters using two or three equations (Ebert 1973, 1975).

Frequently K and S_∞ and n are determined by tagging and so do not have to be estimated from shifts of mean size and, furthermore, mean size might be determined several times during a year or for several years. If one has more equations than unknown parameters, then the analysis moves from a solution of simultaneous equations to a least-squares problem and not only can one obtain an estimate of parameters but also estimates of standard errors. Solving for one or more parameters with *at least* as many mean sizes as parameters to be estimated, is a nonlinear regression problem.

A deep-sea euphausid *Meganyctiphanes norvegica*

Here is an example where both growth and survival parameters are estimated from size data. John Mauchline kindly supplied me with length measurements for a deep-sea euphausid (Table 13.12) gathered at different seasons over a period of several years (Mauchline 1985). There are no growth estimates so not only is it necessary to estimate Z from mean size but also K, S_∞, and the shape parameter n. It is possible to do the analysis by fixing some of the parameters at some reasonable values such as Brody-

TABLE 13.13. Analysis of Growth and Survival Parameters (±1 se) for the Deep-Sea Euphausid *Meganyctiphanes norvegica* with Different Numbers of Fixed Parameters

Fixed	Parameters			
	S_∞	K	n	Z
2	39.3	1.379± 0.069	-1	1.139± 0.074
1	36.650± 2.401	1.733± 0.410	-1	1.144± 0.078
0	33.400± 1.821	3.722± 1.737	+2.515± 0.107	1.091± 0.946

Note: fixed parameters are boldface ; S_R is 10mm; analysis by **ZKAN.BAS** (Ebert 1987b)

-Bertalanffy growth so n = -1 It also may be reasonable to fix S_∞ as 90% or 95% of the largest individual in all of the samples that were gathered for a species.

Analysis of the euphausid data proceeded in several steps. With the first run of **ZKAN.BAS**, I tried to estimate 4 parameters: S_∞, K, n, and Z. I could not obtain convergence, which is not unusual and so for the second attempt I fixed two of the parameters: S_∞ and n, which is shown as the first row in Table 13.13. I then used the parameter estimates to try another run but this time fixing just n at -1. This run is shown as the second row in Table 13.13. I used the parameter estimates as starting values for an additional run and this time did not fix any of the values. The program failed to converge but the direction of n was evermore negative and so another run was made starting with a positive value for n. Recall (Chapter 11) that in the Richards function large negative values for n are the same as large positive values. Starting with a positive value for n, convergence was obtained and results are shown as the last row of Table 13.13.

The positive value for n when four parameters were estimated indicates there may be an exponential phase of the growth curve. It is uncertain whether this is an intrinsic part of growth in this species or represents seasonal variation in

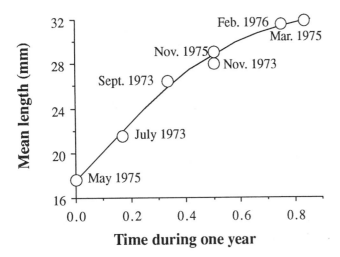

FIGURE 13.21 Change in mean length of the deep-sea euphausid *Meganyctiphanes norvegica* during a year using data from Table 13.12; fitted line is based on estimating 4 parameters by non-linear regression (Ebert 1987b) with **ZKAN.BAS**: S_∞ = 33.400 mm, K = 3.722 yr^{-1}; n = +2.515 and Z = 1.091 yr^{-1}.

food and hence slow growth during summer months. It is not possible to tell with the existing data. The final fitted line is shown in Figure 13.21 together with the data points from Table 13.12.

Diadema setosum from Zanzibar

When growth parameters have been estimated by tagging, Equation 13.29 has just one unknown, Z, which is much better than attempting to estimate growth parameters as well as survival. Growth parameters were provided for *Diadema setosum* jaws in Chapter 11 but what is required is growth in diameter because this is what is used in the size-frequency distribution presented in Figure 13.11.

The grow parameters for the jaw are converted to growth parameters for test diameter first by determining the allometric relationship between test diameter, S, and jaw length, J

$$S = \alpha J^\beta. \tag{13.30}$$

With the Richards function, the parameter K remains unchanged, S_∞ is determined from Equation 13.30 and n, the shape parameter, for test growth is calculated from n_j for jaw growth using the exponent β from Equation 13.30

$$n = \beta n_j.$$

The Richards function parameters for *Diadema setosum* are

$$n = -0.213,$$
$$K = 0.008 \ yr^{-1},$$
$$S_\infty = 9.192 \ cm,$$

and

$$b = 1.00.$$

The size of animals in mode I of the size-frequency analysis shown in Table 13.10 is 1.802cm, which can be used as a reasonable estimate of S_R and mean diameter for the entire distribution is 3.232 cm.
An estimate of Z can be obtained by using **ZKAN.BAS**

$$\boxed{Z = 0.40 \ yr^{-1}}.$$

The annual survival probability, p_x, is e^{-Z} and so

$$p_x = 0.67 \ yr^{-1}.$$

When the Richards function isn't Appropriate: *Strongylocentrotus franciscanus*.

The Tanaka function was introduced in Chapter 11 and data for the red sea urchin *Strongylocentrotus franciscanus* were presented as an example. In order to estimate survival from growth parameters and size-structure, it is necessary to modify Equation 13.29 so that it is appropriate for the Tanaka function. This is accomplished by substituting the Tanaka function into Equation 13.26 to yield Equation 13.31, which looks much worse than it really is. The parameters α and β are the allometric parameters for calculating diameter, S, from jaw size and T is the time since annual recruitment, which, in many cases, can be set equal to 0 because the mean size of the first mode of a size-frequency distribution is used as S_R. The size-frequency histograms for *S. franciscanus* at San Nicolas Island, CA, are shown in Figure 13.22 together with mean sizes for each distribution.

Because just Z is to be estimated with Equation 13.31, it is not necessary to develop a program that would estimate many parameters as is done by **ZKAN.BAS**. The Newton-Raphson algorithm with one parameter is all that is required. The procedure is the same as was introduced in Chapter 2. First change the function so it is equal to 0; that is, subtract the left side of Equation 13.31 from the right side. Next find the derivative with respect to Z, which is easy because Z is

$$\overline{S_T} = (1 - e^{-Z}) \sum_{t=0}^{\omega} e^{-Zt} \ \alpha \left[\frac{1}{\sqrt{f}} ln \left| 2f((t+T)-c) + 2\sqrt{f^2 \ ((t+T) - c)^2 + fa} \right| + d \right]^\beta \tag{13.31}$$

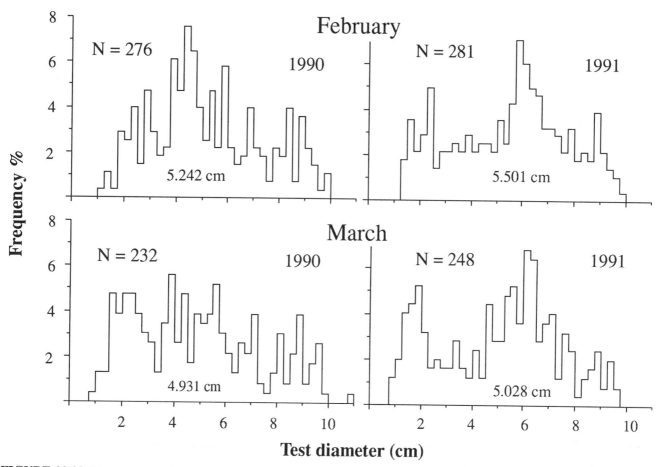

FIGURE 13.22 Size-frequency distributions for *Strongylocentrotus franciscanus* at San Nicolas Island, CA; distributions are for time of tagging (1990) and at the time of collection (1991); means are shown for each distribution (Ebert and Russell 1993).

TABLE 13.14 Summary of Data and Results for Estimating Z for Red Sea Urchins at San Nicolas Is., CA

| year | Mean diameter cm | |
	February	March
1990	5.242	4.931
1991	5.501	5.028
Z =	0.261±0.014	0.303±0.006

Note: Tanaka parameters: f=9.192, d = -0.112, a = 7.396; size at recruitment based on position of first mode, S_R = 1.8 cm; confidence values are ±1se; data from Ebert and Russell (1993).

not in the Tanaka function and so the problem can be thought of in terms of a growth function, G(t), and survival, which reduces Equation 13.31 to

$$\overline{S_T} = (1 - e^{-Z}) \sum_{t=0}^{\omega} e^{-Zt} G(t) \qquad (13.32)$$

and so for the Newton-Raphson algorithm

$$F(Z) = (1 - e^{-Z}) \sum_{t=0}^{\omega} e^{-Zt} G(t) - \overline{S_T} \qquad (13.33)$$

and

$$F'(Z) = e^{-Z} \sum_{t=0}^{\omega} e^{-Zt} G(t) - (1 - e^{-Z}) \sum_{t=0}^{\omega} t e^{-Zt} G(t). \qquad (13.34)$$

G(t) can be any growth function and Z can be estimated by

$$Z_{i+1} = Z_i - \frac{F(Z)}{F'(Z)}.$$

The program for estimating Z using the Tanaka function is given as **ZTAN.BAS** at the end of this chapter.

Table 13.14 shows the summary of data required to use Equation 13.21 to estimate Z. Estimation of parameters for the Tanaka function was shown in Chapter 11. With Z equal to about 0.3 yr^{-1}, the annual survival rate, p_x, is 0.74 which would make the annual mortality rate 26%.

TABLE 13.15 Growth Transitions of *Strongylocentrotus franciscanus* at San Nicolas Island Subtidal

Diameter 1991 ↓	Initial Diameter (cm) in 1990					
	1.1-2.5	2.6-4.0	4.1-5.5	5.6-7.0	7.1-8.5	8.6-10.0
February						
1.1 -2.5						
2.6-4.0	1	1				
4.1-5.5	1	11	1			
5.6-7.0		5	32	6		
7.1-8.5				18	17	
8.6-10.0					8	13
Total	2	17	33	24	25	13
March						
1.1 -2.5						
2.6-4.0	2					
4.1-5.5	3	13	1			
5.6-7.0		9	33	3		
7.1-8.5			2	15	5	
8.6-10,0					6	5
Total	5	22	36	18	11	5
Combined						
1.1 -2.5						
2.6-4.0	0.429	0.026				
4.1-5.5	0.571	0.615	0.029			
5.6-7.0		0.359	0.942	0.214		
7.1-8.5			0.029	0.786	0.611	
8.6-10.0					0.389	1.000

Note: transitions in February and March are numbers of animals from an initial size class (columns) in 1990 that were collected one year later in the initial or larger size class (rows); "Combined" data show fraction of individuals in each transition for the combined February and March samples; data from Ebert and Russell (1993).

SURVIVAL ESTIMATION FROM A SIZE-TRANSITION MATRIX

A final class of methods for estimating survival are to use growth transitions and size-frequency data without a specific growth model and with some relaxation of the restrictions of stable and stationary size structure. In matrix notation

$$N_{T+1} = AN_T \tag{13.35}$$

where A is the transition matrix, N_T is an initial size-frequency distribution and N_{T+1} is the size distribution one time interval later. Elements of N_T are actual counts, not relative frequencies. If relative frequencies are used so that the size distribution sums to 100%, it is necessary to assume that the population was stable and stationary.

The transition matrix A is composed of two sets of probabilities, growth, g_{ij}, and survival, s_j (Equation 13.36).

$$\begin{pmatrix} n_1 \\ n_2 \\ n_3 \\ \vdots \\ n_i \end{pmatrix}(T+1) = \begin{pmatrix} s_1 g_{11} & \cdots & \cdots & \cdots & \cdots \\ s_1 g_{21} & s_2 g_{22} & \cdots & \cdots & \cdots \\ s_1 g_{31} & s_2 g_{32} & s_3 g_{33} & \cdots & \cdots \\ \vdots & \vdots & \vdots & \vdots & \vdots \\ s_1 g_{i1} & s_2 g_{i2} & s_3 g_{i3} & \cdots & s_j g_{ij} \end{pmatrix} \begin{pmatrix} n_1 \\ n_2 \\ n_3 \\ \vdots \\ n_i \end{pmatrix}(T) \tag{13.36}$$

The growth transitions, g_{ij}, generally would be determined from tagged or mapped individuals and so the problem in analysis is to solve for annual survival rates, s_j, which can be done by regression (Equations 13.37).

$$n_i(T+1) = s_1 g_{i1} n_1(T) + s_2 g_{i2} n_2(T) + s_3 g_{i3} n_3(T) + \ldots s_j g_{ij} n_i(T). \tag{13.37}$$

Here is an example of how the method works for the red sea urchin *Strongylocentrotus franciscanus* from San Nicolas Island (Ebert and Russell 1993). Growth transitions, matrix A, are shown as the "Combined" data in

$$n_1(T+1) = s_1g_{11}n_1(T) \tag{13.38}$$
$$n_2(T+1) = s_1g_{21}n_1(T) + s_2g_{22}n_2(T) \tag{13.39}$$
$$n_3(T+1) = s_1g_{31}n_1(T) + s_2g_{32}n_2(T) + s_3g_{33}n_3(T) \tag{13.40}$$
$$\vdots \qquad \vdots \qquad \vdots \qquad \vdots$$
$$n_i(T+1) = s_1g_{i1}n_1(T) + s_2g_{i2}n_2(T) + s_3g_{i3}n_3(T) + \cdots s_jg_{ij}n_i(T) \tag{13.41}$$

TABLE 13.16 Size Structure of *Strongylocentrotus franciscanus* for February and March Samples at San Nicolas Island, CA.

Size (cm)	Feb. '90	Feb. '91	Mar. '90	Mar. '91
≤1.0			1	3
1.1 -2.5	31	43	49	51
2.6-4.0	53	37	46	27
4.1-5.5	83	43	49	46
5.6-7.0	48	85	37	71
7.1-8.5	35	43	23	31
8.6-10.0	26	30	26	19
10.1-11.5				1
Total	276	281	232	248

Table 13.15. The "Combined" data are presented as probabilities of transfer of individuals that lived for the entire time period of one year and so each column of the "Combined" data in Table 13.15 sums to 1.0. For example, in the size class 2.6-4.0, that is the *original* size in 1990, 17 tagged animals were recovered for the February sample and 22 for the March sample for a total of 39 animals. Of these tagged individuals, 11 of the February sample and 13 of the March sample transferred into the 4.1-5.5 cm class for a total of 24. The transfer probability is 24/39 or 0.615, which is the value shown in the "Combined" section of Table 13.15.

The vectors N_T and N_{T+1} of Equation 13.35 are the size-frequency distributions determined in 1990 and 1991 and are shown in Table 13.16. These are the same distributions that are shown in Figure 13.22 except that the interval size is 1.5 cm rather than 0.25 cm as it is in Figure 13.22. The reason for the larger interval is made clear by the small numbers of individuals shown in Table 13.15, all of which are tagged. Small numbers require that larger interval sizes be used.

Starting with Equation 13.36, the regression equations relating growth transitions ("Combined" data in Table 13.15), and s_j, for each size class are shown in Equations 13.38 to 13.41.

The combination of initial size and number in a size class, final size and number in a size class, growth transition probability, and the unknown survival rate is shown in Table 13.17. The column labeled 1991 is the size distribution in 1991 (Table 13.16) and the size distribution of 1990 is included in the body of the table under "Size classes in

1990". For example, there were 31 individuals in the 1.1-2.5 size class in February 1990 (Table 13.16) and so 31 appears in the 1.1-2.5 column for 1990 in Table 13.17. The probability was 0.429 of transfer to the 2.6-4.0 size class and 0.571 for transfer to the 4.1-5.5 size class. What is unknown is the annual survival rate.

What appears in a particular size class in 1991 is determined by the transfers into and out of the size class. For example, the row labeled 4.1-5.5 in February 1991 (Table 13.17) contains animals that transferred into this class from the 1.1-2.5 cm class ($0.571 \times 31s_1$) plus animals that transferred in from the 2.6-4.0 cm class ($0.615 \times 53s_2$) plus individuals that didn't grow and so remained in the 4.1-5.5 class ($0.029 \times 83s_3$). All that is being done is book keeping with animals leaving and entering that can be accounted for by observation and animals disappearing at a rate that is unknown but can be estimated using regression analysis and the following data file:

```
1,37,13.299,1.378,0,0,0,0
1,43,17.701,32.595,2.407,0,0,0
1,85,0,19.027,78.186,10.272,0,0
1,43,0,0,2.407,37.728,21.385,0
1,30,0 0 0 0 13.615,26
2,27,21.021,1.196,0,0,0,0
2,46,27.979,28.290,1.421,0,0,0
2,71,0,16.514,46.158,7.918,0,0
2,31,0,0,1.421,29.082,14.053,0
2,19,0,0,0,0,8.947,26
```

The first number in each row of the file is for time of year: 1=February and 2=March. The second number in each line is the observed number of individuals in the size class in 1991 as shown in Table 13.17. The next 6 numbers in each line are the transitions with unknown survival. For example, in the first line of the file, which is for February, 37 is the number observed in the first size class (2.6-4.0) in 1991, 13.299 is 0.429×31, 1.378 is 0.026×53, and, as shown in Table 13.17 the rest of the transitions for the first line are all 0.

The dependent variable, Y, in the data file is the observed number in 1991 and there are 6 independent variables X1, X2, ... X6 that are the growth probability multiplied by the number in 1990. A model for the regression with 6 survival coefficients and no constant is

Y = X1*s1 + X2*s2 + X3*s3 + X4*s4 + X5*s5 + X6*s6.

TABLE 13.17 Growth Transitions (g_{ij}) for _Strongylocentrotus franciscanus_ from 1990 to 1991 (Table 13.15) Applied to the Size-Distributions of 1990 (Table 13.16) to Produce the Size Structures Observed in 1991 (Table 13.16)

Size classes 1991 February	(n_i) '91 ↓	1.1-2.5	2.6-4.0	4.1-5.5	5.6-7.0	7.1-8.5	8.6-10.0
1.1-2.5	43						
2.6-4.0	37 =	$0.429 \times 31s_1$	$0.026 \times 53s_2$				
4.1-5.5	43 =	$0.571 \times 31s_1$	$0.615 \times 53s_2$	$0.029 \times 83s_3$			
5.6-7.0	85 =		$0.359 \times 53s_2$	$0.942 \times 83s_3$	$0.214 \times 48s_4$		
7.1-8.5	43 =			$0.029 \times 83s_3$	$0.786 \times 48s_4$	$0.611 \times 35s_5$	
8.6-10.0	30 =					$0.389 \times 35s_5$	$1 \times 26s_6$
March							
1.1-2.5	51						
2.6-4.0	27 =	$0.429 \times 49s_1$	$0.026 \times 46s_2$				
4.1-5.5	46 =	$0.571 \times 49s_1$	$0.615 \times 46s_2$	$0.029 \times 49s_3$			
5.6-7.0	71 =		$0.359 \times 46s_2$	$0.942 \times 49s_3$	$0.214 \times 37s_4$		
7.1-8.5	31 =			$0.029 \times 49s_3$	$0.786 \times 37s_4$	$0.611 \times 23s_5$	
8.6-10.0	19 =					$0.389 \times 23s_5$	$1 \times 26s_6$

Note: survival rates, s_i, are the solutions to (n_i in 1991) = $g_{ij} \times$ (n_i in 1990) $\times s_i$; data from Ebert and Russell (1993).

TABLE 13.18 Estimates (±se) of Survival Rates, s_i, for Each of Six Size Classes of Red Sea Urchins (_Strongylocentrotus franciscanus_) from 1990 to 1991

Size class (cm)	Coefficient	Survival yr^{-1}
1.1-2.5	s_1	1.474±0.469
2.6-4.0	s_2	0.350±0.457
4.1-5.5	s_3	1.033±0.312
5.6-7.0	s_4	0.523±1.429
7.1-8.5	s_5	1.008±2.627
8.6-10.0	s_6	0.505±1.155

Note: February and March samples combined; data from Ebert and Russell (1993).

TABLE 13.19 Estimates (±se) of Overall Survival Rate, s_1 for Red Sea Urchins at San Nicolas Is., CA

Sample	Survival (s_1)	Z yr^{-1}
February	0.793±0.092	0.232
March	0.858±0.086	0.153
Combined	0.818±0.061	0.201

Note: annual survival rate is e^{-Z} so $Z = -ln(s_1)$; data from Ebert and Russell (1993).

Survival coefficients, s_j, could not be estimated for the individual samples (February and March) because there are more coefficients (s_j) than dependent variables, Y. As a consequence, the February and March samples were combined as a single analysis so there were 10 dependent variables (10 lines in the data file) and 6 survival

coefficients, s1 to s6. The results of the regression analysis (Table 13.18) are unsatisfactory because survival rates can not be greater than 1.0 and three out of the six estimates of s_j were 1.0 or greater.

The regression was recast so a single survival rate, s_1, was estimated

$$Y = X1*s1 + X2*s1 + X3*s1 + X4*s1 + X5*s1 + X6*s1. \tag{13.42}$$

Because just one coefficient was estimated, it was possible to treat the February and March samples separately as well as estimating an overall survival rate (Table 13.19).

Finally, data were analyzed using a common survival rate for every _two_ size classes so there are survival rates s1, s2, and s3

$$Y = X1*s1 + X2*s1 + X3*s2 + X4*s2 + X5*s3 + X6*s3. \tag{13.43}$$

The results of this analysis are biologically possible because all survival estimates were between 0 and 1 (Table 13.20) and so indicate a possible decrease in survival with increasing size. An alternate explanation is there was size-dependent movement that caused an apparent change in survival rates (Ebert and Russell 1993).

The matrix approach to estimating survival based on transitions of numbers into and out of size categories eliminates the problem of having to specify a particular growth model and also explicitly includes variation in growth. The method also reduces the problems of having to

TABLE 13.20. Annual Size-Class Specific Survival Rates, s_i ($\pm$se), of *Strongylocentrotus franciscanus* at San Nicolas Island, CA

Size classes		Survival (s_i) $\pm$ se year^{-1}		
		February	March	Combined
1.1-2.5 2.6-4.0 }	s_1	0.961$\pm$0.336	0.865$\pm$0.122	0.911$\pm$0.139
4.1-5.5 5.6-7.0 }	s_2	0.740$\pm$0.137	0.986$\pm$0.124	0.817$\pm$0.078
7.1-8.5 8.6-10.0 }	s_3	0.727$\pm$0.383	0.477$\pm$0.229	0.626$\pm$0.173

Note: Estimates based on combining adjacent size classes; data from Ebert and Russell (1993).

have stable and stationary structure. On the negative side, it is necessary to work with a closed system where all individuals can be accounted for or it is necessary to add some additional assumptions such as migration is not size-specific so there are no size-specific additions or losses due to movement. It is best if the system is closed to movement (*e. g.* Sainsbury 1982).

Size-frequency distributions are very rich sources of information about organisms. Contained within the numbers of individuals in size categories is all of recent history for a population including growth, survival, intrinsic variability in these rates, recruitment and year-to-year variation in all of these. It very nearly exceeds one's abilities to cope because frequently the shapes of the growth and survival functions are unknown, yet the shapes of these functions determine the over-all shape of the size-distributions. Teasing information from size distributions can be done and is made much easier if some additional information is known, such as growth transitions. Much of this will become easier as additional information is assembled on patterns of growth and survival across taxonomic groups.

PROBLEMS

1. The simulations of size-frequency distributions shown at the beginning of this chapter used $\sigma = 0.1\mu$, and constant survival. Do a series of simulations where you explore a different growth model or changing mortality rates. For example, explore changes in c of the Weibull function and then choose one value of c and examine the consequences when n or the Richards function is changed. Do sufficient simulations so that you can discuss the effect of the parameter on size-structure.

TABLE 13.21 Size Structure of *Haliotis mariae* on the Dhofari Coast, Sultanate of Oman, 1992

Size	6 Jan.	9 April	25 Sept.	12 Nov.
5	0	21	0	0
10	0	20	0	0
15	1	16	1	1
20	1	12	1	3
25	7	0	8	4
30	11	0	25	13
35	20	1	18	19
40	31	3	11	13
45	29	6	14	12
50	26	7	18	13
55	18	13	21	14
60	31	8	30	23
65	9	21	28	13
70	2	22	26	7
75	11	17	18	5
80	9	7	24	8
85	5	8	10	3
90	2	5	11	2
95	1	1	8	1
100	1	1	3	
105	1			

Note: size class are in millimeters and represent the upper end of the interval (5 means 0.1-5.0, 10 means 5.1-10 etc.); data extracted from Shepherd *et al.* (1995).

2. Shepherd *et al.* (1995) studied the Omani abalone *Haliotis mariae* on the Dhofari coast in 1992. They gathered size-frequency data which are shown in the following table (Table 13.21). Use the size data to estimate parameters of the Brody-Bertalanffy growth equation by finding modes and using the progression of modes to obtain growth-equation parameters.

TABLE 13.22 Mean Length of Female Shrimp Caught in the Kuwait Fishery

Month	1979			1981	
	$\overline{L}$	N	t	$\overline{L}$	N
Jan.	42.533	30			
Feb.	35.458	107			
Mar.	34.672	122			
April	26.191	105			
May	23.639	643	0.000		
June	25.266	924	0.083		
July	27.983	586	0.167	27.835	593
Aug.	28.590	1383	0.250	27.524	542
Sept.	29.467	821	0.333	31.049	553
Oct.	30.513	452	0.417	34.714	392
Nov.	32.327	202	0.500	40.357	370
Dec.	36.275	240	0.583	36.077	26

Note: data from C. P. Mathews.

TABLE 13.23 Length of Tiger Prawns, *Panaeus semisulcatus*, in Kuwait beginning in March (M) 1988 and Ending in September (S) 1989

Size	M	J	J	A	S	O	N	J	F	M	A	M	J	J	A	S
12	1	0	0	0	0	0	0	0	0	0	0	2	0	0	0	0
13	2	0	0	0	0	0	0	0	0	0	0	2	0	0	0	0
14	1	0	0	0	0	0	0	0	0	0	0	4	0	0	0	0
15	2	2	0	0	0	0	0	0	0	0	0	10	2	0	0	0
16	2	1	3	0	0	0	0	0	0	0	0	15	0	0	0	0
17	3	2	7	0	0	0	0	0	0	0	0	11	0	0	0	0
18	4	5	17	0	0	0	0	0	0	0	0	23	0	0	0	0
19	3	6	13	0	0	0	0	0	0	0	0	15	0	0	0	0
20	2	10	7	1	0	0	0	0	0	0	0	6	1	3	11	1
21	1	20	28	1	3	0	0	0	0	0	0	2	0	6	6	1
22	0	43	46	0	7	0	0	0	0	0	0	10	3	11	18	1
23	0	53	68	3	21	0	0	0	0	0	0	6	2	20	20	1
24	1	37	65	8	16	0	0	0	0	0	0	6	1	26	46	3
25	0	51	53	31	37	0	0	0	0	0	0	2	23	28	70	17
26	0	33	69	25	44	0	0	0	1	0	0	2	5	22	47	13
27	1	22	53	37	63	1	0	0	1	0	0	0	18	23	94	16
28	1	15	51	50	42	1	1	0	0	0	0	0	39	28	131	17
29	0	8	23	41	33	3	1	0	0	0	0	0	0	23	92	29
30	0	10	24	60	67	4	3	0	1	0	0	0	8	25	107	38
31	0	4	18	52	84	10	2	0	0	0	0	0	3	17	115	83
32	0	1	12	58	75	9	6	0	0	0	0	1	0	14	107	140
33	0	1	14	34	99	14	9	1	1	1	1	0	0	10	77	128
34	1	0	0	31	81	25	7	2	1	1	2	0	0	14	50	166
35	0	2	0	29	81	36	13	4	1	0	3	0	0	4	23	110
36	1	1	0	16	63	30	9	4	1	2	1	2	1	1	18	138
37	1	1	0	10	74	42	15	4	2	3	3	1	1	1	7	82
38	2	2	4	8	40	49	24	6	2	4	3	0	0	2	2	66
39	1	1	1	3	20	31	34	3	4	3	3	0	0	5	2	20
40	1	1	0	3	19	16	36	8	4	3	3	1	0	1	2	25
41	1	3	0	2	0	21	31	10	4	10	7	1	4	1	5	16
42	3	0	0	0	3	10	29	10	6	14	9	2	3	1	17	0
43	1	0	0	1	4	0	20	7	7	15	12	2	4	1	0	0
44	4	1	1	2	0	4	13	5	4	17	16	4	3	1	1	0
45	2	3	1	1	0	2	9	3	3	15	18	5	5	2	1	0
46	2	1	1	1	0	1	5	2	3	12	11	4	3	4	0	0
47	4	2	7	1	1	1	0	1	3	7	9	4	4	2	0	0
48	2	1	7	0	0	0	1	0	0	4	5	3	2	2	2	2
49	1	0	0	1	2	0	1	0	0	1	1	1	0	1	0	0
50	1	0	0	0	1	1	0	0	0	1	1	1	0	1	0	0
51	0	0	0	0	1	0	0	0	0	0	0	0	0	0	1	0

Note: data from Xu and Mohammed (1996).

3. *Panaeus semisulcatus* is an important shrimp species in the Kuwait fishery. The following data for female shrimp (Table 13.22) were kindly supplied by C. P. Mathews and are discussed in greater detail in Mathews *et al.* (1987); $\overline{L}$ = mean length (mm), N = number in the sample, t = time since recruitment, which is the minimum mean size for the year; recruitment length, L_R, is 22mm for the 1979 data and 26 mm for the 1981 data; L_∞ = 55 mm was estimated for the 1981 data. Estimate K, Z and the shape parameter n of the Richards function using these data.

4. Tiger prawns, *Panaeus semisulcatus* were studied in Kuwait waters during 1988 and 1989 by Xu and Mohammed (1996); the following data (Table 13.23) start with May 1988; size is carapace length measured in millimeters. Estimate K, S_∞, Z and n using this larger data set for *P. semisulcatus* and compare your results with results obtained in problem #3.

BASIC PROGRAMS

FREQ_GENERAL.BAS

This program takes raw size data just as entered from a data sheet and can create three different output files: (1) a file suitable for drawing a histogram using Cricket Graph; (2) a file that is correctly formatted for decomposition into component normal curves using **SIZE_ANALYSIS.BAS**, which employs the Macdonald-Pitcher algorithm; and, (3) a file that contains frequency and cumulative frequency, which, following a bit of editing, is suitable for determining normal deviates from cumulative frequency. This final data file would be used with the program **X.BAS**.

Data for **FREQ_GENERAL.BAS** must be individual measurements separated by a <return> and saved as a text file; use a word processor. Table 13.4 for *Diadema setosum* would have 153 lines because N= 153, and would start

```
0.71
0.88
1.22
etc.
```

Although the data were sorted before Table 13.4 was created, there is no need to sort a data file for **FREQ_GENERAL.BAS**.

You must choose the size of the interval of the frequency histogram. For example, if animals range from 0.25 cm to 10.00 cm, you may want to use an interval of 0.25 cm. You also must specify the tolerance of the measurements. If you measured animals to 0.01 cm then the tolerance is 0.01. If you measured just to the closest millimeter so measurements are, for example, 2.3 cm and 9.6 cm, then the tolerance is 0.1.

Output files repeat the name of the data file with extensions, **.FREQ**, **.OUT** and **.MAC**. Files with the **.FREQ** extension have frequencies and cumulative frequencies. Files with the **.OUT** extension are for making graphs like Figure 3.11 with Cricket Graph; however, files should first be examined with an editor such as Word. Size and frequency values must be separated by a tab stop. If you are using a BASIC that does not do this you will have to do a global replace to insert tabs. A second problem is that Cricket Graph will not accept numbers in scientific notation such as 2.45E-3, which would be considered to be alpha-numeric and would not be graphed as a number. You have to go through and change all numbers in scientific notation to decimal form so 2.45E-3 would be 0.00245. Also, be sure to save the file as a text file so you can open it using Cricket Graph. Or, you can copy and paste the file into Cricket Graph. If the file is very large, you may want to use a spreadsheet application to convert numbers from scientific to decimal notation.

Once you have the file opened in Cricket Graph you can graph using either by plotting points and connecting dots or, better still, create a graph that fills-in a solid color under an area.

The file with the extension **.FREQ** can not be opened by **X.BAS**. Two important changes must be made. First all columns other than the cumulative frequency column must be removed and, second, the number of data values must be entered as the first value in the data file.

Here is a sample of how the program **FREQ_GENERAL.BAS** is run with a data set called diadema.dat, the size data for *Diadema setosum* from Zanzibar.

Want to continue? (Y/N):

    enter y or Y <cr>

Filename with data:

    enter diadema.dat <cr>

Interval or bucket size you wish to use:

    enter .25 <cr>

Measurement accuracy e.g. 0.1 or 0.01:

    enter .01 <cr>

Do you want to save a file with cumulative frequency? (Y/N)

    enter y or Y <cr>

Do you want to save a file for drawing a histogram? (Y/N)

    enter y or Y<cr>

Do you want to save a file for size-frequency analysis? (Y/N)

    enter y or Y <cr>

Want to continue? (Y/N)

    enter n or N <cr>

Files created with the data set diadema.dat will be: diadema.FREQ, diadema.OUT and diadema.MAC.

FREQ_GENERAL.BAS

```
10 dim sz(1010),sx(2001),a(101),b(101),
   p(101),cf(101)
20 print "A general program for creating
   data files for:"
30 print "      1) drawing frequency
   histograms"
40 print "      2) size-frequency
   analysis using Size_analysis.bas."
50 print "      3) calculating normal
   deviates from cumulative frequency
   with"
60 print "             the program x.bas."
70 print ""
80 print "This is another fine product
   from Cornered Rat Software©"
90 print "
   T. A. Ebert     1996"
100 input "Want to continue? (Y/N): ";q$
110 if q$ = "N" or q$ = "n" then goto
    900
120 for i = 0 to 2000
130 sx(i) = 0
140 next i
150 for i = 0 to 100
160 a(i) = 0
170 b(i) = 0
180 p(i) = 0
190 sz(i) = 0
200 cf(i) = 0
210 next i
220 input "Filename with data: ";f$
230 input "Interval or bucket size you
    wish to use: ";iz
240 input "Measurement accuracy e.g. 0.1
    or 0.01: ";tol
250 open f$ for input as #9
260 p = instr(f$,".") : f$ = left$(f$,p)
270 n = 0
280 for i = 1 to 2000
290 input #9,sx(i)
300 if eof(9) then goto 330
310 n = n+1
320 next i
330 close #9
340 for i = 1 to n
350 k = int((sx(i)+(iz-tol))/iz)
360 sz(k) = sz(k)+1
370 next i
380 for k = 1 to 200
390 if sz(k) = 0 then next k
400 for j = 200 to 1 step -1
410 if sz(j) = 0 then next j
420 ir = j-k+1
430 cf(k-1) = 0
440 for i = k to j
450 a(i) = i*iz
460 b(i) = a(i)-iz
470 if b(i) < tol then b(i) = 0
480 p(i) = sz(i)/n*100
490 cf(i) = cf(i-1)+sz(i)/n
500 next i
510 input "Do you want to save a file
    with cumulative frequency? (Y/N)
    ";q$
520 if q$ = "Y" or q$ = "y" then goto
    550
530 if q$ <> "N" and q$ <> "n" then goto
    510
540 goto 620
550 x$ = "FRQ"
560 open f$+x$ for output as #9
570 print #9,j-k
580 for i = k to j-1
590 print #9,a(i)"," sz(i)"," cf(i)
600 next i
610 close #9
620 input "Do you want to save a file
    for drawing a histogram? (Y/N) ";q$
630 if q$ = "Y" or q$ = "y" then goto
    660
640 if q$ <> "N" and q$ <> "n" then goto
    620
650 goto 770
660 x$ = "OUT"
670 open f$+x$ for output as #9
680 print #9,b(k),"0","0"
690 p(j+1) = 0
700 sz(j+1) = 0
710 for i = k to j
720 print #9,b(i),p(i),sz(i)
730 print #9,b(i+1),p(i),sz(i)
740 next i
750 print #9,b(j+1),"0","0"
760 close #9
770 input "Do you want to save a file
    for size-frequency analysis? (Y/N)
    ";q$
780 if q$ = "Y" or q$ = "y" then goto
    810
790 if q$ <> "N" and q$ <> "n" then goto
    770
800 goto 100
```

```
810 x$ = "MAC" : open f$+x$ for output
    as #9
820 print #9,f$
830 print #9,ir
840 for i = k to j
850 a = i*iz
860 print #9,sz(i);",";a
870 next i
880 close #9
890 goto 100
900 end
```

SIZE_ANALYSIS.BAS

The Macdonald and Pitcher method for decomposing a size-frequency distribution (Macdonald and Pitcher 1979) into normal components is the best one that I have found. I have taken a portion of the FORTRAN code they developed for their 1979 paper and have translated it into BASIC. In doing this I have made changes in input and output but have retained what I believe are the essential aspects of one section of their much larger program. I have extracted their subroutine that estimates proportions, means and standard deviation of as many components as you assert are present. The program permits holding means and standard deviations fixed or estimating a common standard deviation for all components. The proportions are always permitted to change.

Input for the program is simple. A file is required that has a size-frequency distribution with number and size. The data have been grouped into classes with some interval width and in the file "size" means the *right* boundary of the size interval. To make life easier, I have written a small program called **FREQ_GENERAL.BAS** that takes raw size data and creates a file that is suitable for the size-frequency analysis program.

Here is an example of output you would obtain from **FREQ_GENERAL.BAS** that is suitable for decomposition into component normal distributions. The data used was in a file called diadema.dat, which is printed as the first line in the output file. The second line is the number of size classes to be read by **SIZE_ANALYSIS.BAS**.

```
diadema.dat

25
1 ,0.75
1 ,1
1 ,1.25
12 ,1.5
17 ,1.75
14 ,2
15 ,2.25
6 ,2.5
2 ,2.75
```

```
0 ,3
0 ,3.25
4 ,3.5
3 ,3.75
18 ,4
20 ,4.25
10 ,4.5
8 ,4.75
3 ,5
4 ,5.25
4 ,5.5
4 ,5.75
2 ,6
1 ,6.25
2 ,6.5
1 ,6.75
```

The first number for each pair is the number of individuals in the 0.25cm interval and the second number is the *right* or upper boundary of this interval. For example, there was one individual in the interval 0.51 - 0.75 and 20 individuals in the interval from 4.01 to 4.25. You now are all set to run the program **SIZE_ANALYSIS.BAS**.

Do you want to continue (Y/N)

Enter Y or y and<cr> to continue. N or n will terminate the program.

Enter file name with data:

Enter the file name for the data that were grouped using the program Frequency. For example, the **diadema.dat** data after grouping was stored as **diadema.MAC** so respond by typing the filename, that is, **diadema.MAC**, and then <cr>. The data will be displayed on the screen

```
diadema.dat
1    1    0.75
2    1    1
3    1    1.25
4    12   1.5
5    17   1.75
6    14   2
7    15   2.25
8    6    2.5
9    2    2.75
10   0    3
11   0    3.25
12   4    3.5
13   3    3.75
14   18   4
15   20   4.25
```

16 10 4.5
17 8   4.75
18 3   5
19 4   5.25
20 4   5.5
21 4   5.75
22 2   6
23 1   6.25
24 2   6.5
25 1   6.75

How many components should be extracted

The number of components is a bit tricky. This is a good reason to have a graph of the data. The *Diadema* data were shown in Figure 13.11.
There certainly are two components and, possibly, one might get away with as many as 4. I might choose one component with a mean of about 1.7, another with a mean of 4 and two more at 5.5 and 6.2. This is not unreasonable and so I will use 4 components and respond
4<cr>

Fraction that is component 1

Make a reasonable guess such as

.45<cr>

Fraction that is component 2

Again, make a reasonable guess such as

.40<cr>

Fraction that is component 3

Make a guess such as

.1<cr>

Fraction that is component 4

Guess

.05<cr>

You could just say .25 for all of them but it helps a bit to have guesses that are not terribly far away from correct values. You now will have to enter guesses for the means.

Mean of component 1

1.8 <cr>

and so on for all four means. Again, try to make reasonable guesses.

Standard deviation of component 1

This will be the hardest to guess but recall that about 67% of the area of a normal curve is contained within ±1 sd so look at the graph and make a guess such as 0.4 or 0.5 or something like that. Actually 0.5 seems a bit large so 0.4 probably would be better.

.4<cr>

Use the same value for all the rest of the responses unless you have some better guesses. After you have made all of your guesses, they will appear on the screen.

Proportion	Mean	sd
0.45000	1.70000	0.40000
0.40000	4.00000	0.40000
0.10000	5.50000	0.40000
0.05000	6.30000	0.40000

Want to continue (Y/N)

Enter Y or y<cr>

Estimate proportions, means and standard errors
Iteration limit =

The number you enter should not matter a great deal, but the program may dither along for some time if you enter a large number. Generally, a solution should be found fairly rapidly but I usually set this at 100 so

100<cr>

Would you like to print a table of Observed and Expected (Y/N)
Getting such a table is useful at the end of the analysis but not now; you have several more passes to make before you will be done. Consequently, enter

N or n followed by <cr>

Would you like to have all means fixed (Y/N)

For the first time through it probably is a good idea to fix these, that is, hold them at the initial values you entered and improve just the estimates of proportions. So enter
Y or y followed by <cr>

Would you like all sd's to be fixed (Y/N)

For the first pass through, keep them fixed at your initial guesses so enter

Y or y followed by <cr>

After what will seem like an unusual length of time

Number of iterations: 6

Do you want results saved in a file (Y/N)

Too soon to save things, so enter

N or n followed by <cr>

You now get to see the results

Proportions and their standard errors
Proportion # 1 .450435 ± 4.035271E-02 (se)
Proportion # 2 .4220504 ± .0409879 (se)
Proportion # 3 .1069763 ± 2.940577E-02 (se)
Proportion # 4 .0205382 ± 1.639948E-02 (se)
Means and their standard errors
Mean # 1 1.7 ±-99 (se)
Mean # 2 4 ±-99 (se)
Mean # 3 5.5 ±-99 (se)
Mean # 4 6.3 ±-99 (se)
Standard deviations and their standard errors
sd # 1 .4 ±-99 (se)
sd # 2 .4 ±-99 (se)
sd # 3 .4 ±-99 (se)
sd # 4 .4 ±-99 (se)
Degrees of freedom = 21
Chi-squared = 28.08466

Notice that the standard errors are -99 for the means and standard deviations. This means that none was estimated because they were held constant. At the bottom of the screen you have the following message:

<return> to continue...

After you have looked at your results, hit <cr> and you are ready to continue the process of improving estimates of parameters>

Want to continue (Y/N)

Enter Y or y<cr>

Estimate proportions, means and standard errors

Iteration limit =

100<cr>

Would you like to print a table of Observed and Expected (Y/N)

N or n followed by <cr>

Would you like to have all means fixed (Y/N)

We could try to improve the means or standard deviations. Let's pick standard deviations rather than means and so enter Y or y followed by <cr>

Would you like all sd's to be fixed (Y/N)

This time enter N or n followed by <cr>

Want to estimate a common standard deviation (Y/N)

This is a safe way to improve things so enter Y or y followed by <cr>

After some time you will get additional results and you will be ready to make some additional improvements such as fixing means 3 and 4 and let means 1 and 2 be estimated. Then fix means 1 and 2 and estimate means 3 and 4. Finally, let all 4 means be fit together with a common standard deviation:

Number of iterations: 16

Expected number	Observed number	Right boundary
0.13104	1	0.75
0.81199	1	1.00
3.49945	1	1.25
9.55865	12	1.50
16.56118	17	1.75
18.20893	14	2.00
12.70608	15	2.25
5.62552	6	2.50
1.58467	2	2.75
0.34874	0	3.00
0.54459	0	3.25
2.45216	4	3.50
7.41384	3	3.75
14.22926	18	4.00
17.34096	20	4.25
13.49908	10	4.50
7.09480	8	4.75
3.68913	3	5.00
3.66986	4	5.25
4.19777	4	5.50
3.48949	4	5.75
2.24930	2	6.00
1.56396	1	6.25
1.24620	2	6.50
1.28336	1	6.75

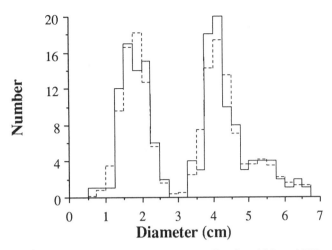

FIGURE 13.23 *Diadema setosum* from Zanzibar 16 June 1976. Solid line is the observed distribution and the dashed line is the expected distribution based on 4 components.

work and so you may be stuck with a common sd or a common sd for components 2-4.

The plot of observed and expected appears to be quite good, which also is the conclusion from the χ^2 value of 16.4, which, with 16 degrees of freedom has p>0.25. This seems like a reasonable fit.

The worst part of this program is that it is possible to do a large number of unreasonable things mostly associated with trying to extract more components than one really should. The best rule is never, never, never try to extract more components than you can reasonably see. If you can't see them then they aren't there. This is not to say that a large mode may not be composed of a large number of age classes. Frequently this surely is the case; however, it serves little useful purpose to try to tease these distributions apart when no hint of modes exists. Such an approach builds stories that may be very wrong and, as a consequence, contributes to pollution of the literature.

Proportions and their standard errors
Proportion # 1 .4509554 ± .0402786 (se)
Proportion # 2 .4213419 ± 4.287867E-02 (se)
Proportion # 3 9.953198E-02 ± 3.067096E-02 (se)
Proportion # 4 .0281707 ± 2.349151E-02 (se)
Means and their standard errors
Mean # 1 1.802155 ± 4.492309E-02 (se)
Mean # 2 4.108355 ± 5.393839E-02 (se)
Mean # 3 5.375691 ± .1935565 (se)
Mean # 4 6.303342 ± .3562988 (se)
Standard deviations and their standard errors
sd # 1 .363511 ± 2.508998E-02 (se)
sd # 2 .363511 ±-99 (se)
sd # 3 .363511 ±-99 (se)
sd # 4 .363511 ±-99 (se)
Degrees of freedom = 16
Chi-squared = 16.40697

There is just one standard error for the standard deviation estimates because it is the standard error for the estimate of the common standard deviation for all four components (Figure 13.23). The fact that a common standard deviation was estimated also accounts for the 16 degrees of freedom: components (4), means (4), and standard deviation (1) equals a loss of 9 degrees of freedom and with 25 categories, 25 - 9 = 16df.

You could continue to improve the estimate of standard deviations if you wanted to. First, hold components 2, 3 and 4 constant and estimate the sd for component 1. Then hold sd for components 1, 3 and 4 constant and estimate the sd for component 2. Next hold 1 and 2 constant and try to estimate sd for components 3 and 4. Finally, try to estimate sd for all four components. It is possible that this may not

SIZE_ANALYSIS.BAS

```
10 dim p(10),ph(70),pi(10),mu(10),
   sg(10),n(70),x(70),gm(70,10)
20 dim bm(70,29),p3(70),dc(29),sd(29),
   pv(29),qv(29),r$(29),nf(70)
30 dim sm(10),np(70),p9(10),m9(10),
   s9(10),m$(10),s$(10),st(29),ss(29)
40 print "Size-frequency decomposition
   based on the method of P. D.
   Macdonald"
50 print "and T. J. Pitcher 1979. J.
   Fish. Res. Bd. Canada 36: 987-1001."
60 print
70 print "BASIC version       January
   1995      T. A. Ebert"
80 print "Another fine product from
   Cornered Rat Software©"
90 print
100 input "Do you want to continue with
    this program  (Y/N)";f$
110 if f$ = "N" or f$ = "n" then goto
    3730
120 if f$ <> "Y" and f$ <> "y" then goto
    90
130 input "Enter file name with data:
    ";f$
140 open f$ for input as #1
150 input #1,n$ : input #1,m
160 print n$
170 mh = m
180 mm = m-1
190 nb = 0
200 for j = 1 to m
```

```
210 input #1,n(j),x(j)
220 print j,n(j),x(j)
230 nb = nb+n(j)
240 next j
250 close #1
260 for j = 1 to m
270 ph(j) = n(j)/nb
280 next j
290 print
300 input "How many components should be
    extracted? ";k9
310 km = k9-1
320 for j = 1 to k9
330 print "Fraction that is component
    ";j,
340 input pi(j)
350 next j
360 print
370 for j = 1 to k9
380 print "Mean of component ";j,
390 input mu(j)
400 next j
410 print
420 for j = 1 to k9
430 print "Standard deviation of
    component ";j,
440 input sg(j)
450 next j
460 print "Proportion            Mean
    sd"
470 for j = 1 to k9
480 print pi(j),mu(j),sg(j)
490 next j
500 gosub 4780
510 print
520 input "Want to continue with this
    data set (Y/N)";a$
530 if a$ <> "Y" and a$ <> "y" then goto
    90
540 rem
550 print
560 print " Estimate proportions, means
    and standard errors"
570 gosub 4780
580 input "Iteration limit = ";mx
590 input "Would you like to print a
    table of Observed and Expected (Y/N)
    ";t$
600 input "Would you like to have all
    means fixed (Y/N) ";a$
610 if a$ <> "Y" and a$ <> "y" then goto
    660
620 for i = 1 to k9
630 m$(i) = "Y"
640 next i
650 goto 700
660 for i = 1 to k9
670 print "Is mean";i;
680 input "fixed? (Y/N) ";m$(i)
690 next i
700 input "Would you like all sd's to be
    fixed (Y/N) ";a$
710 if a$ <> "Y" and a$ <> "y" then goto
    770
720 q$ = "N"
730 for i = 1 to k9
740 s$(i) = "Y"
750 next i
760 goto 900
770 input "Want to estimate a common
    standard deviation? (Y/N)";q$
780 if q$ = "Y" or q$ = "y" then goto
    840
790 for i = 1 to k9
800 print "Is sd";i;
810 input "fixed? (Y/N) ";s$(i)
820 next i
830 goto 900
840 for i = 1 to k9
850 s$(i) = "N"
860 next i
870 for i = 2 to k9
880 sg(i) = sg(1)
890 next i
900 nt = 0
910 if mx <= 0 then let nt = -1
920 if nt >= mx then goto 2440
930 nt = nt+1
940 for i6 = 1 to k9
950 if sg(i6) > 0 then goto 990
960 print "Invalid standard deviation
    ";i,sg(i)
970 gosub 4560
980 goto 510
990 tm = 0
1000 fm = 0
1010 zm = 0
1020 ki = km+i6
1030 kk = km+k9+i6
1040 for j6 = 1 to mm
1050 z = (x(j6)-mu(i6))/sg(i6)
1060 gosub 3770
1070 gm(j6,i6) = fj-fm
1080 bm(j6,ki) = -pi(i6)*(zj-zm)/sg(i6)
1090 bm(j6,kk) = -pi(i6)*(z*zj-
     tm*zm)/sg(i6)
1100 tm = z
1110 fm = fj
```

```
1120 zm = zj
1130 next j6
1140 gm(m,i6) = 1-fm
1150 bm(m,ki) = -pi(i6)*(-zm)/sg(i6)
1160 bm(m,kk) = -pi(i6)*(-tm*zm)/sg(i6)
1170 next i6
1180 for j = 1 to m
1190 gh = gm(j,k9)
1200 for i = 1 to km
1210 bm(j,i) = gm(j,i)-gh
1220 next i
1230 next j
1240 for j = 1 to m
1250 p3(j) = 0
1260 for i = 1 to k9
1270 p3(j) = p3(j)+pi(i)*gm(j,i)
1280 next i
1290 next j
1300 for j = 1 to m
1310 if p3(j) > 0 then goto 1370
1320 l$ = "T"
1330 if mx <= 0 then goto 2770
1340 print "Sorry, a parameter dropped
       to zero ... or even lower."
1350 gosub 4560
1360 goto 510
1370 next j
1380 r = 0
1390 for i = 1 to km
1400 r = r+1
1410 p(r) = pi(i)
1420 next i
1430 for i = 1 to k9
1440 if m$(i) = "Y" or m$(i) = "y" then
     goto 1520
1450 r = r+1
1460 p(r) = mu(i)
1470 ki = km+i
1480 if r = ki then goto 1520
1490 for j = 1 to m
1500 bm(j,r) = bm(j,ki)
1510 next j
1520 next i
1530 if q$ = "Y" or q$ = "y" then goto
     1660
1540 r = 0
1550 for i = 1 to k9
1560 if s$(i) = "Y" or s$(i) = "y" then
     goto 1640
1570 r = r+1
1580 p(r) = sg(i)
1590 kk = km+k9+i
1600 if r = kk then goto 1640
1610 for j = 1 to m
1620 bm(j,r) = bm(j,kk)
1630 next j
1640 next i
1650 goto 1790
1660 kk = km+k9+1
1670 r = r+1
1680 p(r) = sg(1)
1690 if r = kk then goto 1730
1700 for j = 1 to m
1710 bm(j,r) = bm(j,kk)
1720 next j
1730 for i = 2 to k9
1740 kk = km+k9+i
1750 for j = 1 to m
1760 bm(j,r) = bm(j,r)+bm(j,kk)
1770 next j
1780 next i
1790 im = 0
1800 for ir = 1 to r
1810 for ic = ir to r
1820 im = im+1
1830 nf(im) = 0
1840 for j = 1 to m
1850 nf(im) =
     nf(im)+bm(j,ir)*bm(j,ic)*nb/p3(j)
1860 next j
1870 next ic
1880 next ir
1890 kh = km
1900 km = r
1910 gosub 4040
1920 r = km
1930 km = kh
1940 if l$ = "F" then goto 1990
1950 print "Egad! A singular matrix"
1960 if mx <= 0 then goto 2770
1970 gosub 4560
1980 goto 510
1990 if mx <= 0 then goto 2440
2000 for i = 1 to r
2010 dc(i) = 0
2020 for j = 1 to m
2030 dc(i) =
     dc(i)+bm(j,i)*ph(j)*nb/p3(j)
2040 next j
2050 next i
2060 for i = 1 to r
2070 st(i) = 0
2080 next i
2090 im = 0
2100 for ir = 1 to r
2110 for ic = ir to r
2120 im = im+1
2130 st(ir) = st(ir)+dc(ic)*nf(im)
```

```
2140 if ir <> ic then let st(ic) =
     st(ic)+dc(ir)*nf(im)
2150 next ic
2160 next ir
2170 pi(k9) = 1
2180 for i = 1 to km
2190 pi(i) = pi(i)+st(i)
2200 pi(k9) = pi(k9)-pi(i)
2210 next i
2220 kr = km
2230 for i = 1 to k9
2240 if m$(i) = "Y" or m$(i) = "y" then
     goto 2270
2250 kr = kr+1
2260 mu(i) = mu(i)+st(kr)
2270 next i
2280 if q$ = "Y" or q$ = "y" then goto
     2350
2290 for i = 1 to k9
2300 if s$(i) = "Y" or s$(i) = "y" then
     goto 2330
2310 kr = kr+1
2320 sg(i) = sg(i)+st(kr)
2330 next i
2340 goto 2400
2350 kr = kr+1
2360 sg(1) = sg(1)+st(kr)
2370 for i = 2 to k9
2380 sg(i) = sg(1)
2390 next i
2400 for i = 1 to r
2410 if abs(st(i)) > 5.000000E-06 then
     goto 920
2420 next i
2430 goto 2450
2440 print "Limit of iterations has been
     reached"
2450 i = 0
2460 sd = 0
2470 im = 0
2480 for kr = 1 to r
2490 for ic = kr to r
2500 im = im+1
2510 if kr = ic then goto 2540
2520 if ic <= km then let sd =
     sd+2*nf(im)
2530 goto 2570
2540 i = i+1
2550 sd(i) = sqr(nf(im))
2560 if ic <= km then let sd = sd+nf(im)
2570 next ic
2580 next kr
2590 sd = sqr(sd)
2600 kr = km
2610 for i = 1 to k9
2620 sm(i) = -99
2630 if m$(i) = "Y" or m$(i) = "y" then
     goto 2660
2640 kr = kr+1
2650 sm(i) = sd(kr)
2660 next i
2670 if q$ = "Y" or q$ = "y" then goto
     2750
2680 for i = 1 to k9
2690 ss(i) = -99
2700 if s$(i) = "Y" or s$(i) = "y" then
     goto 2730
2710 kr = kr+1
2720 ss(i) = sd(kr)
2730 next i
2740 goto 2770
2750 kr = kr+1
2760 ss(1) = sd(kr)
2770 print "Number of iterations: ";nt
2780 input "Do you want results saved in
     a file (Y/N)";g$
2790 if g$ <> "Y" and g$ <> "y" then
     goto 2820
2800 input "Filename for results: ";f$
2810 open f$ for output as #1
2820 if g$ = "Y" or g$ = "y" then print
     #1,"Number of iterations: ";nt
2830 for j = 1 to m
2840 np(j) = nb*p3(j)
2850 next j
2860 if t$ <> "Y" and t$ <> "y" then
     goto 2980
2870 print "Expected number, observed
     number, and right boundary"
2880 if g$ = "Y" or g$ = "y" then print
     #1,"Expected number, observed
     number, and right boundary"
2890 for j = 1 to mm
2900 print j,np(j),n(j),x(j)
2910 next j
2920 print m,np(m),n(m),2*x(mm)-x(mm-1)
2930 if g$ <> "Y" and g$ <> "y" then
     goto 2980
2940 for j = 1 to mm
2950 print #1,j,np(j),n(j),x(j)
2960 next j
2970 print #1,m,np(m),n(m),2*x(mm)-x(mm-
     1)
2980 if g$ <> "Y" and g$ <> "y" then
     goto 3280
2990 print #1,""
3000 print #1,"Proportions and their
     standard errors"
```

```
3010 for i = 1 to k9
3020 print #1,"Proportion ";i;pi(i)
3030 next i
3040 if l$ = "T" then goto 3080
3050 for i = 1 to k9
3060 print #1,"se for ";i;sd(i)
3070 next i
3080 print #1,""
3090 print #1,"Means and their standard
     errors"
3100 for i = 1 to k9
3110 print #1,"Mean ";i;mu(i)
3120 next i
3130 if l$ = "T" then goto 1670
3140 for i = 1 to k9
3150 print #1,"se for ";i;sm(i)
3160 next i
3170 print #1,""
3180 print #1,"Standard deviations and
     their standard errors"
3190 for i = 1 to k9
3200 print #1,"sd ";i;sg(i)
3210 next i
3220 kh = k9
3230 if q$ = "Y" or q$ = "y" then let kh
     = 1
3240 if l$ = "T" then goto 3280
3250 for i = 1 to kh
3260 print #1,"se for ";i;ss(i)
3270 next i
3280 print "Proportions and their
     standard errors"
3290 for i = 1 to k9
3300 print "Proportion ";i;pi(i)
3310 next i
3320 if l$ = "T" then goto 3360
3330 for i = 1 to k9
3340 print "se for ";i;sd(i)
3350 next i
3360 print
3370 print "Means and their standard
     errors"
3380 for i = 1 to k9
3390 print "Mean ";i;mu(i)
3400 next i
3410 if l$ = "T" then goto 3450
3420 for i = 1 to k9
3430 print "se for ";i;sm(i)
3440 next i
3450 print
3460 print "Standard deviations and
     their standard errors"
3470 for i = 1 to k9
3480 print "sd ";i;sg(i)
3490 next i
3500 kh = k9
3510 if q$ = "Y" or q$ = "y" then let kh
     = 1
3520 if l$ = "T" then goto 3560
3530 for i = 1 to kh
3540 print "se for ";i;ss(i)
3550 next i
3560 cq = 0
3570 for j = 1 to m
3580 if ph(j) <= 0 then goto 3630
3590 if p3(j) > 0 then goto 3620
3600 print "Invalid parameter"
3610 goto 510
3620 cq = cq+ph(j)*log(p3(j)/ph(j))
3630 next j
3640 cq = -2*nb*cq
3650 df = mm-r
3660 print "Degrees of freedom = ";df
3670 print "Chi-squared = ";cq
3680 if g$ = "Y" or g$ = "y" then print
     #1,"Degrees of freedom = ";df
3690 if g$ = "Y" or g$ = "y" then print
     #1,"Chi-squared = ";cq
3700 if g$ = "Y" or g$ = "y" then close
     #1
3710 input "<return> to continue
     .....";x$
3720 goto 510
3730 end
3740 rem Area to the left of x beneath
     the standardized normal
3750 rem probability curve.  This is
     based on the Nov. 1972 FORTRAN
3760 rem version used by Macdonald and
     Pitcher 1972
3770 y = abs(z)
3780 xs = z*z
3790 zj = 0.398942*exp(-xs/2)
3800 if (y-1) > 0 then goto 3840
3810 fj = xs*(xs*(xs*6.659694E-07-
     9.444656E-06)+1.154347E-04)-
     1.187328E-03
3820 fj = z*(xs*(xs*(xs*fj+9.973557E-
     03)-0.06649)+0.398942)+0.5
3830 return
3840 if (y-2.8) > 0 then goto 3890
3850 t = 1/(1+0.231642*y)
3860 qt = t*(t*(t*1.330274-
     1.821256)+1.781478)-0.356564
3870 qt = zj*t*(t*qt+0.319382)
3880 goto 3970
3890 b = xs+7
3900 a = y*(b+6)
```

```
3910 c = y*a+5*b
3920 b = y*c+4*a
3930 a = y*b+3*c
3940 c = y*a+2*b
3950 b = y*c+a
3960 qt = zj*c/b
3970 if z > 0 then goto 4000
3980 fj = qt
3990 return
4000 fj = 1-qt
4010 return
4020 rem Invert a positive definite
     symmetric matrix in upper
4030 rem triangular form
4035 rem
4040 for i = 1 to km
4050 r$(i) = "T"
4060 next i
4070 for i = 1 to km
4080 bg = 0
4090 l4 = 1
4100 for j = 1 to km
4110 if (r$(j) = "T" and (abs(nf(l4)) >
     bg)) then goto 4130
4120 goto 4160
4130 bg = abs(nf(l4))
4140 k3 = j
4150 m3 = l4
4160 l4 = l4+km-j+1
4170 next j
4180 if bg <> 0 then goto 4210
4190 l$ = "T"
4200 return
4210 l4 = k3
4220 r$(k3) = "F"
4230 qv(k3) = 1/nf(m3)
4240 pv(k3) = 1
4250 nf(m3) = 0
4260 if k3 <= 1 then goto 4350
4270 m1 = k3-1
4280 for j = 1 to m1
4290 pv(j) = nf(l4)
4300 nf(l4) = 0
4310 l4 = l4+km-j
4320 qv(j) = qv(k3)*pv(j)
4330 if r$(j) = "T" then let qv(j) =
     -qv(j)
4340 next j
4350 l4 = (k3-1)*(2*km-k3)/2
4360 if k3 >= km then goto 4450
4370 k1 = k3+1
4380 for j = k1 to km
4390 lp = l4+j
4400 pv(j) = -nf(lp)
```

```
4410 if r$(j) = "T" then let pv(j) =
     -pv(j)
4420 qv(j) = -nf(lp)*qv(k3)
4430 nf(lp) = 0
4440 next j
4450 l4 = 1
4460 for j = 1 to km
4470 for k3 = j to km
4480 nf(l4) = nf(l4)+pv(j)*qv(k3)
4490 l4 = l4+1
4500 next k3
4510 next j
4520 next i
4530 l$ = "F"
4540 return
4550 rem Here is a subsoutine that
     restores parameter values
4560 for i = 1 to k9
4570 pi(i) = p9(i)
4580 mu(i) = m9(i)
4590 sg(i) = s9(i)
4600 next i
4610 print "Paramaters have been
     restored to previous values."
4620 print
4630 for i = 1 to k9
4640 print "Proportion ";i;pi(i)
4650 next i
4660 print
4670 for i = 1 to k9
4680 print "Mean ";i;mu(i)
4690 next i
4700 print
4710 for i = 1 to k9
4720 print "sd ";i;sg(i)
4730 next i
4740 print
4750 return
4760 rem
4770 rem Routine to save parameters
4780 for i = 1 to k9
4790 p9(i) = pi(i)
4800 m9(i) = mu(i)
4810 s9(i) = sg(i)
4820 next i
4830 return
```

ZKAN.BAS

This is a program that uses means of size distributions to estimate Z, K, asymptotic size (S_∞), and the shape parameter n of the Richards Function. The method is based on the idea that mean size of individuals in a population is a function of growth and survival parameters. It is assumed that the population is seasonally stable and stationary and there is a single reproductive episode each year.

Data are mean sizes such as lengths or weights together with the fraction of a year since recruitment to the population. For example, if annual recruitment of individuals size S_R takes place each year in June then the mean size of the population in June is assigned T=0 and size data from other months is scaled based on T=0 for June. Mean size in July would have T=1/12 or T=0.0833, August would have T=2/12 or T=0.1667, etc. The data file starts with the *number* of size - time pairs and then that number of mean sizes and times. Here is an example for a euphausid *Meganyctiphanes norvegica*. The data file is taken from Table 13.12. Note that it is not necessary to arrange the data file in any particular order with respect either to size or time.

```
7
21.5373, .1667
26.4146, .3333
27.9873, .5
31.9119, .8333
17.6349, 0
29.1009, .5
31.5436, .75
```

When you run **ZKAN.BAS**, there will be a series of questions that must be answered. Most are obvious but I will provide just a few observations and suggestions.

Are data from a file or entered on the keyboard? (F/K)

If you answer with a K or a k for "keyboard" you will be the following prompt.

Number of data pairs =

Enter the number of observations you have of mean size and time since recruitment at size SR. After this you will get prompts for size and time until you have entered the total number of data pairs.

Average size =

at time =

Be sure that average size is for the entire size-frequency distribution. Don't truncate your data.

Do you want data saved in a file? (Y/N)

If you respond Y or y, you will get the following prompt.

File name for data:

Do you want to continue with this data set? (Y/N)

Answer Y or y if you want to continue. You now will be given the chance to estimate one or more parameters.

What is the variance of function values that can be used to terminate the program (e.g. 0.0001, .000001)?

A value of 0.0001 isn't bad particularly if you are attempting to estimate values for all four parameters. If you have just one parameter and several mean values then 0.00001 or even smaller would be ok.

Do you want to estimate Z? (Y/N)

Do you want to estimate K? (Y/N)

Do you want to estimate asymptotic size? (Y/N)

Do you want to estimate n, the shape parameter? (Y/N)

Answer Y or y for all parameters you wish to estimate but be careful. The more parameters you attempt to estimate, the larger the standard errors and the longer it will take. Often it will be best to fix n at some value such as -1 (Brody-Bertalanffy), +1 (logistic) or +25 (Gompertz).

Initial values of parameters

Z =

K =

Maximum size =

Size at recruitment =

Shape parameter, n =

If convergence does not take place, you will have to fix one or more of the parameters until you discover the general region where a final solution may reside. First fix n at -1 and try to estimate all other parameters, then try n=+1. Which gave a smaller SSE? If +1 gave the smaller SSE then use the parameter estimates from this run as starting values for a run to estimate all 4 parameters. It is possible that the difference in SSE between the run with estimating 3 and the run with estimating 4 parameters is so small that it is not worth estimating 4 because of the increase in se. You must decide what you need.

ZKAN.BAS

```
10 dim t(50),s(50),f(50,6),a(6,6),
   m(6,6),b(6,6),e(6),c(50)
20 print ;tab (7);"A product from
   CORNERED RAT SOFTWARE"
```

```
30 print : print ;tab (12);"ZKAN by T.
   A. Ebert, 1990; revised 1994, 1996"
   : print
40 print " This is a program to estimate
   Z, K, Asymptotic size, "
50 print " and the shape parameter n of
   the Richards Function" : print
60 input "Do you want to continue? (Y/N)
   ";f$
70 if f$ = "N" or f$ = "n" then goto
   3300
80 if f$ <> "Y" and f$ <> "y" then goto
   60
90 rem
100 for i = 1 to 50
110 t(i) = 0
120 s(i) = 0
130 c(i) = 0
140 next i
150 for i = 1 to 50
160 for j = 0 to 6
170 f(i,j) = 0
180 next j
190 next i
200 for i = 1 to 6
210 e(i) = 0
220 for j = 0 to 6
230 a(i,j) = 0
240 m(i,j) = 0
250 b(i,j) = 0
260 next j
270 next i
280 mg = -log(1.000000E-05)
290 input " Are data from a file or
   entered on the keyboard? (F/K)";y$
300 if y$ = "F" or y$ = "f" then goto
   500
310 print
320 print
330 input " Number of data pairs = ";n
340 for i = 1 to n
350 print : input "Average size  =
   ";s(i)
360 input "at time = ";t(i)
370 print
380 next i
390 input "Do you want data saved in a
   file? (Y/N) ";y$
400 if y$ = "N" or y$ = "n" then goto
   580
410 print
420 input "File name for data: ";f$
430 open f$ for output as #1
440 print #1,n
450 for i = 1 to n
460 print #1,s(i),t(i)
470 next i
480 close #1
490 goto 580
500 rem
510 input "Name of file with data: ";f$
520 open f$ for input as #1
530 input #1,n
540 for i = 1 to n
550 input #1,s(i),t(i)
560 next i
570 close #1
580 rem
590 input "Do you want to continue with
   this data set? (Y/N) ";f$
600 if f$ = "Y" or f$ = "y" then goto
   650
610 print
620 input "Do you want to continue with
   a new file? (Y/N) ";f$
630 if f$ = "N" or f$ = "n" then goto
   3300
640 goto 90
650 ic = 0
660 id = 0
670 ie = 0
680 ig = 0
690 for i = 1 to n
700 print t(i);s(i)
710 next i
720 print " What is the variance of
   function values that can be used to
   "
730 input "terminate the program (e.g.
   0.0001, .000001)? ";rq
740 r = 0
750 input "Do you want to estimate Z?
   (Y/N) ";y$
760 if y$ = "Y" or y$ = "y" then ic =
   r+1 : r = r+1
770 print
780 input "Do you want to estimate K?
   (Y/N) ";y$
790 if y$ = "Y" or y$ = "y" then id =
   r+1 : r = r+1
800 print
810 input "Do you want to estimate
   asymptotic size? (Y/N) ";y$
820 if y$ = "Y" or y$ = "y" then ie =
   r+1 : r = r+1
830 print
840 input "Do you want to estimate n,
   the shape parameter? (Y/N) ";y$
```

```
850 if y$ = "Y" or y$ = "y" then ig =
    r+1 : r = r+1
860 print
870 print ;tab (10);"Initial values of
    parameters"
880 input "Z = ";z
890 input "K = ";k
900 input "Maximum size = ";a
910 input "Size at recruitment = ";sr
920 input "Shape parameter, n = ";nr
930 rem f(i,c1) is S(I) -F, the value of
    the function
940 rem f(i,ic) is the partial with
    respect to Z
950 rem f(i,id) is the partial with
    respect to K
960 rem f(i,ie) is the partial with
    respect to a, asymptotic size.
970 rem f(i,ig) is the partial with
    respect to n, the shape parameter
980 rem
990 c1 = r+1
1000 zs = 1000
1010 ks = 1000
1020 asv = 10000
1030 ns = 1000
1040 print "Estimates of parameters"
1050 print "        Z","K","S∞","n"
1060 print
1070 p1 = exp(z)-1
1080 p3 = 1-exp(-z)
1090 d = -1/nr
1100 a9 = log(a/sr)
1110 a8 = (a/sr)^(-d)
1120 for i = 1 to n
1130 for j = 0 to c1
1140 f(i,j) = 0
1150 next j
1160 next i
1170 b = (a^d-sr^d)/a^d
1180 kk = int(mg/z+0.5)
1190 for j = 1 to n
1200 for i = 0 to kk
1210 a1 = exp(-z*i)
1220 a6 = exp(-k*(i+t(j)))
1230 b1 = 1-b*a6
1240 a2 = b1^(-nr)
1250 f(j,c1) = f(j,r+1)+a1*a2
1260 a3 = i*a1
1270 f(j,ic) = f(j,ic)+a3*a2
1280 a4 = b1^(-(nr+1))
1290 a5 = (i+t(j))*a6
1300 f(j,id) = f(j,id)+a1*a4*a5
1310 f(j,ie) = f(j,ie)+a1*a6*a4
1320 a7 = -log(b1)
1330 f(j,ig) =
     f(j,ig)+a1*a2*(a7+a6*a*a9/(nr*b1))
1340 next i
1350 f(j,ic) = a*(f(j,r+1)-
     p1*f(j,ic))/exp(z)
1360 f(j,c1) = p3*f(j,c1)
1370 f(j,id) = -a*nr*b*p3*f(j,id)
1380 f(j,ie) = -a*p3/sr*(a/sr)^(1/nr-
     1)*f(j,ie)+f(j,c1)
1390 f(j,ig) = a*p3*f(j,ig)
1400 f(j,c1) = s(j)-a*f(j,c1)
1410 next j
1420 rem Multiply the matrix by its
     transpose
1430 for i = 1 to r
1440 for j = 1 to c1
1450 a(i,j) = 0
1460 next j
1470 next i
1480 for i = 1 to r
1490 for j = 1 to c1
1500 for l = 1 to n
1510 a(i,j) = a(i,j)+f(l,i)*f(l,j)
1520 next l
1530 next j
1540 next i
1550 gosub 1760
1560 for i = 0 to r
1570 e(i) = 0
1580 next i
1590 for i = 1 to r
1600 for g = 1 to r
1610 e(i) = e(i)+b(i,g)*a(g,c1)
1620 next g
1630 next i
1640 z = z+e(ic)
1650 k = k+e(id)
1660 a = a+e(ie)
1670 nr = nr+e(ig)
1680 print z,k,a,nr
1690 xz = (zs-z)^2+(ks-k)^2+(asv-
     a)^2+(ns-nr)^2
1700 if xz <= rq then goto 2140
1710 zs = z
1720 ks = k
1730 asv = a
1740 ns = nr
1750 goto 1070
1760 for i = 1 to r
1770 for j = 1 to r
1780 b(i,j) = 0
1790 next j
1800 next i
```

```
1810 for j = 1 to r
1820 b(j,j) = 1
1830 next j
1840 for j = 1 to r
1850 for i = j to r
1860 if a(i,j) <> 0 then goto 1900
1870 next i
1880 print "Singular matrix "
1890 goto 580
1900 for m = 1 to r
1910 s = a(j,m)
1920 a(j,m) = a(i,m)
1930 a(i,m) = s
1940 s = b(j,m)
1950 b(j,m) = b(i,m)
1960 b(i,m) = s
1970 next m
1980 t = 1/a(j,j)
1990 print "------------------------------------"
2000 for m = 1 to r
2010 a(j,m) = t*a(j,m)
2020 b(j,m) = t*b(j,m)
2030 next m
2040 for l = 1 to r
2050 if l = j then 2110
2060 t = -a(l,j)
2070 for m = 1 to r
2080 a(l,m) = a(l,m)+t*a(j,m)
2090 b(l,m) = b(l,m)+t*b(j,m)
2100 next m
2110 next l
2120 next j
2130 return
2140 print
2150 print
2160 print ;"          Final values"
2170 print
2180 if ic <> 0 then print "Z =";z
2190 if id <> 0 then print "K =";k
2200 if ie = 0 then goto 2220
2210 print "S∞ = ";a
2220 if ig = 0 then goto 2240
2230 print "Shape parameter, n = ";nr
2240 print
2250 if n-r < 1 then goto 2460
2260 ss = 0
2270 for i = 1 to n
2280 ss = ss+f(i,c1)*f(i,c1)
2290 next i
2300 print "Residual SS = ";ss
2310 ms = ss/(n-r)
2320 print "Mean Square Error = ";ms
2330 for i = 1 to r
2340 m(i,i) = sqr(b(i,i)*ms)
2350 next i
2360 print "Standard deviations of
     parameters with ";n-r;"df"
2370 print
2380 if ic = 0 then goto 2400
2390 print "sd for Z = ",m(ic,ic)
2400 if id = 0 then goto 2420
2410 print ;"sd for K = ",m(id,id)
2420 if ie = 0 then goto 2440
2430 print "sd for S∞ = ",m(ie,ie)
2440 if ig = 0 then goto 2460
2450 print "sd for the shape parameter,
     n = ",m(ig,ig)
2460 rem
2470 input "<RETURN> to continue";x$
2480 print "      Values for plotting" :
     print
2490 print : input "File name for
     results: ";f$
2500 open f$ for output as #1
2510 print #1,n
2520 print "  Time","Mean Size"
2530 print
2540 for j = 0 to 11
2550 f = 0
2560 for i = 0 to kk
2570 a1 = exp(-z*i)
2580 a6 = exp(-k*(i+j/12))
2590 b1 = 1-b*a6
2600 a2 = b1^(-nr)
2610 f = f+a1*a2
2620 next i
2630 f = f*p3*a
2640 print j,f
2650 next j
2660 rem
2670 print
2680 input "<RETURN> to continue...";f$
2690 rem
2700 print "    Residuals" : print
2710 print
     "Observed","Calculated","Residual"
2720 print
2730 for j = 1 to n
2740 r2 = s(j)-f(j,c1)
2750 print s(j),r2,f(j,c1)
2760 next j
2770 print #1,"          Final values"
2780 print #1," "
2790 if ic <> 0 then print #1,"Z = ",z
2800 if id <> 0 then print #1,"K = ",k
2810 if ie = 0 then goto 2830
2820 print #1,"S∞ =",a
```

```
2830 if ig = 0 then goto 2850
2840 print #1,"Shape parameter, n =",nr
2850 print #1," "
2860 if n-r < 1 then goto 2990
2870 print #1,"Residual SS = ",ss
2880 print #1,"Mean Square Error = ",ms
2890 print "Standard deviations of
     parameters with ";n-r;"df"
2900 print #1," "
2910 if ic = 0 then goto 2930
2920 print #1,"sd for Z = ",m(ic,ic)
2930 if id = 0 then goto 2950
2940 print #1,"sd for K = ",m(id,id)
2950 if ie = 0 then goto 2970
2960 print #1,"sd for S∞ = ",m(ie,ie)
2970 if ig = 0 then goto 2990
2980 print #1,"sd for the shape
     parameter, n = ",m(ig,ig)
2990 rem
3000 print #1,"    Values for plotting"
     : print #1," "
3010 print #1,"Time","Mean Size"
3020 for j = 0 to 11
3030 f = 0
3040 for i = 0 to kk
3050 a1 = exp(-z*i)
3060 a6 = exp(-k*(i+j/12))
3070 b1 = 1-b*a6
3080 a2 = b1^(-nr)
3090 f = f+a1*a2 : next i
3100 f = f*p3*a
3110 print #1,j,f
3120 next j
3130 print #1,"   Residuals"
3140 print
     #1,"Observed","Calculated","Residual
     "
3150 print #1," "
3160 for j = 1 to n
3170 r2 = s(j)-f(j,c1)
3180 print #1,s(j),r2,f(j,c1)
3190 next j
3200 close #1
3210 rem
3220 print
3230 goto 590
3240 print "System will not converge. "
3250 print "Try other initial estimates"
3260 print "or fix one or more
     parameters."
3270 print
3280 input "" < return " to
     continue...";y$
3290 goto 580
```

```
3300 end
```

ZTAN.BAS

The following program estimates Z from a size-frequency distribution and growth parameters of the Tanaka function. The structure of the program matches Equation 13.21 and so is specifically written for the example with sea urchins with α and β included for allometry of the diameter *vs.* jaw length. The program can be run with data sets that do not have an allometric adjustment by entering 1.0 for both α and β.

```
10 dim lt(10000),t(10000)
20 print "Estimating Z with Tanaka
   parameters"
30 print "Cornered Rat Software  June
   1992 T. A. Ebert"
40 print : input "Do you want to
   continue? (y/n) ";s$
50 if s$ = "N" or s$ = "n" then goto 470
60 input "Mean size: ";ms
70 input "Size and time at recruitment:
   ";10,t0
80 input "Do you want to use the same
   parameters as the last run?
   (y/n)";s$
90 if s$ = "Y" or s$ = "y" then goto 160
100 print
110 print
120 input "Enter f, d, and a: ";f,d,a
130 input "Enter the alpha and beta for
    dia(y) vs jaw(x) allometry";alp,b
140 rem alpha and beta are ALP and B
150 input "Enter an initial estimate for
    Z:";z
160 g = -log(1.000000E-08)
170 s0 = (10/alp)^(1/b)
180 e = exp(sqr(f)*(s0-d))
190 c = a/e-e/4/f
200 for j = 1 to 10000
210 fz = 0
220 dz = 0
230 fs = 0
240 ds = 0
250 for i = 0 to (g/z+0.5)
260 t = i+t0
270 lt(i) = 1/sqr(f)*log(2*abs(f*(t-
    c)+sqr(f*f*(t-c)^2+f*a)))+d
280 lt(i) = alp*lt(i)^b
290 fz = fz+exp(-z*i)
300 dz = dz-i*exp(-z*i)
310 fs = fs+exp(-z*i)*lt(i)
320 ds = ds-i*exp(-z*i)*lt(i)
```

```
330 next i
340 dz = (fz*ds-fs*dz)/(fz*fz)
350 print "DZ= ";dz
360 fs = fs/fz-ms
370 print "FS= ";fs
380 cr = fs/dz
390 z = z-cr
400 print z,cr
410 if abs(cr) <= 1.000000E-06 then goto
    430
420 next j
430 print "Final Z = ";z
440 print
450 print
460 goto 40
470 end
```

14

Macroparameters

*L*ive fast, love hard, and die young

Country-western song recorded by
Faron Young, 1955

INTRODUCTION

Survival functions such as the Weibull or Gompertz (Chapter 12) and individual growth functions such as the Richards or Tanaka (Chapter 11) can be used to estimate l_x and m_x and smooth values of p_x and f_x. It is possible to combine p_x and f_x values determined from these general functions and to explore how changes in parameters of these functions, which Caswell (1982a) calls "macroparameters", can modify population growth. Survival and growth parameters provide summary statistics of physiological, morphological and behavioral attributes and can be used to explore patterns across species. Caswell (1982a) has provided the necessary mathematical framework for conducting a sensitivity analysis of survival and individual growth parameters.

MODEL DEVELOPMENT

In the present state of development, a restricted range of growth and survival models has been used. The Weibull distribution was introduced in Chapter 12 and is one way of modeling survival

$$n_x = n_0 e^{-(x/b)^c}$$

or

$$l_x = e^{-(x/b)^c}.$$

The parameter b is

$$b = \omega e^{-v/c},$$

where ω is the age at which the last individual is alive out of a cohort, which I am going to choose as 1000 individuals. One out of 1000 also defines the parameter v

$$v = ln(-ln(0.001))$$

or

$$v = 1.932644734.$$

If ω would be defined in some other manner, then v also would have to be changed. For example, if ω would be the last one out of a cohort of 10,000 then

$$v = ln(-ln(0.0001))$$

and so forth.

Starting at age 0, l_x is

$$l_x = \exp(-(x/(\omega*\exp(-1.932644734/c)))^\wedge c) \quad (14.1)$$

and annual survival, p_x, is

$$p_x = e^{[(x/b)^c - ((x+1)/b))^c]}. \qquad (14.2)$$

In the following analyses, growth will be modeled with the Brody-Bertalanffy equation that was introduced in Chapter 11 and, in particular, this equation will be used as a reasonable expression of age-specific fecundity, where m_x is the number of female offspring produced by a female age x

$$m_x = m_\infty(1 - e^{-K(x-\alpha)}). \qquad (14.3)$$

The parameter m_∞ is the maximum number of female offspring produced and α is the age at which gonads begin to develop but m_α still is zero. If first reproduction would take place at age 4 then α would be 3 followed by the first age where $m_x \neq 0$, age 4. Note that Equation 14.3 is appropriate only when age x is equal to or greater than the age when gonads begin to develop. For

$$x < \alpha, \quad m_x = 0$$

and for

$$x \geq \alpha, \quad m_x = m_\infty(1 - e^{-K(x-\alpha)}).$$

The values of p_x are defined in Equation 14.2 and f_x values are defined using Equation 14.3 and an estimate of early survival, p_0

$$f_x = p_0 m_x \qquad (14.4)$$

Because the Weibull function is used, survival during the first time period, p_0, is kept separate from other survival values because mortality often is very much higher during the first time period than in following periods. If a different survival model would be used, such as the 5-parameter Siler equation (Chapter 12), then p_0 would not have to be treated separately but could be part of the survival model. The trade-off for purposes of modeling is that the Weibull plus p_0 represents just 3 parameters whereas the Siler equation has 5. Keeping p_0 separate and starting at age 1 tends to improve the fit of the Weibull for a wide range of species (cf. Pinder et al. 1978).

In creating a transition matrix, most transitions can be calculated from data by using growth and survival functions. Early survival, p_0, often is difficult to calculate or may be subject to substantial error. In many circumstances, it may be best to calculate p_0 by creating the characteristic equation, setting $\lambda = 1.0$ and solving for p_0. For an age-structured population, this is very simple because the coefficients of the characteristic equation are

$$1 - m_1 p_0 - m_2 p_0 p_1 - m_3 p_0 p_1 p_2 - \ldots m_\omega p_0 p_1 p_2 p \cdots p_{\omega-1}$$

so, when $\lambda = 1$,

$$1 - p_0(m_1 + m_2 p_1 + m_3 p_1 p_2 + \ldots m_\omega p_1 p_2 \cdots p_{\omega-1}) = 0$$

or

$$p_0 = 1/(m_1 + m_2 p_1 + m_3 p_1 p_2 + \ldots m_\omega p_1 p_2 p \cdots p_{\omega-1}) \qquad (14.5)$$

This relationship comes directly from how the characteristic equation is constituted from a life-cycle graph (Chapter 4) The key point here is that λ is assumed to equal 1.0.

Nothing is special about doing a sensitivity analysis for individual elements of the matrix (Chapter 7). The new twist comes in doing a sensitivity analysis for the parameters: m_∞, K, ω, c, α, and p_0. Rather than computing, for example, $\partial\lambda/\partial p_2$ we want to calculate partials such as $\partial\lambda/\partial\omega$. The general formula for any parameter, ϕ, is

$$\frac{\partial\lambda}{\partial\phi} = \sum_{x=0}^{\omega}\left[\frac{\partial\lambda}{\partial p_x}\frac{\partial p_x}{\partial\phi} + \frac{\partial\lambda}{\partial f_x}\frac{\partial f_x}{\partial\phi}\right] \qquad (14.6)$$

The calculation of $\partial\lambda/\partial p_x$ and $\partial\lambda/\partial f_x$ comes directly from what we have done before

$$\frac{\partial\lambda}{\partial p_x} = \frac{v_{x+1}c_x}{<v,c>} \qquad (14.7)$$

and

$$\frac{\partial\lambda}{\partial f_x} = \frac{v_1 c_x}{<v,c>} \qquad (14.8)$$

The stable-age vector (c) and reproductive-value vector (v) are the associated vectors of the dominant latent root for the original transition matrix and the transposed matrix respectively. It also is possible to obtain c and v from the Euler equation (Chapter 3) or from the life-cycle graph (Chapter 4). In the computer program associated with this chapter, **MACRO.BAS**, substantial computer time is saved by calculating c_x and v_x as was shown for the Euler equation. This includes the recursive formula for the stable-age distribution, c, and the reproductive value, v

$$c_x = p_{x-1}\lambda^{-1}c_{x-1} \qquad (14.9)$$

with c_1 defined as equal to 1. Equation 14.9 with $c_1 = 1$ is the way that c_x was calculated using life-cycle graphs (Chapter 4).

The recursive formula for the reproductive value starts with the final value of v_x and works backwards, just like reversing the arrows of a life-cycle graph (Chapter 4) or using a transpose of the transition matrix (Chapter 5)

$$v_x = f_x\lambda^{-1} + p_x\lambda^{-1}v_{x+1}. \qquad (14.10)$$

Recall that the final value for p_x *always is equal to zero* and so the final reproductive value, v_ω, is

$$v_\omega = f_\omega\lambda^{-1}.$$

These recursive formulae save some computer time but can only be used with the age-structured Leslie matrix. Basically, everything is the same, so far, as has been presented in previous chapters. What is new is the fact that partials must be evaluated for the growth and survival parameters

$$\frac{\partial p_x}{\partial c} = p_x[(x/b)^c(ln(x) - ln(b) - v/c) - $$
$$((x+1)/b)^c(ln(x+1) - ln(b)-v/c)], \qquad (14.11)$$

$$\frac{\partial p_x}{\partial \omega} = p_x[ce^{v/c}/\omega^2)((x+1)/b)^{c-1}(x+1) - x(x/b)^{c-1}], \qquad (14.12)$$

$$\frac{\partial p_x}{\partial K} = \frac{\partial p_x}{\partial m_\infty} = \frac{\partial p_x}{\partial \alpha} = \frac{\partial p_x}{\partial p_0} = 0, \qquad (14.13)$$

$$\frac{\partial f_x}{\partial p_0} = m_\infty(1-e^{-K(x-\alpha)}) = m_x, \qquad (14.14)$$

$$\frac{\partial f_x}{\partial m_\infty} = p_0(1-e^{-K(x-\alpha)}) = \frac{f_x}{m_\infty}, \qquad (14.15)$$

$$\frac{\partial f_x}{\partial K} = (x-\alpha)p_0m_\infty e^{-K(x-\alpha)}, \qquad (14.16)$$

and

$$\frac{\partial f_x}{\partial \alpha} = -Kp_0m_\infty e^{-K(x-\alpha)}) \qquad \text{for } x \geq \alpha \qquad (14.17a)$$

and

$$\frac{\partial f_x}{\partial \alpha} = 0 \qquad \text{for } x < \alpha. \qquad (14.17b)$$

Elasticities for parameters, ϕ, can be calculated as

$$e_\phi = \frac{\phi}{\lambda}\frac{\partial \lambda}{\partial \phi}. \qquad (14.18)$$

The elasticities are not exclusively associated with single elements of the transition matrix and so Σe_ϕ does not have to equal 1.0.

TABLE 14.1 Age-specific Fecundity (m_x) for the Starfishes *Asterina phylactica* and *A. gibbosa*

Age (yr)	*A. phylactica*	*A. gibbosa*
0	0	0
1	0	0
2	51.0	0
3	66.8	0
4	77.4	443.0
5	84.4	541.0
6	89.2	618.2
7	92.4	678.8
8		726.4
9		764.2
10		792.2
11		816.6
12		835.0

Note: estimates based on Ebert (1996) from data presented by Emson and Crump (1979, 1984).

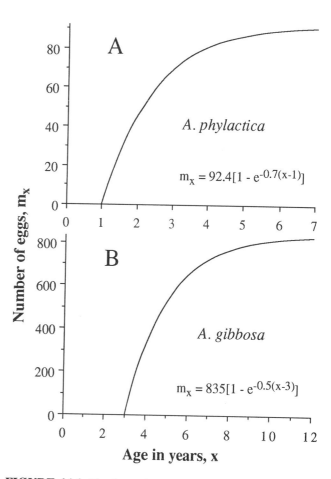

FIGURE 14.1 Numbers of eggs, m_x, as a function of age for two small starfishes, *Asterina gibbosa* and *A. phylactica*; based on data from Emson and Crump (1979, 1984) and previous analysis (Ebert 1996).

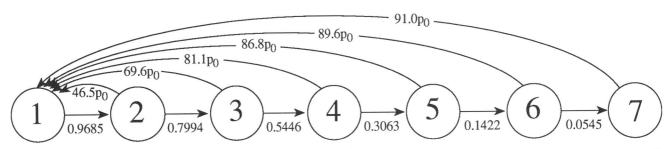

FIGURE 14.2 Life-cycle graph for *Asterina phylactica* showing transitions calculated using the survival, p_x, and fecundity, m_x, values in Table 14.1.

$$
A = \begin{pmatrix}
0 & 0.3184 & 0.4765 & 0.5550 & 0.5940 & 0.6134 & 0.6230 \\
0.9685 & 0 & 0 & 0 & 0 & 0 & 0 \\
0 & 0.7994 & 0 & 0 & 0 & 0 & 0 \\
0 & 0 & 0.5446 & 0 & 0 & 0 & 0 \\
0 & 0 & 0 & 0.3063 & 0 & 0 & 0 \\
0 & 0 & 0 & 0 & 0.1422 & 0 & 0 \\
0 & 0 & 0 & 0 & 0 & 0.0545 & 0
\end{pmatrix}
\qquad (14.19)
$$

$$
A = \begin{pmatrix}
0 & 0 & 0 & 0.2270 & 0.3647 & 0.4482 & 0.4988 & 0.5296 & 0.5482 & 0.5595 & 0.5664 & 0.5705 \\
0.9948 & 0 & 0 & 0 & 0 & 0 & 0 & 0 & 0 & 0 & 0 & 0 \\
0 & 0.9643 & 0 & 0 & 0 & 0 & 0 & 0 & 0 & 0 & 0 & 0 \\
0 & 0 & 0.9061 & 0 & 0 & 0 & 0 & 0 & 0 & 0 & 0 & 0 \\
0 & 0 & 0 & 0.8253 & 0 & 0 & 0 & 0 & 0 & 0 & 0 & 0 \\
0 & 0 & 0 & 0 & 0.7286 & 0 & 0 & 0 & 0 & 0 & 0 & 0 \\
0 & 0 & 0 & 0 & 0 & 0.6236 & 0 & 0 & 0 & 0 & 0 & 0 \\
0 & 0 & 0 & 0 & 0 & 0 & 0.5173 & 0 & 0 & 0 & 0 & 0 \\
0 & 0 & 0 & 0 & 0 & 0 & 0 & 0.4160 & 0 & 0 & 0 & 0 \\
0 & 0 & 0 & 0 & 0 & 0 & 0 & 0 & 0.3243 & 0 & 0 & 0 \\
0 & 0 & 0 & 0 & 0 & 0 & 0 & 0 & 0 & 0.2450 & 0 & 0 \\
0 & 0 & 0 & 0 & 0 & 0 & 0 & 0 & 0 & 0 & 0.1794 & 0
\end{pmatrix}
\qquad (14.20)
$$

APPLICATIONS OF MACROPARAMETER ANALYSIS

Two small starfishes, *Asterina phylactica* and *A. gibbosa*

Asterina gibbosa and *A. phylactica* are two small starfishes that have been studied in some detail (Crump 1978, Emson and Crump 1979, 1984, Strathmann *et al.* 1984). *Asterina gibbosa* lays egg masses and abandons them whereas *A. phylactica* lays egg masses that are brooded. I have used the data gathered on *Asterina gibbosa* and *A. phylactica* to assemble a life table (Ebert 1996). To estimate survival, I used the Weibull function with c = 3 for both *Asterina* species in order to have survival rates decrease as animals increased in size, which is what Emson and Crump contend. I selected ω = 7 years for *A. phylactica* and ω = 12 years for *A. gibbosa* based on the likely growth curves and age to 95% of maximum size (Ebert 1996).

Egg number, m_x, is a function of age (Table 14.1) with ages, α, when gonads begin to develop of 1 and 3 for *A.*

phylactica and *A. gibbosa* respectively. Results of nonlinear regression of egg number, m_x, as a function of age yielded $K = 0.7$ yr^{-1} for *A. phylactica* and $K = 0.5$ yr^{-1} for *A. gibbosa* (Figure 14.1). Age-specific fecundity and survival were assembled as life-cycle graphs to estimate p_0 using Equation 14.5. The calculation of p_0 using Equation 14.5 is shown in detail in Table 14.2. The calculation of

$$p_0 m_{x+1} \prod_{i=1}^{x} p_i$$

is the same as forming the characteristic equation from a life-cycle graph. The life-cycle graph for *Asterina phylactica* is shown in Figure 14.2 which includes p_x and m_x for ages ≥ 1, but p_0 is left as an unknown.

When p_0 is used in Figure 14.2, matrix **A** is produced (Equation 14.19) and the dominant latent root is 1.0 as expected. A similar procedure was used to create the appropriate matrix for *Asterina gibbosa* (Equation 14.20)

The analysis of the two *Asterina* species proceeds by assembling the stable-age and reproductive-value vectors, **c** and **v** (Table 14.3) and the partials for matrix elements, $\partial\lambda/\partial a_{ij}$ as shown in Chapter 7. The formulae are again given in Equations 14.7 and 14.8 and the results for the two starfishes are shown in Tables 14.4 and 14.5. Equations 14.12 to 14.17 were used to calculate the partials of matrix elements with respect to the growth and survival parameters and then these were combined with the partials as shown in Equation 14.6 to show the change in fitness with respect to changes in the parameters (Table 14.6) and comparison of the parameter elasticities for the two *Asterina* species is shown in Figure 14.3.

The pattern of elasticities is similar for the two *Asterina* species. The largest elasticities are associated with survival and life span, c and ω. The smallest values are for age at which gonads begin to develop, α, and the growth-rate constant K. Other than age at first reproduction, the absolute value of the elasticities for *A. phylactica* are larger than the elasticities for *A. gibbosa*. In general, small changes in parameters for *A. gibbosa* will have less of an effect on λ than similar changes in *A. phylactica*. The sum of the absolute value of elasticities, $\Sigma|e_\phi| = 1.22$ for *A. gibbosa* and $\Sigma|e_\phi| = 1.72$ for *A. phylactica*. On this basis, *A. phylactica* could be considered to be evolutionarily more volatile.

The calculated elasticities are reasonably accurate. If changes are made in parameters, one at a time, the changes in λ are approximately correct. In Table 14.7, most of the parameters were changed by 10%; however, both ω and α were changed by integer values. For both of these parameters the increment was by 1 so ω was changed from 6 to 7 and α was changed from 1 to 2. These changes are 16.7% and 100% respectively.

TABLE 14.2 Intermediate Terms Needed to Estimate p_0 for *Asterina phylactica*

x	m_x	p_x	$\prod_{i=1}^{x} p_i$	$m_{x+1}\prod_{i=1}^{x} p_i$
1	0	0.9685	0.9685	$45.0515p_0$
2	46.5155	0.7994	0.7743	$53.9000p_0$
3	69.6144	0.5446	0.4217	$34.1933p_0$
4	81.0850	0.3063	0.1292	$11.2082p_0$
5	86.7812	0.1422	0.0184	$1.6453p_0$
6	89.6098	0.0545	0.0010	$0.0910p_0$
7	91.0144		0.0	
Σ				$146.0893p_0$

Note: $146.0893p_0 = 1$ or $p_0 = 1/146.0893 = 0.0068$ yr^{-1}.

TABLE 14.3 Stable-Age, c, and Reproductive-Value, v, Vectors for *Asterina gibbosa* and *A. phylactica*

Age	*Asterina gibbosa* c	v	*Asterina phylactica* c	v
1	1	1	1	1
2	0.9948	1.0052	0.9685	1.0325
3	0.9593	1.0424	0.7743	0.8933
4	0.8693	1.1504	0.4217	0.7652
5	0.7174	1.1189	0.1292	0.6860
6	0.5227	1.0351	0.0184	0.6473
7	0.3260	0.9412	0.0010	0.6230
8	0.1686	0.8552		
9	0.0701	0.7828		
10	0.0227	0.7233		
11	0.0056	0.6687		
12	0.0010	0.5705		
<v,c>	5.8704		3.1154	

Note: analysis based on data of Emson and Crump (1979, 1984).

TABLE 14.4. Partials of Matrix Elements for *Asterina gibbosa*;

Age(x)	Element, a_{ij}	$\partial\lambda/\partial p_x$	Element, a_{ij}	$\partial\lambda/\partial f_x$
1	$a_{2,1}$	0.1712	$a_{1,1}$	0.1703
2	$a_{3,2}$	0.1767	$a_{1,2}$	0.1695
3	$a_{4,3}$	0.1880	$a_{1,3}$	0.1634
4	$a_{5,4}$	0.1657	$a_{1,4}$	0.1481
5	$a_{6,5}$	0.1265	$a_{1,5}$	0.1222
6	$a_{7,6}$	0.0838	$a_{1,6}$	0.0890
7	$a_{8,7}$	0.0475	$a_{1,7}$	0.0555
8	$a_{9,8}$	0.0225	$a_{1,8}$	0.0287
9	$a_{10,9}$	0.0086	$a_{1,9}$	0.0119
10	$a_{11,10}$	0.0026	$a_{1,10}$	0.0039
11	$a_{12,11}$	0.0005	$a_{1,11}$	0.0010
12			$a_{1,12}$	0.0002

Note: data from Emson and Crump (1979, 1984).

**TABLE 14.5 Partials for Matrix Elements,
$\partial\lambda/\partial a_{ij}$, for *Asterina phylactica***

Age(x)	Element, a_{ij}	$\partial\lambda/\partial p_x$	Element, a_{ij}	$\partial\lambda/\partial f_x$
1	$a_{2,1}$	0.3314	$a_{1,1}$	0.3210
2	$a_{3,2}$	0.2777	$a_{1,2}$	0.3109
3	$a_{4,3}$	0.1902	$a_{1,3}$	0.2485
4	$a_{5,4}$	0.0929	$a_{1,4}$	0.1354
5	$a_{6,5}$	0.0268	$a_{1,5}$	0.0415
6	$a_{7,6}$	0.0037	$a_{1,6}$	0.0059
7			$a_{1,7}$	0.0003

Note: analysis based on data gathered by Emson and Crump (1979, 1984).

**TABLE 14.6 Calculated Partials of λ and Elasticities
with Respect to Growth and Survival Parameters**

ϕ	*Asterina gibbosa*		*Asterina phylactica*	
	$\partial\lambda/\partial\phi$	elasticity	$\partial\lambda/\partial\phi$	elasticity
c	0.0857	0.2571	0.1189	0.3568
ω	0.0359	0.3951	0.0772	0.4629
K	0.1648	0.0824	0.2136	0.1495
m_∞	0.0002	0.1703	0 0035	0.3210
α	-0.0477	-0.1430	-0.1040	-0.1040
p_0	246.5494	0.1703	46.1822	0.3210

Note: c and ω are from the Weibull function; K is the growth-rate constant and m_∞ is maximum reproduction which in this case is number of eggs; α is the age at which gonads begin to develop and is one year *before* the age at first reproduction; p_0 is first-year survival.

TABLE 14.7 Test of Elasticity Values from *Asterina phylactica*

ϕ	Original value	Elasticity	% change	New value	New λ	Expected λ
c	3	0.3568	10	3.3	1.029	1.036
ω	6	0.4629	16.67	7	1.071	1.077
K	0.7	0.1495	10	0.77	1.014	1.015
m_∞	91	0.3210	10	100.1	1.031	1.032
α	1	-0.1040	100	2	0.845	0.896
p_0	0.00695	0.3210	10	0.0076	1.031	1.032

Note: the new value of λ is the one determined by using the program **MACRO.BAS**; expected value of λ is the value calculated by using the elasticity value in Table 14.6.

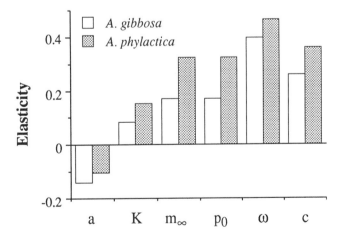

FIGURE 14.3 Elasticities for growth and survival parameters for two small starfishes, *Asterina gibbosa* and *A. phylactica*; data gathered by Emson and Crump (1979, 1984); analysis presented in Ebert (1996).

Lizard Species of the Genus *Sceloporus*

Life history attributes of the lizard genus *Sceloporus* have been studied by Tinkle and Ballinger (1972) for *S. undulatus* and by Ballinger (1973) for *S. poinsetti* and *S. jarrovi*. Values of l_x and m_x of the three species are shown in Table 14.8. Survival was modeled by separating first-year survival, p_0, from subsequent survival and the l_x columns were adjusted to start at 1.00 at age 1, which means starting at 1 and dividing each l_x value in the column by l_1 (Table 14.9). Fecundity as a function of age used the Brody-Bertalanffy model (Equation 14.3) and parameter estimates are shown in Table 14.10; m_∞ was fixed at the maximum number of female offspring and age when gonads begin to develop, α, was taken as one year before first reproduction except for *S. jarrovi* where 65% matured during the first year and so α was selected to be 0.5.

TABLE 14.8 Survival, l_x, and Fecundity, m_x, of Three *Sceloporus* Species

S. undulatus			S. poinsetti			S. jarrovi	
x	l_x	m_x	x	l_x	m_x	l_x	m_x
0	1.00	0	0	1.000	0	1.000	0
1	0.14	0	1	0.131	0	0.180	1.2
1.8	0.11	7.9	2	0.059	4.5	0.086	5.25
2.8	0.03	8.9	3	0.045	6.0	0.032	5.75
3.8	0.01	8.9	4	0.027	6.5	0.012	6.00
			5	0.015	7.0		
			6	0.008	7.5		
			7	0.004	7.5		

Note: data for *S. undulatus* from Tinkle and Ballinger (1972), *S. poinsetti* and *S. jarrovi* data from Ballinger (1973); *S. undulatus* from Colorado; *S. poinsetti* from Texas and *S. jarrovi* from Arizona.

TABLE 14.9 Adjusted l_x Values to Start at Age 1 for Three Lizard Species of the Genus *Sceloporus*;

S. undulatus		S. poinsetti		S. jarrovi
Age, x	l_x	Age, x	l_x	l_x
0 (1.0)	1.000	0 (1)	1.000	1.000
0.8 (1.8)	0.786	1 (2)	0.450	0.478
1.8 (2.8)	0.214	2 (3)	0.343	0.178
2.8 (3.8)	0.071	3 (4)	0.206	0.067
		4 (5)	0.114	
		5 (6)	0.061	
		6 (7)	0.031	
c	2.00		1.27	1.66
ω	4		7	4

Note: ω fixed at one age past the observations; ages used to estimate Weibull parameters must start at age 0; actual ages are shown in parentheses; ω is the age based on the first age being age 0 and so ω = 4 is 4 time periods from the start of the table; for example, ω = 4 in the table is actually age 5.

TABLE 14.10 Reproductive Parameters, ϕ, for Three Lizard Species of the Genus *Sceloporus*

ϕ	S. undulatus	S. poinsetti	S. jarrovi
m_∞	8.9	7.5	6.0
K	2.74	0.82	0.63
α	1	1	0.5

Note: data from Tinkle and Ballinger 1972 and Ballinger 1973.

The analysis was performed for each species using the program **MACRO.BAS**. The transition matrix for *S. undulatus*, with $p_0 = 0.1399$, is shown in Equation 14.21, for *S. poinsetti*, with $p_0 = 0.2071$ in Equation 14.22 and for *S. jarrovi*, with $p_0 = 0.2456$ in Equation 14.23.

Sceloporus undulatus

$$\mathbf{A} = \begin{pmatrix} 0 & 1.1605 & 1.2354 & 1.2403 & 1.2406 \\ 0.6494 & 0 & 0 & 0 & 0 \\ 0 & 0.2738 & 0 & 0 & 0 \\ 0 & 0 & 0.1155 & 0 & 0 \\ 0 & 0 & 0 & 0.0487 & 0 \end{pmatrix}$$

(14.21)

Sceloporus poinsetti

$$\mathbf{A} = \begin{pmatrix} 0 & 0.8691 & 1.2518 & 1.4204 & 1.4947 & 1.5274 & 1.5418 & 1.5481 \\ 0.5579 & 0 & 0 & 0 & 0 & 0 & 0 & 0 \\ 0 & 0.4388 & 0 & 0 & 0 & 0 & 0 & 0 \\ 0 & 0 & 0.3876 & 0 & 0 & 0 & 0 & 0 \\ 0 & 0 & 0 & 0.3539 & 0 & 0 & 0 & 0 \\ 0 & 0 & 0 & 0 & 0.3289 & 0 & 0 & 0 \\ 0 & 0 & 0 & 0 & 0 & 0.3091 & 0 & 0 \\ 0 & 0 & 0 & 0 & 0 & 0 & 0.2928 & 0 \end{pmatrix}$$

(14.22)

Sceloporus jarrovi

$$\mathbf{A} = \begin{pmatrix} 0.3982 & 0.9008 & 1.1685 & 1.3111 & 1.3870 \\ 0.5007 & 0 & 0 & 0 & 0 \\ 0 & 0.2244 & 0 & 0 & 0 \\ 0 & 0 & 0.1226 & 0 & 0 \\ 0 & 0 & 0 & 0.0726 & 0 \end{pmatrix}$$

(14.23)

The calculated l_x and m_x schedules for the three *Sceloporus* species are shown in Table 14.11 and the stable-age and reproductive-value vectors are shown in Table 14.12. Partials with respect to individual matrix elements, a_{ij}, are shown in Table 14.13 and the sensitivities of λ with respect to parameters, ϕ, are shown in Table 14.14. Comparison of the parameter elasticities for the three *Sceloporus* species is shown in Figure 14.4.

The pattern of elasticities is the same for all three species although the magnitude of values differs. All three species have highest elasticities associated with maximum reproduction, m_∞, and first-year survival, p_0. Evolutionary volatility, $\Sigma|e_\phi|$, also differs across the species. *Sceloporus undulatus* and *S. poinsetti* are similar with $\Sigma|e_\phi|$ values of 1.5 - 1.6 but *S. jarrovi* has a value of 2.2. Small changes in parameters α, K, m_∞, and p_0 would have a greater effect on population growth of *S. jarrovi* than on the other two lizard species. On the other hand, *S. jarrovi* would be less sensitive to changes associated with survival, ω and c of the Weibull function than would *S. undulatus* and *S. poinsetti*.

TABLE 14.11 Calculated Survival, l_x, and Fecundity, m_x, Schedules for Three Lizard Species of the Genus *Sceloporus*. Using Parameters Shown in Tables 14.9 and 14.10

Age, x	*S. undulatus* l_x	m_x	*S. poinsetti* l_x	m_x	*S. jarrovi* l_x	m_x
0	1	0	1	0	1	0
1	0.1394	0	0.2071	0	0.2456	1.621
2	0.0905	8.325	0.1155	4.197	0.1230	3.668
3	0.0248	8.863	0.0507	6.045	0.0276	4.758
4	0.0029	8.898	0.0196	6.859	0.0034	5.338
5	0.0001	8.900	0.0069	7.218	0.0002	5.648
6			0.0023	7.376		
7			0.0007	7.445		
8			0.0002	7.476		

TABLE 14.12 Stable-age (c) and Reproductive-value (v) Vectors for Three Lizard Species of the Genus *Sceloporus*

Age, x	*S. undulatus* c	v	*S. poinsetti* c	v	*S. jarrovi* c	v
1	1	1	1	1	1	1
2	0.6494	1.5399	0.5579	1.7923	0.5007	1.2019
3	0.1778	1.3856	0.2448	2.1041	0.1124	1.3416
4	0.0205	1.3007	0.0949	2.1990	0.0138	1.4118
5	0.0010	1.2406	0.0336	2.2000	0.0010	1.3870
6			0.0110	2.1441		
7			0.0034	1.9951		
8			0.0010	1.5481		
<v,c>	2.27435		2.82971		1.77343	

Note: ages start at 1 and so the c columns all end with 0.001 by definition of the manner in which ω was calculated.

TABLE 14.13. Partials, $\partial\lambda/\partial p_x$ and $\partial\lambda/\partial f_x$, of Individual Matrix Elements, a_{ij}, for Three Lizard Species of the Genus *Sceloporus*

Species	Age(x)	Element, a_{ij}	$\partial\lambda/\partial p_x$	Element, a_{ij}	$\partial\lambda/\partial f_x$
S. undulatus	1	$a_{2,1}$	0.6771	$a_{1,1}$	0.4397
	2	$a_{3,2}$	0.3956	$a_{1,2}$	0.2855
	3	$a_{4,3}$	0.1017	$a_{1,3}$	0.0782
	4	$a_{5,4}$	0.0112	$a_{1,4}$	0.0090
	5			$a_{1,5}$	0.0004
S. poinsetti	1	$a_{2,1}$	0.6334	$a_{1,1}$	0.3534
	2	$a_{3,2}$	0.4149	$a_{1,2}$	0.1972
	3	$a_{4,3}$	0.1903	$a_{1,3}$	0.0865
	4	$a_{5,4}$	0.0738	$a_{1,4}$	0.0335
	5	$a_{6,5}$	0.0255	$a_{1,5}$	0.0119
	6	$a_{7,6}$	0.0078	$a_{1,6}$	0.0039
	7	$a_{8,7}$	0.0019	$a_{1,7}$	0.0012
	8			$a_{1,8}$	0.0003
S. jarrovi	1	$a_{2,1}$	0.6777	$a_{1,1}$	0.5639
	2	$a_{3,2}$	0.3788	$a_{1,2}$	0.2824
	3	$a_{4,3}$	0.0895	$a_{1,3}$	0.0634
	4	$a_{5,4}$	0.0108	$a_{1,4}$	0.0078
	5			$a_{1,5}$	0.0006

Note: data from Tinkle and Ballinger (1972) and Ballinger (1973).

TABLE 14.14 Calculated Partials of λ with Respect to Growth and Survival Parameters, φ, Together with Elasticities

φ	*Sceloporus undulatus* ∂λ/∂φ	elasticity	*Sceloporus poinsetti* ∂λ/∂φ	elasticity	*Sceloporus jarrovi* ∂λ/∂φ	elasticity
c	0.1501	0.3002	0.1563	0.1985	0.0982	0.1630
ω	0.0837	0.3346	0.0471	0.3299	0.0631	0.2525
K	0.0237	0.0649	0.2037	0.1670	0.5987	0.3772
m_∞	0.0494	0.4397	0.0471	0.3534	0.0940	0.5639
a	-0.0638	-0.0638	-0.1363	-0.1363	-0.4969	-0.2485
p_0	3.1543	0.4397	1.7066	0.3534	2.2960	0.5639
$\sum|e_\phi|$		1.6429		1.5385		2.1690

Note: c and ω are from the Weibull function; K is the growth-rate constant and m_∞ is maximum value of m_x; a is the age at which gonads begin to develop and is one year *before* the age at first reproduction; p_0 is first-year survival.

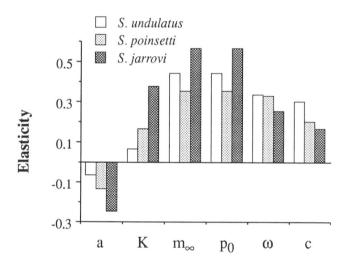

FIGURE 14.4 Elasticity values for growth and survival parameters for three lizard species of the genus *Sceloporus*; data from Tinkle and Ballinger (1972) and Ballinger (1973).

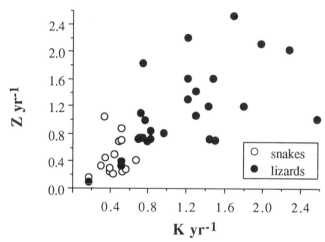

FIGURE 14.5 Relationship between Z of the decaying exponential survivorship curve ($N_t = N_0 e^{-Zt}$) and K of the Brody-Bertalanffy model; data from Shine and Charnov (1992).

GENERAL COMMENTS

In the analysis that has been presented for starfishes and lizards, just two models have been used: the Weibull and the Brody-Bertalanffy and so the analysis is rather procrustean. As shown in previous chapters, there is not a single model for survival and a single model for growth but rather an uncertain number of such models.   If comparisons were to be made across groups of sea urchins, then the Tanaka function should be used rather than Brody-Bertalanffy. Mammal survival possibly should be modeled using the 5-parameter equation of Siler (1979) rather than the 2-parameter Weibull. Differences among groups of organisms means that a program such as **MACRO.BAS** has to be expanded to accommodate different patterns; however, having said all this, an important question is whether such differences really matter in understanding the evolution of

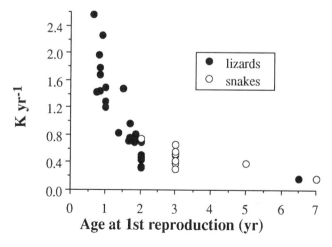

FIGURE 14.6 Relationship between the growth-rate constant K of the Brody-Bertalanffy model and the age at first reproduction for lizards and snakes; data from Shine and Charnov (1992).

life histories. Is it not possible that the broad and important patterns can be understood with just simple models such as exponential survival with just one parameter and Brody-Bertalanffy with 2 parameters? Maybe; the consequences of more complex models, however, have not been explored.

One additional factor in interpreting the elasticities can not be explored in detail here and is a good place to stop this book because it points in a direction for a great deal of further work. There are trade-offs among the parameters and a growing interest in exploring the implications of covariance between survival and fecundity for a wide range of organisms (e.g. Pease and Bull 1988, Niewiarowski and Dunham 1994, Weiser 1994, Boyd *et al.* 1995, van Tienderen 1995, Elle 1996).

Trade-offs are evident by patterns that emerge in comparative studies and one approach to the exploration of patterns has been an examination of the relationship between growth and survival by using the decaying exponential to model survival and the Brody-Bertalanffy equation to model growth. Charnov (1993) has provided a number of data sets that show these relationships for certain groups of vertebrates and invertebrates. For example, Shine and Charnov (1992) show a number of relationships among growth and survival parameters for snakes and lizards. A scatter diagram of Z *vs.* K (Figure 14.5) shows a positive relationship, which means that the faster maximum size is approached (increasing K), the higher the mortality rate (increased Z).

Other patterns exist and Figure 14.6 shows a relationship between age at first reproduction and the growth-rate constant K. In Figure 14.6 K is the constant for growth of the entire body rather than of m_x as I have been using in this chapter. The pattern shown in Figure 14.6 suggests that rapid growth to maximum size is associated with early first reproduction and delay in first reproduction is associated with a longer period of growth to a maximum size.

The relationship between survival and growth (Figure 14.5) certainly makes sense as a trade-off but the relationship between growth and age at first reproduction (Figure 14.6) seems less like a trade-off and more like just a correlation with a possible link through survival. There are other patterns shown by Charnov (1993) and clearly all, in some way, are related to λ. Examination and interpretation of patterns created using different growth and survival models related to fitness has never been done but it would be very interesting and, I suspect, will provide new insights into the evolution of life histories.

PROBLEMS

In the following data sets, determine growth and survival parameters using the Weibull and Brody-Bertalanffy models. For the Weibull be sure to start with age 1 and save p_0 as a separate parameter. The Brody-Bertalanffy model should be used for m_x. In a nonlinear regression, fix m_∞ as the maximum observed value for m_x and fix the age when gonads begin to develop, a, as one time unit before the age at first reproduction. Assume that λ = 1 and do an elasticity analysis of the parameters.

1. The following data set for waterbuck (*Kobus ellipsiprymnus*) in the Umfolozi Game Reserve in South Africa (Melton 1983) was introduced in Chapter 2. Here is the life table and fecundity schedule (Table 14.15). Do the analysis two ways. First use age at which gonads begin to develop as 1 and then redo the analysis with the age equal to 1.5.

TABLE 14.15 Life Table for Waterbuck

Age	l_x	m_x	Age	l_x	m_x
0	1000	0	7	125.5	0.42
1	193	0	8	111.3	0.42
2	180.8	0.25	9	89.0	0.42
3	180.8	0.42	10	64.6	0.42
4	174.2	0.42	11	17.9	0.42
5	166.1	0.42	12	0	
6	145.8	0.42			

Note: age in years; data from Melton (1983).

2. Use the data of Strijbosch and Creemers (1988) for a lizard, *Lacerta vivipara*, which were gathered from 1976-1982 at Bergen in the Netherlands (Table 14.16). Compare your results with the analysis of *Sceloporus* spp as presented in this chapter.

TABLE 14.16 Life Table for *Lacerta vivipara*

x	l_x	m_x	x	l_x	m_x
0	1000	0.00	4	57	4.88
1	424	0.08	5	10	6.50
2	308	2.94	6	7	6.50
3	158	4.13	7	2	6.50

Note: data from Strijbosch and Creemers (1988).

3. Turner *et al.* (1969) present an estimated age-specific survival and a fecundity schedule for a lizard *Xantusia vigilis* (Table 14.17). This problem was introduced in chapter 4. They begin their table at the age of first reproduction, 3 years, and so it in necessary to give a bit of thought to how to separate-out p_0.

TABLE 14.17 Life Table for *Xantusia vigilis*

Age	l_x	m_x
3	0.45	0.5
4	0.32	0.5
5	0.23	0.5
6	0.16	0.5
7	0.08	0.5
8	0.02	0.5
9	0.001	0.5

Note: data from Turner *et al.* (1969).

4. Survival and reproductive data (Table 14.18) for a freshwater snail, *Viviparus georgianus*, have been taken from Buckley (1986) and were analyzed in Chapter 7. Note that the reproductive output increases with size and survival rate decreases with age. Compare your analysis with results for *Asterina* spp. in this chapter.

TABLE 14.18 Life Table for a Freshwater Snail
Viviparus georgianus

Age (x)	p_x	m_x
0	0.321	0
1	0.639	0
2	0.565	1.24
3	0.307	4.80
4	0.0	8.35

Note: data from Buckley (1986)

BASIC PROGRAMS

MACRO.BAS uses parameters from the Weibull and Brody-Bertalanffy models to create a transition matrix with an option of estimating p_0 so that the population growth-rate, λ, is equal to 1. Parameter values that are required are m_∞, K, and α for increase of m_x with age and c and ω for survival as a function of age. These parameters must be estimated from some other program such as **SIMPLEX.BAS** or a commercial software package that has a nonlinear regression module. In general, determination of ω and c will be done starting with age 1; however, for purposes of estimating parameters, it is necessary to reassign ages so the first age is 0. In general, all that this requires is subtracting 1 from all ages.

```
10 dim a(32,31),b(32),f(31,31),q(961),
   c(31),p(33),v(31,31),lx(31)
20 dim mx(31),px(31),fx(31),s(33)
30 print : print
```

```
40 print "Program for creating a
   transition matrix using
   macroparameters." : print
50 print "Original version developed
   using Applesoft BASIC"
60 print "Modified for Microsoft BASIC,
   May 1988, and revised 1996"
70 print:print "Another fine product
   from Cornered Rat Software©"
80 print : print : input "Do you want to
   continue? (Y/N) ";f$
90 if f$ = "Y" or f$ = "y" then goto 130
100 if f$ = "N" or f$ = "n" then goto
    1550
110 goto 80
130 print : print : print "Reproductive
    information" : print : print
140 input "maximum number of offspring
    (m∞) ";mi : print : input "growth
    rate constant (K) ";k1
150 print : input "age when gonads begin
    to develop (a) ";a : print
160 input "Do you want to estimate p(0)
    so lambda = 1.0? (y/n) ";f$
170 if f$ = "Y" or f$ = "y" then goto
    190
180 input "Survival rate for the first
    time period (p0) ";p0
190 print : print : print ;"    Weibull
    parameters" :
200 input "age at which lx is 0.001 (w):
    ";w : print : input "shape parameter
    (c): ";c
210 input "Do you want results saved in
    a file? (y/n) ";p$
220 if p$ <> "Y" and p$ <> "y" then goto
    250
230 input "File name for output: ";g$
240 open g$ for output as #1
250 n = w+1
260 for i = 0 to n+1 : for j = 0 to n :
    a(i,j) = 0 : next j : lx(i) = 0 :
    px(i) = 0 : fx(i) = 0 : mx(i) = 0 :
    next i
270 v = log(-log(0.001))
280 b = exp(log(w)-1/c*v)
290 for i = 1 to n
300 lx(i) = exp(-(((i-1)/b)^c))
310 px(i) = ((i-1)/b)^c-((i)/b)^c
320 px(i) = exp(px(i))
330 a(i+1,i) = px(i) : next i
340 for i = 0 to n
350 mx(i) = mi*(1-exp(-k1*(i-a)))
360 if mx(i) < 0 then mx(i) = 0
```

```
370 next i
380 if f$ <> "Y" and f$ <> "y" then goto
    440
390 s(1) = px(1)
400 for i = 2 to n : s(i) = s(i-1)*px(i)
    : next i
410 p0 = mx(1)
420 for i = 2 to n : p0 = p0+mx(i)*s(i-
    1) : next i
430 p0 = 1/p0 : print "p0 = ";p0
440 lx(0) = 1 : for i = 1 to n : lx(i) =
    lx(i)*p0 : next i
450 print : print "x"," lx","mx"
460 for i = 0 to n : print i,lx(i),mx(i)
    : next i
465 print:print
470 if p$ <> "Y" and p$ <> "y" then goto
    520
480 print #1,"p0 = ",p0
490 print #1,"x","lx","mx"
500 for i = 0 to n : print
    #1,i,lx(i),mx(i) : next i
510 print #1," "
520 for i = 1 to n : a(1,i) = p0*mx(i) :
    fx(i) = a(1,i) : next i
530 for i = 1 to n : for j = 1 to n
540 print a(i,j), : next j : print
550 next i
560 if p$ <> "Y" and p$ <> "y" then goto
    600
570 for i = 1 to n : for j = 1 to n
580 print #1,a(i,j), : next j : print
    #1," "
590 next i
600  print : input "<return> to
    continue...";w$
610  print "Coefficients of the
    characteristic equation" : print
620 if p$ <> "Y" and p$ <> "y" then goto
    640
630 print #1,"Coefficients of the
    characteristic equation" : print
    #1," "
640 q(1) = 1 : q(n) = 0
650 for k = n to 1 step -1
660 q(k+1) = -a(1,k)
670 if k = 1 then goto 710
680 for i = 1 to k-1
690 q(k+1) = q(k+1)*a(i+1,i)
700 next i
710 next k
720 for i = 1 to n+1 : print q(i) : next
    i
725 if p$<>"Y" and p$<>"y" then goto 740
```

```
730 for i = 1 to n+1 : print #1,q(i) :
    next i
740 for i = 1 to 31 : p(i) = 0 : b(i) =
    0 : next i
750 for i = 1 to n+1 : p(i) = q(n+2-i)
760 next i
770 for i = 1 to n+1 : b(i) = p(i+1)*i :
    next i : x = 10
780 q = 0
790 u = 1 : f1 = 0 : f0 = 0
800 q = q+1 : for i = 1 to n+1 : f0 =
    f0+p(i)*u : f1 = f1+b(i)*u : u = u*x
    : next i
810 u = x-f0/f1 : if abs(x-u) < 1E-08
    then goto 850
820 x = u
830 if q > 500 then goto 890
840 goto 790
850 print : print "lambda = ",x
860 if p$ <> "Y" and p$ <> "y" then goto
    930
870 print #1," " : print #1," lambda =
    ",x
880 goto 930
890 print "500 iterations completed:"
900 print "Root = ";x;" function = ";f0
910 if p$ <> "Y" and p$ <> "y" then goto
    930
920 print #1,"Root = ";x;" function =
    ";f0
930 for i = 0 to n+2 : b(i) = 0 : q(i) =
    0 : next i
940 b(1) = 1 : for i = 1 to n-1 : b(i+1)
    = b(i)*px(i)/x : next i
950 print : print "Stable-age
    distribution" : print
960 for i = 1 to n : print b(i) : next i
970 for i = n to 1 step -1
980 q(i) = fx(i)/x+px(i)/x*q(i+1)
990 next i
1000 print : print "Reproductive-value
    vector"
1010 for i = 1 to n : print i;tab
    (10);q(i) : next i
1020 if p$ <> "Y" and p$ <> "y" then
    goto 1050
1030 print #1," " : print #1,"
    Age","Stable-age
    distribution","Reproductive value
    vector"
1040 for i = 1 to n : print
    #1,i,b(i),q(i) : next i
1050 print "Partials and elasticities" :
    print
```

```
1060 sp = 0
1065 for ,i = 1 to 25 : p(i) = 0 : c(i) =
     0 : next i
1070 for i = 1 to n : sp = sp+q(i)*b(i)
     : next i
1080 print : print "<v,c> = ";sp : print
1090 for i = 1 to n : p(i) =
     q(i+1)*b(i)/sp
1100 c(i) = q(1)*b(i)/sp : next i
1110 print
     "d(lambda)/d(p)","d(lambda)/d(f)"
1120 for i = 1 to n : print p(i),c(i) :
     next i : print : print
1130 if p$ <> "Y" and p$ <> "y" then
     goto 1180
1140 print #1,"Partials and
     elasticities" : print #1," "
1150 print #1,"<v,c> = ",sp : print #1,"
     "
1160 print
     #1,"d(lambda)/d(p)","d(lambda)/d(f)"
1170 for i = 1 to n : print #1,p(i),c(i)
     : next i : print #1," "
1180 print : input " <return> to
     continue..";w$
1190 p2 = c*exp(v/c)/w/w : pw =
     (1/b)^(c-1)
1200 pc = -((1/b)^c)*(-log(b)-v/c)*px(1)
1210 pw = px(1)*p2*pw
1220 fk = 0 : fm = 0 : fa = 0 : fl = 0 :
     fc = 0 : pk = 0 : pm = 0
1225 pa = 0 : pp = 0 : fc = 0 : rw = 0:
     rc=0
1230 for i = 1 to n
1240 pc = (i/b)^c*(log(i)-log(b)-v/c)
1250 pc = pc-((i+1)/b)^c*(log(i+1)-
     log(b)-v/c)
1260 pc = pc*px(i)
1270 rc = rc+p(i+1)*pc
1280 pw = ((i+1)/b)^(c-1)*(i+1)-
     (i/b)^(c-1)*i : pw = pw*px(i+1)*p2
1290 rw = rw+p(i+1)*pw
1300 next i
1310 fk = 0 : fm = 0 : fa = 0 : fp = 0
1320 for i = 1 to n
1330 if mx(i) = 0 then goto 1380
1340 fk = fk+p0*(i-a)*mi*exp(-k1*(i-
     a))*c(i)
1350 fm = fm+fx(i)/mi*c(i)
1360 fa = fa-p0*mi*k1*exp(-k1*(i-
     a))*c(i)
1370 fp = fp+mx(i)*c(i)
1380 next i : print : print
1390 print "d(lambda)/d(c) = ";rc,"e(c)
     = ";rc*c/x
1400 print "d(lambda)/d(w) = ";rw,"e(w)
     =";rw*w/x
1410 print "d(lambda)/d(K) = ";fk,"e(K)
     = ";fk*k1/x
1420 print "d(lambda)/d(m∞) =
     ";fm,"e(m∞) = ";fm*mi/x
1430 print "d(lambda)/d(a) = ";fa,"e(a)
     = ";fa*a/x
1440 print "d(lambda)/d(p0) =
     ";fp,"e(p0) = ";fp*p0/x
1445 if p$<>"Y" and p$<>"y" then goto
     1530
1450 print #1," "
1460 print #1,"d(lambda)/d(c) =
     ";rc,"e(c) = ";rc*c/x
1470 print #1,"d(lambda)/d(w) =
     ";rw,"e(w) =";rw*w/x
1480 print #1,"d(lambda)/d(K) =
     ";fk,"e(K) = ";fk*k1/x
1490 print #1,"d(lambda)/d(m∞) = ";fm,"
     e(m∞) = ";fm*mi/x
1500 print #1,"d(lambda)/d(a) =
     ";fa,"e(a) = ";fa*a/x
1510 print #1,"d(lambda)/d(p0) =
     ";fp,"e(p0) = ";fp*p0/x
1520 close #1
1530 input "<return> to continue...";w$
1540 goto 80
1550 end
```

Bibliography

Åberg, P. 1992a. A demographic study of two populations of the seaweed *Ascophyllum nodosum*. Ecology 73: 1473-1487.

Åberg, P. 1992b. Size-based demography of the seaweed *Ascophyllum nodosum* in stochastic environments. Ecology 73: 1488-1501.

Abramowitz, M. and I.A. Stegun [eds.] 1972. *Handbook of Mathematical Functions with Formulas, Graphs, and Mathematical Tables (10th Edition)*. National Bureau of Standards, Washington, DC. Reprinted by Dover Publications, Inc., New York. 1046pp.

Ainley, D.G., R.E. LeResche, and W.J.L. Sladen 1983. *Breeding Biology of the Adélie Penguin*. U. Calif. Press, Berkeley, CA 240pp.

Alvarez-Buylla, E.R. and M. Slatkin 1991. Finding confidence limits on population growth rates. Trends in Ecology and Evolution 6: 221-224.

Alvarez-Buylla, E.R. and M. Slatkin 1993. Finding confidence limits on population growth rates: Monte Carlo test of a simple analytic method. Oikos 68: 273-282.

Arambasic, M. ,M. Pasic, L. Kojic, A. Kalauzi and V. Markovic 1987. The growth of pond snail *Lymnaea stagnalis* L. in laboratory conditions. Zoologische Jarbucher Abteilung für Anatomie und Ontogenie der Tiere 116:119-128.

Astheimer, H. 1986. A length class model of the population dynamics of the Antarctic krill *Euphausia superba* Dana. Polar Biology 6:227-232.

Astheimer, H., H. Krause, and S. Rakusa-Suszczewski. 1985. Modelling individual growth of the Antarctic krill *Euphausia superba* Dana. Polar Biology 4: 65-73.

Bachelet, G. 1980. Growth and recruitment of the tellinid bivalve *Macoma balthica* at the southern limit of its geographical distribution, the Gironde estuary (SW France) Marine Biology 59: 105-117.

Bairstow, L. 1914. Appendix to *Investigations Relating to the Stability of the Aeroplane*, Advisory Committee for Aeronautics. Reports and Memoranda, No.154, London, 51-64 (Cited by Grattan-Guinness 1994)

Baker, S.L. 1973. Growth of the red sea urchin *Strongylocentrotus franciscanus* in two natural habitats. MS Thesis, San Diego State University, San Diego, California, 83pp.

Baker, T.T., R. Lafferty and T.J. Quinn II. 1991. A general growth model for mark-recapture data. Fisheries Research 11: 257-281.

Ballinger, R.E. 1973. Comparative demography of two viviparous iguanid lizards (*Sceloporus jarrovi* and *Sceloporus poinsetti*) Ecology 54: 269-283.

Barkalow, F.S., R.B. Hamilton and R.F. Soots. 1970. The vital statistics of an unexploited gray squirrel population. Journal of Wildlife Management 34: 489-500.

Barry, J.P. and M.J. Tegner 1989. Inferring demographic processes from size-frequency distributions: simple models indicate specific patterns of growth and mortality. Fishery Bulletin, U. S. 88: 13-19.

Bartlett, N.R. and J.C. Noble. 1985. The population biology of plants with clonal growth. III. Analysis of tiller mortality in *Carex arenaria*. Journal of Ecology 73:1-10.

Benton, T.G., A. Grant and T.H. Clutton-Brock 1995. Does environmental stochasticity matter? Analysis of red deer life-histories on Rum. Evolutionary Ecology 9: 559-574.

Bernardelli, H. 1941. Population waves. Journal of the Burma Research Society 31: 1-18.

Bertalanffy, L. von. 1934. Untersuchungen über die Gesetzlichkeit des Wachstums. I. Teil Allgemeine Grundlagen der Theorie: mathematische und physiologische Gesetzlichkeiten des Wachstums bei Wassertieren. Wilhelm Roux' Archiv für Entwicklungsmechanik der Organismen 131: 613-652.

Bertalanffy, L. von. 1938. A quantitative theory of organic

growth. Human Biology 10:181-213.

Beverton, R.J. and S.J. Holt 1956 A review of methods for estimating mortality rates in exploited fish populations, with special reference to sources of bias in catch sampling. Conseil Permanent International pour l'Exploration de la Mer. Rapports et Procè-Verbaux des Réunions 140 (part I): 67-83.

Bhattacharya, C.G. 1967. A simple method of resolution of a distribution into Gaussian components. Biometrics 23: 115-135.

Bierzychudek, P. 1982. The demography of jack-in-the-pulpit, a forest perennial that changes sex. Ecological Monographs 52: 335-351.

Birch, L.C. 1948. The intrinsic rate of natural increase of an insect population. Journal of Animal Ecology 17: 15-26.

Bliss, C.I. 1967. *Statistics in Biology. Vol. 1*. McGraw-Hill Book Co., New York. 558p

Bowman, R.S. and J.R. Lewis 1977. Annual fluctuations in the recruitment of *Patella vulgata* L. Journal of the Marine Biological Association of the U.K. 57: 793-815.

Boyce, M.S. 1977. Population growth with stochastic fluctuations in the life table. Theoretical Population Biology 12: 366-373.

Boyd, I.L., J.P. Croxall, N.J. Lunn and K. Reid 1995. Population demography of Antarctic fur seals: the costs of reproduction and implications for life-histories. Journal of Animal Ecology 64: 505-518.

Brewer, J.W. 1976. The analytical solution of a time-varying logistic growth equation. IEEE Transactions on systems, man, and cybernetics , May 1976: 384-386.

Brock, R.D. and D. Thissen 1976. Fitting multicomponent models for growth in stature. Proceedings of the 9th International Biometrics Conference 1: 431-442.

Brody, S. 1927. Growth and development with special reference to domestic animals III. Growth rates, their evaluation and significance. University of Missouri Agricultural Experiment Station Research Bulletin 97. Columbia, Missouri 70pp.

Brody, S. 1945. *Bioenergetics and Growth with Special Reference to the Efficiency Complex in Domestic Animals*. Reinhold Publ. Corp., New York, 1023p

Brousseau, D.J. and J.A. Baglivo 1988. Life tables for two field populations of soft-shell clam, *Mya arenaria*, (Mollusca: Pelecypoda) from Long Island Sound. Fishery Bulletin, U. S. 86: 567-579

Brown, J.R. 1988. Multivariate analyses of the role of environmental factors in seasonal and site-related growth variation in the Pacific oyster *Crassostrea gigas*. Marine Ecology Progress Series 45: 225-236.

Buckley, D.E. 1986. Bioenergetics of age-related versus size-related reproductive tactics in female *Viviparus georgianus*. Biological Journal of the Linnean Society 27:293-309.

Busso, C.A. and J.H. Richards 1995. Drought and clipping effects on tiller demography and growth of two tussock grasses in Utah. Journal of Arid Environments 29: 239-251.

Bustamante, J. 1996. Population viability analysis of captive and released bearded vulture populations. Conservation Biology 10: 822-831.

Cain, M.L. 1990. Patterns of *Solidago altissima* ramet growth and mortality - The role of below-ground ramet connections. Oecologia 82: 201-209.

Calkins, D.G. and K.W. Pitcher 1982. Population assessment, ecology, and trophic relationships of Steller sea lions in the Gulf of Alaska. Alaska Department of Fish and Game, Final Report RU243. Alaska Department of Fish and Game, 333 Raspberry Rd., Anchorage, AK 99502, 128pp.

Canales, J., M.C. Trevisan, J.F. Silva and H. Caswell 1994. A demographic study of an annual grass (*Andropogon brevifolius* Schwarz) in burnt and unbrunt savanna. Acta Œcologica 15: 261-273.

Carey, J.R. 1982. Demography and population dynamics of the Mediterranean fruit fly. Ecological Modeling, 16:125-150.

Carlander, K.D. 1950. *Handbook of Freshwater Fishery Biology*. Wm. C. Brown Co., Dubuque, Iowa, 281pp.

Carlsson, B.Å. and T.V Callaghan 1991. Simulation of fluctuating populations of *Carex bigelowii* tillers classified by type, age and size. Oikos 60: 231-240.

Carroll, L. 1865. *Alice's Adventures in Wonderland* (reprinted in M. Gardner, 1960. The Annotated Alice, Bramhall House, New York, 352pp.)

Carroll, L. 1876 *The Hunting of the Snark* (reprinted in J. Tanis and J. Dooley [eds.] *Lewis Carroll's The Hunting of the Snark*, William Kaufmann, Inc., Los Altos, California, in cooperation with Bryn Mawr College Library, 129pp + xliv plates)

Cassie, R.M. 1950. The analysis of polymodal frequency distributions by the probability paper method. New Zealand Science Review 8: 90-91.

Cassie, R.M. 1954. Some uses of probability paper in the analysis of size frequency distributions. Australian Journal of Marine and Freshwater Research 5: 513-522.

Caswell, H. 1982a. Life history theory and the equilibrium status of populations. American Naturalist 120: 317-339.

Caswell, H. 1982b. Stable population structure and reproductive value for populations with complex life cycles. Ecology 63: 1223-1231.

Caswell, H. 1985. The evolutionary demography of clonal reproduction. pp 187-224 *In:* J.B.C. Jackson, L.W. Buss and R.E. Cook [eds.] *Population Biology and Evolution of Clonal Organisms*. Yale U. Press, New Haven, Connecticut

Caswell, H. 1986. Life cycle models for plants. Lectures on Mathematics in the Life Sciences 18: 171-233.

Caswell, H. 1989. *Matrix Population Models*. Sinauer

Associates. Inc., Sunderland, Massachusetts. 328 pp.

Caswell, H., R.J. Naiman, and R. Morin 1984. Evaluating the consequences of reproduction in complex salmonid life cycles. Aquaculture 43: 123-134.

Caswell, H. and P.A. Werner 1978. Transient behavior and life history analysis of teasel (*Dipsacus sylvestris* Huds.). Ecology 59: 53-66.

Caughley, G. 1967. Parameters for seasonally breeding populations. Ecology 48: 834-839.

Caughley, G. 1977. *Analysis of Vertebrate Populations.* Wiley, New York, 234pp.

Causton, D.R. 1969. A computer program for fitting the Richards function. Biometrics 25: 401-409.

Causton, D.R., C.O. Elias, and P. Hadley. 1978. Biometrical studies of plant growth. I. The Richards function, and its application in analyzing the effects of temperature on leaf growth. Plant Cell Environment 1: 163-184.

Charlesworth, B. 1980. *Evolution in Age-Structured Populations.* Cambridge Studies in Mathematical Biology 1., Cambridge U. Press, New York, 300pp.

Charnov, E.L. 1993. *Life History Invariants: Some Explorations of Symmetry in Evolutionary Ecology,* Oxford University Press, New York. 167pp.

Chen, Y., D.A. Jackson and H.H. Harvey 1992. A comparison of von Bertalanffy and polynomial functions in modeling fish growth data. Canadian Journal of Fisheries and Aquatic Sciences 49: 1228-1235.

Clarke, P.J. 1992. Predispersal mortality and fecundity in the grey mangrove (*Avicennia marina*) in south-eastern Australia. Australian Journal of Ecology 17: 161-168.

Clarke, P.J. 1995. The population dynamics of the mangrove *Avicennia marina*; demographic synthesis and predictive modelling. Hydrobiologia 295: 83-88.

Clarke, P.J. and W.G. Allaway 1993. The regeneration niche of the grey mangrove (*Avicennia marina*); effects of salinity, light and sediment factors on establishment, growth and survival in the field. Oecologia 93: 548-556.

Cloern, J.E., and F.H. Nichols. 1978. A von Bertalanffy growth model with a seasonally varying coefficient. Journal of the Fisheries Research Board of Canada 35:1479-1482.

Cohen, J.E. 1995. *How Many People Can the Earth Support.* W.W. Norton, New York, 532pp.

Cole, L.C. 1954. The population consequences of life history phenomena. Quarterly Review of Biology 29: 103-137.

Collett, D. 1994. *Modelling Survival Data in Medical Research.* Chapman & Hall, New York, 347pp.

Congdon, J.D., A.E. Dunham and R.C. van Loben Sels. 1993. Delayed sexual maturity and demographics of Blanding's turtles (*Emydoidea blandingii*): Implications for conservation and management of long-lived organisms. Conservation Biology 7: 826-833.

Congdon, J.D., A.E. Dunham and R.C. van Loben Sels.

1994. Demographics of common snapping turtles (*Chelydra serpentina*): Implications for conservation and management of long-lived organisms. American Zoologist 34: 397-408.

Cormack, R.M. 1964. Estimates of survival from the sighting of marked animals. Biometrika 51: 429-438.

Cortés, E. and G.R. Parsons 1996. Comparative demography of two populations of the bonnethead shark (*Sphyrna tiburo*). Canadian Journal of Fisheries and Aquatic Sciences 53: 709-718.

Craig, G.Y. and G. Oertel 1966. Deterministic models of living and fossil populations of animals. The Quarterly Journal of the Geological Society of London 122: 315-355.

Crivelli, A.J. and D. Mestre 1988. Life history traits of pumpkinseed, *Lepomis gibbosus*, introduced into the Camargue, a Mediterranean wetland. Archive für Hydrobiologie 111: 449-466.

Crouse, D.T., L.B. Crowder, and H. Caswell. 1987. A stage-based population model for loggerhead sea turtles and implications for conservation. Ecology 68: 1412-1423.

Croxall, J.P., P. Rothery, S.P.C. Pickering and P.A. Prince 1990. Reproductive performance, recruitment and survival of wandering albatrosses *Diomedea exulans* at Bird Island, South Georgia. Journal of Animal Ecology 59: 775-796.

Crump, R.G. (1978). Some aspects of the population dynamics of *Asterina gibbosa* (Asteroidea). Journal of the marine biological Association U.K. 58: 451-466.

de Kroon, H., A. Plaisier, J. van Groenendael and H. Caswell 1986. Elasticity: the relative contribution of demographic parameters to population growth rate. Ecology 67: 1427-1431.

DeAngelis, D.L. and C.C. Coutant 1982. Genesis of bimodal size distributions in species cohorts. Transactions of the American Fisheries Society 111: 384-388.

DeAngelis, D.L. and L.J. Gross [eds.] 1992. *Individual-Based Models and Approaches in Ecology.* Chapman & Hall, New York, 525pp.

Deevey, E.S. 1947, Life tables for natural populations of animals. The Quarterly Review of Biology 22: 283-314.

Dobson, A.P. and A.M. Lyles 1989. The population dynamics and conservation of primate populations. Conservation Biology 3: 362-380.

Doyle, R.W. and W. Hunte 1981. Genetic changes in "fitness" and yield of a crustacean population in a controlled environment. Journal of Experimental Marine Biology and Ecology 52: 147-156.

Drake, D.A. 1992. Dynamic Aquaria [review of W.H. Adey and K. Loveland 1991. *Dynamic Aquaria: Building Living Ecosystems*]. Ecology 73:1133.

Dumont, H.J. and S.S.S. Sarma 1995. Demography and population growth of *Asplanchna girodi* (Rotifera) as a function of prey (*Anuraeopsis fissa*) density.

Hydrobiologia 306: 97-107.

Dyke, B., T.B. Gage, P.L. Alford, B. Swenson and S. Williams-Blangero 1995. Model life table for captive chimpanzees. American Journal of Primatology 37: 25-37.

Ebert, T.A. 1973. Estimating growth and mortality rates from size data. Oecologia 11: 281-298.

Ebert, T.A. 1975. Growth and mortality in postlarval echinoids. American Zoologist 15: 755-775.

Ebert, T.A. 1980. Estimating parameters in a flexible growth equation, the Richards function. Canadian Journal of Fisheries and Aquatic Sciences 37: 687-692.

Ebert, T.A. 1981 Estimating mortality from growth parameters and a size distribution when recruitment is periodic. Limnology and Oceanography 26: 764 -769.

Ebert, T.A. 1982. Longevity, life history, and relative body wall size in sea urchins. Ecological Monographs 52: 353-394.

Ebert, T.A. 1985. Sensitivity of fitness to macroparameter changes: an analysis of survivorship and individual growth in sea urchin life histories. Oecologia 65: 461-467.

Ebert, T.A. 1987a. Morphological data for tropical and subtropical sea urchins: 1968-1982. Technical Report. 87-1. Center for Marine Studies, San Diego State University, San Diego, California. 104p

Ebert, T.A. 1987b. Estimating growth and survival parameters by nonlinear regression using average size in catches. pp35-44 In: D. Pauly and G. R. Morgan [eds.] Theory and Application of Length-based Methods in Stock Assessment. ICLARM Conference Proceedings 13.

Ebert, T.A. 1988. Allometry, design and constraint of body components and of shape in sea urchins. Journal of Natural History 22: 1407-1425.

Ebert, T.A. 1996. The consequences of broadcasting, brooding, and asexual reproduction in echinoderm metapopulations. Oceanologica Acta 19: 217-226.

Ebert, T. A. 1998. An analysis of the importance of Allee effects in management of the red sea urchin Strongylocentrotus franciscanus. pp 619-627 In: R. Mooi and M. Telford [eds.] Echinoderms: San Francisco. Proceedings, 9th International Echinoderm Conference, A. A. Balkema, Brookfield, Vermont

Ebert, T.A. and C.A. Ebert 1989. A method for studying vegetation dynamics when there are no obvious individuals: Virtual-population analysis applied to the tundra shrub Betula nana L. Vegetatio 85: 33-44.

Ebert, T.A. and R.F. Ford. 1986. Population ecology and fishery potential of the spiny lobster Panulirus penicillatus at Enewetak Atoll, Marshall Islands. Bulletin of Marine Science 38: 56-67.

Ebert, T.A. and M.P. Russell 1993. Growth and mortality of subtidal red sea urchins (Strongylocentrotus franciscanus) at San Nicolas Island, California, USA: problems with models. Marine Biology 117: 79-89.

Ebert, T.A. and M.P. Russell 1994. Allometry and Model II nonlinear regression. Journal of Theoretical Biology 168: 367-372.

Ebert, T.A., S.C. Schroeter and J.D. Dixon 1993. Inferring demographic processes from size-frequency distributions: Effect of pulsed recruitment on simple models. Fishery Bulletin, US 91: 237-243.

Ebert, T.A., S.C. Schroeter, J.D. Dixon and P. Kalvass 1994. Settlement patterns of red and purple sea urchins (Strongylocentrotus franciscanus and S. purpuratus) in California, USA. Marine Ecology Progress Series 111:41-52.

Ebert., T.A. and D.C. Lees 1996. Growth and loss of tagged individuals of the predatory snail Nucella lamellosa in areas within the influence of the Exxon Valdez oil spill in Prince William Sound. American Fisheries Society Symposium 18: 349-361.

Ebert, T.A., J.D. Dixon, S.C. Schroeter, P.E. Kalvass, N.T. Richmond, W.A. Bradbury and D.A. Woodby (in review) Growth and mortality of red sea urchins (Strongylocentrotus franciscanus) across a latitudinal gradient.. Ecological Monographs

Edmondson, W.T. 1945. Ecological studies of sessile rotatoria, Part II. Dynamics of populations and social structures. Ecological Monographs 15: 141-192.

Elle, E. 1996. Reproductive trade-offs in genetically distinct clones of Vaccinium macrocarpon, the American cranberry. Oecologia 107: 61-70.

Emlen, J. 1970. Age-specificity and ecological theory. Ecology 51: 588-601.

Emlet, R.B. 1989. Apical skeletons of sea urchins Echinodermata: Echinoidea: Two methods for inferring mode of larval development. Paleobiology 15: 223-254.

Emson, R.H. and R.G. Crump. 1979. Description of a new species of Asterina (Asteroidea), with an account of its ecology. Journal of the Marine Biological Association U.K. 59, 77-94.

Emson, R.H., and R.G. Crump. 1984. Comparative studies on the ecology of Asterina gibbosa and A. phylactica at Lough Ine. Journal of the Marine Biological Association U.K. 64: 35-53.

Emson, R.H. and P.V. Mladenov. 1987. Studies of the fissiparous holothurian Holothuria parvula (Selenka) (Echinodermata: Holothuroidea). Journal of Experimental Marine Biology and Ecology 111: 195-211.

Enright, N.J. 1992. Factors affecting reproductive behavior in the New Zealand nikau palm, Rhopalostylis sapida Wendl. et Drude. New Zealand Journal of Botany 30: 69-80.

Enright, N. and J. Ogden. 1979. Applications of transition matrix models in forest dynamics: Araucaria in Papua

New Guinea and *Nothofagus* in New Zealand. Australian Journal of Ecology 4: 3-23.

Enright, N.J. and A.D. Watson 1992. Population dynamics of the nikau palm, *Rhopalostylis sapida* (Wendl. et Drude), in a temperate forest remnant near Auckland, New Zealand. New Zealand Journal of Botany 30: 29-43.

Eriksson, O. 1988. Ramet behaviour and population growth in the clonal herb *Potentilla anserina*. Journal of Ecology 76: 522-536.

Euler, L. 1760. Researches générales sur la mortalité et la multiplication du genre humain. Mémoires de l'Académie Royale des Sciences et Belles Lettres (Belgium) 16: 144-164 (Translated by N. Keyfitz and B. Keyfitz 1970. A general investigation into the mortality and multiplication of the human species. Theoretical Population Biology 1: 307-314.)

Evans, F.C. and F.E. Smith 1952. The intrinsic rate of natural increase for the human louse, *Pediculus humanus* L. American Naturalist 86: 299-310.

Fabens, A.J. 1965. Properties and fitting of the von Bertalanffy growth curve. Growth 29: 265-289.

Fabricius, K.E. 1995. Slow population turnover in the soft coral genera *Sinularia* and *Sarcophyton* on mid- and outer-shelf reefs of the Great Barrier Reef. Marine Ecology Progress Series 126: 145-152.

Fancy, S.G., K.R. Whitten and D.E. Russell 1994. Demography of the Porcupine caribou herd, 1983-1992. Canadian Journal of Zoology 72: 840-846.

Fetcher, N. and G.R. Shaver. 1983. Life histories of tillers of *Eriophorum vaginatum* in relation to tundra disturbance. Journal of Ecology, 71:131-147.

Fiedler, P.L. 1987. Life history and population dynamics of rare and common mariposa lilies (*Calochortus* Pursh: Liliaceae). Journal of Ecology 75: 977-995.

Finch, C.E. 1990. *Longevity, Senescence, and the Genome*. U. Chicago Press, Chicago, Illinois, 922pp.

Fisher, R.A. 1930. *The Genetical Theory of Natural Selection*. Clarendon Press, Oxford (reprinted and revised, 1958. Dover, New York, 291pp.)

Fletcher, W.J. 1984. Intraspecific variation in the population dynamics and growth of the limpet, *Cellana tramoserica*. Oecologia 63: 110-121.

Ford, E. 1933. An account of the herring investigations conducted at Plymouth during the years from 1924-1933. Journal of the Marine Biological Association U.K. 19: 305-384.

Ford, H. 1981. The demography of three populations of dandelion. Biological Journal of the Linnean Society 15: 1-11.

Fournier, D.A., and P.A. Breen 1983. Estimation of abalone mortality rates with growth analysis. Transactions of the American Fisheries Society 112: 403-411.

Fournier, D.A., J.R. Sibert, and M. Terceiro 1991. Analysis of length frequency samples with relative abundance data for the Gulf of Marine northern shrimp (*Pandalus borealis*) by the MULTIFAN method. Canadian Journal of Fisheries and Aquatic Sciences 48: 591-598.

Fournier, D.A., J.R. Sibert, J. Majkowski, and J. Hampton 1990. MULTIFAN a likelihood-based method for estimating growth parameters and age composition from multiple length frequency data sets illustrated using data for southern bluefin tuna (*Thunnus maccoyii*). Canadian Journal of Fisheries and Aquatic Sciences 47: 301-317.

Fox, G.A. 1993. Failure-time analysis: Emergence, flowering, survivorship and other waiting times. pp 253-289 *In:* S. M. Scheiner and J. Gurevitch [eds.] *Design and Analysis of Ecological Experiments*. Chapman & Hall, New York

Francis, R.I.C.C. 1995. An alternative mark-recapture analogue of Schnute's growth model. Fisheries Research 23: 95-111.

Frank, P.W. 1965. Shell growth in a natural population of the turban snail, *Tegula funebralis*. Growth 29: 395-403.

Fraser, F.C. 1936. On the development and distribution of the young stages of krill (*Euphausia superba*). Discovery Reports 14: 3-192.

Frazer, N.B. 1983a. Demography and life history evolution of the Atlantic loggerhead sea turtle, *Caretta caretta*. Dissertation, University of Georgia, Athens, 233pp.

Frazer, N.B. 1983b. Survivorship of adult female loggerhead sea turtles, *Caretta caretta*, nesting on Little Cumberland Island, Georgia, USA. Herpetologica 39: 436-447.

Frazer, N.B. 1984. A model for assessing mean age-specific fecundity in sea turtle populations. Herpetologica 40: 281-291.

Frazer, N.B. 1987. Preliminary estimates of survivorship for wild juvenile loggerhead sea turtles (*Caretta caretta*). Journal of Herpetology 21: 232-235.

Gage, J.D. 1987. Growth of the deep-sea irregular sea urchins *Echinosigra phiale* and *Hemiaster espergitus* in the Rockall Trough (N. E. Atlantic Ocean). Marine Biology 96: 19-30.

Gage, T.B. 1988. Mathematical hazard models of mortality: An alternative to model life tables. American Journal of Physical Anthropology 76: 429-441.

Gage, T.B. and B. Dyke 1988. Model life tables for the larger Old World monkeys. American Journal of Primatology 16: 305-320.

Ghiselin, M.T. 1987. Evolutionary aspects of marine invertebrate reproduction. pp 609-665 *In:* A. C. Giese, J. S. Pearse and V. B. Pearse [eds.] *Reproduction of Marine Invertebrates Vol. IX General Aspects: Seeking Unity in Diversity*. Boxwood Press, Pacific Grove, CA.

Gómez, A., M. Temprano and M. Serra 1995. Ecological genetics of a cyclical parthenogen in temporary habitats. Journal of Evolutionary Biology 8: 601-622.

Gompertz, B. 1825. On the nature of the function

expressive of the law of human mortality, and on a new mode of determining the value of life contingencies. Philosophical Transactions of the Royal Society of London 115: 513-585.

Goodman, D. 1987. The demography of chance extinction pp 11-34 In: M. E. Soulé [ed.] *Viable Populations for Conservation*. Cambridge University Press, New York

Gonick, L. and W. Smith 1993. *The Cartoon Guide to Statistics*. Harper Collins, New York, 231pp.

Gould, S.J. and E.S. Vrba 1982. Exaptation -- a missing term in the science of form. Paleobiology 8: 4-15.

Grant, P.R. and B.R. Grant 1992. Demography and the genetically effective sizes of two populations of Darwin's finches. Ecology 73: 766-784.

Grattan-Guinness, I. 1994. The roots of equations: Detection and approximation. pp 563-566 *In:* I. Grattan-Guinness [ed.] *Companion Encyclopedia of the History and Philosophy of the Mathematical Sciences. Vol. 1.* Routledge, New York.

Grattan-Guinness, I. and W. Ledermann 1994. Matrix theory. pp 775-786 *In:* I. Grattan-Guinness [ed.] *Companion Encyclopedia of the History and Philosophy of the Mathematical Sciences. Vol. 1.* Routledge, New York.

Graunt, J. 1662. Natural and political observations mentioned in a following index, and made upon the bills of mortality. Printed by Tho. Roycroft for John Martin, James Allestry, and Tho. Dicas, at the Sign of the Bell in St. Paul's Church-yard. [Reprinted 1964. Journal of the Institute of Actuaries 90: 1-61.]

Green, R.H. 1970 Graphical estimation of rates of mortality and growth. Journal of the Fisheries Research Board of Canada 27: 204-208.

Green, R.H. 1979. Matrix population models applied to living populations and death assemblages. American Journal of Science 279: 481-487.

Hairston, N.G., Jr. and C.E. Cáceres 1996. Distribution of crustacean diapause: micro- and macroevolutionary pattern and process. Hydrobiologia 320: 27-44.

Hairston, N.G., Jr. and B.T. De Stasio, Jr. 1988. Rate of evolution slowed by a dormant propagule pool. Nature 336: 239-242.

Hald, A. 1952. *Statistical Theory with Engineering Applications*. John Wiley & Sons, Inc., New York, 783pp.

Halley, E. 1693a. An estimate of the degrees of the mortality of mankind, drawn from curious tables of the births and funerals at the city of Breslaw; with an attempt to ascertain the price of annuities upon lives. Philosophical Transactions of the Royal Society of London 17: 596-610, 654-656. [Reprinted in facsimile 1985, Journal of the Institute of Actuaries 112: 278-298.]

Halley, E. 1693b. Some further considerations on the Breslaw bills of mortality. By the same hand, &c.

Philosophical Transactions of the Royal Society of London 17: 654-656. [Reprinted in facsimile 1985, Journal of the Institute of Actuaries 112: 299-301.]

Harding, J.P. 1949. The use of probability paper for the graphical analysis of polymodal frequency distributions. Journal of the Marine Biological Association of the United Kingdom 28: 141-153.

Harper, J.L. 1977. *Population Biology of Plants*. Academic Press, New York, 892pp.

Hartman, G. D. 1995. Age determination, age structure, and longevity in the mole, *Scalopus aquaticus* (Mammalia: Insectovora). Journal of Zoology, London 237: 107-122.

Hartnoll, R.G., and A.D. Bryant 1990. Size-frequency distributions in decapod crustacea-The quick, the dead, and the cast-offs. Journal of Crustacean Biology 10: 14-19.

Hasselblad, V. 1966. Estimation of parameters for a mixture of normal distributions. Technometrics 8: 431-444.

Hastings, C. Jr. 1955. *Approximations for Digital Computers*, Princeton Univ. Press, Princeton, New Jersey, 201pp.

Hearn, W.S. and G.M. Leigh 1994. Comparing polynomial and von Bertalanffy growth functions for fitting tag-recapture data. Canadian Journal of Fisheries and Aquatic Sciences 51: 1689-1691.

Henderson, P.A. and R.N. Bamber 1987. On the reproductive biology of the sand smelt *Atherina boyeri* Risso (Pisces: Atherinidae) and its evolutionary potential. Biological Journal of the Linnean Society 32: 395-415.

Hendler, G. 1991. Echinodermata: Ophiuroidea pp 255-511 *In:* Giese, A.C. J.S. Pearse and V. B. Pearse [eds.] *Reproduction of Marine Invertebrates Vol. VI Echinoderms and Lophophorates*, Boxwood Press, Pacific Grove, CA.

Heyde, C.C. and J.E. Cohen 1985. Confidence intervals for demographic projections based on products of random matrices. Theoretical Population Biology 27: 120-153.

Hickey, F. 1960. Death and reproductive rate of sheep in relation to flock culling and selection. New Zealand Journal of Agricultural Research 3: 332-344.

Hickey, F. 1963. Sheep mortality in New Zealand. New Zealand Agriculturist 15: 1-3.

Hiraldo, F., J. J. Negro, J. A. Donázar and P. Gaona 1996. A demographic model for a population of the endangered lesser kestrel in southern Spain. Journal of Applied Ecology 33: 1085-1093.

Hoagland, K.E. and R. Robertson 1988. An assessment of poecilogony in marine invertebrates: phenomenon or fantasy? Biological Bulletin 174: 109-125.

Hoddle, M.S. 1991. Lifetable construction for the gorse seed weevil, *Apion ulicis* (Forster) (Coleoptera: Apionidae) before gorse pod dehiscence, and life history strategies of the weevil. New Zealand Journal of Zoology 18: 399-

404.

Hoff, B. 1992. *The Te of Piglet*. Dutton (Penguin Books), New York, 257pp.

Holla, T.A. and P. Knowles 1988. Age structure analysis of a virgin white pine, *Pinus strobus*, population. Canadian Field-Naturalist 102: 221-226.

Hoenig, J. M. 1983. Empirical use of longevity data to estimate mortality rates. Fishery Bulletin, U. S. 82: 898-903.

Horvitz, C.C. and D.W. Schemske 1995. Spatiotemporal variation in demographic transitions of a tropical understory herb: projection matrix analysis. Ecological Monographs 65: 155-192.

Hubbell, S.P. and P.A. Werner 1979. On measuring the intrinsic rate of increase of populations with heterogeneous life histories. American Naturalist 113: 277-293.

Huenneke, L.F. and P.L. Marks 1987. Stem dynamics of the shrub *Alnus incana* ssp. *rugosa*: transition matrix models. Ecology 68: 1234-1242.

Hughes, R.N. and D.J. Roberts 1981. Comparative demography of *Littorina rudis*, *L. nigrolineata* and *L. neritoides* on three contrasted shores in North Wales. Journal of Animal Ecology 50: 251-268.

Hughes, T.P. 1984. Population dynamics based on individual size rather than age: A general model with a reef coral example. American Naturalist. 123:778-795.

Humphrey, R.R. and A.B. Humphrey 1990. *Idria columnaris*: age as determined by growth rate. Desert Plants 10: 51-54.

Hunt, R. 1982. *Plant Growth Curves. The Functional Approach to Plant Growth Analysis*. Edward Arnold, London, 248pp.

Huxley, J.S. 1932. *Problems of Relative Growth*. Methuen & Co., London, 265pp.

Jacobsen, T. and J.A. Kushlan 1989. Growth dynamics in the American alligator (*Alligator mississippiensis*). Journal of Zoology, London 219: 309-328.

Jensen, A. L. 1971. The effect of increased mortality on the young in a population of brook trout, a theoretical analysis. Transactions of the American Fisheries Society 100: 456-459.

Johnson, D.H., A.B. Sargent, and S.H. Allen. 1975. Fitting Richards' curve to data of diverse origins. Growth 39: 315-330.

Jolicoeur, P. 1975. Linear regressions in fisheries research: some comments. Journal of the Fisheries Research Board of Canada. 32: 1491-1494.

Jolicoeur, P. 1990. Bivariate allometry: Interval estimation of the slopes of the ordinary and standardized normal major axes and structural relationship. Journal of Theoretical Biology 144: 275-285.

Jolicoeur, P., J. Pontier and H. Abidi 1992. Asymptotic models for the longitudinal growth of human stature. American Journal of Human Biology 4: 461-468.

Jolicoeur, P., J. Pontier, M.-O. Perin and M. Sempé 1988. A lifetime asymptotic growth curve for human height. Biometrics 4: 995-1003.

Jolly, G.M. 1965. Explicit estimates from capture-recapture data with both death and immigration - stochastic model. Biometrika 38: 301-321.

Kaliz, S. and M.A. McPeek 1992. Demography of an age-structured annual: Resampling projection matrices, elasticity analyses and seed bank effects. Ecology 73: 1082-1093.

Kanefuji, K. and T. Shohoji 1990. On a growth model of human height. Growth, Development and Aging 54: 155-165.

Kaur, A., C.O. Ha, K. Jong, V.E. Sands, H.T. Chan, E. Soepadmo and P.S. Ashton 1978. Apomixis may be widespread among trees of the climax rain forest. Nature 271: 440-442.

Keith, L.B. and L.A. Windberg 1978 A demographic analysis of the snowshoe hare cycle. Wildlife Monographs 58: 1-70.

Kendall, M. and A. Stuart 1977. *The Advanced Theory of Statistics. Vol. 1. Distribution Theory*. Charles Griffin & Co., Ltd., London, 472pp.

Kermack, K.A. 1954. A biometrical study of *Micraster coranguinum* and *M . (Isomicraster) senonensis*. Philosophical Transactions of the Royal Society, London Series B. 237: 375-428.

Keyfitz, N. 1968. *Introduction to the Mathematics of Population*. Addison-Wesley Publ. Co., Inc., Menlo Park, California, 450pp.

Keyfitz, N. and W. Flieger 1968. *World Population. An Analysis of Vital Data*. University of Chicago Press, Chicago, Illinois, 672pp.

Kikuno, T. 1982 Observations of early developments of the Antarctic krill, *Euphausia superba* Dana. pp 38-43 In: T. Hoshiai and Y. Naito (eds.) Proceedings of the 5th Symposium on Antarctic Biology. Memoirs of the National Institute of Polar Research, Special Issue, Tokyo.

Kingsley, M. 1989 Population dynamics of the narwhal *Monodon monoceros*: an initial assessment (Odontoceti: Monodontidae). Journal of Zoology, London 219: 201-208.

Kiritani, K. and F. Nakasuji 1967. Estimation of the stage-specific survival rate in the insect population with overlapping stages. Researches on Population Ecology 9: 143-152.

Klinger, T. 1993. The persistence of haplodiploidy in algae. Trends in Ecology and Evolution 8(7): 256-258

Knight, R.R. and L.L. Eberhardt 1985. Population dynamics of Yellowstone grizzly bears. Ecology 66: 323-334.

Knox, R.B. 1967. Apomixis: seasonal and population differences in a grass. Science 157: 325-326.

Kruger, G. 1978. Zur Mathematik des tierischen Washstums III. Testung der Gompertz-Funktion als Washstumsformel am Beispiel von *Siliqua patula* (Bivalvia) und *Thunnus thynnus* (Pisces). Helgoländer wissenschftliche Meeresuntersuchungen 31: 499-526.

Kuo, S.S. 1965. *Numerical Methods and Computers.* Addison-Wesley, Palo Alto, California, 341pp.

LaBarbera, M. 1989. Analyzing body size as a factor in ecology and evolution. Annual Review of Ecology and Systematics 20: 97-117.

Lande, R. 1979. Quantitative genetic analysis of multivariate evolution, applied to brain:body size allometry. Evolution 33: 402-416.

Lande, R. 1988. Demographic models of the northern spotted owl (*Strix occidentalis caurina*). Oecologia 75: 601-607.

Laws, E.A. and J.W. Archie 1981. Appropriate use of regression analysis in marine biology. Marine Biology 65: 13-16.

Lazaridou-Dimitriadou, M. 1995. The life cycle, demographic analysis, growth and secondary production of the snail *Helicella* (*Xerothracia*) *pappi* (Schütt, 1962) (Gastropoda Pulmonata) in E. Macedonia (Greece). Malacologia 37: 1-11.

Lebreton, J.-D., K.P. Burnham, J. Colbert and D.R. Anderson 1992. Modeling survival and testing biological hypotheses using marked animals: A unified approach with case studies. Ecological Monographs 62: 67-118.

Lee, E.T. 1992. *Statistical Methods for Survival Data Analysis.* John Wiley & Sons, Inc., New York, 482pp.

Lefkovitch, L.P. 1965. The study of population growth in organisms grouped by stages. Biometrics 21: 1-18.

Lenski, R.E. and P.M. Service 1982. The statistical analysis of population growth rates calculated from schedules for survivorship and fecundity. Ecology 63: 655-662.

Lesica, P. 1995. Demography of *Astragalus scaphoides* and effects of herbivory on population growth. Great Basin Naturalist 55: 142-150.

Leslie, P.H. 1945. On the use of matrices in certain population mathematics. Biometrika 33: 183-212.

Leslie, P.H. 1948. Some further notes on the use of matrices in population dynamics. Biometrika 35: 213-245.

Leverich, W.J. and D.A. Levin 1979. Age-specific survivorship and reproduction in *Phlox drummondii.* American Naturalist 113: 881-903.

Levin, L.A., H. Caswell, K.D. DePatra and E.L. Creed 1987. Demographic consequences of larval development mode: planktotrophy vs. lecithotrophy in *Streblospio benedicti.* Ecology 68: 1877-1886.

Levitan, D.R. 1993. The importance of sperm limitation to the evolution of egg size in marine invertebrates. American Naturalist 141: 517-536.

Libertini, G. 1988. An adaptive theory of the increasing mortality with increasing chronological age in populations in the wild. Journal of Theoretical Biology 132: 145-162.

Lodal, J. and H. Grue 1985. Age determination and age distribution in populations of moles (*Talpa europaea*) in Denmark. Acta Zoologica Fennica 173: 279-281.

Łomnicki, A. 1988. *Population Ecology of Individuals.* Princeton University Press, New Jersey, 223pp.

Lotka, A.J. 1907a. Relation between birth rates and death rates. Science 26: 21-22.

Lotka, A.J. 1907b. Studies on the mode of growth of material aggregates. American Journal of Science 4th Ser. 24 (whole number 174): 199-216.

Lotka, A.J. 1945. Population analysis as a chapter in the mathematical theory of evolution. pp 355-385 *In:* W. E. LeGros Clark and P. B. Medawar [eds.] *Essays on Growth and Form.* Oxford University Press, Oxford

Lovejoy, C.O., R.S. Meindl, T.R. Pryzbeck, T.S. Barton, K.G. Heiple and D. Kotting. 1977. Paleodemography of the Libben site, Ottawa County, Ohio. Science 198: 291-293.

Lumer, H. 1937. The consequences of sigmoid growth for relative growth functions. Growth 1: 140-154.

Lynch, M. 1985. Spontaneous mutations for life-history characters in an obligate parthenogen. Evolution, 39(4):804-818.

MacArthur, R.H. 1960. Review of *Population Studies: Animal Ecology and Demography.* Cold Spring Harbor Symposia on Quantitative Biology Volume XXII. Quarterly Review of Biology 35: 82-83.

Macdonald, P.D. and P.E.J. Green 1988. User's guide to program MIX: an interactive program for fitting mixtures of distributions. Ichtus Data Systems, Hamilton, Ontario.

Macdonald, P.D. and T.J. Pitcher. 1979. Age-groups from size-frequency data: A versatile and efficient method of analyzing distribution mixtures. Journal of the Fisheries Research Board of Canada 36: 987-1001.

Manly, B.F.J. 1976. Extensions to Kiritani and Nakasuji's method for analysing insect stage frequency data. Researches on Population Ecology 17: 191-199.

Manly, B.F.J. 1977. A further note on Kiritani and Nakasuji's model for stage-frequency data including comments on the use of Tukey's jackknife technique for estimating variances. Researches on Population Ecology 18: 177-186.

Manly, B.F.J. 1985. Further improvements to a method for analysing stage-frequency data. Researches on Population Ecology 27: 325-332.

Manly, B.F.J. 1990. *Stage-Structured Populations. Sampling, Analysis and Simulation.* Chapman and Hall, New York, 187pp.

Marzolin, G. 1988. Polygynie du Cincle plongeur (*Cinclus cinclus*) dans les côtes de Lorraine. L'Oiseau et la Revue Française d'Ornithologie 58: 277-286.

Mathews, C.P., M. Al-Hossaini, A.R. Abdul Ghaffar, and

M. Al-Shoushani 1987. Assessment of short-lived stocks with special reference to Kuwait's shrimp fisheries: A contrast of the results obtained from traditional and recent size-based techniques. pp 147-166 *In*: D. Pauly and G. R. Morgan [eds.] Theory and Application of Length-based Methods in Stock Assessment. ICLARM Conference Proceedings 13.

Mauchline, J. 1965. The larval development of the euphausid *Thysanoessa raschii* (M. Sars). Crustaceana 9: 31-40.

Mauchline, J. 1980. The biology of mysids and euphausids. Advances in Marine Biology 7: 1-454.

Mauchline, J. 1985 Growth and production of Euphausiacea (Crustacea) in the Rockall Trough. Marine Biology 90: 19- 26.

May, R.M. 1976. Estimating r: a pedagogical note. American Naturalist 110: 496-499.

McArdle, B. H. 1987. The structural relationship: regression in biology. Canadian Journal of Zoology 66: 2329-2339.

McCarthy, D. P. and D. J. Smith 1995. Growth curves for calcium-tolerant lichens in the Canadian Rocky Mountains. Arctic and Alpine Research 27: 290-297.

McFadden, C.S. 1991. A comparative demographic analysis of clonal reproduction in a temperate soft coral. Ecology 72: 1846-1866.

McNew, R.W. and R.C. Summerfelt 1978. Evaluation of a maximum-likelihood estimator for analysis of length-frequency distributions. Transactions of the American Fisheries Society 107: 730-736.

McPeek, M.A. and S. Kalisz 1993. Population sampling and bootstrapping in complex designs: demographic analysis. pp 232-252 *In:* S. M. Scheiner and J. Gurevitch [eds.] *Design and Analysis of Ecological Experiments.* Chapman & Hall, New York.

Medawar, P.B. 1981. *The Uniqueness of the Individual.* (revised 2nd edition). Dover Publications, Inc., New York, 162pp.

Melton, D.A. 1983. Population dynamics of waterbuck (*Kobus ellipsiprymnus)* in the Umfolozi Game Reserve. African Journal of Ecology 21: 77-91.

Menges, E. 1986. Predicting the future of rare plant populations: Demographic monitoring and modeling. Natural Areas Journal 6: 13-25.

Messerton-Gibbons, M. 1993. Why demographic elasticities sum to one: a postscript to de Kroon *et al.* Ecology 74: 2467-2468.

Meyer, J.S., C.G. Ingersoll, L.L. McDonald and M.S. Boyce 1986. Estimating uncertainty in population growth rates: jackknife vs. bootstrap techniques. Ecology 67: 1156-1166.

Metz, J.A.J. and O. Diekmann [eds.] 1986. *The Dynamics of Physiologically Structured Populations.* Lecture Notes in Biomathematics vol. 68. Springer- Verlag, New York, 511pp.

Michaels, H.J. and F.A. Bazzaz 1986. Resource allocation and demography of sexual and apomictic *Antenanaria parlinii.* Ecology 67: 27-36.

Michod, R.E. and W.W. Anderson 1980. On calculating demographic parameters from age frequency data. Ecology 61: 265-269.

Mode, C.J. and R.C. Busby 1982. An eight-parameter model of human mortality -- the single decrement case. Bulletin of Mathematical Biology 44: 647-659.

Mode, C.J. and M.E. Jacobson 1984. A parametric algorithm for computing model period and cohort human survival functions. International Journal of Bio-medical Computing 15: 341-356.

Moloney, K.A. 1986. A generalized algorithm for determining category size. Oecologia 69: 176-180.

Momot, W.T. 1967. Population dynamics and productivity of the crayfish *Orconectes virilis* in a marl lake. American Midland Naturalist 78: 55-81.

Morand, S., F. Robert and V.A. Connors 1995. Complexity in parasite life cycles: population biology of cestodes in fish. Journal of Animal Ecology 64: 256-264.

Morrison, P., R. Dietrich and D. Preston 1977. Body growth in sixteen rodent species and subspecies maintained in laboratory colonies. Physiological Zoology 50: 294-310.

Murie, A. 1944. *The Wolves of Mount McKinley.* Fauna of the National Parks of the United States. Fauna Series No. 5. U.S. Department of the Interior, National Park Service. Government Printing Office, Washington, DC, 238pp.

Nakaoka, M. and S. Matsui 1994. Annual variation in the growth rate of *Yoldia notabilis* (Bivalvia: Nuculanidae) in Otsuchi Bay, northeastern Japan, analyzed using shell microgrowth patterns. Marine Biology 119: 397-404.

Nelder, J.A. 1961. The fitting of a generalization of the logistic curve. Biometrics 17: 89-110.

Nelder, J.A. and R. Mead 1965. A simplex method for function minimization. Computer Journal 7: 308-313.

Neville, A.C. 1963. Daily growth layers for determining the age of grasshopper populations. Oikos 14: 1-8.

Niewiarowski, P.H. and A.E. Dunham 1994. The evolution of reproductive effort in squamate reptiles: costs, trade-offs, and assumptions reconsidered. Evolution 48: 137-145.

Noda, T. and S. Nakao 1996. Dynamics of an entire population of the subtidal snail *Umbonium costatum*: the importance of annual recruitment fluctuation. Journal of Animal Ecology 65: 196-204.

Norton, P.M. 1994. Simple spreadsheet models to study population dynamics, as illustrated by a mountain reedbuck model. South African Journal of Wildlife Research 24: 73-81.

O'Neill, R. 1985. Function minimization using a simplex procedure. pp 79-87 *In*: P. Griffiths and I. D. Hill (eds.) *Applied Statistics Algorithms.* Ellis Horwood Ltd,

Chichester, England.

Orzack, S.H. 1985. Population dynamics in variable environments V. the genetics of homeostasis revisited. American Naturalist 125: 550-572.

Ottesen, P.O. and J.S. Lucas 1982. Divide or broadcast: Interrelation of asexual and sexual reproduction in a population of fissiparous hermaphroditic seastar *Nepanthia belcheri* (Asteroidea: Asterinidae). Marine Biology 69: 223-233.

Packard, J.M. 1985. Preliminary assessment of uncertainty involved in modeling manatee populations. Manatee Population Research Report No. 9. Tech. Rpt. No 8-9. Florida Cooperative Fish and Wildlife Research Unit. University of Florida, Gainesville, Florida. 19pp.

Pake, C.E. and D.L. Venable 1996. Seed banks in desert annuals: Implications for persistence and coexistence in variable environments. Ecology 77: 1427-1435.

Paloheimo, J.E. and W.D. Taylor 1987. Comments on life table parameters with reference to *Daphnia pulex*. Theoretical Population Biology 32: 289-302.

Pauly, D. 1981. The relationships between gill surface area and growth performance in fish: a generalization of von Bertalanffy's theory of growth. Meeresforschung 28:251-282.

Pauly, D. 1980. On the interrelationships between natural mortality, growth parameters and mean environmental temperature in 175 fish stocks. Journal Conseil Permanent International pour l'Exploration de la Mer. 39: 175-192.

Pauly, D. 1987. A review of the ELEFAN system for analysis of length-frequency data in fish and aquatic invertebrates. pp 7-34 *In*: D. Pauly and G. R. Morgan [eds.] Theory and Application of Length-based Methods in Stock Assessment. ICLARM Conference Proceedings 13.

Pearl, R. and J.R. Miner 1935. Experimental studies on the duration of life. XIV. The comparative mortality of certain lower organisms. Quarterly Review of Biology 10: 60-79.

Pease, C.M. and J.H. Bull 1988. A critique of methods for measuring life history trade-offs. Journal of Evolutionary Biology 1: 293-303.

Pennington, R.H. 1965. *Introductory Computer Methods and Numerical Analysis*. Macmillan & Co., New York, New York, 452pp.

Pielou, E.C. 1977. *Mathematical Ecology*. Wiley-Interscience, New York, 385pp.

Pienaar, L.V., and J.A. Thomson 1973. Three programs used in population dynamics WVONB -- ALOMA -- BHYLD (FORTRAN 1130). Fisheries Research Board of Canada Technical Report 137: 1-33.

Pinder, J.E. III, J.G. Wiener, and M.H. Smith 1978. The Weibull distribution: A new method of summarizing survivorship data. Ecology 59: 175-179.

Poole, L. and M. Borchers 1979. *Some Common BASIC Programs, 3rd Edition*. OSBORNE/McGraw-Hill, Berkeley, California, 195pp.

Powell, R.A., J.W. Zimmerman, D.E. Seaman, and J.F. Gilliam 1996. Demographic analyses of a hunted black bear population with access to a refuge. Conservation Biology 10: 224-234.

Prairie, Y.T., R.H. Peters and D.F. Bird 1995. Natural variability and the estimation of empirical relationships: a reassessment of regression methods. Canadian Journal of Fisheries and Aquatic Sciences 52: 788-798.

Press, W.H., B.P. Flannery, S.A. Teukolsky and W.T. Vetterling 1986. *Numerical Recipes. The art of Scientific Computing*. Cambridge University Press, New York, 818pp.

Rayner, J.M.V. 1985. Linear relations in biomechanics: the statistics of scaling functions. Journal of Zoology, London (A) 206: 415-439.

Reeve, E.C.R. and J.S. Huxley 1945. Some problems in the study of allometric growth. pp 121-156. *In:* W. E Le Gros Clark and P. B. Medawar [eds] *Essays on "Growth and Form" Presented to D'Arcy Wentworth Thompson*. Claredon, Oxford.

Richards, F.J. 1959. A flexible growth function for empirical use. Journal of Experimental Botany 10:290-300.

Richardson, J.I. 1982. A population model for adult female loggerhead sea turtles (*Caretta caretta*) nesting in Georgia. Dissertation, University of Georgia, Athens. 204p

Ricker, W.E. 1973. Linear regression in fishery research. Journal of Fisheries Research Board of Canada 30: 409-434.

Ricker, W.E. 1975a. Computation and interpretation of biological statistics of fish populations. Bulletin of the Fisheries Research Board of Canada 191. 382p

Ricker, W.E. 1975b. A note concerning Professor Jolicoeur's comments. Journal of Fisheries Research Board of Canada 32: 1494-1498.

Ricker, W.E. 1982. Letter to Editor on 'Linear regressions for naturally variable data.' Biometrics 38: 859-860.

Rideout, R. S. 1978. Asexual reproduction as a means of population maintenance in the coral reef asteroid *Linckia multifora* on Guam. Marine Biology 47: 287-295.

Robb, R.C. 1929. On the nature of hereditary size-limitation. II. The growth of parts in relation to the whole. British Journal of Experimental Biology 6: 311-324.

Roff, D.A. 1992. *The Evolution of Life Histories. Theory and Analysis*. Chapman & Hall, New York, 535pp.

Rosen, P., A.D. Woodhead and K.H. Thompson 1981. The relationship between the Gompertz constant and maximum potential lifespan; its relevance to theories of aging. Experimental Gerontology 16: 131-135.

Rosenbrock, H. 1960. An automatic method for finding the greatest or least value of a function. Computer Journal 3: 175-184.

Rowley, I. and G. Chapman 1991. The breeding biology, food, social organisation, demography and conservation of the Major Mitchell or Pink Cockatoo, *Cacatua leadbeateri*, on the margin of the Western Australian wheatbelt. Australian Journal of Zoology 39: 211-261.

Russell, M.P. and J.P. Huelsenbeck 1989. Seasonal variation in brood structure of *Transennella confusa* Bivalvia: Veneridae. The Veliger 32: 288-295.

Sager, G. 1978. Integrationen von Zuwachsansätzen nach der Auffassung von L. v. Bertalanffy. Anatomischer Anzeiger 144: 147-157.

Sager, G. 1979a. Zuwachsfunktionen vom Typ $dW/dt=kt^{p-1}(t_E^p - t^p)^q$ und ihre Integrale. Anatomischer Anzeiger 145: 268-275.

Sager, G. 1979b. Die Wachstumsfunktion
$$W=E\left\{\sin\left[\frac{\pi}{2}\left(\frac{t}{t_E}\right)^p\right]\right\}^{2q}$$
und ihre Eigenschaften. Anatomischer Anzeiger 145: 369-379.

Sager, G. 1979c. A generalized form of the Bertalanffy functions of organic growth. Anatomischer Anzeiger 146: 188-200.

Sager, G. 1979d. Die Wachstumsfunktion
$$W=E\left\{1-\sin^{2q}\left[\frac{\pi}{2}\left(\frac{t_E-t}{t_E}\right)^p\right]\right\}$$
und ihre Eigenschaften. Anatomischer Anzeiger 146: 270-276.

Sager, G. 1980a. Die Erprobung von Wachstumsfunktionen am Beispiel *Siliqua patula* (Bivalvia). Anatomischer Anzeiger 148: 446-461.

Sager, G. 1980b. Die Funktion $W = (a-be^{-ct})^n$ als Verallgemeinerung der klassischen Wachstumsfunktionen. Anatomischer Anzeiger 148: 274-286.

Sager, G. 1980c. Zuwachsfunktionen vom Typ $dW/dt = kW^m/(t+t_0)^p$ und ihre Integrale. Anatomischer Anzeiger 147: 445-457.

Sager, G. 1982. Das Längenwachstum der Nordsee Seezunge (*Solea vulgaris* Quensel) und die Problematik der Jahresschwankungen. Anatomischer Anzeiger 151:160-178.

Sager, G. 1984a. Annuale Wachstumsschwankungen der Muschel *Macoma balthica* nach Daten von Bachelet (1980). Beitrage zur Meereskunde 50:43-50.

Sager, G. 1984b. Saisonal-modifizierte Formen der abgewandelten Janoschek-Funktion. Gegenbaurs Morphologisches Jahrbuch 130: 659-669.

Sager, G. 1986. Wachstumsspezifische Approximationen für die Herzmuschel *Cerastoderma edule* L. in der Waddenzee nach Daten von Beukema (1975-78). Bietrage zur Meereskunde. 55: 55-66.

Sager, G. and R. Sammler. 1984. Seasonal length growth of the Norwegian Krill (*Meganyctiphanes norvegica*) after data from Wiborg (1966-1969). Zoologische Jarbucher Abteilung für Anatomie und Ontogenie der Tiere 112:79-84.

Sager, G. and F. Gosselck. 1986. Investigation into seasonal growth of *Branchiostoma lanceolatum* off Heligoland, according to data by Courtney (1975). Internationale Revue gesamten Hydrobiologie. 71:701-707.

Saila, S.B. and R.G. Lough 1981 Mortality and growth estimation from size data -- an application to some Atlantic herring larvae. Conseil Permanent International pour l'Exploration de la Mer. Rapports et Procè-Verbaux des Réunions 178: 7-14.

Sainsbury, K.J. 1982. Population dynamics and fishery management of the Paua, *Haliotis iris*. II. Dynamics and management as examined using a size class population model. New Zealand Journal of Marine and Freshwater Research 16: 163-173.

Saunders, W.B. 1984. *Nautilus* growth and longevity: evidence from marked and recaptured animals. Science 224: 990-992.

Schaaf, W.E., D.S. Peters, D.S. Vaughan, L. Coston-Clements and C.W. Krouse 1987. Fish population responses to chronic and acute pollution: The influence of life history strategies. Estuaries 10: 267-275.

Schmidt, K.P. and L.R. Lawlor. 1983. Growth rate projection and life history sensitivity for annuals with a seed bank. American Naturalist 121: 525-539.

Schnute, J. 1981. A versatile growth model with statistically stable parameters. Canadian Journal of Fisheries and Aquatic Sciences 38: 1128-1140.

Schnute, J. and D. Fournier 1980. A new approach to length-frequency analysis: growth structure. Canadian Journal of Fisheries and Aquatic Sciences 37: 1337-1351.

Searles, R.B. 1980. The strategy of the red algal life history. American Naturalist 115: 113-120.

Sebens, K.P. 1982. Asexual reproduction in *Anthopleura elegantissima* (Anthozoa: Actiniaria): Seasonality and spatial extent of clones. Ecology 63:434-444.

Seber, G.A F. 1965. A note on the multiple recapture census. Biometrika 52: 249-259.

Seber, G.A.F. 1982. *The Estimation of Animal Abundance and Related Parameters*. Macmillan, New York, 654pp.

Seim, E. and B.-E. Sæther 1983. On rethinking allometry: which regression model to use? Journal of Theoretical Biology 104: 161-168.

Shepherd, J.G. 1987. A weakly parametric method for estimating growth parameters from length composition data. pp 113-119 *In*: D. Pauly and G. R. Morgan [eds.] Theory and Application of Length-based Methods in Stock Assessment. ICLARM Conference Proceedings

13.

Shepherd, S.A. 1990. Studies on southern Australian abalone (genus *Haliotis*). XII. Long-term recruitment and mortality dynamics of an unfished population. Australian Journal of Marine and Freshwater Research 41: 475-492.

Shepherd, S.A., D. Al-Wahaibi and A.R. Al-Azri 1995. Shell growth checks and growth of the Omani abalone *Haliotis mariae*. Marine and Freshwater Research 46: 575-582.

Shine, R. and E.L. Charnov 1992. Patterns of survivorship, growth and maturation in snakes and lizards. American Naturalist 139: 1257-1269.

Shirakihara, M., A. Takemura and K. Shirakihara 1993. Age, growth, and reproduction of the finless porpoise, *Neophocaena phocaenoides*, in the coastal waters of western Kyushu, Japan. Marine Mammal Science 9: 392-406.

Shoukry, A. and M. Hafez 1979. Studies on the biology of the Mediterranean fruit fly *Ceratitis capitata*. Entomologia Experimentalis et Applicata 26: 33-39.

Shreve, F. 1917. The establishment of desert perennials. Journal of Ecology 5: 210-216.

Siler, W. 1979. A competing-risk model for animal mortality. Ecology 60: 750-757.

Silvertown, J., M. Franco, I. Pisanty, and A. Mendoza 1993. Comparative plant demography — relative importance of life-cycle components to the finite rate of increase in woody and herbaceous perennials. Journal of Ecology 81: 465-476.

Sims, S.E. 1985. Selected computer programs in FORTRAN for fish stock assessment. FAO Technical Paper 259. 183p

Sinclair, A.R.E. 1974. The natural regulation of buffalo populations in East Africa. East African Wildlife Journal 12: 169-183.

Smith, G.T. and I.C.R. Rowley 1995. Survival of adult and nestling western long-billed corellas, *Cacatua pastinator*, and Major Mitchell cockatoo, *C. leadbeateri*, in the wheatbelt of Western Australia. Wildlife Research 22: 155-162.

Smith, R.J. 1980. Rethinking allometry. Journal of Theoretical Biology 87: 97-111.

Smith, S.H. 1974. The growth and mortality of the littleneck clam *Protothaca staminea* in Tia Juana Slough. MS Thesis, San Diego State University, San Diego, California, 116p.

Smith, S.V. 1972. Production of calcium carbonate on the mainland shelf of southern California. Limnology and Oceanography 17: 28-41.

Solbrig, O.T. and B.B. Simpson 1974. Components of regulation of a population of dandelions in Michigan. Journal of Ecology 62: 473-486.

Soukupová, L. 1988. Short life-cycles in two wetland sedges. Aquatic Botany 30: 49-62

Sparre, P. 1987. A method for the estimation of growth, mortality and gear selection/recruitment parameters from length-frequency samples weighted by catch per effort. pp 75-102 *In*: D. Pauly and G. R. Morgan [eds.] Theory and Application of Length-based Methods in Stock Assessment. ICLARM Conference Proceedings 13.

Spinage, C.A. 1972. African ungulate life tables. Ecology 53: 645-652.

Sprent, P. and G.R. Dolby. 1980. Response to Query: the geometric mean functional relationship. Biometrics 36: 547-550.

Sprent, P. and G.R. Dolby. 1982. Response to letter from W. E. Ricker. Biometrics 38: 860.

Staikou, A., M. Lazaridou-Dimitriadou and E. Pana 1990 The life cycle, population dynamics, growth and secondary production of the snail *Bradybaena fruticum* (Müller, 1774) (Gastropoda Pulmonata) in northern Greece. Journal of Molluscan Studies 56: 137-146.

Starfield, A.M. and A.L. Bleloch 1986. *Building Models for Conservation and Wildlife Management*. Macmillan, New York, 253pp.

Stearns, S.C. 1992. The Evolution of Life Histories. Oxford University Press, New York, 249pp.

Steenbergh, W.F. and C.H. Lowe 1983. Ecology of the saguaro: III Growth and demography. Scientific Monograph Series, Number 17. U.S. National Park Service. U.S. Government Printing Office, Washington, DC 228p

Stocker, L.J. 1991. Effects of size and shape of colony on rates of fission, fusion, growth and mortality in a subtidal invertebrate. Journal of Experimental Marine Biology and Ecology 149: 161-175.

Stohlgren, T.J. and P.W. Rundel. 1986. A population model for a long-live, resprouting chaparral shrub: *Adenostoma fasciculatum*. Ecological Modeling, 34:245-257.

Stoner, D.S. 1989. Fragmentation: A mechanism for the stimulation of genet growth rates in an encrusting colonial ascidian. Bulletin of Marine Science 45: 277-287.

Strathmann, R.R., M.F. Strathmann and R.H. Emson (1984). Does limited brood capacity link adult size, brooding, and simultaneous hermaphroditism? A test with the starfish *Asterina phylactica*. American Naturalist 123: 796-818.

Strijbosch, H. and R.C.M. Creemers 1988. Comparative demography of sympatric populations of *Lacerta vivipara* and *Lacerta agilis*. Oecologia 76: 20-26.

Surlyk, F. 1972. Morphological adaptations and population structures of the Danish chalk brachiopods (Maastrichtiam, upper Cretaceous). Det Kongelige Danske Videnskabernes Selskab Biologiske Skrifter 19:1-57.

Sylvester, J.J. 1883. On the equation to the secular inequalities in the planetary theory. Philosophical

Magazine (Ser. 5) 16: 267-269.

SYSTAT 1992. *SYSTAT. Statistics, Version 5.2 Edition* SYSTAT, Inc., Evanston, Illinois, 724pp.

Tahan, M. 1972. *The Man Who Counted: a Collection of Mathematical Adventures.* W. W. Norton & Co., New York, (translated by L. Clark and A. Reid and illustrated by P. R. Baquero. 1993), 244pp.

Tanaka, M. 1982. A new growth curve which expresses infinite increase. Publications Amakusa Marine Biological Laboratory. Kyushu University 6:167-177.

Tanaka, M. 1988. Eco-physiological meaning of parameters of ALOG growth curve. Publications Amakusa Marine Biological Laboratory. Kyushu University 9: 103-106.

Tanaka, S. 1962. A method of analysing a polymodal frequency distribution and its application to the length distribution of the porgy, *Taius tumifrons* (T. and S.). Journal of the Fisheries Research Board of Canada 19: 1143-1159.

Taylor, B.J.R. 1965. The analysis of polymodal frequency distributions. Journal of Animal Ecology 34: 445-452.

Templeton, A.R. 1982. The prophecies of parthenogenesis. pp 75-101. *In:* H. Dingle and J. P. Hegmann [eds.] *Evolution and Genetics of Life Histories.* Springer-Verlag, New York

Templeton, A.R. and D.A. Levin 1979. Evolutionary consequences of seed pools. American Naturalist 114: 232-249.

Tessier, G. 1948. La relation d'allométrie: sa signification statistique et biologique. Biometrics 4: 14-48.

Tinkle, D.W. and R.E. Ballinger 1972. *Sceloporus undulatus*: A study of the intraspecific comparative demography of a lizard. Ecology 53: 570-584.

Tinkle, D.W., J.D. Congdon and P.C. Rosen 1981. Nesting frequency and success: Implications for the demography of painted turtles. Ecology 62: 1426-1432.

Tuljapurkar, S.D 1982. Population dynamics in variable environments. II. Correlated environments, sensitivity analysis and dynamics. Theoretical Population Biology 21: 114-140.

Tuljapurkar, S. 1990. *Population Dynamics in Variable Environments.* Lecture notes in Biomathematics 85. Springer-Verlag, New York, 154pp.

Tunnicliffe, V.J. 1980. Biological and physical processes affecting the survival of a stony coral, *Acropora cervicornis.* Ph.D. Dissert. Yale University

Turner, F.B., G.A. Hoddenbach, P.A. Medica and J.R. Lannom 1970. The demography of the lizard, *Uta stansburiana* Baird and Girard, in southern Nevada. Journal of Animal Ecology 39:505-519.

Turnpenny, A.W.H., R.N. Bamber and P.A. Henderson 1981. Biology of the sand smelt (*Atherina presbyter*) around Fawley Power Station. Journal of Fish Biology 18: 417-427.

van Aarde, R.J. 1987. Demography of a Cape porcupine,

Hystrix africaneaustrialis, population. Journal of Zoology, London 213: 205-212.

van Groenendael, J., H. de Kroon, S. Kalisz and S. Tuljapurkar 1994. Loop analysis: Evaluating life history pathways in population projection matrices. Ecology 75: 2410-2415.

Van Sickle, J. 1977a Mortality rates from size distributions. The application of a conservation law. Oecologia 27: 311-318.

Van Sickle, J. 1977b Mortality estimates from size distributions: A critique of Smith's model. Limnology and Oceanography 22: 774-775.

van Tienderen, P.H. 1995. Life cycle trade-offs in matrix population models. Ecology 76: 2482-2489.

Vandermeer, J. 1978. Choosing category size in a stage projection matrix. Oecologia 32: 79-84.

Vanzolini, P.E., J.W. Wright, C.J. Cole and O. Cuellar 1978. Parthenogenetic lizards. Science 201: 1152-1155.

Verzon, D.J. and S.H. Moolgavkar 1988. A method for computing profile-likelihood based confidence intervals. Applied Statistics 37: 87-94.

Walford, L.A. 1946. A new graphic method of describing the growth of animals. Biological Bulletin 90: 141-147.

Watanabe, Y., J.L. Butler and T. Mori 1988. Growth of Pacific saury, *Cololabis saira*, in the northeastern and northwestern Pacific Ocean. Fishery Bulletin, U.S. 86: 489-498.

Waters, I., and J.M. Shay. 1991. A field study of the effects of water depth, order of emergence and flowering on the growth of *Typha glauca* shoots using the Richards model. Aquatic Botany 39:231-242.

Watkinson. A.R. and J. White 1985. Some life-history consequences of modular construction in plants. Philosophical Transactions of the Royal Society, London Series B 313:31-51.

Weibull, W. 1951. A statistical distribution function of wide applicability. Journal of Applied Mechanics 18: 293-296.

Weinberg, J.R., H. Caswell and R.B. Whitlatch 1986. Demographic importance of ecological interactions: how much do statistics tell us? Marine Biology 93: 305-310.

Weiser, W. 1994. Cost of growth in cells and organisms: General rules and comparative aspects. Biological Reviews of the Cambridge Philosophical Society 68: 1-33.

Williams, G.C. 1957. Pleiotropy, natural selection, and the evolution of senescence. Evolution 11: 398-411.

Winsor, C.P. 1932. The Gompertz curve as a growth curve. Proceedings of the National Academy of Science 18: 1-8.

Winsor, F. 1958. *The Space Child's Mother Goose.* Simon and Schuster, New York.

Wood, J. 1958. Age structure and productivity of a gray fox population. Journal of Mammalogy 39: 74-86.

Wood, S.N. 1994. Obtaining birth and mortality patterns

from structural population trajectories. Ecological Monographs 64: 23-44.

Woodley, J.D., E.A. Chornesky, P.A. Clifford, J.B.C. Jackson, L.S. Kaufman, N. Knowlton, J.C. Lang, M.P. Pearson, J.W. Porter, M.C. Rooney, K.W. Rylaarsdam, V.J. Tunnicliffe, C.M. Wahle, J.L. Wulff, A.S.G. Curtis, M.D. Dallmeyer, B.P. Jupp, M.A.R. Koehl, J. Neigel and E.M. Sides 1981. Hurricane Allen's impact on Jamaican coral reefs. Science 214: 749-755.

Woodward, F.I. 1980. Review of Hunt (1978): Plant Growth Analysis. Journal of Applied Ecology 17: 516.

Wright, R.G. and G.M. Van Dyne 1981. Population age structure and its relationship to the maintenance of a semidesert grassland undergoing invasion by mesquite. Southwestern Naturalist 26: 13-22.

Xu, X. and H.M.A. Mohammend 1996. An alternative approach to estimating growth parameters from length-frequency data, with application to green tiger prawns. Fishery Bulletin, U.S. 94: 145-155.

York, A.E. 1994. The population dynamics of northern sea lions, 1975 - 1985. Marine Mammal Science 10: 38-51.

Young, M.Y.Y. and R.A. Skillman 1975 A computer program for analysis of polymodal frequency distributions (ENORMSEP), FORTRAN IV. Fishery Bulletin, U.S. 73: 681.

Zammuto, R.M. 1987. Life histories of mammals: analyses among and within *Spermophilus columbianus* life tables. Ecology 68: 1351-1363.

Zar, J.H. 1974. *Biostatistical Analysis*. Prentice-Hall, Inc., Englewood Cliffs, New Jersey 620p

Zedler, P.H., K. Guehlstorff, C. Scheidlinger, and C.R. Gautier 1998. The population ecology of a dune thistle, *Crisium rhothophilum* (Asteraceae). American Journal of Botany 70: 1516-1527.

Zweifel, R.G. and C.H. Lowe 1966. The ecology of a population of *Xantusia vigilis*, the desert night lizard. American Museum Novitates 2247: 1-57.

Index

Some terms and symbols, such as r, λ and f_x, are used throughout this book. I have included them in the index for the first few appearances for purposes of definition and some initial calculations and then I have ignored them. No one really wants to know every page that mentions λ.